Lecture Notes in Computer S<

Commenced Publication in 1973
Founding and Former Series Editors:
Gerhard Goos, Juris Hartmanis, and Jan van Leeuwen

Øyvind Ytrehus (Ed.)

Coding
and Cryptography

International Workshop, WCC 2005
Bergen, Norway, March 14-18, 2005
Revised Selected Papers

 Springer

Volume Editor

Øyvind Ytrehus
University of Bergen
Department of Informatics
N-5020 Bergen, Norway
E-mail: oyvind@ii.uib.no

Library of Congress Control Number: 2006927431

CR Subject Classification (1998): E.3-4, G.2.1, C.2, J.1

LNCS Sublibrary: SL 4 – Security and Cryptology

ISSN   0302-9743
ISBN-10   3-540-35481-6 Springer Berlin Heidelberg New York
ISBN-13   978-3-540-35481-9 Springer Berlin Heidelberg New York

Springer is a part of Springer Science+Business Media

springer.com

© Springer-Verlag Berlin Heidelberg 2006
Printed in Germany

Typesetting: Camera-ready by author, data conversion by Scientific Publishing Services, Chennai, India
Printed on acid-free paper   SPIN: 11779360   06/3142   5 4 3 2 1 0

Preface

This volume contains refereed papers devoted to coding and cryptography. These papers are the full versions of a selection of the best extended abstracts accepted for presentation at the International Workshop on Coding and Cryptography (WCC 2005) held in Bergen, Norway, March 14–18, 2005. Each of the 118 extended abstracts originally submitted to the workshop were reviewed by at least two members of the Program Committee. As a result of this screening process, 58 papers were selected for presentation, of which 52 were eventually presented at the workshop together with four invited talks.

The authors of the presented papers were in turn invited to submit full versions of their papers to the full proceedings. Each of the full-version submissions were once again thoroughly examined and commented upon by at least two reviewers. This volume is the end result of this long process.

I am grateful to the reviewers who contributed to guaranteeing the high standards of this volume, and who are named on the next pages. It was a pleasure for me to work with my program co-chair Pascale Charpin, whose experienced advice I have further benefited greatly from during the preparation of this volume. Discussions with Tor Helleseth and Ángela Barbero were also useful in putting the volume together. Finally, I would like to thank all the authors and all the other participants of the WCC 2005 for making it in every sense a highly enjoyable event.

March 2006 — Øyvind Ytrehus

Organization

WCC 2005 was organized by the Selmer Center at the Department of Informatics, University of Bergen, Norway, in cooperation with INRIA Rocquencourt.

Conference Chair	Tor Helleseth (University of Bergen, Norway)
Program Co-chairs	Pascale Charpin (INRIA Rocquencourt, France)
	Øyvind Ytrehus (University of Bergen, Norway)
Program Committee	

D. Augot (INRIA Rocquencourt, France)
C. Carlet (Université Paris VIII, France)
P. Charpin (**Co-chair**, INRIA Rocquencourt, France)
C. Ding (Hong Kong University of Science and
   Technology, China)
H. Dobbertin (University of Bochum, Germany)
S. Dodunekov (Institute of Mathematics, Sofia,
   Bulgaria)
I. Dumer (UC Riverside, USA)
G. Gong (University of Waterloo, Canada)
T. Helleseth (University of Bergen, Norway)
I. Honkala (University of Turku, Finland)
T. Hholdt (DTU, Denmark)
T. Johansson (Lund University, Sweden)
G. Kabatianski (IPIT, Moscow, Russia)
T. Lange (University of Bochum, Germany)
J. Massey (Lund University, Sweden)
M. Mihaljevic (Serbian Acad. of Sciences and Art,
   Serbia and Montenegro)
K. Nyberg (Nokia, Finland)
M.G. Parker (University of Bergen, Norway)
K. Paterson (Royal Holloway, UK)
I. Semaev (University of Bergen, Norway)
N. Sendrier (INRIA, France)
D. Stinson (University of Waterloo, Canada)
H. van Tilborg (Eindhoven University of
   Technology, The Netherlands)
S. Vladuts (Université de Marseille, France)
Ø. Ytrehus (**Co-chair**, University of Bergen,
   Norway)
G. Zémor (ENST, France)
V. Zinoviev (IPIT, Moscow, Russia)

Other Referees for WCC 2005

In addition to the members of the Program Committee, the following were also involved as reviewers in the WCC 2005 review process:

Alexei Ashikhmin
Roberto M. Avanzi
Leonid Bassalygo
Peter Beelen
Raghav Bhaskar
Peter Boyvalenkov
Dario Catalano
Yang Cui
Ernst M. Gabidulin
Philippe Gaborit
Clemens Heuberger
Claude-Pierre Jeannerod
Shaoquan Jiang

Ellen Jochemsz
Antoine Joux
Alexander Kholosha
Emil Kolev
Kristine Lally
Ivan Landjev
Vladimir Levenshtein
Simon Litsyn
Pierre Loidreau
Nikolai Manev
Marine Minier
Thomas Mittelholzer
Kiril Morozov

Wakaha Ogata
Kalle Ranto
Petri Rosendahl
Berry Schonemakers
SoongHan Shin
Andrey Sidorenko
Faina Solov'eva
Jeremy Thorpe
Jean-Pierre Tillich
Jing Ying
Nam Yul Yu

Sponsoring Institutions

The Selmer Center, University of Bergen
The Norwegian Research Council (NFR)

Table of Contents

Second Support Weights for Binary Self-dual Codes

Keisuke Shiromoto

Department of Information Systems
Aichi Prefectural University
Nagakute, Aichi 480-1198, Japan
keisuke@ist.aichi-pu.ac.jp

Abstract. In this work, we investigate the second generalized Hamming weights for binary doubly-even self-dual codes from the point of view of corresponding t-designs by the Assmus-Mattson theorem. In particular, for extremal doubly-even self-dual codes, we shall give a bound on the weights and determine the weights by using the block intersection numbers of corresponding t-designs. Moreover we study the support weight enumerators for binary doubly-even self-dual codes and determine the second support weight enumerators for binary extremal doubly-even self-dual codes of length 56 and 96.

1 Introduction

Generalized Hamming weights for linear codes over finite fields were introduced by Wei as an application in keyless cryptography ([16]). He also gave the characterization of the performance of a linear code on the wire-tap channel II from its weight hierarchy. The support weight enumerators for linear codes over finite fields were first introduced in [5] as a generalization of the Hamming weight enumerators, and many researchers have investigated the generalized Hamming weights for various classes of linear codes (e.g. [15]).

As for the self-dual codes, Dougherty and Gulliver determined the second and third generalized Hamming weights for binary self-dual codes of length up to 28 and of length 48 and 72 in [3]. Milenkovic, Coffey and Compton ([11]) determined the third support weight enumerator for binary extremal doubly-even self-dual codes of length 32. Chen and Coffey ([2]) studied the connection between the trellis structures and the generalized Hamming weights of some binary extremal self-dual codes.

The purpose of this work is to study the second generalized Hamming weights for binary extremal doubly-even self-dual codes. For that purpose, we first consider t-designs obtained from the codes by the Assmus-Mattson theorem. By investigating the block intersection numbers of these t-designs, we shall give a bound on the weights and determine the weights for some extremal self-dual codes.

Ø. Ytrehus (Ed.): WCC 2005, LNCS 3969, pp. 1–13, 2006.

2   Notation and Terminology

Let C be an $[n, r]$ code over a finite field $\mathbb{F}_q$ of q elements. For a vector $\boldsymbol{x} = (x_1, \ldots, x_n) \in \mathbb{F}_q$ and a subset $D \subseteq \mathbb{F}_q^n$, we define the *support* of $\boldsymbol{x}$, the *support* of D, and the *Hamming weight* of $\boldsymbol{x}$ respectively as follows:

$$\mathrm{supp}(\boldsymbol{x}) = \{i \mid x_i \neq 0\},$$
$$\mathrm{Supp}(D) = \bigcup_{\boldsymbol{x} \in D} \mathrm{supp}(\boldsymbol{x}),$$
$$\mathrm{wt}(\boldsymbol{x}) = |\mathrm{supp}(\boldsymbol{x})|.$$

We denote the set of $[n, m]$ subcodes of C by $\mathcal{D}_m(C)$. For each g, $1 \leq g \leq r$, the g-th *generalized Hamming weight* (GHW) d_g of C is defined by Wei ([16]) as follows:

$$d_g = d_g(C) = \min\{|\mathrm{Supp}(D)| \; : \; D \in \mathcal{D}_g(C)\}.$$

In particular, if $g = 1$, then $d_1 = d$ is the *minimum Hamming weight* of C. And the following bound is known (cf. [16]):

$$(q^r - 1)d_{r-1}(C) \leq (q^r - q)d_r(C). \tag{1}$$

For each g, $1 \leq g \leq r$, the g-th *support weight enumerator* of C is defined as

$$W_C^{(g)}(x, y) = \sum_{D \in \mathcal{D}_g(C)} x^{n - |\mathrm{Supp}(D)|} y^{|\mathrm{Supp}(D)|}$$
$$= \sum_{i=0}^{n} A_i^{(g)} x^{n-i} y^i,$$

where

$$A_i^{(g)} = |\{D \in \mathcal{D}_g(C) \; : \; |\mathrm{Supp}(D)| = i\}|.$$

In particular, if $r = 1$, then $W_C^{(1)}(x, y) + 1 = W_C(x, y)$ is the *Hamming weight enumerator* of C and $A_i^{(1)} = A_i(C) = A_i$, $i = 1, 2, \ldots, n$ is the *Hamming weight distribution* of C.

Let $C^\perp$ be the dual code of C. A *self-dual* code C is an $[n, n/2]$ code such that $C = C^\perp$. If C is a binary code and the Hamming weights of all codewords of C are divisible by 4, C is called a *doubly-even* code. It is well-known that the length of any binary doubly-even self-dual code is divisible by 8. The following bound is the most famous bound on the minimum Hamming weight for binary $[n, n/2, d]$ doubly-even self-dual code C ([9], [8]):

$$d \leq 4 \left\lfloor \frac{n}{24} \right\rfloor + 4.$$

If C meets the bound, that is, $d = 4\lfloor n/24 \rfloor + 4$, then C is called an *extremal* code.

A t-(v, k, λ) *design* is a collection $\mathcal{B}$ of k-subsets (called *blocks*) of a set V of v *points*, such that any t-subset of V is contained in exactly λ blocks. For t-(v, k, λ) design $(V, \mathcal{B})$, there are

$$\lambda_s = \frac{\lambda \binom{v-s}{t-s}}{\binom{k-s}{t-s}}$$

blocks in $\mathcal{B}$ that contain all the points in any s-subset of V, $0 \leq s \leq t$. In [1], Assmus and Mattson proved the following theorem, which is called the *Assmus-Mattson Theorem*.

Theorem 1. *Let C be an $[n, r, d]$ code over $\mathbb{F}_q$, and let $d^{\perp}$ denote the minimum Hamming weight of $C^{\perp}$. Let $w = n$ when $q = 2$ and otherwise the largest integer w satisfying*

$$w - \left(\frac{w + q - 2}{q - 1}\right) < d,$$

defining $w^{\perp}$ similarly. Suppose there is an integer t with $0 < t < d$ that satisfies the following condition: the number s of i $(1 \leq i \leq n - t)$ such that $A_i^{\perp} \neq 0$ is at most $d - t$. Then for each i with $d \leq i \leq w$, the supports of codewords in C of weight i, provided there are any, yield a t-design. Similarly, for each j with $d^{\perp} \leq j \leq \min\{w^{\perp}, s\}$, the supports of codewords in $C^{\perp}$ of weight j, provided there are any, form a t-design.

As a consequence of the above theorem, the binary doubly-even self-dual codes hold t-designs (cf. [6]).

Theorem 2. *Let C be a binary $[24m + 8\mu, 12m + 4\mu, 4m + 4]$ extremal doubly-even code for $\mu = 0, 1$ or 2. Then the supports of codewords in C of any fixed weight except 0 hold t-designs for the following parameters:*

(a) $t = 5$ *if* $\mu = 0$ *and* $m \geq 1$,
(b) $t = 3$ *if* $\mu = 1$ *and* $m \geq 0$, *and*
(c) $t = 1$ *if* $\mu = 2$ *and* $m \geq 0$.

3 Second Support Weights for Extremal Codes

For a binary $[n, r]$ code C and its $[n, 2]$ subcode $D = \{\mathbf{0}, \boldsymbol{x}, \boldsymbol{y}, \boldsymbol{x} + \boldsymbol{y}\}$, we have

$$|\mathrm{Supp}(D)| = |\mathrm{supp}(\boldsymbol{x}) \cup \mathrm{supp}(\boldsymbol{y}) \cup \mathrm{supp}(\boldsymbol{x} + \boldsymbol{y})| = |\mathrm{supp}(\boldsymbol{x}) \cup \mathrm{supp}(\boldsymbol{y})|.$$

Thus we have that

$$d_2(C) = \min\{|\mathrm{supp}(\boldsymbol{x}) \cup \mathrm{supp}(\boldsymbol{y})| \ : \ \boldsymbol{x}, \boldsymbol{y} \in C \setminus \{\mathbf{0}\}, \ \boldsymbol{x} \neq \boldsymbol{y}\}.$$

Let C be a binary $[n, n/2, d = 4\lfloor n/24 \rfloor + 4]$ doubly-even self-dual code. For C, set $V = \{1, 2, \ldots, n\}$ and let $\mathcal{B}_C = \mathcal{B}$ be the set of the supports of all the codewords of weight d in C. From Theorem 2, we see easily that $(V, \mathcal{B})$ is a t-$(n, d, \lambda = A_d \binom{d}{t}/\binom{n}{t})$ design. For any two codewords $\boldsymbol{x}, \boldsymbol{y} \in C$, since the inner product $x \cdot y$ is 0 in $\mathbb{F}_2$, the intersection between $\text{supp}(x)$ and $\text{supp}(y)$ should be even. For a block $B \in \mathcal{B}$, we denote m_{2j}^B by the number of blocks in $\mathcal{B}$ which intersect B with $2j$ elements. We sometimes simply denote m_{2j} for m_{2j}^B when the numbers are independent of the choice of a block B. The system of equations was proved in [10] for $t = 1, 3$ or 5 (see also [14] and [4]):

$$\sum_{j=0}^{d/2} \binom{2j}{s} m_{2j}^B = \lambda_s \binom{d}{s} \quad (s = 0, 1, \ldots, t).$$

For any of two distinct codewords $x, y \in C$ of weight d,

$$\text{wt}(x + y) = \text{wt}(x) + \text{wt}(y) - 2|\text{supp}(x) \cap \text{supp}(y)|$$
$$= 2d - 2|\text{supp}(x) \cap \text{supp}(y)| \geq d.$$

We have that $|\text{supp}(x) \cap \text{supp}(y)| \leq d/2$ and thus $m_{2j}^B = 0$ for $j = d/4 + 1, \ldots, d/2 - 1$ and $m_d^B = 1$. Therefore it follows immediately from the above system of equations:

$$\sum_{j=0}^{d/4} \binom{2j}{s} m_{2j}^B = (\lambda_s - 1) \binom{d}{s} \quad (s = 0, 1, \ldots, t). \tag{2}$$

Let $m(C)$ be the size of the largest intersection between any two blocks, that is,

$$m(C) = \max\{2j \in \{0, 2, \ldots, d/4\} \ : \ m_{2j}^B \neq 0, \ B \in \mathcal{B}\}.$$

From the above argument and the bound (1), we have the following bounds on the second generalized Hamming weights for binary extremal doubly-even self-dual codes.

Theorem 3. *Let C be a binary $[n, n/2, d = 4\lfloor n/24 \rfloor + 4]$ doubly-even self-dual code C. Then we have that*

$$6 \left\lfloor \frac{n}{24} \right\rfloor + 6 \leq d_2(C) \leq 8 \left\lfloor \frac{n}{24} \right\rfloor + 8 - m(C).$$

We set that $\mathcal{B} = \{B_1, B_2, \ldots, B_{A_d}\}$. The *intersection matrix* of C on weight d is defined by the $A_d \times A_d$ matrix $M = (m_{i,j})$ such that

$$m_{i,j} = 2\alpha \text{ if } |B_i \cap B_j| = 2\alpha, \ \alpha = 0, 1, \ldots, d/2.$$

We note that the matrix M is a symmetric matrix, every ith row of M contains exactly $m_{2t}^{B_i}$ "$2t$"s and all of the diagonal elements are d. For example, let C be a binary $[8, 4, 4]$ doubly-even self-dual code having generator matrix:

$$G = \begin{bmatrix} 1\,0\,0\,0\,1\,1\,1\,0 \\ 0\,1\,0\,0\,1\,1\,0\,1 \\ 0\,0\,1\,0\,1\,0\,1\,1 \\ 0\,0\,0\,1\,0\,1\,1\,1 \end{bmatrix}.$$

Then the supports of all codewords of weight 4 form a 3-$(8, 4, 1)$ design. From the system of equations (2), we have that $m_0 = 1$ and $m_2 = 12$. Therefore the intersection matrix M of C is as follows:

$$M = \begin{bmatrix}
4 & 2 & 2 & 2 & 2 & 2 & 2 & 2 & 2 & 2 & 0 & 2 & 2 \\
2 & 4 & 2 & 2 & 2 & 2 & 2 & 2 & 2 & 0 & 2 & 2 & 2 \\
2 & 2 & 4 & 2 & 2 & 2 & 2 & 2 & 0 & 2 & 2 & 2 & 2 \\
2 & 2 & 2 & 4 & 2 & 2 & 2 & 0 & 2 & 2 & 2 & 2 & 2 \\
2 & 2 & 2 & 2 & 4 & 2 & 2 & 0 & 2 & 2 & 2 & 2 & 2 \\
2 & 2 & 2 & 2 & 2 & 4 & 0 & 2 & 2 & 2 & 2 & 2 & 2 \\
2 & 2 & 2 & 2 & 2 & 0 & 4 & 2 & 2 & 2 & 2 & 2 & 2 \\
2 & 2 & 2 & 2 & 0 & 2 & 2 & 4 & 2 & 2 & 2 & 2 & 2 \\
2 & 2 & 2 & 0 & 2 & 2 & 2 & 2 & 4 & 2 & 2 & 2 & 2 \\
2 & 2 & 0 & 2 & 2 & 2 & 2 & 2 & 2 & 4 & 2 & 2 & 2 \\
2 & 0 & 2 & 2 & 2 & 2 & 2 & 2 & 2 & 2 & 4 & 2 & 2 \\
0 & 2 & 2 & 2 & 2 & 2 & 2 & 2 & 2 & 2 & 2 & 4 & 2 & 2 \\
2 & 2 & 2 & 2 & 2 & 2 & 2 & 2 & 2 & 2 & 2 & 4 & 0 \\
2 & 2 & 2 & 2 & 2 & 2 & 2 & 2 & 2 & 2 & 2 & 0 & 4
\end{bmatrix}.$$

Define the set Δ_s as follows:

$$\Delta_s = \{S \subseteq V \ : \ |S| = s, \ S = B_i \cup B_j, \ B_i, B_j \in \mathcal{B}, \ B_i \neq B_j\}.$$

Using the intersection matrix of a binary extremal doubly-even self-dual code, we have the following theorems.

Theorem 4. *Let C be a binary $[n, n/2, d = 4\lfloor n/24 \rfloor + 4]$ doubly-even self-dual code. If $m_0, m_1, \ldots, m_{d/2}$ are uniquely determined by the system of equations (2), then*

(a) $|\Delta_{2d-2\alpha}| = A_d m_{2\alpha}/2$ *if $\alpha = 1, \ldots, d/4 - 1$, and*
(b) $|\Delta_{2d-2\alpha}| = A_d m_{2\alpha}/6$ *if $\alpha = d/4$.*

Proof. From the intersection matrix of C, for any α, $\alpha = 1, \ldots d/4$, the number of all the pairs of two distinct blocks whose intersection is 2α points is $A_d m_{2\alpha}/2$.

We first consider the case (a) and show that every union of two distinct blocks whose intersection is 2α points is different from others for $\alpha = 1, \ldots d/4 - 1$. We assume that there exist distinct blocks $B_1, B_2, B_3, B_4 \in \mathcal{B}$ such that $|B_1 \cap B_2| = |B_3 \cap B_4| = 2\alpha$ and $B_1 \cup B_2 = B_3 \cup B_4$. Let A be the set of the intersection of B_3 and B_4. Since the symmetric difference $B_1 \triangle (B_3 \triangle B_4)$ is also the support of a codeword in C,

$$|B_1 \triangle (B_3 \triangle B_4)| = 3d - 4\alpha - 2|B_1 \cap (B_3 \triangle B_4)| \geq d.$$

Thus we have that $d - 2\alpha \geq |B_1 \cap (B_3 \triangle B_4)|$ and so $A \subseteq B_1$. Because of $|B_1 \cap (B_i \setminus A)| \geq d/2 - \alpha$ for $i = 3$ or 4, we have that $|B_1 \cap B_i| \geq d/2 + \alpha$. Therefore, it follows that

$$|B_1 \triangle B_i| = |B_1| + |B_i| - 2|B_1 \cap B_i| \leq d - 2\alpha(< d).$$

A contradiction. Therefore the number of cardinality $2d - 2\alpha$ distinct unions of two distinct blocks in $\mathcal{B}$ is $A_d m_{2\alpha}/2$.

Next we consider the case (b) and show that, for any cardinality $3d/2$ union U of two distinct blocks in $\mathcal{B}$, the number of pairs of two distinct blocks B_i and B_j in $\mathcal{B}$ with $B_i \cup B_j = U$ is three. Let B_1 and B_2 be two distinct blocks in $\mathcal{B}$ such that $|B_1 \cap B_2| = d/2$. Since $B = B_1 \triangle B_2$ is also a block in $\mathcal{B}$ and $|B_1 \cap B| = |B_2 \cap B| = d/2$, we have that

$$B_1 \cup B_2 = B_1 \cup B = B_2 \cup B.$$

Conversely, we assume that there exist two distinct blocks $B_3, B_4 \in \mathcal{B} \setminus \{B_1, B_2, B\}$ such that $B_1 \cup B_2 = B_3 \cup B_4$. If B_3 intersects B_1 with at most $d/2 - 2$ points, then B_3 intersects B_2 at least $d/2 + 2$ and so $|B_2 \triangle B_3| \leq d - 2$. Thus each B_j, $j = 3$ and 4, intersects each B_i, $i = 1$ and 2 with $d/2$ points. Suppose that $|B_3 \cap (B_1 \cap B_2)| = c \geq 1$. Because of $|B_3 \cap B_1| = |B_3 \cap B_2| = d/2$, $|B_3| = d/2 + d/2 - c = d - c$. A contradiction. So there are no such pairs of two distinct blocks in $\mathcal{B}$. Therefore the number of cardinality $3d/2$ distinct unions of two distinct blocks in $\mathcal{B}$ is $A_d m_{d/2}/6$. □

Theorem 5. *Let C be a binary $[n, n/2, d = 4\lfloor n/24 \rfloor + 4]$ doubly-even self-dual code. If $m_0, m_1, \ldots, m_{d/2}$ are uniquely determined by the system of equations (2), then*

(a) $A_{3d/2}^{(2)}(C) = A_d m_{d/2}/6$,

(b) $A_{3d/2+2}^{(2)}(C) = A_d m_{d/2-2}/2$, *and*

(c) $A_{2d-2\alpha}^{(2)}(C) \geq A_d m_{2\alpha}/2$ *for $\alpha = 1, 2, \ldots, d/4 - 2$.*

Proof. (c) From Theorem 4, there are at least $A_d m_{2\alpha}/2$ cardinality $2d - 2\alpha$ unions of the supports of two distinct codewords in C for $\alpha = 1, 2, \ldots, d/4 - 2$.

(a), (b) Suppose that there exist two distinct codewords $x, y \in C$ such that $\mathrm{wt}(x) \geq d + 4$, $\mathrm{wt}(y) \geq d + 4$ and $|\mathrm{supp}(x) \cup \mathrm{supp}(y)| \leq 3d/2 + 2$. Then we have that

$$|\mathrm{supp}(x) \cap \mathrm{supp}(y)| = \mathrm{wt}(x) + \mathrm{wt}(y) - |\mathrm{supp}(x) \cup \mathrm{supp}(y)| \geq \frac{d}{2} + 6,$$

and so

$$\begin{aligned} \mathrm{wt}(x + y) &= |\mathrm{supp}(x) \cup \mathrm{supp}(y)| - |\mathrm{supp}(x) \cap \mathrm{supp}(y)| \\ &\leq (3d/2 + 2) - (d/2 + 6) = d - 4. \end{aligned}$$

Thus we may assume that $\mathrm{wt}(x) = d$, $\mathrm{wt}(y) = d + 4$ and $|\mathrm{supp}(x) \cup \mathrm{supp}(y)| \leq 3d/2 + 2$. If $|\mathrm{supp}(x) \cup \mathrm{supp}(y)| = 3d/2 + 2$, then $\mathrm{wt}(x + y) = d$ and so there exist two distinct blocks in $\mathcal{B}$ whose union is $\mathrm{supp}(x) \cup \mathrm{supp}(y)$. If $|\mathrm{supp}(x) \cup \mathrm{supp}(y)| = 3d/2$, then $\mathrm{wt}(x + y) = d - 2$. Therefore it finds that, for any two dimensional subcode D of C such that $|\mathrm{Supp}(D)| = 3d/2$ or $3d/2 + 2$, there exist at least two distinct blocks in $\mathcal{B}$ which correspond to the supports of codewords in D. □

For a linear code C over $\mathbb{F}_q$ having generator matrix G, let $C^{(m)}$ be the linear code over $\mathbb{F}_{q^m}$ having generator matrix G. The following equation is proved in [7] (cf. [13]).

Lemma 6

$$W_{C^{(m)}}(x,y) = \sum_{j=0}^{m} [m]_j \, W_C^{(j)}(x,y),$$

where $[m]_j = \prod_{i=0}^{j-1}(q^m - q^i)$.

We denote the elements of $\mathbb{F}_4$ by $0, 1, \omega, \bar{\omega}$. For vectors $\boldsymbol{x} = (x_1, \ldots, x_n), \boldsymbol{y} = (y_1, \ldots, y_n) \in \mathbb{F}_4^n$, the *Hermitian inner product* between $\boldsymbol{x}$ and $\boldsymbol{y}$ is defined by

$$\langle \boldsymbol{x}, \boldsymbol{y} \rangle = \sum_{i=1}^{n} x_i \bar{y}_i,$$

where $^-$ is given by $\bar{0} = 0, \bar{1} = 1$ and $\bar{\omega} = \omega$. For an $[n, k]$ code C over $\mathbb{F}_4$, the *Hermitian dual* code is defined by

$$C^{\perp_H} = \{ \boldsymbol{y} \in \mathbb{F}_4^n \ : \ \langle \boldsymbol{y}, \boldsymbol{x} \rangle = 0, \text{ for all } \boldsymbol{x} \in C \}.$$

If $C = C^{\perp_H}$, then C is called a *Hermitian self-dual* code.

Lemma 7. *Let C be a binary $[n, n/2, d]$ self-dual code with generator matrix G. Then $C^{(2)}$ is an $[n, n/2, d]$ Hermitian self-dual code over $\mathbb{F}_4$.*

Proof. Let $\boldsymbol{g}_1, \boldsymbol{g}_2, \ldots, \boldsymbol{g}_{n/2}$ be the rows of G. For any (not necessarily distinct) two codewords $\boldsymbol{x} = \sum_{i=1}^{n/2} \alpha_i \boldsymbol{g}_i, \boldsymbol{y} = \sum_{j=1}^{n/2} \beta_j \boldsymbol{g}_j \in C^{(2)}$, $\alpha_i, \beta_j \in \mathbb{F}_4$ for all i and j, the Hermitian inner product

$$\langle \boldsymbol{x}, \boldsymbol{y} \rangle = \sum_{i=1}^{n/2} \alpha_i \langle \boldsymbol{g}_i, \sum_{j=1}^{n/2} \beta_j \boldsymbol{g}_j \rangle$$

$$= \sum_{i=1}^{n/2} \sum_{j=1}^{n/2} \alpha_i \bar{\beta}_j \langle \boldsymbol{g}_i, \boldsymbol{g}_j \rangle$$

$$= \sum_{i=1}^{n/2} \sum_{j=1}^{n/2} \alpha_i \bar{\beta}_j (\boldsymbol{g}_i \cdot \boldsymbol{g}_j)$$

is 0 in $\mathbb{F}_4$, where $(\boldsymbol{g}_i \cdot \boldsymbol{g}_j)$ denotes the inner product in $\mathbb{F}_2$. Thus it finds that $C^{(2)} \subseteq (C^{(2)})^{\perp_H}$. On the other hand, since the dimension of $C^{(2)}$ is also $n/2$, we have that $|C^{(2)}| = 4^{n/2}$. Therefore it follows that $C^{(2)} = (C^{(2)})^{\perp_H}$. From Lemma 6, we have that $A_1(C^{(2)}) = \cdots = A_{d-1}(C^{(2)}) = 0$ and $A_d(C^{(2)}) \neq 0$. So the minimum Hamming weight of $C^{(2)}$ is d. □

The following result is well-known as the Gleason theorem (cf. [12], [6] and [8]).

Lemma 8. *Let C be an $[n, n/2]$ code over $\mathbb{F}_q$. Then if $q = 2$ and C is a doubly-even self-dual code, then*

$$W_C(x, y) = \sum_{i=0}^{\lfloor n/24 \rfloor} a_i(x^8 + 14x^4y^4 + y^8)^{n/8-3i}(x^4y^4(x^4 - y^4)^4)^i,$$

and if $q = 4$ and C is a Hermitian self-dual code, then

$$W_C(x, y) = \sum_{i=0}^{\lfloor n/6 \rfloor} b_i(x^2 + 3y^2)^{n/2-3i}(y^2(x^2 - y^2)^2)^i,$$

where the a_i and b_i are integers.

Proposition 9

$$W_C^{(2)}(x, y) = \frac{1}{6}\left\{W_{C^{(2)}}(x, y) - 3W_C(x, y) + 2\right\}.$$

Proof. Applying Lemma 6 with $q = 2$ and $m = 2$, the equation immediately follows.   □

By combining Lemma 7, Lemma 8 and Proposition 9, we have the following result.

Theorem 10. *If C is a binary $[n, n/2]$ doubly-even self-dual code, then*

$$W_C^{(2)}(x, y) = \frac{1}{6}\left\{\sum_{i=0}^{\lfloor n/6 \rfloor} b_i(x^2 + 3y^2)^{n/2-3i}(y^2(x^2 - y^2)^2)^i \right.$$
$$\left. -3\sum_{i=0}^{\lfloor n/24 \rfloor} a_i(x^8 + 14x^4y^4 + y^8)^{n/8-3i}(x^4y^4(x^4 - y^4)^4)^i + 2\right\}.$$

4   Second Generalized Hamming Weights for Some Extremal Codes

Using the system of the equations (2), we can calculate the block intersection numbers m_i for some extremal doubly-even self-dual codes of length n. Therefore, we can also determine the second generalized Hamming weights for these codes by considering the numbers $m(C)$.

Lemma 11. *Let C be a binary $[n, n/2, d = 4\lfloor n/24 \rfloor + 4]$ doubly-even self-dual code. Then $d_2(C) = 6$ for $n = 8, 16$ and $d_2(C) = 12$ for $n = 24, 32$.*

Proof. See [3], [6] and [2].   □

Lemma 12. *The second generalized Hamming weight for any $[40, 20, 8]$ doubly-even self-dual code C is 12.*

Proof. The supports of codewords of weight 12 in C form a 1-$(40, 12, 57)$ design. Because of $\lambda_0 = 285$ and $\lambda_1 = 57$, we have that

$$m_0 + m_2 + m_4 = 284$$
$$2m_2 + 4m_4 = 448,$$

from the system of equations (2). Suppose that $m_4 = 0$. Then it follows that $m_2 = 224$ and $m_0 = 60$. Thus, there are two points v_1 and v_2 in a block B such that at least $m_2/\binom{8}{2} = 8$ blocks of $\mathcal{B}$ that intersect B in exactly v_1 and v_2. Since there are only $40 - 8 = 32$ points in V which are not contained in B, at least two of these blocks intersect in more than two points. Therefore $m_4 \neq 0$, and so $d_2(C) = 12$ from Theorem 3.   □

Lemma 13. *The second generalized Hamming weight for the* $[48, 24, 12]$ *doubly-even self-dual code* C *is 18.*

Proof. The set of all supports of codewords of weight 12 in C forms a 5-$(48, 12, 8)$ design. So we can uniquely determine that $m_0 = 630, m_2 = 8316, m_4 = 7425$ and $m_6 = 924 \neq 0$ from the system of equations (2).   □

Lemma 14. *The second generalized Hamming weight for any* $[56, 28, 12]$ *doubly-even self-dual code* C *is 18.*

Proof. The set of all supports of codewords of weight 12 in C forms a 3-$(56, 12, 65)$ design. Thus we can uniquely determine that $m_0 = 621, m_2 = 4800, m_4 = 2580$ and $m_6 = 188 \neq 0$.   □

Lemma 15. *The second generalized Hamming weight for any* $[64, 32, 12]$ *doubly-even self-dual code* C *is 18 or 20.*

Proof. The set of all supports of codewords of weight 12 in C forms a 1-$(64, 12, 558)$ design. From the system of equations (2), if $m_6 = 0$, then $367 \leq m_4 \leq 1671$ since $m_0 \geq 0$ and $m_2 \geq 0$. Thus we have that $m_6 \neq 0$ or $m_4 \neq 0$.   □

Remark 16. We have not found any code C whose second generalized Hamming weight is 20 by computer search. We still conjecture that $d_2(C) = 18$.

Lemma 17. *If there exists a* $[72, 36, 16]$ *doubly-even self-dual code* C, *then the second generalized Hamming weight for* C *is 24.*

Proof. The set of all supports of codewords of weight 16 in C forms a 5-$(72, 16, 78)$ design. Then we have that $m_8 = 2310 \neq 0$ by the similar argument to Lemma 13 (see also, [4]).   □

Proposition 18. *The second generalized Hamming weight for any* $[80, 40, 16]$ *doubly-even self-dual code* C *is 24 or 26.*

Proof. The set of all supports of codewords of weight 16 in C forms a 3-$(80, 16, 665)$ design. Suppose that $m_8 = 0$. Then it follows that $m_0 = 3132, m_2 = 43200, m_4 = 40800$ and $m_6 = 10432 \neq 0$. Thus we have that $m_8 \neq 0$ or $m_6 \neq 0$.   □

Lemma 19. *The second generalized Hamming weight for any* $[88, 44, 16]$ *doubly-even self-dual code* C *is* 24 *or* 26.

Proof. The set of all supports of codewords of weight 16 in C forms a 1-$(88, 16, 5848)$ design. Suppose that $m_8 = 0$. Moreover we assume that $m_6 = 0$. Then it follows that $14598 \leq m_4 \leq 23388$ from $m_0 \geq 0$ and $m_2 \geq 0$. Thus it finds that there are at least $\lceil m_4 / \binom{16}{4} \rceil \geq \lceil 14598 / \binom{16}{4} \rceil = 9$ blocks in $\mathcal{B}$ which contain any 4 points v_1, v_2, v_3, v_4 in a block B. Since there are only $88 - 16 = 72$ points in V which are not contained in B, these blocks intersect others in except for v_1, v_2, v_3 and v_4. Therefore we have that $m_6 \neq 0$. □

Lemma 20. *If there exists a* $[96, 48, 20]$ *doubly-even self-dual code* C, *then the second generalized Hamming weight for* C *is* 30.

Proof. The set of all supports of codewords of weight 20 in C forms a 5-$(96, 20, 816)$ design. Then we can uniquely determine that $m_0 = 32505, m_2 = 708300, m_4 = 1561845, m_6 = 792900, m_8 = 116025, m_{10} = 5480 \neq 0$. □

Lemma 21. *If there exists a* $[104, 52, 20]$ *doubly-even self-dual code* C, *then the second generalized Hamming weight for* C *is* 30 *or* 32.

Proof. The set of all supports of codewords of weight 20 in C forms a 3-$(104, 20, 7125)$ design. Suppose that $m_{10} = 0$. Then it follows that $9891 \leq m_8 \leq 82470$ since $m_0 \geq 0$ and $m_6 \geq 0$. Therefore we have that $m_8 \neq 0$. □

Lemma 22. *If there exists a* $[112, 56, 20]$ *doubly-even self-dual code* C, *then the second generalized Hamming weight for* C *is* 30, 32 *or* 34.

Proof. The set of all supports of codewords of weight 20 in C forms a 1-$(112, 20, 63525)$ design. Suppose that $m_{10} = 0$ and $m_8 = 0$. Moreover we assume that $m_6 = 0$. Then it follows that $279501 \leq m_4 \leq 317620$ from $m_0 \geq 0$ and $m_2 \geq 0$. Thus it finds that there are at least $\lceil m_4 / \binom{20}{4} \rceil \geq \lceil 279501 / \binom{20}{4} \rceil = 58$ blocks in $\mathcal{B}$ which contain any 4 points v_1, v_2, v_3, v_4 in a block B. Since there are only $112 - 20 = 92$ points in V which are not contained in B, these blocks intersect others in except for v_1, v_2, v_3 and v_4. Therefore we have that $m_6 \neq 0$. □

Lemma 23. *If there exists a* $[120, 60, 24]$ *doubly-even self-dual code* C, *then the second generalized Hamming weight for* C *is* 36 *or* 38.

Proof. The set of all supports of codewords of weight 24 in C forms a 5-$(120, 24, 8855)$ design. Suppose that $m_{12} = 0$. Then we can uniquely determine that $m_{10} = 419520 \neq 0$. □

Lemma 24. *If there exists a* $[128, 64, 24]$ *doubly-even self-dual code* C, *then the second generalized Hamming weight for* C *is* 36, 38 *or* 40.

Proof. The set of all supports of codewords of weight 24 in C forms a 3-$(128, 24, 78430)$ design. Suppose that $m_{12} = 0$ and $m_{10} = 0$. Then it follows that $1048255 \leq m_8 \leq 1713855$ and so $m_8 \neq 0$. □

Table 1.

n	d	t-(v,k,λ) designs	d_2	references
8	4	3-(8,4,1)	6	[3], etc.
16	4	1-(16,4,7)	6	[3], etc.
24	8	5-(24,8,1)	12	[6], [3], etc.
32	8	3-(32,8,7)	12	[2]
40	8	1-(40,8,57)	12	
48	12	5-(48,12,8)	18	[3], [2]
56	12	3-(56,12,65)	18	
64	12	1-(64,12,558)	18 or 20	
72	16	5-(72,16,78)	24	[3]
80	16	3-(80,16,665)	24 or 26	
88	16	1-(88,16,5848)	24 or 26	
96	20	5-(96,20,816)	30	
104	20	3-(104,20,7125)	30 or 32	
112	20	1-(112,20,63525)	30, 32 or 34	
120	24	5-(120,24,8855)	36 or 38	
128	24	3-(128,24,78430)	36, 38, or 40	
136	24	1-(136,24,705510)	≤ 42	
144	28	5-(144,28,98280)	42 or 44	

Lemma 25. *If there exists a* $[136, 68, 24]$ *doubly-even self-dual code* C, *then the second generalized Hamming weight for* C *is at most* 42.

Proof. The set of all supports of codewords of weight 24 in C forms a 1-$(136, 24, 705510)$ design. Suppose that $m_{12} = 0$, $m_{10} = 0$ and $m_8 = 0$. Moreover we assume that $m_6 = 0$. Then it follows that $4468219 \leq m_4 \leq 4233054$. Thus it finds that there are at least $\lceil m_4/\binom{24}{4} \rceil \geq \lceil 4468219/\binom{24}{4} \rceil = 421$ blocks in $\mathcal{B}$ which contain any 4 points v_1, v_2, v_3, v_4 in a block B. Since there are only $136 - 24 = 112$ points in V which are not contained in B, these blocks intersect others in except for v_1, v_2, v_3 and v_4. Therefore we have that $m_6 \neq 0$. □

Lemma 26. *If there exists a* $[144, 72, 28]$ *doubly-even self-dual code* C, *then the second generalized Hamming weight for* C *is* 42 *or* 44.

Proof. The set of all supports of codewords of weight 28 in C forms a 5-$(144, 28, 98280)$ design. Suppose that $m_{14} = 0$. Then we have that $53222 \leq m_{12} \leq 7521276$ and so $m_{12} \neq 0$. □

We summarize the results in the following theorem on the second generalized Hamming weight for each extremal code of length n.

Theorem 27. *If there exists a binary* $[n, n/2, d = 4\lfloor n/24 \rfloor + 4]$ *doubly-even self-dual code* C *for* $8 \leq n \leq 144$, *then the second generalized Hamming weight* d_2 *of* C *is given in Table 1.*

The second support weight enumerators for a $[48, 24, 12]$ and a putative $[72, 36, 16]$ binary doubly-even self-dual code were found in [3]. So we shall focus on the second support weight enumerators for the other binary doubly-even self-dual codes. Since $m_0, \ldots, m_6$ for a binary $[56, 28, 12]$ doubly-even self-dual codes

Table 2.

$A_i^{(2)}(C_{96})$	i
2938244480	30
186629461200	32
7045835998400	34
199291263806160	36
4574744637832000	38
87439630245125320	40
1407670349998923200	42
19162822065941055600	44
220954897445587326528	46
2159581817482038356700	48
17897349267021111990720	50
125726758328799425384400	52
748033353865421165001280	54
3763971884492563244750520	56
15984923398507198732221120	58
57139317625768079355422960	60
171327331504711070574027840	62
429083154858222192050233225	64
892973054496261690811488832	66
1534674972545912449195676400	68
2161889920680257626892426688	70
2473993764265941647156926200	72
2275231592226313773450849600	74
1659721559142483086102143440	76
945096197886559621218329280	78
411834641673293162669223180	80
133929962826423251810998080	82
31465473895466131097318000	84
5113677593380965527379136	86
541023784763938828424200	88
34041973190568146168384	90
1097865261838779477200	92
13562254143959639360	94
26894375014056762	96

$A_i^{(2)}(C_{56})$	i
256620	18
10565100	20
300998880	22
5632389945	24
77887280016	26
810987952320	28
6329994271776	30
37317157650045	32
165332448557640	34
545437564471800	36
1326807463581600	38
2342437785741690	40
2938066233999120	42
2543737485612960	44
1459873538104800	46
524139462502110	48
107830495153836	50
10978269398460	52
413623584640	54
2530030237	56

and $m_0, \ldots, m_{10}$ for a putative binary $[96, 48, 20]$ doubly-even self-dual codes are uniquely determined by the system of equations (2), we have the following results by combining Theorem 5 and Theorem 10.

Corollary 28. *If C_{56} is a binary $[56, 28, 12]$ doubly-even self-dual code and C_{96} is a putative binary $[96, 48, 20]$ doubly-even self-dual code, then the support weight enumerators for C_{56} and C_{96} are determined in Table 2, respectively.*

Proof. From Theorem 5, it follows that $A_{2i}^{(2)}(C_{56}) = 0$ for $i = 0, \ldots, 8$, $A_{18}^{(2)}(C_{56}) = 256620$ and $A_{20}^{(2)}(C_{56}) = 10565100$, and $A_{2i}^{(2)}(C_{96}) = 0$ for $i = 0, \ldots,$

14, $A_{30}^{(2)}(C_{96}) = 2938244480$ and $A_{32}^{(2)}(C_{96}) = 186629461200$. So we can uniquely determine the coefficients b_i in Theorem 10 for C_{56} and C_{96}.   □

References

1. E.F. Assmus and H.F. Mattson, New 5-designs, *Journal of Combinatorial Theory* **6** (1969), pp. 122–151.
2. H. Chen and J.T. Coffey, Trellis structure and higher weights of extremal self-dual codes, *Designs, Codes and Cryptography* **24** (2001), pp. 15–36.
3. S.T. Dougherty and T.A. Gulliver, Higher weights and binary self-dual codes, *Proceedings of the International Workshop on Coding and Cryptography* (2001), Paris (France), pp. 177–188.
4. M. Harada, M. Kitazume and A. Munemasa, On a 5-designs related to an extremal doubly even self-dual code of length 72, *Journal of Combinatorial Theory* A **107** (2004), pp. 143–146.
5. T. Helleseth, T. Kløve and J. Mykkeltveit, The weight distribution of irreducible cyclic codes with block length $n_1((q^l - 1)/N)$, *Discrete Mathematics* **18** (1977), pp. 179–211.
6. W.C. Huffman and V.S. Pless, *Fundamentals of Error-Correcting Codes*, Cambridge, 2003.
7. T. Kløve, The weight distribution of linear codes over $GF(q^l)$ having generator matrix over $GF(q)$, *Discrete Mathematics* **106/107** (1992), pp. 311–316.
8. F.J. MacWilliams and N.J. A. Sloane, *The Theory of Error-Correcting Codes*, North-Holland, Amsterdam 1977.
9. C.L. Mallows and N.J.A. Sloane, An upper bound for self-dual codes, *Inform. Contr.* **22** (1973), pp. 188–200.
10. N. S. Mendelsohn, Intersection numbers of t-designs, *Studies in Pure Mathematics*, pp. 145–150, Academic Press, London, 1971.
11. O. Milenkovic, S.T. Coffey and K.J. Compton, The third support weight enumerators of the doubly-even, self-dual $[32, 16, 8]$ codes, *IEEE Trans. Inform. Theory* **49** (2003), pp. 740–746.
12. V.S. Pless and W.C. Huffman (Editors), *Handbook of Coding Theory*, North-Holland, Amsterdam 1998.
13. K. Shiromoto, The weight enumerator of linear codes over $GF(q^m)$ having generator matrix over $GF(q)$, *Designs, Codes and Cryptography* **16** (1999), pp. 87–92.
14. V.D. Tonchev, A characterization of designs related to the Witt system $S(5, 8, 24)$, *Math. Z.* **191** (1986), pp. 225–230.
15. M.A. Tsfasman and S.G. Vlǎdut, Geometric approach to higher weights, *IEEE Trans. Inform. Theory* **41** (1995) pp. 1564–1588.
16. V.K. Wei, Generalized Hamming weights for linear codes, *IEEE Trans. Inform. Theory* **37** (1991) pp. 1412–1418.

On Codes Correcting Symmetric Rank Errors

Nina I. Pilipchuk and Ernst M. Gabidulin

Moscow Institute of Physics and Technology,
141700 Dolgoprudnyi, Russia
{nina.pilipchuk, gab}@mail.mipt.ru

Abstract. We study the capability of rank codes to correct so-called symmetric errors beyond the $\lfloor \frac{d-1}{2} \rfloor$ bound. If $d \geq \frac{n+1}{2}$, then a code can correct symmetric errors up to the maximal possible rank $\lfloor \frac{n-1}{2} \rfloor$. If $d \leq \frac{n}{2}$, then the error capacity depends on relations between d and n. If $(d+j) \nmid n$, $j = 0, 1, \ldots, m-1$, for some m, but $(d+m) \mid n$, then a code can correct symmetric errors up to rank $\lfloor \frac{d+m-1}{2} \rfloor$. In particular, one can choose codes correcting symmetric errors up to rank $d-1$, i.e., the error capacity for symmetric errors is about twice more than for general errors.

1 Introduction

Let F_q be a base field and let F_{q^n} be an extension of degree n of F_q.

The *rank* norm rank(M) of a *matrix* $M \in F_q^{n \times n}$ is defined as the algebraic rank of this matrix, i.e., the *maximal number* of rows (or, columns) which are linearly independent over F_q. The *rank distance* between M_1 and M_2 is defined as $d(M_1, M_2) = \text{rank}(M_1 - M_2)$. A *matrix code* $\mathcal{M} \subset F_q^{n \times n}$ is any set of matrices with code distance $d(\mathcal{M}) = d = \min\{d(M_1, M_2) | M_1, M_2 \in \mathcal{M}; \ M_1 \neq M_2\}$.

The *rank* norm $r(\mathbf{g})$ of a *vector* $\mathbf{g} = g_1, g_2, \ldots, g_n$, $\mathbf{g} \in F_{q^n}^n$, is defined as the *maximal number* of coordinates g_j which are linearly independent over the base field F_q.

A *vector code* $\mathcal{V} \subset F_{q^n}^n$ is any set of vectors with code distance $d(\mathcal{V}) = d = \min\{r g_1 - g_2 \mid g_1, g_2 \in \mathcal{V}; \ g_1 \neq g_2\}$.

Let $\mathbf{g}_0 = g_1, g_2, \ldots, g_n$, $g_j \in F_{q^n}$, be a basis of F_{q^n} over F_q. Then any vector $\mathbf{m} = (m_1, m_2, \ldots, m_n) \in F_{q^n}^n$ can be uniquely represented as

$$\mathbf{m} = (m_1, m_2, \ldots, m_n) = \mathbf{g}_0 M = g_1, g_2, \ldots, g_n M, \tag{1}$$

where M is the $n \times n$-matrix in F_q. One refers to the matrix M as the matrix $\mathbf{g}_0$-representation of the vector $\mathbf{m}$. Note that $r(\mathbf{m}) = \text{rank}(M)$.

Let the vector $\mathbf{m}$ and the matrix M be defined by Eq. (1). Let M^t be the transposed matrix. Then the vector

$$\mathbf{m}^t = (\tilde{m}_1, \tilde{m}_2, \ldots, \tilde{m}_n) = \mathbf{g}_0 M^t = g_1, g_2, \ldots, g_n M^t \tag{2}$$

is called $\mathbf{g}_0$-transposed of $\mathbf{m}$.

If $\mathbf{m} = \mathbf{m}^t$, or, equivalently, $M = M^t$, then $\mathbf{m}$ is called the $\mathbf{g}_0$-symmetric vector.

Ø. Ytrehus (Ed.): WCC 2005, LNCS 3969, pp. 14–21, 2006.

Given a vector code $\mathcal{V}$ and a basis $\mathbf{g}_0$, one can get a corresponding matrix code $\mathcal{M}$ in the $\mathbf{g}_0$-representation as $\mathcal{V} = \mathbf{g}_0 \mathcal{M}$, and vice versa.

Let $\mathcal{V}_k$ be an (n, k, d) linear vector code with maximal rank distance $d = n - k + 1$ (an MRD code). Let $\mathcal{M}_k$ be be the corresponding matrix code in the $\mathbf{g}_0$-representation. Codes $\mathcal{V}_k$ containing a subcode of $\mathbf{g}_0$-*symmetric* vectors (respectively, matrix codes $\mathcal{M}_k$ containing a subcode of *symmetric* matrices) are of particular interest. It is known that such codes can correct not only all the errors of rank up to $\lfloor \frac{d-1}{2} \rfloor$ but also many $\mathbf{g}_0$-*symmetric* errors of rank beyond this bound [1].

The number of correctable symmetric errors depends on n and d. In this paper, we investigate the error capacity of codes with respect to different relations between n and d.

If $k \leq (n+1)/2$, equivalently, $d \geq \frac{n+1}{2}$, then all the $\mathbf{g}_0$-symmetric errors up to rank $\lfloor \frac{n-1}{2} \rfloor$ can be corrected, i.e., beyond the $\lfloor (d-1)/2 \rfloor$ bound.

If $d \leq \frac{n}{2}$, then the error capacity depends on relations between d and n. If, for some m, $(d+m) \mid n$, but $(d+j) \nmid n$, $j = 0, 1, \ldots, m-1$, then a code can correct symmetric errors up to rank $\lfloor \frac{d+m-1}{2} \rfloor$. In particular, one can choose codes correcting symmetric errors up to rank $d - 1$, i.e., the error capacity for symmetric errors is about twice the error capacity for general errors.

2 Codes Containing Symmetric Subcodes

Let

$$\mathbf{g}_0 = g_1, g_2, \ldots, g_n, \quad g_j \in F_{q^n}, \tag{3}$$

be a basis of F_{q^n} over F. Associate with $\mathbf{g}_0$ the $n \times n$-matrix

$$\mathbf{G}_n = \begin{bmatrix} g_1 & g_2 & \cdots & g_n \\ g_1^{[1]} & g_2^{[1]} & \cdots & g_n^{[1]} \\ g_1^{[2]} & g_2^{[2]} & \cdots & g_n^{[2]} \\ \cdots & \cdots & \cdots\cdots \\ g_1^{[n-1]} & g_2^{[n-1]} & \cdots & g_n^{[n-1]} \end{bmatrix}, \tag{4}$$

We use the notation $[i] := q^i$, if $i \geq 0$ and $[i] := q^{n+i}$, if $i < 0$. It is well known (see, e.g., [4]) that the matrix $\mathbf{G}_n$ is non singular.

Definition 1. *A basis* $\mathbf{g}_0 = g_1, g_2, \ldots, g_n$ *is called a **weak self-orthogonal** basis if*

$$\mathbf{G}_n \mathbf{G}_n^T = \mathbf{\Lambda},$$

where $\mathbf{\Lambda}$ *is a diagonal matrix in* F_{q^n}, *not necessarily a multiple of the identity matrix* $\mathbf{I}_n$.

Lemma 1 ([3]). *Weak self-orthogonal bases always exist.*

Lemma 2 ([3]). *A linear (n, k, d) MRD code $\mathcal{V}_k$ with the standard generator matrix of the form*

$$
\mathbf{G}_k = \begin{bmatrix}
g_1 & g_2 & \cdots & g_n \\
g_1^{[1]} & g_2^{[1]} & \cdots & g_n^{[1]} \\
g_1^{[2]} & g_2^{[2]} & \cdots & g_n^{[2]} \\
\cdots & \cdots & \cdots \cdots \\
g_1^{[k-1]} & g_2^{[k-1]} & \cdots & g_n^{[k-1]}
\end{bmatrix}. \tag{5}
$$

contains an $(n, 1, n)$ subcode consisting of symmetric matrices in $\mathbf{g}_0$-representation if $\mathbf{g}_0 = g_1, g_2, \ldots, g_n$ is a weak self-orthogonal basis. This subcode is generated by the first row of the matrix $\mathbf{G}_k$.

The code $\mathcal{V}_k$ is said to be generated by the basis $\mathbf{g}_0$.

Let $\mathcal{M}_k$ be the matrix $\mathbf{g}_0$-representation of the vector code $\mathcal{V}_k$, $\mathcal{M}_k^t$ be the transposed matrix code. The corresponding vector code $\mathcal{V}_k^t$ consists of vectors which are $\mathbf{g}_0$-transposed vectors of $\mathcal{V}_k$.

Lemma 3 ([3]). *The parity check matrix of the code $\mathcal{V}_k$ is given by the last $n - k$ rows of the matrix (4), i. e.,*

$$
\mathbf{H}_{n-k} = \begin{bmatrix}
g_1^{[k]} & g_2^{[k]} & \cdots & g_n^{[k]} \\
g_1^{[k+1]} & g_2^{[k+1]} & \cdots & g_n^{[k+1]} \\
\cdots & \cdots & \cdots \cdots \\
g_1^{[n-1]} & g_2^{[n-1]} & \cdots & g_n^{[n-1]}
\end{bmatrix}. \tag{6}
$$

The parity check matrix $\widehat{\mathbf{H}}_{n-k}$ of the transposed code $\mathcal{V}_k^t$ is as follows:

$$
\widehat{\mathbf{H}}_{n-k} = \begin{bmatrix}
g_1^{[1]} & g_2^{[1]} & \cdots & g_n^{[1]} \\
g_1^{[2]} & g_2^{[21]} & \cdots & g_n^{[2]} \\
\cdots & \cdots & \cdots \cdots \\
g_1^{[n-k]} & g_2^{[n-k]} & \cdots & g_n^{[n-k]}
\end{bmatrix}. \tag{7}
$$

3   Correcting Symmetric Rank Errors

Let $\mathbf{g} \in \mathcal{V}_k$ be a code vector. Assume that a received vector is

$$
\mathbf{y} = \mathbf{g} + \mathbf{e}, \tag{8}
$$

where $\mathbf{e}$ is an error in vector representation. The code $\mathcal{V}_k$ can correct errors $\mathbf{e}$ of rank up to $\lfloor (d-1)/2 \rfloor$. Usually one calculates a syndrome

$$\mathbf{s}_1 = \mathbf{y}\mathbf{H}_{n-k}^t = (\mathbf{g}+\mathbf{e})\mathbf{H}_{n-k}^t = \mathbf{e}\mathbf{H}_{n-k}^t \tag{9}$$

and applies one of the fast decoding algorithms (e.g., see, [5, 6]).

We start just with this procedure. If decoding is successful, then end. This means that the rank $r(\mathbf{e})$ of an error $\mathbf{e}$ is not greater than $\lfloor \frac{d-1}{2} \rfloor$. Otherwise we conclude that $r(\mathbf{e}) > \lfloor \frac{d-1}{2} \rfloor$.

Nevertheless we can *assume* that an error vector $\mathbf{e}$ is a g_0-symmetric one, i. e., $\mathbf{e} = \mathbf{e}^t$, and continue decoding. Transpose the received vector (8) and calculate the syndrome for the transposed code V_k^t:

$$\mathbf{s}_2 = \mathbf{y}^t\widehat{\mathbf{H}}_{n-k}^t = (\mathbf{g}^t + \mathbf{e}^t)\widehat{\mathbf{H}}_{n-k}^t = \mathbf{e}^t\widehat{\mathbf{H}}_{n-k}^t = \mathbf{e}\widehat{\mathbf{H}}_{n-k}^t. \tag{10}$$

Then we use both $\mathbf{s}_1$ and $\mathbf{s}_2$ to find an error $\mathbf{e}$ possible beyond the $\lfloor \frac{d-1}{2} \rfloor$ bound. If $\mathbf{e}$ is a g_0-symmetric error, then end. Otherwise the decision is made that an error is uncorrectable.

From the point of view of coding theory, this means that one considers a rank code with an equivalent parity check matrix of the form

$$\mathbf{H}_{equ} = \begin{pmatrix} \mathbf{H}_{n-k} \\ \widehat{\mathbf{H}}_{n-k} \end{pmatrix}. \tag{11}$$

We have to find the rank distance of a code $\mathcal{V}_{equ}$ defined by the parity check matrix (11). If rank distance of $\mathcal{V}_{equ}$ is D, then one can correct symmetric errors up to rank $\lfloor \frac{D-1}{2} \rfloor$.

The value of D depends on relations between the code rank distance d of the code $\mathcal{V}_k$ and code length n.

The case $k \le (n+1)/2$, or, $d \ge \frac{n+1}{2}$, was investigated completely in [1]. The equivalent parity check matrix can be rewritten after a permutation of rows and deletion of identical rows in the form

$$\mathbf{H}_{equ} = \begin{bmatrix} h_1 & h_2 & \cdots & h_n \\ h_1^{[1]} & h_2^{[1]} & \cdots & h_n^{[1]} \\ \cdots & \cdots & \cdots & \cdots \\ h_1^{[n-2]} & h_2^{[n-2]} & \cdots & h_n^{[n-2]} \end{bmatrix}, \tag{12}$$

where $h_j = g_j^{[1]}$, $j = 1, 2, \ldots, n$. Hence the code rank distance of $\mathcal{V}_{equ}$ is exactly $D = n$. Therefore all the g_0-symmetric errors up to rank $\lfloor \frac{n-1}{2} \rfloor$ can be corrected, i.e., beyond the $\lfloor (d-1)/2 \rfloor$ bound.

Consider the case $k \ge \frac{n+2}{2}$, or, $d \le \frac{n}{2}$. Denote $g_j^{[k]} = h_j$, $j = 1, 2, \ldots, n$. Then the equivalent parity check matrix can be rewritten after a permutation of rows in the form

$$
\mathbf{H}_{equ} = \begin{bmatrix}
h_1 & h_2 & \cdots & h_n \\
h_1^{[1]} & h_2^{[1]} & \cdots & h_n^{[1]} \\
\cdots & \cdots & \cdots \cdots \\
h_1^{[d-2]} & h_2^{[d-2]} & \cdots & h_n^{[d-2]} \\
h_1^{[d]} & h_2^{[d]} & \cdots & h_n^{[d]} \\
h_1^{[d+1]} & h_2^{[d+1]} & \cdots & h_n^{[d+1]} \\
\cdots & \cdots & \cdots \cdots \\
h_1^{[2d-2]} & h_2^{[2d-2]} & \cdots & h_n^{[2d-2]}
\end{bmatrix}.
\tag{13}
$$

It was pointed out in [1] that this parity check matrix defines a code $\mathcal{V}_{equ}$ of rank distance D, where $d = n - k + 1 \leq D \leq 2d - 1$. The precise value of D depends on d and n. Here we find these precise values.

We recall the following necessary and sufficient conditions that a parity check matrix $\mathbf{H}$ defines a code with rank distance d.

Lemma 4 ([5]). *Let $\mathbf{H}$ be an $r \times n$ parity check matrix in $GF(q^n)$. Let $\mathcal{Y}_s$ be the set of $n \times s$ matrices in the base field $GF(q)$ of full rank s. The matrix $\mathbf{H}$ defines a code with rank distance d if and only if, for any matrix $\mathbf{Y}_s \in \mathcal{Y}_s$, $s = 1, 2, \ldots, d - 1$, we have*

$$\mathrm{rank}(\mathbf{HY}_s) = s,$$

and there exists a matrix $\mathbf{Y}_d \in \mathcal{Y}_d$ such that

$$\mathrm{rank}(\mathbf{HY}_d) = d - 1 < d.$$

Proof. Let $\mathbf{H}$ be a parity check matrix of a code with rank distance d. Since for any code $d - 1 \leq r$, then for any matrix $\mathbf{Y}_s \in \mathcal{Y}_s$, $s = 1, 2, \ldots, d - 1$, $\mathrm{rank}(\mathbf{HY}_s) \geq s$. Assume that $\mathrm{rank}(\mathbf{HY}_s) > s$ for some $\mathbf{Y}_s$. Then there exists a non zero s-vector $\mathbf{y}$ such that $\mathbf{y}\mathbf{Y}_s^t\mathbf{H}^t = 0$. But a non zero n-vector $\mathbf{y}\mathbf{Y}_s^t$ has rank at most $s < d$ and can not be a code vector. Thus $\mathrm{rank}(\mathbf{HY}_s) = s$. On the other hand, a code vector of rank d exists. Hence, the relation $\mathrm{rank}(\mathbf{HY}_d) = d - 1 < d$ must be satisfied.

The inverse statement is evident. $\square$

Apply this Lemma to the parity check matrix $\mathbf{H}_{equ}$.

First, we consider simple examples. Let $k = n - 1$. Thus the original rank code $\mathcal{V}_k$ is of rank distance $d = 2$. The equivalent parity check matrix is as follows:

$$
\mathbf{H}_{equ} = \begin{bmatrix}
h_1 & h_2 & \cdots & h_n \\
h_1^{[2]} & h_2^{[2]} & \cdots & h_n^{[2]}
\end{bmatrix}.
\tag{14}
$$

Lemma 5. *Let n be **odd**. Then the parity check matrix (14) defines a code of rank distance $D = 3$.*

Proof. Multiply to the right the matrix (14) by a $n \times 2$ matrix $\mathbf{Y}_2 \in \mathcal{Y}_2$ in F_q of rank 2. Then we obtain the matrix

$$\mathbf{Z} = \begin{bmatrix} f_1 & f_2 \\ f_1^{q^2} & f_2^{q^2} \end{bmatrix}, \tag{15}$$

where $f_1 \in F_{q^n}$ and $f_2 \in F_{q^n}$ are linearly independent over F_q. Calculate the determinant of $\mathbf{Z}$:

$$\det(\mathbf{Z}) = f_1 f_2^{q^2} - f_2 f_1^{q^2} = f_1 f_2^{q^2} \left(1 - \left(\frac{f_1}{f_2} \right)^{q^2-1} \right) \tag{16}$$

Note that $f_1/f_2 \neq 1$ and the field F_{q^n} does not contain elements of order $q^2 - 1$ because n is odd. Therefore, the matrix $\mathbf{Z}$ is nonsingular and by Lemma 4 the matrix (14) defines a code of rank distance $D = 3$. □

In this case, the original code of rank distance 2 can correct symmetric errors of rank 1.

Let $n = 2m$ be **even** and $k = n - 1$. Then it is possible to choose a matrix (15) in such a manner that f_1 and f_2 are linearly independent over F_q but an element f_1/f_2 has order $q^2 - 1$. Thus the conditions of Lemma 4 do not satisfy. The equivalent parity check matrix (14) defines a code of rank distance $D = 2$. Nevertheless one can show that *list* decoding gives a list of errors of rank 1 and a symmetric error of rank 1 is always in this list.

In general, we have for some s and $\mathbf{Y}_s \in \mathcal{Y}_s$:

$$\mathbf{H}(\mathbf{Y}_s) = \mathbf{H}_{equ} \mathbf{Y}_s = \begin{bmatrix} f_1 & f_2 & \cdots & f_s \\ f_1^{[1]} & f_2^{[1]} & \cdots & f_s^{[1]} \\ \cdots & \cdots & \cdots \cdots \\ f_1^{[d-2]} & f_2^{[d-2]} & \cdots & f_s^{[d-2]} \\ f_1^{[d]} & f_2^{[d]} & \cdots & f_s^{[d]} \\ f_1^{[d+1]} & f_2^{[d+1]} & \cdots & f_s^{[d+1]} \\ \cdots & \cdots & \cdots \cdots \\ f_1^{[2d-2]} & f_2^{[2d-2]} & \cdots & f_s^{[2d-2]} \end{bmatrix}, \tag{17}$$

where $f_1, f_2, \cdots, f_s$ are linearly independent over the base field $GF(q)$. We examine $\mathrm{rank}(\mathbf{H}_{equ} \mathbf{Y}_s)$ for s from 1 to $2d - 2$.

1. We can consider the matrix $\mathbf{H}(\mathbf{Y}_s)$ as a parity check matrix of a code $\mathcal{V}(\mathbf{Y}_s)$ of length s. Dimension of $\mathcal{V}(\mathbf{Y}_s)$ depends on rank of $\mathbf{H}(\mathbf{Y}_s)$. If conditions of Lemma 4 are satisfied, then $\mathrm{rank}(\mathbf{H}_{equ} \mathbf{Y}_s) = s$ and $\dim(\mathcal{V}(\mathbf{Y}_s)) = 0$.
2. It is clear that rank distance D of the code $\mathcal{V}_{equ}$ satisfy $D \geq d$ since already the first $d - 1$ rows of the parity check matrix (13) provide rank distance at least d. Hence we have to consider only cases $s \geq d$.

3. Let $s = d$. Suppose that $d \mid n$. Then $GF(q^d) \subset GF(q^n)$. We can choose a matrix $\mathbf{Y}_d$ such that elements $f_1, f_2, \cdots, f_d$ form a basis of $GF(q^d)$. Then $f_j^{[d]} = f_j$, $j = 1, 2, \ldots, d$. Hence we see that $\mathrm{rank}(\mathbf{H}(\mathbf{Y}_d)) = d - 1$, $\dim(\mathcal{V}(\mathbf{Y}_d)) = 1$. So by Lemma 4 the the parity check matrix $\mathbf{H}_{equ}$ defines a code with rank distance $D = d$.

On the other hand, if $d \nmid n$, then $\dim(\mathcal{V}(\mathbf{Y}_d)) = 0$. Otherwise the upper part of $\mathbf{H}(\mathbf{Y}_d)$, namely

$$\begin{bmatrix} f_1 & f_2 & \cdots & f_d \\ f_1^{[1]} & f_2^{[1]} & \cdots & f_d^{[1]} \\ \cdots & \cdots & \cdots \cdots \\ f_1^{[d-2]} & f_2^{[d-2]} & \cdots & f_d^{[d-2]} \end{bmatrix} \tag{18}$$

and the lower part of $\mathbf{H}(\mathbf{Y}_d)$

$$\begin{bmatrix} f_1^{[d]} & f_2^{[d]} & \cdots & f_d^{[d]} \\ f_1^{[d+1]} & f_2^{[d+1]} & \cdots & f_d^{[d+1]} \\ \cdots & \cdots & \cdots \cdots \\ f_1^{[2d-2]} & f_2^{[2d-2]} & \cdots & f_d^{[2d-2]} \end{bmatrix} \tag{19}$$

would define the same one-dimensional code that it is impossible. Thus the matrix $\mathbf{H}_{equ}$ defines the MRD code of distance $D \geq d + 1$.

4. In a similar manner we can prove.

Lemma 6. *Let* $(d + m) \mid n$, *but* $(d + j) \nmid n$, $j = 0, 1, \ldots, m - 1$, *then the parity check matrix* $\mathbf{H}_{equ}$ *defines a code with rank distance* $D \geq d + m$.

5. Finally, the following statement is valid.

Lemma 7. *Let* $(d + j) \nmid n$, $j = 0, 1, \ldots, 2d - 2$, *then the parity check matrix* $\mathbf{H}_{equ}$ *defines a code with maximal rank distance* $D = 2d - 1$.

Corollary 1. *If* n *is a prime, then the parity check matrix* $\mathbf{H}_{equ}$ *from (13) for all* d *defines a MRD code of rank distance* $D = 2d - 1$.

4   Conclusion

We have investigated the error capacity of linear (n, k, d) MRD codes generated by weak self-orthogonal bases. These codes allow to correct not only all errors of rank not greater than $\lfloor (d - 1)/2 \rfloor$ but also many specific (namely, symmetric) errors beyond this bound.

In particular, if $d \geq \frac{n+1}{2}$, then codes can correct symmetric errors up to rank $\frac{n-1}{2}$.

If $d \leq \frac{n}{2}$, one can choose codes correcting symmetric errors up to rank $d - 1$, i.e., the error capacity for symmetric errors is about twice more than for general errors.

References

1. E.M.Gabidulin, N.I. Pilipchuk, "Transposed Rank Codes Based on Symmetric Matrices," in: *Proc. of the WCC'2003*. P. 203-211. 24-28 March 2003. Versailles (France)
2. E.M.Gabidulin, N.I. Pilipchuk, "Symmetric rank codes ," *Problems of Information Transmission*. V. 40. No. 2. Pp. 3-18, 2004.
3. E.M.Gabidulin, N.I. Pilipchuk, "Symmetric matrices and codes correcting rank errors beyond the $\lfloor \frac{d-1}{2} \rfloor$ bound," *Discrete Applied Mathematic* . 2005 (to be published).
4. F.J. MacWilliams, N.J.A. Sloane, "The Theory of Error Correcting Codes," 8th ed, North Holland Press, Amsterdam, 1993.
5. E.M.Gabidulin, "Theory of Codes with Maximum Rank Distance," *Problems of Information Transmission*, v. 21, No. 1, pp. 3-14, 1985.
6. E. M. Gabidulin, "A Fast Matrix Decoding Algorithm For Rank-Error-Correcting Codes." In: (Eds G. Cohen, S. Litsyn, A. Lobstein, G. Zemor), *Algebraic coding* , pp. 126-132, Lecture Notes in Computer Science No. 573, Springer-Verlag, Berlin, 1992.

Error and Erasure Correction of Interleaved Reed–Solomon Codes

Georg Schmidt, Vladimir R. Sidorenko*, and Martin Bossert

Department Telecommunications
and Applied Information Theory
University of Ulm, Germany
{georg.schmidt, vladimir.sidorenko, martin.bossert}@uni-ulm.de

Abstract. We present an algorithm for error and erasure correction of interleaved Reed–Solomon codes. Our algorithm is based on an algorithm recently proposed by Bleichenbacher et al. This algorithm is able to correct many error patterns beyond half the minimum distance of the interleaved Reed–Solomon code. We extend the algorithm in a way, that it is not only able to correct errors, but can correct both, errors and erasures simultaneously. Furthermore we present techniques to describe the algorithm in an efficient way. This can help to reduce the complexity when implementing the algorithm.

1 Introduction

Recently, a decoding algorithm for interleaved Reed–Solomon codes has been introduced in [1]. For constructing an interleaved Reed–Solomon code, l codewords of a Reed–Solomon code $\mathcal{RS}\,(q; n, k, d)$ with length n, dimension k, and minimum distance $d = n - k + 1$ consisting of symbols from the Galois field $\mathbb{F}_q$ are used. These codewords are arranged row-wise into an $l \times n$ matrix $\boldsymbol{C}$. All matrices obtainable in this way constitute a code of length $N = l \cdot n$, dimension $K = l \cdot k$, and minimum distance d with symbols from $\mathbb{F}_q$. Equivalently, these matrices can be interpreted as codewords from a code of length n, dimension k, and minimum distance d with symbols from the field $\mathbb{F}_{q^l}$. We denote such a code by $\mathcal{IRS}(q^l; n, k, d)$. Assume that a codeword $\boldsymbol{c} \in \mathcal{IRS}(q^l; n, k, d) \subset \mathbb{F}_{q^l}^n$ has been corrupted by an additive error vector $\boldsymbol{e} \in \mathbb{F}_{q^l}^n$ of Hamming weight $\theta = \mathrm{wt}(\boldsymbol{e})$. If we would decode the resulting vector $\boldsymbol{y} = \boldsymbol{c} + \boldsymbol{e}$ by using a *Bounded Minimum Distance* (BMD) decoder for all l codewords of the underlying Reed–Solomon codewords independently, we would be able to correct the errors as long as $\theta \leq \lfloor \frac{d-1}{2} \rfloor$. With the decoder from [1], we will be able with high probability to decode $\boldsymbol{y}$, as long as

$$\theta \leq \tfrac{l}{l+1} \cdot (n - k) \triangleq \theta_{\max} \,. \tag{1}$$

---

* The work of Vladimir R. Sidorenko is supported by Deutsche Forschungsgemeinschaft (DFG), Germany

Ø. Ytrehus (Ed.): WCC 2005, LNCS 3969, pp. 22–35, 2006.

From (1) we see, that depending on the choice of l, the maximum correcting radius θ_{max} lies within the range $\lfloor \frac{d-1}{2} \rfloor \leq \theta_{max} < d-1$ (except for the trivial case $d = 1$). However, it should be mentioned, that this decoding algorithm corrects errors with respect to the field $\mathbb{F}_{q^l}$, not with respect to $\mathbb{F}_q$. Consequently, the errors should not occur independently in the symbols of the underlying Reed–Solomon codes, but affect columns of C with symbols from $\mathbb{F}_q$. We have such a situation, e.g., with concatenated codes using interleaved outer codes [2]. The decoding algorithm proposed in [1] can be seen as generalization of the Welch–Berlekamp algorithm described in [3]. The basic idea is to localize the errors in all words of the underlying Reed–Solomon codes simultaneously instead of searching them independently in any word. After the positions of the errors are determined, the values of the transmitted symbols are calculated for all l Reed–Solomon codewords independently. This results in a linear system of equations which is uniquely solvable with high probability, as long as (1) is fulfilled. In the following section we describe, how such a linear system of equations can be obtained. Then we present a technique to decompose this system of equations into two parts, one to locate the errors and another to calculate the error values. This enables us to solve the decoding problem more efficiently and describe it in a concise way. Based on this description, we propose an extension of the algorithm, capable of decoding erroneous symbols and erased symbols simultaneously. Since the positions of the erased symbols have not to be located, erasures can be processed with respect to the field $\mathbb{F}_q$ instead of $\mathbb{F}_{q^l}$. This means, that erasures can be corrected even if they occur *independently* in the several words of the underlying Reed–Solomon codes.

2   Interleaved Reed–Solomon Codes

In order to obtain a decoder for correcting errors and erasures, we start with defining Reed–Solomon codes and briefly describing the algorithm proposed in [1].

Definition 1 (Reed–Solomon (RS) code). *Let*

$$\{C(x)\} = \left\{ \sum_{i=0}^{k-1} C_i x^i, \ C_i \in \mathbb{F}_q \right\}$$

be the set of all polynomials of degree smaller than k with coefficients C_i from $\mathbb{F}_q$. Further, let $\alpha_1, \ldots, \alpha_n$ be n distinct elements from $\mathbb{F}_q$, i.e., $\alpha_i, \alpha_j \in \mathbb{F}_q$, $i \neq j \rightarrow \alpha_i \neq \alpha_j$. Then, a Reed–Solomon code $\mathcal{C} = \mathcal{RS}(q; n, k, d)$ can be defined as the set of vectors

$$\mathcal{C}_{RS} = \{ \boldsymbol{c} \, | \, c_i = C(\alpha_i), \ C(x) \in \{C(x)\}, \ i = 1, \ldots, n \} \ .$$

The minimum Hamming distance of such a code is $d = n - k + 1$.

Assume that a codeword $\boldsymbol{c} \in \mathcal{C}$ is corrupted by adding some error vector $\boldsymbol{e} = (e_1, \ldots, e_n)$ with weight $wt(\boldsymbol{e}) = \theta$. In oder to correct this error, the decoder has to reconstruct the vector $\boldsymbol{c}$ from the observed vector $\boldsymbol{y} = \boldsymbol{c} + \boldsymbol{e}$. There is a one-to-one relation between c and the polynomial $C(x)$. Therefore, the decoder can alternatively reconstruct the polynomial $C(x)$ instead of c. Since the decoder observes θ corrupted and $n - \theta$ uncorrupted symbols, we have

$$C(\alpha_i) \neq y_i \ \forall \, i \in \mathrm{supp}(e) \,,$$
$$C(\alpha_i) = y_i \ \forall \, i \in \{1, \ldots, n\} \setminus \mathrm{supp}(e) \,,$$

where $\mathrm{supp}(e)$ is the set of indices of the non-zero components in e. Now, let

$$\Lambda(x) = \beta \cdot \prod_{i \in \mathrm{supp}(e)} (x - \alpha_i) = 1 + \Lambda_1 x + \cdots + \Lambda_\theta x^\theta$$

be some polynomial which is zero at all positions where the vector y differs from c, and let β be chosen such, that $\Lambda(0) = 1$. Since $|\mathrm{supp}(e)| = \mathrm{wt}(e)$, the degree of $\Lambda(x)$ is $\deg(\Lambda(x)) = \theta$. Further let $\Gamma(x) = C(x) \cdot \Lambda(x) = \Gamma_0 + \Gamma_1 x + \cdots + \Gamma_{\theta+k-1} x^{\theta+k-1}$. With this, we can write

$$\Gamma(\alpha_i) = y_i \cdot \Lambda(\alpha_i)|_{i=1}^n \,. \tag{2}$$

In this way, we obtain a linear system of n equations with $2\theta + k$ unknowns which are the coefficients of $\Gamma(x)$ and $\Lambda(x)$. If θ is small enough, i.e., $\theta \leq \frac{n-k}{2}$, this linear system of equations has a unique solution and $\Gamma(x)$ and $\Lambda(x)$ can uniquely be determined. Since the roots of $\Lambda(x)$ correspond to the positions of the erroneous symbols in y, we can locate the erroneous positions by determining $\Lambda(x)$. Therefore $\Lambda(x)$ is called *error locator polynomial*. The calculation of $\widehat{C}(x) = \frac{\Gamma(x)}{\Lambda(x)}$ gives rise to $\hat{c}$, i.e., an estimation for the uncorrupted codeword c. As proposed in [1], this decoding method can be generalized for decoding interleaved Reed–Solomon codes.

Definition 2 (Interleaved Reed–Solomon code). *Let $c^{(i)}$, $i = 1, \ldots, l$ be l codewords from a Reed–Solomon code $\mathcal{RS}\,(q; n, k, d)$ arranged row-wise in the $l \times n$ matrix*

$$C = \begin{pmatrix} c^{(1)} \\ \vdots \\ c^{(l)} \end{pmatrix}$$

with elements from $\mathbb{F}_q$. With it, we define the set of matrices

$$\mathcal{C}_{IRS} = \left\{ C \,|\, c^{(i)} \in \mathcal{RS}\,(q; n, k, d)\,, \ i = 1, \ldots, l \right\} \,.$$

We interpret any matrix in this set as row vector of length n with elements from the field $\mathbb{F}_{q^l}$, and call $\mathcal{C}_{IRS}$ an interleaved Reed–Solomon code of length n, dimension k and minimum distance d with symbols from $\mathbb{F}_{q^l}$ and denote it by $\mathcal{IRS}(q^l; n, k, d)$.

Assume that some codeword $C \in \mathcal{C}_{IRS}$ is corrupted by some error pattern E consisting of symbols from $\mathbb{F}_{q^l}$, i.e., assume that the decoder observes a word

$\boldsymbol{Y} = \boldsymbol{C} + \boldsymbol{E}$ with elements from $\mathbb{F}_{q^l}$. Equivalently, $\boldsymbol{Y}$ can be considered as matrix with elements from $\mathbb{F}_q$ containing the received words $\boldsymbol{y}^{(\ell)} = \boldsymbol{c}^{(\ell)} + \boldsymbol{e}^{(\ell)} = (y_1^{(\ell)}, \ldots, y_n^{(\ell)})$ in its rows. Since a non-zero element E_i in $\boldsymbol{E} = (E_1, \ldots, E_n)$ affects a column of symbols of the underlying Reed–Solomon codewords, we can perform decoding based on this underlying code but use a common error locator polynomial $\Lambda(x)$ for all l codewords. In this way, we obtain a linear system of equations

$$\Gamma^{(1)}(\alpha_i) = y_i^{(1)} \cdot \Lambda(\alpha_i)\Big|_{i=1}^n \;, \; \ldots \;, \; \Gamma^{(l)}(\alpha_i) = y_i^{(l)} \cdot \Lambda(\alpha_i)\Big|_{i=1}^n \tag{3}$$

with $\Gamma^{(\ell)}(x) = C^{(\ell)}(x) \cdot \Lambda(x)$. This system of equations has $l \cdot n$ equations and $l \cdot (\theta + k) + \theta$ unknowns. It cannot have a unique solution, if the number of equations is smaller than the number of unknowns. Consequently, since we want to have a unique decoding result, the maximum error correcting radius is upper bounded by (1). Furthermore, since the equations obtained from the l codewords of the Reed–Solomon code could be linearly dependent, we cannot ensure a unique solution, if the number of errors is in the range $\lfloor \frac{d-1}{2} \rfloor < \theta \le \frac{l}{l+1}(n-k)$. However, we will demonstrate later that a unique solution exists with high probability.

3   Error Location and Correction

The linear system of equations specified by (3) can be stated as matrix equation

$$\boldsymbol{Ax} = \boldsymbol{b} \,. \tag{4}$$

The vector $\boldsymbol{x} = \left(\boldsymbol{\Gamma}^{(1)}, \ldots, \boldsymbol{\Gamma}^{(l)}, \boldsymbol{\Lambda}\right)^T$ consists of the unknown coefficients $\boldsymbol{\Gamma}^{(\ell)} = (\Gamma_0^{(\ell)}, \ldots, \Gamma_{\theta+k-1}^{(\ell)})$ of the polynomials $\Gamma^{(\ell)}(x)$ and the unknown coefficients $\boldsymbol{\Lambda} = (\Lambda_1, \ldots, \Lambda_\theta)$ of the polynomial $\Lambda(x)$. The matrix $\boldsymbol{A}$ and the vector $\boldsymbol{b}$ can be represented by

$$\boldsymbol{A} = \begin{pmatrix} \boldsymbol{G} & \boldsymbol{0} & \ldots & \boldsymbol{0} & -\boldsymbol{L}^{(1)} \\ \boldsymbol{0} & \boldsymbol{G} & \ldots & \boldsymbol{0} & -\boldsymbol{L}^{(2)} \\ \vdots & \vdots & \ddots & \vdots & \vdots \\ \boldsymbol{0} & \boldsymbol{0} & \ldots & \boldsymbol{G} & -\boldsymbol{L}^{(l)} \end{pmatrix}, \quad \text{and } \boldsymbol{b} = \begin{pmatrix} \boldsymbol{y}^{(1)\,T} \\ \boldsymbol{y}^{(2)\,T} \\ \vdots \\ \boldsymbol{y}^{(l)\,T} \end{pmatrix},$$

where the $n \times (\theta + k)$ matrix $\boldsymbol{G}$ and the l different $n \times \theta$ matrices $\boldsymbol{L}^{(\ell)}$ are of the form $\boldsymbol{G} = \left(g_{\mu,\nu}\right) = \left(\alpha_\mu^{\nu-1}\right)$ and $\boldsymbol{L}^{(\ell)} = \left(l_{\mu,\nu}\right) = \left(y_\mu^{(\ell)} \alpha_\mu^\nu\right)$. The vector $\boldsymbol{\Lambda}$ is uniquely determined by (4), provided that (1) holds and $\mathrm{rank}(\boldsymbol{A}) = l \cdot (\theta+k) + \theta$.

To reduce the decoding complexity, we describe a method to extract a linear system of equations with $l \cdot (n - \theta - k)$ equations and θ unknowns to calculate $\Lambda(x)$. For this purpose, we have a closer look at the matrix $\boldsymbol{G}$. We observe, that it consists of the first $\theta + k$ columns of the $n \times n$ Vandermonde matrix

$$\boldsymbol{V}_n = \left(v_{\mu,\nu}\right) = \left(\alpha_\mu^{\nu-1}\right). \tag{5}$$

Let I denote an identity matrix and let V_n^{-1} be an $n \times n$ matrix, such that $V_n^{-1} \cdot V_n = I$. Then, the matrix product $V_n^{-1} \cdot G$ has the structure

$$V_n^{-1} \cdot G = \begin{pmatrix} I \\ 0 \end{pmatrix}$$

with an $(\theta + k) \times (\theta + k)$ identity matrix I and an $(n - \theta - k) \times (\theta + k)$ zero matrix 0. Consequently, by defining the matrix

$$U = \begin{pmatrix} V_n^{-1} & 0 & \cdots & 0 \\ 0 & V_n^{-1} & \ddots & \vdots \\ \vdots & \ddots & \ddots & 0 \\ 0 & \cdots & 0 & V_n^{-1} \end{pmatrix}$$

and multiplying it from the left to both sides of (4), we obtain

$$\tilde{A}x = \tilde{b} \tag{6}$$

with

$$\tilde{A} = UA = \left(\begin{array}{cccc|c} I & 0 & \cdots & 0 & -W_1 \cdot L^{(1)} \\ 0 & 0 & \cdots & 0 & -W_2 \cdot L^{(1)} \\ \hline 0 & I & \cdots & 0 & -W_1 \cdot L^{(2)} \\ 0 & 0 & \cdots & 0 & -W_2 \cdot L^{(2)} \\ \hline \vdots & \vdots & \ddots & \vdots & \vdots \\ \hline 0 & 0 & \cdots & I & -W_1 \cdot L^{(l)} \\ 0 & 0 & \cdots & 0 & -W_2 \cdot L^{(l)} \end{array} \right), \quad \text{and} \quad \tilde{b} = Ub = \begin{pmatrix} W_1 \cdot y^{(1)^T} \\ W_2 \cdot y^{(1)^T} \\ \hline W_1 \cdot y^{(2)^T} \\ W_2 \cdot y^{(2)^T} \\ \hline \vdots \\ \hline W_1 \cdot y^{(l)^T} \\ W_2 \cdot y^{(l)^T} \end{pmatrix}.$$

Here, the matrix W_1 consists of the first $(\theta + k)$ rows of V_n^{-1} and W_2 consists of the last $(n - \theta - k)$ rows of V_n^{-1}, i.e., $V_n^{-1} = \left(\begin{smallmatrix} W_1 \\ W_2 \end{smallmatrix} \right)$. Considering the structure of $\tilde{A}$ we observe, that it consists of l stripes. The last $n - \theta - k$ rows in any stripe have non-zero entries only in the last θ columns, which correspond to the unknown coefficients of $\Lambda(x)$. Consequently, we can use these rows to create a smaller linear system of equations

$$B \cdot \Lambda^T = s \tag{7}$$

with

$$B = - \begin{pmatrix} W_2 \cdot L^{(1)} \\ W_2 \cdot L^{(2)} \\ \vdots \\ W_2 \cdot L^{(l)} \end{pmatrix}, \quad \text{and} \quad s = \begin{pmatrix} W_2 \cdot y^{(1)^T} \\ W_2 \cdot y^{(2)^T} \\ \vdots \\ W_2 \cdot y^{(l)^T} \end{pmatrix}.$$

This system of equations has $l \cdot (n - \theta - k)$ equations and θ unknowns. If (1) holds and $\text{rank}(\boldsymbol{B}) = \theta$, the coefficient vector $\boldsymbol{\Lambda}$ can be determined from it. After this, the coefficients $\boldsymbol{\Gamma}^{(\ell)}$ can be calculated by

$$\boldsymbol{\Gamma}^{(\ell)} = \left(\boldsymbol{W}_1 \cdot \boldsymbol{y}^{(\ell)^T} + \boldsymbol{W}_1 \cdot \boldsymbol{L}^{(\ell)} \cdot \boldsymbol{\Lambda}^T \right)^T. \tag{8}$$

It is shown in [4], that the inverse $\boldsymbol{V}_n^{-1}$ of a Vandermonde matrix $\boldsymbol{V}_n$ can be given in a closed form by

$$\boldsymbol{V}_n^{-1} = \left(v_{\mu,\nu} \right) = n^{-1} \left(\alpha_\nu^{-(\mu-1)} \right), \tag{9}$$

if $\alpha_1, \ldots, \alpha_n \in \mathbb{F}_q$ are the roots of $x^n - 1$ and n is relatively prime to q. This condition holds for Reed–Solomon codes of length $n = q - 1$. Whenever we have $\boldsymbol{V}_n^{-1}$ in the form of (9), we can use this structure to further simplify the calculation of $\boldsymbol{\Lambda}$ and $\boldsymbol{\Gamma}^{(\ell)}$ without ever calculating $\boldsymbol{V}_n^{-1}$ explicitly. We use $\rho = n - k - \theta$ and $\boldsymbol{\alpha}^{(i)} = (\alpha_1^i, \alpha_2^i, \ldots, \alpha_n^i)$ to write the matrix product $\boldsymbol{W}_2 \cdot \boldsymbol{L}^{(\ell)}$ in the form

$$\boldsymbol{W}_2 \cdot \boldsymbol{L}^{(\ell)} = \begin{pmatrix} \boldsymbol{\alpha}^{(\rho+1)} \boldsymbol{y}^{(\ell)^T} & \boldsymbol{\alpha}^{(\rho+2)} \boldsymbol{y}^{(\ell)^T} & \cdots & \boldsymbol{\alpha}^{(\rho+\theta)} \boldsymbol{y}^{(\ell)^T} \\ \boldsymbol{\alpha}^{(\rho)} \boldsymbol{y}^{(\ell)^T} & \boldsymbol{\alpha}^{(\rho+1)} \boldsymbol{y}^{(\ell)} & \cdots & \boldsymbol{\alpha}^{(\rho+\theta-1)} \boldsymbol{y}^{(\ell)^T} \\ \vdots & \vdots & \ddots & \vdots \\ \boldsymbol{\alpha}^{(2)} \boldsymbol{y}^{(\ell)^T} & \boldsymbol{\alpha}^{(3)} \boldsymbol{y}^{(\ell)^T} & \cdots & \boldsymbol{\alpha}^{(\theta+1)} \boldsymbol{y}^{(\ell)^T} \end{pmatrix},$$

where $\boldsymbol{\alpha}^{(i)} \boldsymbol{y}^{(\ell)^T}$ is the scalar product between the vector $\boldsymbol{\alpha}^{(i)}$ and the received vector $\boldsymbol{y}^{(\ell)^T}$. We observe, that the matrix

$$\boldsymbol{Z} = \boldsymbol{Y} \cdot \boldsymbol{V}_n = \left(z_{\mu,\nu} \right) = \left(\boldsymbol{\alpha}^{(\nu-1)} \boldsymbol{y}^{(\mu)^T} \right) \tag{10}$$

contains all these scalar products. Consequently, the matrix product $\boldsymbol{W}_2 \cdot \boldsymbol{L}^{(\ell)}$ can be represented as *Toeplitz Matrix* composed from elements of the ℓ-th row of $\boldsymbol{Z}$. With this, we can represent

$$\boldsymbol{B} = - \left(\begin{array}{ccc} z_{1,\rho+1} \; z_{1,\rho+2} \; \cdots \; z_{1,\rho+\theta} \\ \vdots \quad \vdots \quad \ddots \quad \vdots \\ z_{1,2} \quad z_{1,3} \quad \cdots \; z_{1,\theta+1} \\ \hline \vdots \quad \vdots \qquad \vdots \\ \hline z_{l,\rho+1} \; z_{l,\rho+2} \; \cdots \; z_{l,\rho+\theta} \\ \vdots \quad \vdots \quad \ddots \quad \vdots \\ z_{l,2} \quad z_{l,3} \quad \cdots \; z_{l,\theta+1} \end{array} \right) \tag{11}$$

as a matrix composed of l Toeplitz matrices with elements from $\boldsymbol{Z}$. The vector

$$\boldsymbol{s} = \left(z_{1,\rho}, \ldots, z_{1,1} \mid \ldots \mid z_{l,\rho}, \ldots, z_{l,1} \right)^T \tag{12}$$

can also be written by means of elements from $\boldsymbol{Z}$. To calculate the coefficients $\Gamma^{(\ell)}$, we consider the structure of the matrix product

$$
\boldsymbol{W}_1 \cdot \boldsymbol{L}^{(\ell)} =
\begin{pmatrix}
\boldsymbol{\alpha}^{(1)} \boldsymbol{y}^{(\ell)T} & \boldsymbol{\alpha}^{(2)} \boldsymbol{y}^{(\ell)T} & \cdots & \boldsymbol{\alpha}^{(\theta)} \boldsymbol{y}^{(\ell)T} \\
\boldsymbol{\alpha}^{(0)} \boldsymbol{y}^{(\ell)T} & \boldsymbol{\alpha}^{(n-1)} \boldsymbol{y}^{(\ell)T} & \cdots & \boldsymbol{\alpha}^{(n-\theta+1)} \boldsymbol{y}^{(\ell)T} \\
\vdots & \vdots & \cdots & \vdots \\
\boldsymbol{\alpha}^{(\rho+2)} \boldsymbol{y}^{(\ell)T} & \boldsymbol{\alpha}^{(\rho+1)} \boldsymbol{y}^{(\ell)T} & \cdots & \boldsymbol{\alpha}^{(\rho+\theta+1)} \boldsymbol{y}^{(\ell)T}
\end{pmatrix},
$$

and the structure of the product

$$
\boldsymbol{W}_1 \cdot \boldsymbol{y}^{(\ell)T} =
\begin{pmatrix}
\boldsymbol{\alpha}^{(0)} \boldsymbol{y}^{(\ell)T} \\
\boldsymbol{\alpha}^{(n-1)} \boldsymbol{y}^{(\ell)T} \\
\vdots \\
\boldsymbol{\alpha}^{(\rho+1)} \boldsymbol{y}^{(\ell)T}
\end{pmatrix}.
$$

Carefully examining (8) having this special structure in mind and regarding the fact that $\Lambda_0 = 1$, we observe, that the i-th coefficient of the polynomial $\Gamma^{(\ell)}$ can be calculated by

$$
\Gamma_i^{(\ell)} = \sum_{j=0}^{\theta} \Lambda_j \boldsymbol{\alpha}^{(n-i+j)} \boldsymbol{y}^{(\ell)T} = \sum_{j=0}^{\theta} \Lambda_j z_{\ell, n-i+j} \, . \tag{13}
$$

With this considerations, we are now ready to describe a decoding procedure, which is able to reconstruct a codeword $\boldsymbol{c}$ from an observed vector $\boldsymbol{y}$, provided that θ satisfies (1) and the linear system of equations (4) has a unique solution. Uniqueness is guaranteed for $\theta \leq \lfloor \frac{n-k}{2} \rfloor$ by the properties of the underlying Reed–Solomon code. If θ is larger, a unique solution cannot longer be guaranteed but exists with high probability.

For the sake of simplicity we assume here, that $\alpha_1, \ldots, \alpha_n \in \mathbb{F}_q$ are the roots of $x^n - 1$ and n is relatively prime to q. We have this case, if we consider Reed–Solomon codes of length $n = q - 1$. In a first step, we use the observed word $\boldsymbol{Y}$ to calculate the matrix $\boldsymbol{Z}$ according to (10). The elements of this matrix can be used to create the matrix $\boldsymbol{B}$ and the vector $\boldsymbol{s}$ according to (11) and (12) respectively. In the next step, we have to solve the linear system of equations (7) constituted by $\boldsymbol{B}$ and $\boldsymbol{s}$ with respect to $\boldsymbol{\Lambda}$. The problem here is, that the dimensions of $\boldsymbol{B}$ and $\boldsymbol{s}$ depend on the number of errors θ. The system is solvable, if $\mathrm{rank}(\boldsymbol{B}) = \mathrm{rank}((\boldsymbol{B}|\boldsymbol{s}))$, where $(\boldsymbol{B}|\boldsymbol{s})$ is the matrix obtained by concatenating the matrix $\boldsymbol{B}$ and the column vector $\boldsymbol{s}$. If there are no errors, i.e., $\theta = 0$, the vector $\boldsymbol{s}$ has to be the all-zero vector. Hence, whenever we observe $\boldsymbol{s} = \boldsymbol{0}$, we do not have to decode anything. If we have $\boldsymbol{s} \neq \boldsymbol{0}$, we know, that $\theta > 0$. In this case, we assume $\theta = 1$ and check the rank of $\boldsymbol{B}$ and $(\boldsymbol{B}|\boldsymbol{s})$. If we have $\mathrm{rank}(\boldsymbol{B}) \neq \mathrm{rank}((\boldsymbol{B}|\boldsymbol{s}))$, we know, that the system has no solution and our assumption is wrong. In this case, we increase θ by one and inspect the ranks again. We repeat this procedure, until we find the smallest possible θ for which $\mathrm{rank}(\boldsymbol{B}) = \mathrm{rank}((\boldsymbol{B}|\boldsymbol{s}))$. Furthermore, if rank($\boldsymbol{B}$) = θ holds, we know that the solution of (7) will be unique. Therefore, if rank($\boldsymbol{B}$) < θ, there does not exist a unique solution and consequently our decoding attempt fails. If we have rank($\boldsymbol{B}$) = θ, we solve (7) to obtain $\boldsymbol{\Lambda}$, which in turn gives rise to the error locator polynomial $\Lambda(x)$. Once we found $\Lambda(x)$ we can use (13) to calculate the coefficents of the polynomials $\Gamma^{(\ell)}(x)$ for $\ell = 1, \ldots, l$. In the last step, we calculate the polynomials

$$\widehat{C}^{(\ell)}(x) = \frac{\Gamma^{(\ell)}(x)}{\Lambda(x)} \ \forall \ \ell = 1, \ldots, l \, , \tag{14}$$

and map them to the corresponding codeword $\widehat{\boldsymbol{C}} \in \mathcal{C}_{\mathrm{IRS}}$ to obtain our decoding result.

4 Error and Erasure Correction

Now we assume, that the received codewords are not only corrupted by errors, but also by erasures. Unlike errors, the positions of the erasures can be detected in the received word. Hence, we only need to locate the errors, not the erasures. Consequently, we can correct erasures not only with respect to the field $\mathbb{F}_{q^l}$, but also with respect to the smaller field $\mathbb{F}_q$. In the following, we describe a method for correcting errors in the field $\mathbb{F}_{q^l}$ and erasures in the field $\mathbb{F}_q$ simultaneously. Let ξ denote an erased symbol, i.e., let $\xi \notin \mathbb{F}_q$ be some special symbol not included in $\mathbb{F}_q$. Furthermore, let the addition of ξ and some $\alpha \in \mathbb{F}_q$ be defined by $\alpha + \xi = \xi + \alpha = \xi$. Now, assume that the decoder observes a word $\boldsymbol{Y}$, which is corrupted by some error pattern $\boldsymbol{E}$ consisting of symbols from $\mathbb{F}_{q^l}$. In addition to this, assume that $\boldsymbol{Y}$ is also corrupted by some erasure pattern

$$\mathcal{E} = \begin{pmatrix} \epsilon^{(1)} \\ \vdots \\ \epsilon^{(l)} \end{pmatrix} \, ,$$

where $\epsilon^{(1)}, \ldots, \epsilon^{(l)} \in \{0, \xi\}^n$ are l vectors with zeros and erased symbols. In other words, the decoder observes the word $\boldsymbol{Y} = \boldsymbol{C} + \boldsymbol{E} + \mathcal{E}$, which is the sum of a valid codeword $\boldsymbol{C}$, an error pattern $\boldsymbol{E}$, and an erasure pattern $\mathcal{E}$. As before, we also can interpret

$$\boldsymbol{Y} = \begin{pmatrix} \boldsymbol{y}^{(1)} \\ \vdots \\ \boldsymbol{y}^{(l)} \end{pmatrix}$$

as matrix with elements from $\{\mathbb{F}_q \cup \xi\}$ containing the received words $\boldsymbol{y}^{(\ell)} = (y_1^{(\ell)}, \ldots, y_n^{(\ell)})$.

To find a decoding method for errors and erasures, we again look at the linear system of equations stated in (4). We consider the ℓ-th stripe containing the matrix $\boldsymbol{L}^{(\ell)}$ and the corresponding observed vector $\boldsymbol{y}^{(\ell)}$. For any symbol $y_i^{(\ell)}$ there exists exactly one row in the ℓ-th stripe which depends on this symbol.

If we observe an erasure, i.e., a symbol $y_i^{(\ell)}$ with the value ξ, we do not know anything about the transmitted symbol. Therefore we just erase the i-th row in the ℓ-th stripe of $\boldsymbol{A}$ and the corresponding symbol in $\boldsymbol{b}$. To describe this formally, we first define the operator $\mathcal{N}(\boldsymbol{y}) = \{i : y_i \in \mathbb{F}_q\}$ as set of indices of the non-erasure positions in some vector $\boldsymbol{y}$. Furthermore, let

$$\boldsymbol{M} = \begin{pmatrix} \boldsymbol{m}_1 \\ \vdots \\ \boldsymbol{m}_n \end{pmatrix}$$

be a matrix with n rows and $\boldsymbol{y}$ some vector of length n. Define the projection

$$P_{\mathcal{N}(\boldsymbol{y})}\left[\boldsymbol{M}\right] = \left(\boldsymbol{m}_{i_1}{}^T, \ldots, \boldsymbol{m}_{i_\kappa}{}^T\right)^T, \; \{i_1, \ldots, i_\kappa\} = \mathcal{N}(\boldsymbol{y})$$

to be a matrix composed of the $\kappa = |\mathcal{N}(\boldsymbol{y})|$ rows of $\boldsymbol{M}$ corresponding to $\mathcal{N}(\boldsymbol{y})$. With this projection, we obtain the matrices $\boldsymbol{G}'^{(\ell)} = P_{\mathcal{N}(\boldsymbol{y}^{(\ell)})}\left[\boldsymbol{G}\right]$ and $\boldsymbol{L}'^{(\ell)} = P_{\mathcal{N}(\boldsymbol{y}^{(\ell)})}\left[\boldsymbol{L}\right]$, i.e.,

$$\boldsymbol{G}'^{(\ell)} = \left(g_{\mu,\nu}\right) = \left(\alpha_{i_\mu}^{\nu-1}\right) \text{ and } \boldsymbol{L}'^{(\ell)} = \left(l_{\mu,\nu}\right) = \left(y_{i_\mu}^{(\ell)} \alpha_{i_\mu}^{\nu}\right). \tag{15}$$

Furthermore, we define the punctured vectors $\boldsymbol{y}'^{(\ell)} = (y_{i_1}^{(\ell)}, \ldots, y_{i_\kappa}^{(\ell)})$ containing all non-erased symbols of the corresponding observed vectors $\boldsymbol{y}^{(\ell)}$. Now we can state our error and erasure correcting problem as a linear system of equations

$$\boldsymbol{A}'\boldsymbol{x} = \boldsymbol{b}' \tag{16}$$

with

$$\boldsymbol{A}' = \begin{pmatrix} \boldsymbol{G}'^{(1)} & \boldsymbol{0} & \cdots & \boldsymbol{0} & -\boldsymbol{L}'^{(1)} \\ \boldsymbol{0} & \boldsymbol{G}'^{(2)} & \cdots & \boldsymbol{0} & -\boldsymbol{L}'^{(2)} \\ \vdots & \vdots & \ddots & \vdots & \vdots \\ \boldsymbol{0} & \boldsymbol{0} & \cdots & \boldsymbol{G}'^{(l)} & -\boldsymbol{L}'^{(l)} \end{pmatrix}, \text{ and } \boldsymbol{b} = \begin{pmatrix} \boldsymbol{y}'^{(1)\,T} \\ \boldsymbol{y}'^{(2)\,T} \\ \vdots \\ \boldsymbol{y}'^{(l)\,T} \end{pmatrix}.$$

The vector $\boldsymbol{x} = \left(\boldsymbol{\Gamma}^{(1)}, \ldots, \boldsymbol{\Gamma}^{(l)}, \boldsymbol{\Lambda}\right)^T$ coincides with the one defined previously in (4). Now, let $\varepsilon^{(\ell)} = n - |\mathcal{N}(\boldsymbol{y}^{(\ell)})|$ be the number of erased positions in the ℓ-th observed vector $\boldsymbol{y}^{(\ell)}$. In order to determine the $\theta + k$ unknown coefficients of the polynomial $\Gamma^{(\ell)}(x)$, the rank of $\boldsymbol{G}'^{(\ell)}$ has to be at least $\theta + k$. Therefore, to be able to obtain a unique decoding result, the necessary condition

$$\theta + \varepsilon^{(\ell)} \leq d - 1 \; \forall \, \ell = 1, \ldots, l \tag{17}$$

has to hold. Furthermore, (16) cannot have a unique solution, when the number of rows of $\boldsymbol{A}'$ is smaller than the number of columns. Therefore in order to have a unique solution we additionally have to state the necessary condition

$$\theta + \frac{\varepsilon}{l+1} \leq \frac{l}{l+1}(n - k) \tag{18}$$

with $\varepsilon = \sum_{\ell=1}^{l} \varepsilon^{(\ell)}$. Considering the structure of the matrices $\boldsymbol{G}'^{(\ell)}$, we observe that they again have a Vandermonde like structure, i.e., $\boldsymbol{G}'^{(\ell)}$ consists of the first $\theta + k$ columns of the $\kappa \times \kappa$ Vandermonde matrix

$$\boldsymbol{V}_{\mathcal{N}(\boldsymbol{y}^{(\ell)})} = \left(v_{\mu,\nu}\right) = \left(\alpha_{i_\mu}^{\nu-1}\right) \tag{19}$$

with $\kappa = |\mathcal{N}(\boldsymbol{y}^{(\ell)})|$. To emphasize that the corresponding Vandermonde matrix is composed of the elements $\alpha_{i_1}, \ldots, \alpha_{i_\kappa}$, $\{i_1, \ldots, i_\kappa\} = \mathcal{N}(\boldsymbol{y}^{(\ell)})$, we denote it in slightly misusing the notation by $\boldsymbol{V}_{\mathcal{N}(\boldsymbol{y}^{(\ell)})}$. Let $\boldsymbol{V}_{\mathcal{N}(\boldsymbol{y}^{(\ell)})}^{-1}$ be the inverse of $\boldsymbol{V}_{\mathcal{N}(\boldsymbol{y}^{(\ell)})}$, then we can use the matrix

$$\boldsymbol{U}' = \begin{pmatrix} \boldsymbol{V}_{\mathcal{N}(\boldsymbol{y}^{(1)})}^{-1} & \boldsymbol{0} & \cdots & \boldsymbol{0} \\ \boldsymbol{0} & \boldsymbol{V}_{\mathcal{N}(\boldsymbol{y}^{(2)})}^{-1} & \ddots & \vdots \\ \vdots & \ddots & \ddots & \boldsymbol{0} \\ \boldsymbol{0} & \cdots & \boldsymbol{0} & \boldsymbol{V}_{\mathcal{N}(\boldsymbol{y}^{(l)})}^{-1} \end{pmatrix}$$

to calculate

$$\tilde{\boldsymbol{A}}' = \boldsymbol{U}'\boldsymbol{A}' = \begin{pmatrix} \boldsymbol{I} & 0 & \cdots & 0 & -\boldsymbol{W}_1^{(1)} \cdot \boldsymbol{L}'^{(1)} \\ 0 & 0 & \cdots & 0 & -\boldsymbol{W}_2^{(1)} \cdot \boldsymbol{L}'^{(1)} \\ 0 & \boldsymbol{I} & \cdots & 0 & -\boldsymbol{W}_1^{(2)} \cdot \boldsymbol{L}'^{(2)} \\ 0 & 0 & \cdots & 0 & -\boldsymbol{W}_2^{(2)} \cdot \boldsymbol{L}'^{(2)} \\ \vdots & \vdots & \ddots & \vdots & \vdots \\ 0 & 0 & \cdots & \boldsymbol{I} & -\boldsymbol{W}_1^{(l)} \cdot \boldsymbol{L}'^{(l)} \\ 0 & 0 & \cdots & 0 & -\boldsymbol{W}_2^{(l)} \cdot \boldsymbol{L}'^{(l)} \end{pmatrix}, \text{ and } \tilde{\boldsymbol{b}}' = \boldsymbol{U}'\boldsymbol{b}' = \begin{pmatrix} \boldsymbol{W}_1^{(1)} \cdot \boldsymbol{y}^{(1)^T} \\ \boldsymbol{W}_2^{(1)} \cdot \boldsymbol{y}^{(1)^T} \\ \boldsymbol{W}_1^{(2)} \cdot \boldsymbol{y}^{(2)^T} \\ \boldsymbol{W}_2^{(2)} \cdot \boldsymbol{y}^{(2)^T} \\ \vdots \\ \boldsymbol{W}_1^{(l)} \cdot \boldsymbol{y}^{(l)^T} \\ \boldsymbol{W}_2^{(l)} \cdot \boldsymbol{y}^{(l)^T} \end{pmatrix},$$

and obtain the linear system of equations

$$\tilde{\boldsymbol{A}}' \cdot \boldsymbol{x} = \tilde{\boldsymbol{b}}' . \tag{20}$$

As in the previous section, $\boldsymbol{W}_1^{(\ell)}$ consists of the first $(\theta + k)$ rows of $\boldsymbol{V}_{\mathcal{N}(\boldsymbol{y}^{(\ell)})}^{-1}$ and $\boldsymbol{W}_2^{(\ell)}$ of the remaining $n - \theta - k - \varepsilon^{(\ell)}$ rows. Consequently, we obtain again a matrix with l stripes, where the last $n - \theta - k - \varepsilon^{(\ell)}$ rows of the ℓ-th stripe only have non-zero entries in the last θ columns. However, unlike in the case of decoding without erasures, the number of this rows can be different in any stripe, since the size of $\boldsymbol{W}_2^{(\ell)}$ depends on $\varepsilon^{(\ell)}$. Nevertheless, we can take these rows to form the linear system of equations

$$\boldsymbol{B}' \cdot \boldsymbol{\Lambda}^T = \boldsymbol{s}' \tag{21}$$

with

$$\boldsymbol{B}' = -\begin{pmatrix} \boldsymbol{W}_2^{(1)} \cdot \boldsymbol{L}'^{(1)} \\ \vdots \\ \boldsymbol{W}_2^{(l)} \cdot \boldsymbol{L}'^{(l)} \end{pmatrix}, \text{ and } \boldsymbol{s}' = \begin{pmatrix} \boldsymbol{W}_2^{(1)} \cdot \boldsymbol{y}'^{(1)^T} \\ \vdots \\ \boldsymbol{W}_2^{(l)} \cdot \boldsymbol{y}'^{(l)^T} \end{pmatrix} . \tag{22}$$

Note, that if $\theta + \varepsilon^{(\ell)} = d - 1$, the corresponding matrix $\boldsymbol{W}_2^{(\ell)}$ and also $\boldsymbol{W}_2^{(\ell)} \cdot \boldsymbol{L'}^{(\ell)}$ do not consist of any lines. However, even in this case Equation (21) can have a unique solution as long as Condition (18) is still fulfilled. More precisely, we find a solution for $\boldsymbol{\Lambda}$ in any case, in which we have rank$(\boldsymbol{B'}) = \theta$. If we find a solution, the coefficients $\boldsymbol{\Gamma}^{(\ell)}$ can be obtained from the equation

$$\boldsymbol{\Gamma}^{(\ell)} = \left(\boldsymbol{W}_1^{(\ell)} \cdot \boldsymbol{y'}^{(\ell)^T} + \boldsymbol{W}_1^{(\ell)} \cdot \boldsymbol{L'}^{(\ell)} \cdot \boldsymbol{\Lambda}^T \right)^T \tag{23}$$

provided that Condition (17) holds. Unfortunately, in the presence of erasures, the $\kappa = \left| \mathcal{N}(\boldsymbol{y}^{(\ell)}) \right|$ elements $\alpha_{i_1}, \ldots, \alpha_{i_\kappa}$ constituting the Vandermonde matrix $\boldsymbol{V}_{\mathcal{N}(\boldsymbol{y}^{(\ell)})}$ mostly do not fulfill the properties to give its inverse in the form of (9). Therefore, we need a more general way to obtain the inverse $\boldsymbol{V}_{\mathcal{N}(\boldsymbol{y}^{(\ell)})}^{-1}$. For the sake of a concise notation, let $\beta_\mu = \alpha_{i_\mu} \forall i_\mu \in \mathcal{N}(\boldsymbol{y}^{(\ell)})$, and let $\kappa = \left| \mathcal{N}(\boldsymbol{y}^{(\ell)}) \right|$. It is described in [5], that the inverse of the Vandermonde matrix $\boldsymbol{V}_{\mathcal{N}(\boldsymbol{y}^{(\ell)})} = (v_{\mu,\nu}) = (\beta_\mu^{\nu-1})$ is given by

$$\boldsymbol{V}_{\mathcal{N}(\boldsymbol{y}^{(\ell)})}^{-1} = (v_{\mu,\nu}) = \left(\frac{(-1)^{\nu-1} \Pi_{\kappa-\nu}^\mu (\beta_1, \ldots, \beta_\kappa)}{\prod\limits_{\substack{k=1 \\ k \neq \mu}}^{\kappa} (\beta_k - \beta_\mu)} \right)^T, \tag{24}$$

where $\Pi_r^\mu (\beta_1, \ldots, \beta_\kappa)$ is defined in analogy to an elementary symmetric function by

$$\Pi_r^\mu (\beta_1, \ldots, \beta_\kappa) = \sum_{\substack{1 \leq k_1 < \cdots < k_r \leq \kappa \\ k_1, \ldots, k_r \neq \mu}} \beta_{k_1} \ldots \beta_{k_r} .$$

The numerator $(-1)^{\nu-1} \Pi_{\kappa-\nu}^i (\beta_1, \ldots, \beta_\kappa)$ of (24) is just the coefficient $p_{\mu,\nu}$ of $x^{\nu-1}$ in the polynomial

$$p_\mu(x) = \prod_{\substack{k=1 \\ k \neq \mu}}^{\kappa} (\beta_k - x) = p_{\mu,1} + p_{\mu,2} x + \cdots + p_{\mu,\kappa} x^{\kappa-1} .$$

Furthermore we observe, that the denominator of (24) is just $p_\mu(x)$ evaluated for $x = \beta_\mu$. Consequently, we can write (24) in the less cumbersome form

$$\boldsymbol{V}_{\mathcal{N}(\boldsymbol{y}^{(\ell)})}^{-1} = (v_{\mu,\nu}) = \left(\frac{p_{\mu,\nu}}{p_\mu(\beta_\mu)} \right)^T . \tag{25}$$

From this equation we can easily show, that $\boldsymbol{V}_{\mathcal{N}(\boldsymbol{y}^{(\ell)})}^{-1}$ is an inverse for $\boldsymbol{V}_{\mathcal{N}(\boldsymbol{y}^{(\ell)})}$. For this purpose, we verify, that

$$\boldsymbol{V}_{\mathcal{N}(\boldsymbol{y}^{(\ell)})}^{-1} \boldsymbol{V}_{\mathcal{N}(\boldsymbol{y}^{(\ell)})} = \left[\boldsymbol{V}_{\mathcal{N}(\boldsymbol{y}^{(\ell)})}^{-1} \right]^T \left[\boldsymbol{V}_{\mathcal{N}(\boldsymbol{y}^{(\ell)})} \right]^T = \boldsymbol{I} = (i_{i,j})$$

yields the identity matrix. At this matrix product, the element $i_{i,j}$ is obtained by calculating the scalar product from the i-th row of $[\mathbf{V}_{\mathcal{N}(\mathbf{y}^{(\ell)})}^{-1}]^T$ and the j-th column of $[\mathbf{V}_{\mathcal{N}(\mathbf{y}^{(\ell)})}]^T$:

$$i_{i,j} = \sum_{k=1}^{\nu} \frac{p_{i,k}}{p_i(\beta_i)} \cdot \beta_j^{k-1} = \sum_{k=1}^{\nu} \frac{p_{i,k}\beta_j^{k-1}}{p_i(\beta_i)} = \frac{p_i(\beta_j)}{p_i(\beta_i)}.$$

Due to the fact, that $p_i(x)$ is zero for $x \in \{\beta_1 \dots, \beta_{i-1}, \beta_{i+1}, \dots, \beta_\nu\}$ and non-zero only for $x = \beta_i$, we have

$$i_{i,j} = \begin{cases} 1 & i = j \\ 0 & i \neq j \end{cases}.$$

From this we conclude, that $\mathbf{V}_{\mathcal{N}(\mathbf{y}^{(\ell)})}^{-1}$ inverts $\mathbf{V}_{\mathcal{N}(\mathbf{y}^{(\ell)})}$.

To complete this section, we deduce an error and erasure decoding algorithm from the discussed facts. First, we create for all $\ell = 1, \dots, l$ the punctured vectors $\mathbf{y}'^{(\ell)}$ from the observed word $\mathbf{Y}$. Then we use (25) to obtain the matrices $\mathbf{W}_2^{(\ell)}$. With this, we can calculate the vector $\mathbf{s}'$ according to (22). If we observe $\mathbf{s}' = \mathbf{0}$ we know, that $\mathbf{Y}$ is not corrupted by any errors and we set $\Lambda(x) = 1$. If $\mathbf{s}' \neq \mathbf{0}$, we assume $\theta = 1$ and use (15) to calculate $\mathbf{L}'^{(\ell)}$. Then, we create $\mathbf{B}'$ according to (22). In the same way as before, we inspect $\operatorname{rank}(\mathbf{B}')$ and $\operatorname{rank}((\mathbf{B}'|\mathbf{s}'))$ and increase θ if necessary until we have $\operatorname{rank}(\mathbf{B}') = \operatorname{rank}((\mathbf{B}'|\mathbf{s}'))$. Then we check, whether $\operatorname{rank}(\mathbf{B}') = \theta$ holds, i.e., whether there exists a unique solution. If no unique solution exists, we cancel the decoding attempt with a failure. Otherwise, we obtain Λ as solution of (21), which gives rise to the error locator polynomial $\Lambda(x)$. After this, we use (23) to obtain the coefficients $\mathbf{\Gamma}^{(\ell)}$ of the polynomials $\Gamma^{(\ell)}(x)$. The last step is performed in the same way as before. We use (14) to calculate the polynomials $\widehat{C}(x)^{(\ell)}$ and map them to the corresponding codeword $\widehat{C}$, which is our decoding result.

5   Probability for a Unique Decoding Result

The probability for a decoding failure is upper bounded in [1] by $P_f \leq \theta/q$. In [6], this bound has recently been improved to $P_f \leq \exp\left(1/q^{l-2}\right)/q$. The problem with this bound is, that it does not depend on θ, the number of errors actually occurred. Consequently, it only gives us a tight bound in the case $\theta = \theta_{\max}$. In [7] the following bound on P_f is given, which depends on θ:

$$P_f(\theta) \leq \left(\frac{q^l - \frac{1}{q}}{q^l - 1}\right)^\theta \cdot \frac{q^{-l \cdot (n-k-\theta)+\theta}}{q-1} = \left(\frac{q^l - \frac{1}{q}}{q^l - 1}\right)^\theta \cdot \frac{q^{-(l+1)(\theta_{\max}-\theta)}}{q-1}. \quad (26)$$

We use Monte Carlo methods, to verify the tightness of these bounds. For this purpose, we randomly generate error patterns $\mathbf{E} \in \mathbb{F}_{q^l}^n$ with fixed Hamming

Table 1. Probability P_f to get a decoder failure

	$\mathcal{IRS}((2^4)^3;15,7,9))$			$\mathcal{IRS}((2^5)^3;31,15,17))$		
θ_{max}	6			12		
θ	5	6	7	11	12	13
#experiments w/o unique solution	8	$6.62 \cdot 10^5$	$1 \cdot 10^7$	0	$3.21 \cdot 10^5$	10^7
P_f	$\approx 8 \cdot 10^{-7}$	$6.62 \cdot 10^{-2}$	1	$< 1 \cdot 10^{-7}$	$3.21 \cdot 10^{-2}$	1
Bound from [6]		$6.65 \cdot 10^{-2}$			$3.22 \cdot 10^{-2}$	
Bound (26)	$1.02 \cdot 10^{-6}$	$6.67 \cdot 10^{-2}$	1	$3.08 \cdot 10^{-8}$	$3.23 \cdot 10^{-2}$	1

weights $\text{wt}(\boldsymbol{E}) = \theta$. For any weight θ, 10^7 experiments are performed. In each experiment we randomly generate a codeword $\boldsymbol{C} \in \mathcal{IRS}(q, n, k, d)$ and add some random error vector $\boldsymbol{E}$, $\text{wt}(\boldsymbol{E}) = \theta$. Then we count the number of experiments, in which our decoder is not able to find a unique solution. This yields an estimation for the probability P_f. We perform these experiments for the two different codes $\mathcal{IRS}((2^4)^3; 15, 7, 9)$, and $\mathcal{IRS}((2^5)^3; 31, 15, 17)$. The results are shown in Table 1. We observe, that the bounds from [6] and [7] are quite tight if θ equals to the maximum error correction radius $\theta_{max} = \frac{l}{l+1} \cdot (n - k)$. For $\theta < \lfloor \theta_{max} \rfloor$, we are hardly able to observe any decoding failures, as predicted by (26).

6   Conclusions

In this paper, we extended the algorithm described in [1] to be able to correct errors and erasures simultaneously. Furthermore, we presented techniques for improving the efficiency of interleaved Reed–Solomon decoding compared to [1]. We achieve this improvement by reducing the linear system of equations used in [1] to state the decoding problem to a smaller system of equations for locating the errors. In this way we can reduce the computational complexity of the decoding algorithm. This reduction is obtained by multiplying a transformation matrix based on inverse Vandermonde matrices. In most cases, this multiplication has not to be carried out explicitly, because the structure of the problem allows it to directly give closed forms for the desired system of equations and also for the auxiliary polynomials $\Gamma^{(\ell)}(x)$. This is possible by applying the results from [4].

Based on this, we describe an algorithm which is not only able to correct errors, but both, errors and erasures simultaneously. Unlike errors, which affect a complete symbol $y_i \in \mathbb{F}_{q^l}$ of the interleaved Reed–Solomon code, we can have erased positions $y_i^{(\ell)} \in \mathbb{F}_q$ with respect to the symbols of the underlying Reed–Solomon code. This is explained by the fact, that the positions of the erasures are known, i.e., they do not have to be located first. In presence of erasures, the structure of our linear system of equations is modified such, that obtaining the required inverse Vandermonde matrices gets a little more involved. Therefore, we also describe a more general method to obtain the required inverse matrix based on [5]. This method is suited for our error and erasure decoder.

Acknowledgment

The authors would like to thank Gerd Richter for many helpful comments on this manuscript.

References

1. D. Bleichenbacher, A. Kiayias, and M. Young. Decoding of interleaved Reed Solomon codes over noisy data. In *Springer Lecture Notes in Computer Science*, volume 2719, pages 97–108, January 2003.
2. J. Justesen, C. Thommesen, and T. Høholdt. Decoding of concatenated codes with interleaved outer codes. In *Proc. of IEEE Intern. Symposium on Inf. Theory*, page 329, Chicago, USA, 2004.
3. L. R. Welch and E. R. Berlekamp. Error correction for algebraic block codes. U. S. Patent 4 633 470, issued Dec. 30 1986.
4. H.L. Althaus and R. J. Leake. Inverse of a finite-field Vandermonde matrix. *IEEE Trans. Inform. Theory*, 15:173, January 1969.
5. A. Klinger. The Vandermonde matrix. *Amer. Math. Monthly*, 74:571–574, 1967.
6. A. Brown, L. Minder, and A. Shokrollahi. Probabilistic decoding of interleaved rs-codes on the q-ary symmetric channel. In *Proc. of IEEE Intern. Symposium on Inf. Theory*, page 327, Chicago, USA, 2004.
7. G. Schmidt, V. R. Sidorenko, and M. Bossert. Interleaved Reed–Solomon codes in concatenated code designs. In *Proc. of IEEE ISOC ITW2005 on Coding and Complexity*, pages 187–191, Rotorua, New Zealand, 2005.

A Welch–Berlekamp Like Algorithm for Decoding Gabidulin Codes

Pierre Loidreau

Ecole Nationale Supérieure de Techniques Avancées
Pierre.Loidreau@ensta.fr

Abstract. In this paper, we present a new approach of the decoding of Gabidulin codes. We show that, in the same way as decoding Reed-Solomon codes is an instance of the problem called *polynomial reconstruction*, the decoding of Gabidulin codes can be seen as an instance of the problem of reconstruction of linearized polynomials. This approach leads to the design of two efficient decoding algorithms inspired from the Welch–Berlekamp decoding algorithm for Reed–Solomon codes. The first algorithm has the same complexity as the existing ones, that is cubic in the number of errors, whereas the second has quadratic complexity in $2.5n^2 - 1.5k^2$.

1 Introduction

Gabidulin codes are the analogs for rank metric of Reed–Solomon codes for Hamming metric. Namely, they consist of evaluation of q–polynomials of bounded degree over a set of elements of a finite field, [3]. These codes are optimal codes, both in Hamming and in rank metric and can be used in building cryptosystems, with a much smaller public-key size than McEliece type cryptosystems whose security relies on the difficulty of decoding in Hamming metric [5]. Several polynomial-time decoding algorithms were designed until now enabling to decode Gabidulin codes up to their rank error-correcting capability. It is interesting to note that all of them have an equivalent decoding algorithm in Hamming metric for Reed–Solomon codes, such as *extended Euclidian*, and *Berlekamp–Massey* algorithms, [3, 4, 11, 10].

Concerning Reed-Solomon codes there is still another decoding algorithm based on the analogy between decoding Reed–Solomon codes and solving some instances of the polynomial reconstruction problem [12]. Inspired by such an analogy we reformulated the problem of decoding Gabidulin codes into the problem of q–*polynomial reconstruction*. In the following, we show that the problem of decoding Gabidulin codes can be related to this problem in a simple way. We then derive two polynomial-time decoding algorithms solving this problem. They can be seen as the analogs in rank metric of Welch–Berlekamp algorithms, [1].

2 Rank Metric and Gabidulin Codes

Rank metric was introduced in 1985 by E.M. Gabidulin [3]. Given a vector $\mathbf{c} = (c_1, \ldots, c_n)$ of elements of a finite field $GF(q^m)$, the rank over $GF(q)$ of $\mathbf{c}$

Ø. Ytrehus (Ed.): WCC 2005, LNCS 3969, pp. 36–45, 2006.

is defined as the rank of the $n \times m$ q-ary matrix obtained by expanding each coordinate of $\mathbf{c}$ over a basis of $GF(q^m)/GF(q)$. It is denoted $\mathrm{Rk}(\mathbf{c} \mid GF(q))$.

In the same way, given a code over $GF(q^m)$, the minimum rank distance of the code is the quantity

$$d = Min_{\mathbf{c} \in C \setminus \{0\}}(\mathrm{Rk}(\mathbf{c} \mid GF(q)))$$

Let C be a linear code with parameters (n, k), and minimum rank distance d over $GF(q^m)$. In rank metric the problem of bounded distance decoding of a code can be formulated as such

Decoding$(\mathbf{y}, C, t)$
Find, when it exists, $\mathbf{c} \in C$, and $\mathbf{e}$ where $Rk(\mathbf{e} \mid GF(q)) \leq t$ such that $\mathbf{y} = \mathbf{c} + \mathbf{e}$,

where $\mathbf{y}$ is the received vector over $GF(q^m)$, C is a code over $GF(q^m)$, and t is a positive integer. Provided t is less than or equal to the rank error-correcting capability of the code C, either there is no solution or the solution is unique.

Some general purpose decoding algorithms were constructed, for example in [2] but the best ones were designed by Ourivski and Johannson in [9]. Both are based on writing a set of quadratic equations satisfied by the error-vector, and linearizing a part of it by some extended search over a definite vector space. Provided one wants to correct t rank errors over $GF(q^m)/GF(q)$ in a code of length n, dimension k, their complexity is given by:

- *First strategy:* $O((mt)^3 q^{(t-1)(k+1)})$ operations in $GF(q)$.
- *Second strategy:* $O((k + t)q^{(t-1)(m-t)})$ operations in $GF(q)$.

It is highly exponential. Therefore, given a code C, we are not generally able to solve the **Decoding** problem for the code C, even for small parameters. This property enables to design Public-Key cryptosystems based on codes with theoretically a smaller public-key size than in Hamming metric [5].

In the seminal paper, Gabidulin presented a new family of codes defined by a vector $\mathbf{g} = (g_1, \ldots, g_n)$ of elements of $GF(q^m)$ linearly independent over $GF(q)$. A generating matrix of such a code $Gab_k(\mathbf{g})$ is the matrix G such that

$$G = \begin{pmatrix} g_1 & \cdots & g_n \\ \vdots & \ddots & \vdots \\ g_1^{q^{k-1}} & \cdots & g_n^{q^{k-1}} \end{pmatrix},$$

These codes are called Gabidulin codes and are denoted $Gab_k(\mathbf{g})$. They have minimum rank distance $d = n - k + 1$ and possess fast-polynomial time decoding algorithm. Namely, if we instantiate the problem **Decoding$(\mathbf{y}, C, t)$** with a Gabidulin code of minimum distance d and with $t \leq \lfloor (d-1)/2 \rfloor$, there are fast polynomial time decoding algorithms solving the problem. They are similar to corresponding decoding algorithms for Reed-Solomon codes:

- *Extended Euclidian like :* $\approx t(m + 2n + t^2)$ multiplications in $GF(q^m)$, see [11, 4];
- *Berlekamp–Massey like:* $\approx t(m + 2n + 6t + t^2/2)$ multiplications in $GF(q^m)$, see [10].

3   The Reconstruction of q–Polynomials

q–polynomials (also called *linearized polynomials*) are polynomials of the form

$$P(x) = \sum_{i=0}^{t} p_i x^{q^i}, \quad \forall i, \; p_i \in GF(q^m), \; p_t \neq 0.$$

the integer t is called q–degree of P and is denoted $deg_q(P)$.

Gabidulin codes play the same role in rank metric as Reed-Solomon codes in Hamming metric. Namely, they are evaluation codes of q–polynomials, as defined by Øre [7, 8], on a set of n elements taken from $GF(q^m)$, linearly independent over the base field $GF(q)$. Therefore it is natural to link a so-called *Reconstruction Problem* for q–polynomials to the decoding problem in rank metric. Here is the statement of the problem as presented in [6].

Reconstruction($\mathbf{y} = (y_1, \ldots, y_n), \mathbf{g} = (g_1, \ldots, g_n), k, t$)
Find the set (V, f) where V is a non-zero q–polynomial of q-degree $\leq t$ and where f is a q-polynomial of q-degree $< k$, such that

$$V(y_i) = V[f(g_i)], \text{ for all } i = 1, \ldots, n.$$

This problem can be related to the problem of bounded distance decoding Gabidulin codes, by the following theorem.

Theorem 1. *From any solution to* **Reconstruction**($\mathbf{y}, \mathbf{g}, k, t$), *where the g_i's are linearly independent over $GF(q)$ one gets a solution to* **Decoding**($\mathbf{y}, Gab_k(\mathbf{g}), t$) *in polynomial time.*

Proof. Let $\mathcal{L}$ be the set of solutions of **Reconstruction**($\mathbf{y}, \mathbf{g}, k, t$). Let $(V_1, f_1) \in \mathcal{L}$. Then for all $i = 1 \ldots, n$ we have $V_1(y_i) = V_1[f_1(g_i)]$. By linearity of V_1, we get $V_1(y_i - f_1(g_i)) = 0$, for all $i = 1 \ldots, n$. This is equivalent to the fact that for all $i = 1 \ldots, n$, the field element $e_i \stackrel{def}{=} y_i - f_1(g_i)$ belong to a vector space over $GF(q)$ of dimension at most the q-degree of V_1, that is t. Therefore, the vector $\mathbf{e} = (e_1, \ldots, e_n)$ is of rank at most t and $(\mathbf{c}, \mathbf{e})$ where $\mathbf{c} = (f_1(g_1), \ldots, f_1(g_n))$ is a solution of **Decoding**($\mathbf{y}, Gab_k(\mathbf{g}), t$). All these transformation can clearly be computed in polynomial time. $\square$

Therefore, designing algorithms for reconstructing q–polynomials will enable us to solve the decoding problem in rank metric.

4   Solving the Reconstruction Problem

Suppose we are given,

- A vector $\mathbf{y} = (y_1, \ldots, y_n)$ of elements taken over the field $GF(q^m)$;
- A vector $\mathbf{g} = (g_1, \ldots, g_n)$ of elements taken over the field $GF(q^m)$, that are linearly independent over $GF(q)$;
- Integers k, t;

To solve **Reconstruction**$(\mathbf{y}, \mathbf{g}, k, t)$, we need to find the q-polynomials V of q–degree less than or equal to t, and f of q–degree less than k such that

$$V(y_i) = V[f(g_i)], \quad \text{for all } i = 1, \ldots, n. \tag{1}$$

It is a quadratic system of n equations in $t+1+k$ variables. Basically we have no clue on how to solve this system. A way would be to compute the Gröbner basis of the system by adding the field equations, and then extract the finite number of solutions by computing the number of points of the obtained variety. However we have no precise complexity results on the difficulty the computation.

It is the reason why we consider the following system: Find (V, N), a pair of q-polynomials, such that

$$\begin{cases} V(y_i) = N(g_i), & \forall i = 1, \ldots, n \\ deg_q(V) \leq t, \\ deg_q(N) \leq k + t - 1, \end{cases} \tag{2}$$

This system is a linear system whose unknowns are the $k + 2t + 1$ coefficients of N and V. The following proposition gives a relation between the sets of solutions of the two systems

Proposition 1. *Any solution (V, p) of (1) provides a solution $(V, N = V \circ p)$ to (2).*

Proof. Let (V_0, p_0) be a solution of (1), then the pair $(V_0, N_0 = V_0 \circ p_0)$ is a solution of (2). $\quad\blacksquare$

Moreover, in some cases there is reciprocity.

Proposition 2. *If $t \leq (n - k)/2$ and if there is at least a non-zero solution to 1), then the dimension of the vector space of solutions of (2) has dimension equal to 1, and any non zero solution to (2) provides a solution to (1).*

Proof. Suppose that the dimension of the vector space of solutions of (2) is 0. Then the unique solution to the system is $(0, 0)$. But from Proposition 1 it implies that the only solution to (1) is equally $(0, 0)$.

Now let us consider a non-zero solution (V_0, p_0) of 1) then any solution V, N of (2) satisfies the following system of equations:

$$V_0 [N(g_i) - V \circ p_0(g_i)] = 0, \forall i = 1, \ldots, n$$

the q–polynomial $V_0 [N - V \circ p_0] (x)$ has q–degree less than or equal to $k + 2t - 1$. Since $t \leq (n - k)/2$, this implie that it has degree less than or equal to $n - 1$. Therefore as q–polynomials, we have $V_0 [N - V \circ p_0] (x) = 0$, and since q-polynomials form an integral domain for composition, we get that $N = V \circ p_0$. Moreover, this gives easily that there is some $\alpha \in GF(q^m)$ such that $(V, N) = \alpha(V_0, V_0 \circ p_0)$. Hence the set of solutions to (1) has the form $(\alpha V_0, p_0)$.

5   New Decoding Algorithms

Suppose we receive a vector $\mathbf{y} = \mathbf{c} + \mathbf{e}$ where $\mathbf{c} \in Gab_k(\mathbf{g})$ and $\mathbf{e}$ has rank less than or equal to the error-correcting capability of the code. From Proposition 2 it follows that, we only need to find one solution of the linear system (2) to get the unique solution of **Reconstruction**$(\mathbf{y}, \mathbf{g}, k, t)$. Once we get this solution we can decode easily by merely computing a Euclidian division of q-polynomials.

Namely the decoding algorithm can be described as such:

1. Find a two q-polynomials (V_0, N_0) which are solution of (2);
2. Compute the Euclidian division of N_0 by V_0 and set $f = N_0/V_0$. We have

$$y_i = f(g_i) + e_i,$$

for all $i = 1, \ldots, n$.

The rest of the section is devoted to the description two different algorithms solving system (2).

The second step of the algorithm is not considered here since it was already shown by Øre that the division could be computed in polynomial time. In [7], he designed an algorithmic way of computing the Euclidian division of q-polynomials.

The complexity of computing the Euclidian division between N_0 and V_0 is $(k - 1)t$ multiplications in $GF(q^m)$.

5.1   A Natural Algorithm

Let $\mathcal{V} \overset{def}{=} (v_0, \ldots, v_t)^T$, where the v_i's are the coefficients of the q-polynomial V and $\mathcal{N} \overset{def}{=} (n_0, \ldots, n_{k+t-1})^T$ where the n_i are the coefficients of the q-polynomial N. Set

$$S = \left. \left(\begin{array}{ccc|ccc} g_1 & \cdots & g_1^{[k+t-1]} & y_1 & \cdots & y_1^{[t]} \\ \vdots & \ddots & \vdots & \vdots & \ddots & \vdots \\ g_n & \cdots & g_n^{[k+t-1]} & y_n & \cdots & y_n^{[t]} \end{array} \right) \right\} n$$

Solving (2) is equivalent to solving the system

$$S \times \begin{pmatrix} \mathcal{N} \\ \mathcal{V} \end{pmatrix} = 0. \tag{3}$$

In the unknowns $\mathcal{N}$ and $\mathcal{V}$. Therefore it costs roughly $(k + 2t)^3$ operations over $GF(q)$. It is far too much to be efficiently implemented, compared to the already existing decoding algorithms.

By considering (3), it is clear that a part of the matrix S is independent of the received word, depending only on the parameters of the Gabidulin code. Let us write

$$S = \left(\begin{array}{c|c} G_1 & Y_1 \\ \hline G_2 & Y_2 \end{array} \right),$$

where $G_1 = \left(g_i^{[j]}\right)_{i=1,j=0}^{k+t,k+t-1}$ is the upper left $(k+t) \times (k+t)$ matrix of S. Since, by definition, the g_i's are linearly independent, G_1 an invertible matrix. Therefore solving (3) is equivalent to solving

$$\begin{cases} \mathcal{N} = U \times (Y_1 \mathcal{V}), \\ ((T \times Y_1) + Y_2)\mathcal{V} = 0, \end{cases} \qquad (4)$$

where $U = -G_1^{-1}$ and $T = -G_2 G_1^{-1}$ can be precomputed. The complexity of this algorithm is thus $(k+t)(k+t^2+2t) + t^3/2$ operations over $GF(q^m)$. Even this complexity is not satisfactory compared to the complexity of the existing algorithms, see section 2.

5.2   A Trickier Algorithm

We will now design another algorithm solving the polynomial reconstruction problem. Although less natural it is also more efficient. Our goal consists in finding q–polynomials $V(y)$ of q–degree less than or equal to t and $N(x)$ of q–degree less than $k+t$ satisfying system (2), i.e.

$$V(y_i) - N(g_i) = 0, \quad \forall i = 1, \ldots, n.$$

The idea is to construct two sequences of polynomials $(V_0^{(i)}(y), N_0^{(i)}(x))$ and $(V_1^{(i)}(y), N_1^{(i)}(x))$, satisfying for $i \leq n$ the following property denoted by $\mathcal{P}(i)$

$$\forall k \leq i, \begin{cases} V_0^{(i)}(y_k) - N_0^{(i)}(g_k) = 0, \\ V_1^{(i)}(y_k) - N_1^{(i)}(g_k) = 0, \end{cases}$$

If we manage to bound the degrees of the polynomials such that

$$\begin{cases} deg_q\left(V_0^{(n)}\right) \leq t \\ deg_q\left(N_0^{(n)}\right) \leq k-1+t \end{cases} \quad \text{or} \quad \begin{cases} deg_q\left(V_1^{(n)}\right) \leq t \\ deg_q\left(N_1^{(n)}\right) \leq k-1+t \end{cases}$$

then we have won.

Since the label i runs over n positions, if we increase the degrees of the polynomials at each step then we will not be able to satisfy the condition on the degrees. Therefore a way must be found to keep the degrees as low as possible.

Suppose that we have constructed a sequence of polynomials satisfying $\mathcal{P}(j)$, for all $j = 0, \ldots, i < n$. We show how to build polynomials satisfying $\mathcal{P}(i+1)$. First we evaluate the following quantities.

$$s_0^{(i)} \stackrel{def}{=} V_0^{(i)}(y_{i+1}) - N_0^{(i)}(g_{i+1}),$$
$$s_1^{(i)} \stackrel{def}{=} V_1^{(i)}(y_{i+1}) - N_1^{(i)}(g_{i+1}).$$

These quantities correspond to some defect in what we expect. Namely, if both of them is equal to zero, then $\mathcal{P}(i+1)$ is immediately satisfied.

There are basically two manners to build polynomials satisfying $\mathcal{P}(i+1)$.

– First and most simple is to evaluate

$$N_0^{(i+1)}(x) = N_0^{(i)}(x)^p - s_0^{(i)} N_0^{(i)}(x),$$
$$V_0^{(i+1)}(y) = V_0^{(i)}(y)^p - s_0^{(i)} V_0^{(i)}(y),$$

This corresponds for q–polynomials to the interpolation of the *multivariate* polynomial $Q(x,y) \stackrel{def}{=} V(y) - N(x)$ on the point $[(y_{i+1}, g_{i+1}), 0]$. We check that for all $k = 1 \ldots i+1$, we have $V_0^{(i+1)}(y_k) - N_0^{(i+1)}(x_k) = 0$. It is important to note that this method increases the q–degree of non-zero polynomials by 1.

– The second one corresponds to cross evaluation. We set

$$N_1^{(i+1)}(x) = s_0^{(i)} N_1^{(i)}(x) - s_1^{(i)} N_0^{(i)}(x),$$
$$V_1^{(i+1)}(y) = s_0^{(i)} V_1^{(i)}(y) - s_1^{(i)} V_0^{(i)}(y).$$

This transformation implies that $deg_q(N_1^{(i+1)}) \leq Max(deg_q(N_1^{(i)}), deg_q (N_0^{(i)}))$, with equality if the degrees of $N_1^{(i)}$ and $N_0^{(i)})$ are different. Therefore this does not increase the degrees and one can check that for all $k = 1 \ldots i + 1$, $V_1^{(i+1)}(y_k) - N_1^{(i+1)}(x_k) = 0$.

This is the heart of the decoding algorithm we design. Basically there will be steps where we increase the degrees of the polynomials by maintaining the degrees of the others constant.

Description of the Algorithm. The algorithm is described in Table **??**. We chose not to build the sequences $(N_0^{(i)}, V_0^{(i)})$ and $(N_1^{(i)}, V_1^{(i)})$, but to modify the considered polynomials. Hence we can save space. This implies that at every step i both pairs of polynomials (N_0, V_0) and (N_1, V_1) satisfy the property $\mathcal{P}(i)$.

The algorithm consists of three steps:

– *Precomputation step*:
  • Compute $Int(g_1, \ldots, g_k)$, where $Int(g_1, \ldots, g_k)$ denotes the unique monic polynomial of q–degree k such that $(Int(g_1, \ldots, g_k)(g_i) = 0$, for all $i = 1, \ldots, k$.
  • Compute the list $\mathcal{P}_i$, $i = 1, \ldots, k$ of the k Lagrange interpolation polynomials of q–degree $k - 1$, that is

$$\forall i = 1, \ldots, k, \begin{cases} \mathcal{P}_i(g_j) = 0, \ mboxif j \neq i, \\ \mathcal{P}_i(g_i) = 1. \end{cases}$$

  This set of q–polynomials form a basis of the vector space of q–polynomials of q–degree $k - 1$.

For computation, we can use algorithms described by Øre in his paper for example.

– *Initialisation step*:
  • Set $V_0 = 0$, and $N_0 = Int(g_1, \ldots, g_k)$.
  • Set $V_1(y) = y$ and

$$N_1 = \sum_{i=1}^{k} y_i \mathcal{P}_i.$$

From the properties of the polynomials $\mathcal{P}_i$, the polynomial N_1 has q-degree $k-1$ and satisfies the relations

$$\forall\, i = 1,\ldots,k, \quad N_1(g_i) = y_i.$$

– *Alternate increasing degree step*: This is the most delicate part of the algorithm. Indeed this part consists of checking the degrees of the pairs of polynomials. We now exchange the roles of N_0 and N_1 and V_0 and V_1, so that we will always increase the degree of N_0 and V_0 by one at each step. If we set $s = \lfloor (i-k)/2 \rfloor$, after the ith step we have
  - $deg_q(N_0) = k + s$;
  - $deg_q(V_0) = s$ if $i-k$ is even and $deg_q(V_0) = s+1$ if $i-k$ is odd;
  - $deg_q(N_1) = k + s - 1$ if $i-k$ is even and $deg_q(N_1) = k+s$ if $i-k$ is odd;
  - $deg_q(V_1) = s$.
  Therefore after the final step n the pair of polynomials (N_1, V_1) satisfy the condition for being a solution to system (2), since $deg_q(N_1) = k + \lfloor (n-k)/2 \rfloor - 1$ and $deg_q(V_1) = \lfloor (n-k)/2 \rfloor$.

5.3 Complexity Analysis of the Algorithm

The most complex operation is multiplying elements in finite fields compared to squaring and additioning.

– *Initialisation step*: the only polynomial that cannot be precomputed is N_1 consisting of a linear combination of interpolation polynomials. Hence, the complexity of computing N_1 is k^2 multiplications in $GF(q^m)$.
– *Alternate Incresing Degree step*: Let us evaluate the complexity of the algorithm at step $i \geq k+1$
  - Computation of s_0 and s_1: In any case, it is easy to check that either in the even of in the odd case, the computation it takes exactly $2i-1$ multiplications.
  - Computing $s_0 N_1(x) - s_1 N_0(x)$ and $s_0 V_1(y) - s_1 V_0(y)$ costs equally $2i-1$ multiplications.
  - Computing $N_0(x)^q - s_0 N_0(x)$, and $V_0(x)^q - s_0 V_0(x)$ costs i multiplications.
  Therefore, at every step $k+1 \leq i \leq n$, one has to compute $5i - 2$ multiplications. Hence the total number of multiplications for this step is:

$$\sum_{i=k+1}^{n} 5i - 2 = \frac{5}{2}(n^2 - k^2) + \frac{n-k}{2} - 2,$$

multiplications in $GF(q^m)$.

The overall complexity gives about $\frac{5}{2}n^2 - \frac{3}{2}k^2 + \frac{n-k}{2}$ multiplications.

Table 1. Algorithm for solving the linear system

INPUT: A Gabidulin code $Gab_k(\mathbf{g})$ of length n, and a vector $\mathbf{y} = (y_1, \ldots, y_n)$ at rank distance less than or equal to $t = \lfloor (d-1)/2 \rfloor$ from $Gab_k(\mathbf{g})$.
OUTPUT: A pair of polynomials (N_1, V_1) satisfying system (2)

1. *Initialisation step:*
   - $V_0(y) \leftarrow 0$ and $V_1(y) \leftarrow y$,
   - $N_0(x) \leftarrow Int(g_1, \ldots, g_k)$ and $N_1(x) \leftarrow \sum_{i=1}^{k} y_i \mathcal{P}_i$.
2. *Alternate increasing degree step*
   For $i \in \{k+1, \ldots, n\}$ do
   - $s_0 \leftarrow V_0(y_i) - N_0(g_i)$ and $s_1 \leftarrow V_1(y_i) - N_1(g_i)$,
   - Exchange N_0 and N_1, V_0 and V_1, s_0 and s_1
   - Compute
     (a) $N_1(x) \leftarrow s_0 N_1(x) - s_1 N_0(x)$,
     (b) $V_1(y) \leftarrow s_0 V_1(y) - s_1 V_0(y)$,
     (c) $N_0(x) \leftarrow N_0(x)^q - s_0 N_0(x)$,
     (d) $V_0(y) \leftarrow V_0(y)^q - s_0 V_0(y)$.
3. Return (N_1, V_1).

6   Conclusion

We implemented both algorithms as well as the *extended Euclidian* algorithm in Magma language. It appears, that the first approach is not faster than the extended Euclidian, and has approximately the same complexity, a little less efficient nevertheless.

Computer simulations made in MAGMA show that our second algorithm with complexity $5/2n^2 - 3/2k^2$ is almost always faster than the *extended Euclidian*. The thing is that the complexity of the latter is roughly in $O(t^3 + 2nt)$. This implies that whenever, t is great, the complexity is cubic, whereas when t is small, then the dimension k can be high, Thus reducing the complexity of our algorithm.

References

1. E. R. Berlekamp and L. Welch. Error correction of algebraic block codes. US Patent, Number 4,633,470, 1986.
2. F. Chabaud and J. Stern. The cryptographic security of the syndrome decoding problem for rank distance codes. In K. Kim and T. Matsumoto, editors, *Advances in Cryptology - ASIACRYPT '96*, volume 1163 of *LNCS*. Springer-Verlag, November 1996.
3. E. M. Gabidulin. Theory of codes with maximal rank distance. *Problems of Information Transmission*, 21:1–12, July 1985.
4. E. M. Gabidulin. A fast matrix decoding algorithm for rank-error correcting codes. In G. Cohen, S. Litsyn, A. Lobstein, and G. Zémor, editors, *Algebraic coding*, volume 573 of *LNCS*, pages 126–133. Springer-Verlag, 1991.

5. E .M. Gabidulin, A. V. Paramonov, and O. V. Tretjakov. Ideals over a non-commutative ring and their application in cryptology. In D. W. Davies, editor, *Advances in Cryptology – EUROCRYPT'91*, volume 547 of *LNCS*, pages 482–489. Springer-Verlag, 1991.

6. P. Loidreau. Sur la reconstruction des polynômes linéaires : un nouvel algorithme de décodage des codes de Gabidulin. *Comptes Rendus de l'Académie des Sciences: Série I*, 339(10):745–750, 2004.

7. Ø. Ore. On a special class of polynomials. *Transactions of the American Mathematical Society*, 35:559–584, 1933.

8. Ø. Ore. Contribution to the theory of finite fields. *Transactions of the American Mathematical Society*, 36:243–274, 1934.

9. A. Ourivski and T. Johannson. New technique for decoding codes in the rank metric and its cryptography applications. *Problems of Information Transmission*, 38(3):237–246, September 2002.

10. G. Richter and S. Plass. Error and erasure decoding of rank-codes with a modified Berlekamp-Massey algorithm. In *5th Int. ITG Conference on Source and Channel Coding (SCC 04)*, 2004.

11. R. M. Roth. Maximum-Rank array codes and their application to crisscross error correction. *IEEE Transactions on Information Theory*, 37(2):328–336, March 1991.

12. M. Sudan. Decoding Reed-Solomon codes beyond the error-correction diameter. In *Proceedings of the 35th Annual Allerton Conference on Communication, Control and Computing*, pages 215–224, 1997.

On the Weights of Binary Irreducible Cyclic Codes

Yves Aubry and Philippe Langevin

Université du Sud Toulon-Var, Laboratoire GRIM F-83270 La Garde, France
{langevin, yaubry}@univ-tln.fr
http://{langevin, yaubry}.univ-tln.fr

Abstract. This paper is devoted to the study of the weights of binary irreducible cyclic codes. We start from McEliece's interpretation of these weights by means of Gauss sums. Firstly, a dyadic analysis, using the Stickelberger congruences and the Gross-Koblitz formula, enables us to improve McEliece's divisibility theorem by giving results on the multiplicity of the weights. Secondly, in connection with a Schmidt and White's conjecture, we focus on binary irreducible cyclic codes of *index two*. We show, assuming the generalized Riemann hypothesis, that there are an infinite of such codes. Furthermore, we consider a subclass of this family of codes satisfying the quadratic residue conditions. The parameters of these codes are related to the class number of some imaginary quadratic number fields. We prove the non existence of such codes which provide us a very elementary proof, without assuming G.R.H, that any two-weight binary irreducible cyclic code $c(m, v)$ of index two with v prime greater that three is semiprimitive.

1 Introduction

In a recent paper [9], Wolfmann has proved that a two-weight binary cyclic code is necessarily irreducible. On the other hand, it is well-known that there exist two infinite classes of irreducible cyclic codes with at most two nonzero weights: the subfield codes and the semiprimitive ones. Apart from these two families, 11 exceptional codes have been found by Langevin (see [4]) and, Schmidt and White (see [6]). It has been conjectured in the later paper that this is the whole story. This question is investigated in this paper in the case of the characteristic two.

In the first part of this article, we recall the McEliece interpretation of the weights of an irreducible cyclic code by means of linear combinations of Gauss sums. McEliece's divisibility theorem plays a significant role in the study of weight distributions of irreducible cyclic codes. In particular, Schmidt and White deduce a necessary and sufficient condition for an irreducible cyclic code to be a two-weight code.

In the second part, we use the Stickelberger congruences and the Gross-Koblitz formula to obtain two new results that improve McEliece's theorem. We study the Boolean functions that appear in the dyadic expansion of the weight of a

Ø. Ytrehus (Ed.): WCC 2005, LNCS 3969, pp. 46–54, 2006.

codeword. The estimation of their algebraic degree leads us to results on the divisibility concerning multiplicities by means of Ax's and Katz's theorems.

In the last part, we are interested in Schmidt and White's conjecture on irreducible cyclic $c(m,v)$ codes. Since they proved that it holds for codes of index two (conditionally on the Generalized Riemann Hypothesis), we focus our attention on this class of codes. We prove that, conditionally on G.R.H., there are an infinity of binary irreducible cyclic $c(m,v)$ codes of index two with v prime. This result can be seen as an analogue of the Artin conjecture on primitive roots. Thus, this family of codes seems to be interesting in view Schmidt and White's result. Then, we use a result of Langevin in [4] to prove that there does not exist any two-weight irreducible cyclic $c(m,v)$ code of index two with $v > 3$ prime and $v \equiv 3 \pmod 4$. This provides us an elementary proof, without assuming G.R.H, of a particular instance of the Schmidt and White conjecture, namely that any two-weight binary irreducible cyclic $c(m,v)$ code of index two with v prime greater that 3 is semiprimitive.

2   McEliece's Theorem

Let L be a finite field of order $q = 2^m$. Let n be a divisor of $q-1$ and write $v = (q-1)/n$. Let ζ be a primitive n-th root of unity in L. Consider the following map Φ:

$$\Phi : L \longrightarrow \mathbf{F}_2^n$$
$$a \longmapsto \left(\mathrm{Tr}_{L/\mathbf{F}_2}(a\zeta^{-i})\right)_{i=0}^{n-1}$$

where $\mathrm{Tr}_{L/\mathbf{F}_2}$ is the trace of the field L over $\mathbf{F}_2$. The image $\Phi(L)$ of L by Φ is an irreducible cyclic code of length n, denoted $c(m,v)$, see [6] for the material about these codes. Its dimension is equal to the multiplicative order of 2 modulo n, denoted $\mathrm{ord}_n(2)$. Any binary irreducible cyclic code can be viewed as a $c(m,v)$ code, so let us consider such codes. For an element t of L, let us denote by $w(t)$ the weight of $\Phi(t)$. The well-known McEliece formula gives the weight of $\Phi(t)$ in term of Gauss sums

$$w(t) = \frac{n}{2(q-1)}\left(q + \sum_{\chi \in \Gamma \setminus \{1\}} \tau_L(\chi)\bar{\chi}(t)\right) \tag{1}$$

where Γ is the subgroup of multiplicative characters of L^* that are orthogonal to ζ, see [6]. The Gauss sum $\tau_L(\chi)$ is implicitly defined with respect to the canonical additive character, say μ_L, of L. By definition,

$$\tau_L(\chi) = -\sum_{x \in L^*} \chi(x)\mu_L(x).$$

Note that a change of additive character produces a permutation of weights. As in [6], let us denote by θ the greatest integer such that, for all non trivial $\chi \in \Gamma$, 2^θ divides $\tau_L(\chi)$. The famous Stickelberger theorem (see next section) claims

$$\theta = \min_{0 < j < v} S(jn)$$

where $S(k)$ denotes the sum of the bits in the binary expansion of the natural integer k.

Theorem 1 (McEliece). *All the weights of the irreducible cyclic code $c(m,v)$ are divisible by $2^{\theta-1}$. Moreover, one of them is not divisible by 2^θ.*

Sketch of the proof. It suffices to group together the terms of minimal 2-adic valuation in (1) to get the first part of the theorem. The second part comes from the independence (modulo 2) of the multiplicative characters of L.   □

A two-weight code is a code with two nonzero Hamming weights. The McEliece formula appears as the Fourier inversion formula of the map $t \mapsto f(t) = qz(t) - n$, where $z(t)$ denotes the number of zero components of the codeword $\Phi(t)$. Moreover if G denotes the group of order n in L^*, the map $f(t)$ is defined over the quotient group $V = L^*/G$. Let us set $f := \mathrm{ord}_v(2)$, and since $nv = 2^m - 1$, f divides m and we set $m = fs$.

Theorem 2 (Schmidt-White). *The irreducible cyclic code $c(m,v)$ is a two-weight code if and only if there exists an integer k satisfying the three conditions*

   (i) k divides $v - 1$
   (ii) $k2^{s\theta} \equiv \pm1 \pmod{v}$
   (iii) $k(v - k) = (v - 1)2^{s(f-2\theta)}$

Sketch of the proof. Using Fourier analysis, one can prove that

$$D = \{t \in V \mid 2^\theta \text{ divides } w(t)\}$$

is a difference set of order $2^{f-2\theta}$ implying (iii). This set or its complementary is a (v, k, λ) difference set satisfying (i) & (ii). Surprisingly, the three conditions are sufficient.   □

Traditionally, one says that 2 is semiprimitive modulo v when -1 is in the group generated by 2 in $(\mathbf{Z}/v\mathbf{Z})^*$. In this case, all the Gauss sums are rationals, equal to $\sqrt{q}$ whence $\theta = f/2$, and the code $c(m,v)$ is a two-weight code with $k = 1$. Each of these assertions characterizes the semiprimitivity.

3   Dyadic Weight Formula

In this section, we analyse dyadicaly the function

$$f(t) = \sum_{1 \neq \chi \in \Gamma} \tau_L(\chi)\bar{\chi}(t) = 2^\theta \sum_{i=0}^{+\infty} f_i(t)2^i \tag{2}$$

where the f_i are Boolean functions i.e. map L into $\{0,1\}$. By definition, see [7], the degree of a Boolean function f defined over a $\mathbf{F}_2$-space E of dimension m is equal to the smallest degree of a polynomial $p \in \mathbf{F}_2[X_1, X_2, \ldots, X_m]$ such that

$$\forall(x_1, x_2, \ldots, x_m) \in \mathbf{F}_2^m$$
$$p(x_1, x_2, \ldots, x_m) \equiv f(x_1\beta_1 + x_2\beta_2 + \cdots + x_m\beta_m) \pmod{2},$$

where $(\beta_1, \beta_2, \ldots, \beta_m)$ is any basis of L considered has a vector space over $\mathbf{F}_2$.

In the first part of this section, we use the Stickelberger's congruences to determine the algebraic degree of f_0. In the second part, we will use the Gross-Koblitz formula to give an upper bound on the degree of f_1. For this, we realize the finite field L as the quotient ring $\mathbf{Z}_2[\xi]/(2)$, where ξ is a $(q-1)$-root of unity in an algebraic extension of $\mathbf{Q}_2$ the field of 2-adic numbers. The Teichmüller character of L, denoted by w, is the multiplicative character of L defined by the relation

$$w(\xi \pmod 2) = \xi.$$

It is important to remark that $t \mapsto w(t) \pmod 2$ is nothing but the identity of L^*. The Gross-Koblitz formula below (see [3]) claims the existence of an additive character ψ such that, for any residue a modulo $q-1$, the following holds:

$$\tau_L(\bar{w}^a, \psi) = (-2)^{S(a)} \prod_{j=0}^{f-1} \Gamma_2\left(1 - \langle \frac{2^j a}{q-1} \rangle\right) \tag{3}$$

where $S(a) = a_0 + a_1 + \ldots + a_{f-1}$ is the sum of the bits of $a = \sum_{i=0}^{f-1} a_i 2^i$, $\langle x \rangle$ is the fractional part of x, and Γ_p the 2-adic gamma function defined by

$$\forall k \in \mathbf{N}, \quad \Gamma_2(k) = (-1)^k \prod_{j<k,\, 2 \nmid j} j, \qquad \forall s \in \mathbf{Z}_2, \quad \Gamma_2(s) = \lim_{k \to s} \Gamma_2(k).$$

3.1   The Function f_0

The first approximation of the 2-adic gamma function gives the famous Stickelberger's congruences

$$\tau_L(\bar{w}^a, \psi) \equiv 2^{S(a)} \pmod{2^{1+S(a)}}.$$

We introduce the set

$$J = \{j \mid S(jn) = \theta\},$$

so that

$$f_0(t) \equiv \sum_{j \in J} t^{jn} \pmod 2.$$

Using any $\mathbf{F}_2$-basis of L, the function f_0 becomes a mapping from $\mathbf{F}_2^f$ into $\mathbf{F}_2$. Since all the exponents jn have a constant 2-ary weight equal to θ, the algebraic degree of f_0 is less or equal to θ. The previous McEliece theorem claims that the weights are divisible by $2^{\theta-1}$. The next result gives precisions concerning the multiplicities of the weights. Let us recall that by Ax's theorem (see [1]), for any polynomial $f \in \mathbf{F}_2[X_1, X_2, \ldots, X_m]$ of degree k, the number of solutions in $\mathbf{F}_2^m$ of the equation :

$$f(x_1, x_2, \ldots, x_m) = 0$$

is divisible by $2^{\lceil \frac{m}{k} \rceil - 1}$ where $\lceil r \rceil$ denotes the smallest integer greater or equal to r.

Theorem 3. *The number of codewords of weight of dyadic valuation $\theta - 1$ is divisible by $2^{\lceil f/\theta \rceil - 1}$.*

Proof. The weight of $\Phi(t)$ has valuation $\theta - 1$ if and only if $f_0(t) = 1$. By Ax's theorem the number of solutions is divisible by $2^{\lceil f/\theta \rceil - 1}$ since the degree of the Boolean function f_0 is less or equal to θ.

Example 1. The weights of the binary $[23, 11]$ (subcode of the Golay code) are : $0, 8, 12$ and 16 whence $\theta = 3$ and Theorem 3 claims that the number of codewords of weight 12 is divisible by $2^{\lceil 11/3 \rceil - 1} = 8$. According to [8], this number is $56 \times 23 = 8 \times 161$.

Remark 1. In the case of a two-weight code, the condition (3) of the theorem of Schmidt and White implies a divisibility by a large power of 2. It seems very interesting to study more precisely the function f_0.

3.2   The Function f_1

The first values of the 2-adic gamma function are: $\Gamma_2(0) = 1$, $\Gamma_2(1) = -1$, $\Gamma_2(2) = +1$, $\Gamma_2(3) = -1$, and $\Gamma_2(4) = 3 \equiv -1 \pmod 4$. In particular,

$$\Gamma_2\left(\left(1 - \left(\frac{a}{q-1}\right)\right) \equiv \Gamma_2(1 + a_0 + a_1 2) \equiv (-1)^{1 + a_0 + a_0 a_1} \pmod 4$$

and we get the congruence

$$\tau_L(\bar{\omega}^a, \psi) \equiv (-1)^{Q(a)} 2^{S(a)} \pmod{2^{2 + S(a)}} \tag{4}$$

where $Q(a) = f + a_0 a_1 + a_1 a_2 + \ldots + a_{f-1} a_0$. To improve our approximation of $f(t)$, we introduce the set $K = \{k \in \mathbf{N} \mid 1 \leq k < v, \quad S(kn) = \theta + 1\}$ and the partition $J_\epsilon = \{j \in J \mid Q(jn) \equiv \epsilon \pmod 2\}$. We have

$$f_0(t) + 2f_1(t) \equiv \sum_{j \in J_0} \omega^{jn}(t) - \sum_{j \in J_1} \omega^{jn}(t) + 2 \sum_{k \in K} \omega^{kn}(t) \pmod 4.$$

The Boolean function f_1 depends on the sets K and J_1 but also of the "carry function" $g(t)$ corresponding to the relation

$$\sum_{j \in J} \omega^{jn}(t) \equiv f_0(t) + 2g(t) \pmod 4.$$

By classical 2-adic tricks, we get:

$$g(t) = \frac{1}{2}\left(\sum_{j \in J} \omega^{jn}(t) - \left(\sum_{j \in J} \omega^{jn}(t)\right)^2\right)$$

$$\equiv \sum_{j < j'} \omega^{(j+j')n}(t) \pmod 2.$$

Reducing modulo 2, gluing all pieces together, we get:

$$f_1(t) = \sum_{j<j'} t^{(j+j')n} + \sum_{j\in J_1} t^{jn} + \sum_{k\in K} t^{kn}.$$

Let us recall that Katz's divisibility theorem (see [2]) implies that for any pair of polynomials f_1 and $f_2 \in \mathbf{F}_2[X_1, X_2, \ldots, X_m]$ of degree $k_1 \leq k_2$, the number of solutions in $\mathbf{F}_2^m$ of the system of equations :

$$\begin{cases} f_1(x_1, x_2, \ldots, x_m) = 0 \\ f_2(x_1, x_2, \ldots, x_m) = 0 \end{cases}$$

is divisible by $2^{\lfloor \frac{m-k_1-k_2}{k_2} \rfloor}$ where $\lfloor r \rfloor$ denotes the largest integer smaller or equal to r.

Theorem 4. *Let w_0 be an integer. The number of codewords with weight of the form $w\, 2^{\theta-1}$ with $w \equiv w_0 \pmod 4$ is divisible by $2^{\lfloor \frac{f-3\theta}{2\theta} \rfloor}$.*

Proof. Let $a + 2b + \cdots$ be the 2-adic decomposition of w_0. The weight of $\Phi(t)$ is of the form $w\, 2^{\theta-1}$ if and only if t is a solution of the system

$$f_0(t) = a, \qquad f_1(t) = b.$$

The result is a consequence of the above Katz divisibility theorem since the algebraic degrees of f_0 and f_1 are respectively less or equal to θ and 2θ.

Example 2. A sufficient condition to obtain a non trivial result is $n > 1$ and $5\theta < f$. The first instance is the $[11, 10]$-code ($v = 93$, $\theta = 2$) and the second one is the $[6765, 20]$-code ($n = 6765$, $v = 155$, $\theta = 4$). According to [8], the weight distribution is given by Tab. (1). All the weight are divisible by 8, and the number A_w of codewords of weight w satisfy:

$$\sum_{w\equiv 0 \ (\mathrm{mod}\ 4)} A_w = 1 + 25n \qquad\qquad \equiv 0 \mod 2,$$

$$\sum_{w\equiv 1 \ (\mathrm{mod}\ 4)} A_w = (5 + 45)n \qquad\qquad \equiv 0 \mod 2,$$

$$\sum_{w\equiv 2 \ (\mathrm{mod}\ 4)} A_w = (5 + 20 + 20 + 5)n \qquad \equiv 0 \mod 2,$$

$$\sum_{w\equiv 3 \ (\mathrm{mod}\ 4)} A_w = (4 + 25 + 1)n \qquad\qquad \equiv 0 \mod 2.$$

Table 1. Weight distribution of the $[6765, 20]$ irreducible cyclic code. The number of codewords of weight w is equal to $\mu \times n$, $\bar{w}$ denotes the congruence of $\frac{w}{8}$ modulo 4.

w	3272	3280	3320	3352	3376	3392	3400	3408	3448	3504
$\bar{w}$	1	2	3	3	2	0	1	2	3	2
μ	5	5	4	25	20	25	45	20	1	5

4   Two-Weight Binary Irreducible Cyclic Codes

4.1   Primes Which Generate Squares and Index 2 Codes

Is there infinitely many primes v such that 2 generates the squares modulo v ? Before answering this question, recall that the Artin conjecture asserts that 2 is a primitive root for infinitely many primes (the conjecture is proved by Hooley assuming the Generalized Riemann Hypothesis). In other words, there is infinitely many primes v such that the order of 2 modulo v is equal to $v - 1$.

We consider here an analogue question : is there infinitely many primes v such that 2 generates exactly the squares modulo v ? We can give an another formulation of this question : is there infinitely many primes v such that the order of 2 modulo v is equal to $\frac{v-1}{2}$ or equivalently such that 2 has index 2 modulo v ? Indeed, these problems are equivalent since the group $(\mathbf{Z}/v\mathbf{Z})^*$ is cyclic and the subgroup of squares has index 2 (v odd).

For a positive integer x, let $H(x)$ be the cardinality of the set

$$\{v \le x \mid v \text{ prime and } \operatorname{ord}_v(2) = \frac{v-1}{2}\}.$$

Murata has proved (see [5]) that G.R.H. implies that for every $\varepsilon > 0$,

$$H(x) = \frac{3}{8}\delta\pi(x) + O\left(\frac{2^\varepsilon x \log \log x}{\log^2 x}\right),$$

where

$$\delta = \prod_{\ell \text{ prime}} \left(1 - \frac{1}{\ell(\ell - 1)}\right)$$

is the Artin constant.

Then, under G.R.H., we can use the previous result of Murata to conclude positively to our question: there is infinitely many primes v such that 2 has index 2 modulo v.

Recall that a code $c(m, v)$ is said to have index 2 if the multiplicative order of 2 modulo v is equal to $\varphi(v)/2$, where φ is the Euler function. In particular, we have shown that:

Proposition 1. *Conditionally on G.R.H., there are infinitely many index 2 binary irreducible cyclic codes $c(m, v)$ with v prime.*

Remark 2. Recall that an index 2 binary irreducible cyclic codes $c(m, v)$ with v prime has at most three different nonzero weights. Thus, these codes are good candidates to be two-weight codes. By the way, we can state that, conditionally on G.R.H., there are infinitely many binary cyclic codes with at most three different nonzero weights.

4.2   The Residue Quadratic Case and the Semiprimitivity

For the study of a special class of three-weight codes, Langevin in [4] introduced more restrictive conditions on our integer v which lead us to the quadratic residue case for v, namely the index 2 case with the additional conditions that v is an odd prime greater than 3 with $v \equiv 3 \pmod 4$. In other words, the integer v satisfies the quadratic residue conditions if:

(i) v is a prime greater than 3,
(ii) $\mathrm{ord}_v(2) = \frac{v-1}{2}$,
(iii) $v \equiv 3 \pmod 4$.

This case is of particular interest because of an explicit relation between the class number h of the imaginary quadratic number field $\mathbf{Q}(\sqrt{-v})$ and the Gauss sums (see [4]).

Proposition 2. *There does not exist a two-weight binary irreducible cyclic code satisfying the quadratic residue conditions.*

Proof. Let s be the integer introduced in the section (2). By theorem 3.3 of [4], we know that the code $c(m, v)$ has at most two weights if and only if

$$\frac{v+1}{4} = 2^{hs}. \tag{5}$$

The previous relation implies that:

$$2^{hs+2} \equiv 1 \pmod v.$$

This implies that the order of 2 modulo v divides $hs + 2$. But, by hypothesis, we have $\mathrm{ord}_v(2) = (v-1)/2$. Then, taking the logarithm in (5), we have the inequalities:

$$\frac{v-1}{2} \le hs + 2 = \log(\frac{v+1}{4}) + 2 \tag{6}$$

implying $v = 7$. But this leads to a code with only one nonzero weight: the proposition follows.

The conjecture of Schmidt and White in even characteristic states that an irreducible cyclic code $c(m, v)$ is a two-weight code if and only if it is a semiprimitive code. They proved it, conditionally on G.R.H. for all index 2 codes. We can now prove it also for all index 2 codes with v prime greater than 3 but without assuming G.R.H.

Theorem 5. *A binary irreducible cyclic code $c(m, v)$ of index 2 with v prime greater than 3 is a two-weight code if and only if it is a semiprimitive code.*

Proof. The odd prime v is congruent to 1 or 3 modulo 4. The last congruence comes to the quadratic residue case and the previous proposition implies that the code has three weights. The first congruence $v \equiv 1 \pmod 4$ implies that -1 is a square modulo v and thus is a power of 2 modulo v since 2 has index 2 modulo v and then generates the squares. Thus, the code is semiprimitive.

The converse is a well-known result: the semiprimitivity implies that the code has two weights.

References

1. J. Ax, Zeroes of polynomial over finite fields, *Amer. J. Math.*, Vol. 86, (1964), pp 255–261.
2. N. Katz, On a theorem of Ax, *Amer. J. Math.*, Vol.93, (1971), pp 485–499.
3. N. Koblitz, *p*-adic Analysis: a Short Course on Recent Work, *LMS*, LNS-46, 1980.
4. Ph. Langevin, A new class of two weight codes, *Finite Fields and Their Applications, Glasgow 1995, London Math. Soc.* Lecture Note Ser. 233, 181-187 (1996).
5. L. Murata, A problem analogous to Artin's conjecture for primitive roots and its applications, *Arch. Math.* **57** (1991), 555–565.
6. B. Schmidt and C. White, All two-weight irreducible cyclic codes ?, *Finite Fields and Their Applications* **8** (2002), 1-17.
7. F.J. MacWilliams and N.J.A. Sloane. The Theory of Error-Correcting Codes, Elsevier Science Publisher, 1991.
8. J. MacWilliams and J. Seery, The weight distributions of some minimal cyclics codes, *IEEE transactions on information theory* , IT-27:6, (1981).
9. J. Wolfmann, Are 2-weight projective cyclic codes irreducible ? *IEEE transactions on information theory*, Vol. **51**, N. 2 (2005), 733-737.

3-Designs from $\mathbf{Z}_4$-Goethals-Like Codes and Variants of Cyclotomic Polynomials[*]

Jyrki Lahtonen, Kalle Ranto[**], and Roope Vehkalahti

Department of Mathematics, University of Turku
and Turku Centre for Computer Science TUCS
FIN-20014 Turku, Finland
{lahtonen, kara, roiive}@utu.fi

Abstract. We construct a family of simple 3-$(2^m, 8, 14(2^m - 8)/3)$ designs, with odd $m \geq 5$, from all $\mathbf{Z}_4$-Goethals-like codes $\mathcal{G}_k$ with $k = 2^l$ and $l \geq 1$. In addition, these designs imply also the existence of the other design families constructed from the $\mathbf{Z}_4$-Goethals codes $\mathcal{G}_1$ by Ranto. In the existence proofs we count the number of solutions to certain systems of equations over finite fields and use Dickson polynomials and variants of cyclotomic polynomials and identities connecting them.

1 Introduction

A t-(v, k, λ) *design* is a pair (X, B), where X is a v-element set of *points* and B is a collection of k-element subsets of X (called *blocks*) with the property that every t-element subset of X is contained in exactly λ blocks. A design is *simple* if all the blocks are distinct. In this paper all designs considered are simple.

From the $\mathbf{Z}_4$-Goethals code $\mathcal{G}_1$ Shin, Kumar, and Helleseth [14] constructed a 3-$(2^m, 7, 14(2^m - 8)/3)$ design for odd $m \geq 5$ by taking the supports of codewords of Hamming weight 7. The supports of codewords of Hamming weight 8 in $\mathcal{G}_1$ were analyzed by Ranto [12] and he constructed several families of 3-designs from the different subsets of these supports.

In [13] Ranto verified partly with computer calculations that the designs with the same parameters as introduced in [12] can be also found from the $\mathbf{Z}_4$-Goethals-like codes $\mathcal{G}_k$ with $k \in \{2, 4, 8, 16\}$.

In this paper we prove that for all the designs constructed from $\mathcal{G}_1$ so far we can find a design with the same parameters from $\mathbf{Z}_4$-Goethals-like code $\mathcal{G}_k$ with $k = 2^l$, $l \geq 1$. In addition, we conjecture that they are pairwise nonequivalent.

For a survey on t-designs and $\mathbf{Z}_4$-codes, see [8].

Let $\mathbf{F}$ be the finite field with $q = 2^m$ elements where $m \geq 5$ is odd. The parameter k for the codes $\mathcal{G}_k$ should satisfy $\gcd(k, m) = 1$ (see Definition 5).

Theorem 1 (Main theorem). *Supports in $\mathcal{G}_k$ which are disjoint unions of two nonparallel 2-flats form a* 3-$\left(q, 8, \frac{14}{3}(q - 8)\right)$ *design for all $k = 2^l$, $l \geq 1$.*

---

[*] Part of the results have been published in the dissertation of the second author [13].
[**] Research supported in part by the Academy of Finland (grant 108238).

Ø. Ytrehus (Ed.): WCC 2005, LNCS 3969, pp. 55–68, 2006.

Proof of the main theorem is postponed to Sect. 5. The proofs for the next corollaries are similar to those described in [12] for the case $k = 1$. More details can be found in [13].

Corollary 1. *The supports of size 7 in $\mathcal{G}_{2^l}$ form a 3-$\left(q, 7, \frac{14}{3}(q - 8)\right)$ design.*

Corollary 2. *Certain subsets of supports of size 8 in $\mathcal{G}_{2^l}$ form 3-$(q, 8, \lambda)$ designs where λ has values*

$$\frac{32q^2 - 985q + 5892}{60}, \quad \frac{(q - 8)(q - 32)(q - 49)}{120}, \quad and \quad \frac{56}{15}(q - 8)(q - 12) \ .$$

Corollary 3. *The supports of size 8 in $\mathcal{G}_{2^l}$ form a 3-$(q, 8, \lambda)$ design with*

$$\lambda = \frac{q^4 - 25q^3 + 693q^2 - 10030q + 44712}{120} \ .$$

All the evidence known to us so far validate the following conjecture. Interested reader is referred to [13, Sect. 5.3] for some partial results concerning it.

Conjecture 1. All the designs above are pairwise nonequivalent for every k.

In Sect. 2 we present some useful properties of Dickson polynomials. The key part of this paper is in Sect. 3 where we define recursively some polynomial sets and find identities relating them. In Sect. 4 we describe some preliminaries of $\mathbf{Z}_4$-codes needed to read the proof of the main theorem in Sect. 5. Finally, we make some concluding remarks.

2   Dickson Polynomials

Definition 1. *A Dickson polynomial (of the first kind) of degree n in indeterminate x and with parameter u is*

$$D_n(x, u) = \sum_{i=0}^{\lfloor n/2 \rfloor} \frac{n}{n - i} \binom{n - i}{i} (-u)^i x^{n - 2i} \ .$$

Let $\sigma_1 = x_1 + x_2$, $\sigma_2 = x_1 x_2$, and $S_n = x_1^n + x_2^n$ be the first and second elementary symmetric polynomials and the sum of nth powers in two variables. Dickson polynomials arise from Waring's formula [10, Theorem 1.1] in the following manner:

$$S_n = x_1^n + x_2^n = \sum_{i=0}^{\lfloor n/2 \rfloor} \frac{n}{n - i} \binom{n - i}{i} (-\sigma_2)^i \sigma_1^{n - 2i} = D_n(\sigma_1, \sigma_2) \ .$$

All the polynomials studied in this paper have their coefficients in $\mathbf{F}$ or its algebraic closure $\overline{\mathbf{F}}$ containing $\mathbf{F}_{2^{2k}}$ and hence the primitive $(2^k + 1)$-th root ρ of unity. We need a special case where $n = 2^k + 1$ and by [2, Lemma 2.1] we know that

$$D_{2^k+1}(x, u) = x^{2^k+1} + ux^{2^k-1} + u^2 x^{2^k-3} + u^4 x^{2^k-7} + \cdots + u^{2^{k-1}} x . \qquad (1)$$

With this identity it is quite clear that we have

$$D_{2^k+1}(x, u + v) = D_{2^k+1}(x, u) + D_{2^k+1}(x, v) + x^{2^k+1} . \qquad (2)$$

Clearly, when $u = 0$ the Dickson polynomial $D_n(x, 0) = x^n$ is a permutation polynomial of $\mathbf{F}$ if and only if $\gcd(n, q - 1) = 1$. The following theorem (see e.g. [10, Theorem 3.2]) settles the cases when $u \in \mathbf{F}^*$.

Theorem 2. *If $u \in \mathbf{F}^*$, the Dickson polynomial $D_n(x, u)$ is a permutation polynomial of $\mathbf{F}$ if and only if $\gcd\left(n, q^2 - 1\right) = 1$.*

We need a factorization result of $D_{2^k+1}(x, u)$ in the next section and therefore specialize [10, Theorem 3.12 (i)].

Theorem 3. *Let $\beta_i = \rho^i + \rho^{-i}$. Then we have in $\mathbf{F}_{2^{2k}}[x, y, u]$*

$$D_{2^k+1}(x, u) + D_{2^k+1}(y, u) = (x + y) \prod_{i=1}^{2^{k-1}} (x^2 + \beta_i xy + y^2 + \beta_i^2 u) .$$

We conclude this section with one separate well known lemma. The usual trace function $\mathrm{Tr}_m : \mathbf{F}_{2^m} \to \mathbf{F}_2$ is defined by

$$\mathrm{Tr}_m(x) = x + x^2 + x^4 + \cdots + x^{2^{m-1}}.$$

Lemma 1. *The quadratic equation $x^2 + x = a$ with $a \in \mathbf{F}$ has two roots in $\mathbf{F}$, if $\mathrm{Tr}_m(a) = 0$, and no roots in $\mathbf{F}$ if $\mathrm{Tr}_m(a) = 1$.*

Equation $x^2 + bx = a$, where $b \neq 0$, can be transformed to $(x/b)^2 + x/b = a/b^2$, and the condition in the previous lemma becomes $\mathrm{Tr}_m\left(a/b^2\right) = 0$.

3   Variants of Cyclotomic Polynomials

In order to count the number of solution to one specific equation we need to introduce several polynomials.

Definition 2. *We define three sets of polynomials in $\mathbf{F}_2[s]$ from which the last ones are the usual cyclotomic polynomials.*

$$w_0(s) = s \qquad \text{and} \qquad w_{n+1}(s) = w_n(s)^2 + \left(s^{2^n} + 1\right) w_n(s) + s^{2^{n+1}} + 1 ,$$

$$f_0(s) = s + 1 \qquad \text{and} \qquad f_{n+1}(s) = f_n(s)^2 + f_n(s) + 1 ,$$

$$Q_n(s) = \prod_{\substack{1 \le i \le n \\ \gcd(i,n)=1}} (s - \rho_n^i) \quad \text{where } \rho_n \text{ is a primitive } n\text{-th root of unity.}$$

Table 1. Examples of polynomials

(n, i)	$f_n(s)$	$w_n(s)$	$Q_i(s)$
$(0, 1)$	$1 + s$	s	$1 + s$
$(1, 3)$	$1 + s + s^2$	$1 + s + s^2$	$w_1(s)$
$(2, 5)$	$1 + s + s^4$	$1 + s + s^2 + s^3 + s^4$	$w_2(s)$
$(3, 15)$	$1 + s + s^2 + s^4 + s^8$	$1 + s + s^3 + s^4 + s^5 + s^7 + s^8$	$w_3(s)$
$(4, 17)$	$1 + s + s^{16}$	$1 + s + s^2 + \cdots + s^{16}$	$w_4(s)$
$(5, 51)$	$1 + s + s^2 + s^{16} + s^{32}$	$1 + \cdots + s^5 + \cdots + s^{32}$	$1 + \cdots + s^4 + \cdots + s^{32}$

As an example we give some polynomials in Table 1. For the proof of the next lemma, see [13, Lemma 5.6].

Lemma 2. *The polynomials $w_n(s)$ have the following properties:*

1. $w_n(s) = \sum_{i=1}^{n} \binom{n}{i} t_i(s)^{2^{n-i}}$ *where* $t_i(s) = \sum_{j=0}^{2^i} s^j$;
2. $w_n(s) \neq 0$ *for every* $s \in \mathbf{F}$;
3. $\mathrm{Tr}_m \left(w_{n+1}(s)/w_n(s)^2 \right) = 1$ *for every* $s \in \mathbf{F}$;
4. $w_{2^l-1}(s)^2 \cdot (s+1) = D_{2^{2^l}+1}(s+1, 1)$ *is a permutation polynomial of* $\mathbf{F}$.

These properties are needed in the proof of Theorem 4. Actually, the polynomials $w_n(s)$ and $f_n(s)$ are related to each other with the following transformation.

Lemma 3. *Let* $p(s) \in \overline{\mathbf{F}}[s]$ *and define* $\overline{p(s)}$ *to be the resiprocal polynomial of* $p(s+1)$, *i.e.,* $\overline{p(s)} = s^{\deg(v(s))} v(1/s)$ *with* $v(s) = p(s+1)$. *Then we have the following identities:*

1. $\overline{w_n(s)} = f_n(s)$ *for all* $n \geq 0$;
2. $\overline{p(s)} = 0$ *if and only if* $p(s) = 0$;
3. $\overline{p(s)r(s)} = \overline{p(s)} \; \overline{r(s)}$;
4. $\overline{p(s) + r(s)} = \overline{p(s)} + \overline{r(s)}$ *if* $\deg(p(s)) = \deg(r(s))$;
5. $\overline{(s+1)^n} = 1$ *for all* $n \geq 0$;
6. $\deg \left(\overline{p(s)} \right) = \deg(p(s))$ *if and only if* $p(1) \neq 0$.

Definition 3. *Let $W_k(s, T)$ and $F_k(s, T)$ be compositions (with respect to the variable T and from right to left) of $k - 1$ quadratic linearized polynomials and $P_k(s, T)$ a linearized polynomial given below*

$$W_k(s, T) = \left(T^2 + w_{k-1}(s)T \right) \circ \left(T^2 + w_{k-2}(s)T \right) \circ \cdots \circ \left(T^2 + w_1(s)T \right)$$

$$F_k(s, T) = \left(T^2 + f_{k-1}(s)T \right) \circ \left(T^2 + f_{k-2}(s)T \right) \circ \cdots \circ \left(T^2 + f_1(s)T \right)$$

$$P_k(s, T) = \sum_{i=0}^{k-1} \left(s^{2^k + 1 - 2^{i+1}} + 1 \right) T^{2^i} .$$

Remark 1. Let $\tau = x^2$ and $\tau^0 = x$. The linearized polynomials in $\overline{\mathbf{F}}[x]$ form an algebra $\overline{\mathbf{F}}\{\tau\}$ where the multiplication is the composition in $\overline{\mathbf{F}}[x]$, see e.g. [3]. We could write $W_k(s, T)$ as $(\tau + w_{k-1}(s)\tau^0) \dots (\tau + w_1(s)\tau^0)$ (with $x = T$) and polynomials $f_n(s)$ as $(\tau + \tau^0)^n + 1$ (with $x = s$). From this τ-form we see easily that $f_{2^l-1}(s) + 1 = (\tau + \tau^0)^{2^l-1} = \sum_{i=0}^{2^l-1} \tau^i = \text{Tr}_{2^l}(s)$ in $\overline{\mathbf{F}}[s]$.

In the proof of the main theorem we end up with (9), i.e. the equation $P_k(s, T) = \sum_{i=1}^{k} a^{2^i}$. We should count the solutions satisfying the conditions mentioned in the next theorem. The key idea here is that in the cases $k = 2^l$ the polynomials $P(s, T)$ and $(s + 1)W_k(s, T)$ are identical and the composition structure of $W_k(s, T)$ makes counting the number of solutions possible.

Let $\mathbf{F}^a = \mathbf{F} \setminus \{0, 1, a, a + 1\}$ for any $a \in \mathbf{F} \setminus \{0, 1\}$.

Theorem 4. *For every $l \geq 1$ and $a \in \mathbf{F} \setminus \{0, 1\}$ the equation $(s+1)W_{2^l}(s, T) = \sum_{i=1}^{2^l} a^{2^i}$ has exactly $(q - 8)/2$ solutions $(s, T) \in \mathbf{F}^a \times \mathbf{F}$ with $\text{Tr}_m\left(T/s^2\right) = 0$.*

Proof. We give only a sketch of proof and an interested reader can find the whole proof from [13, Theorem 5.10].

By the definition of $W_{2^l}(s, T)$ the equation splits into a chain of $2^l - 1$ nested equations

$$(s + 1)\left[U_{2^l-1}^2 + w_{2^l-1}(s)U_{2^l-1}\right] = \sum_{i=1}^{2^l} a^{2^i}$$

$$U_{2^l-2}^2 + w_{2^l-2}(s)U_{2^l-2} = U_{2^l-1}$$

$$\vdots$$

$$U_2^2 + w_2(s)U_2 = U_3$$

$$U_1^2 + w_1(s)U_1 = U_2 \ .$$

The first equation has two roots in $\mathbf{F}$ for $q/2 - 4$ values of $s \in \mathbf{F}^a$ because of two facts: for every $s \in \{0, 1, a, a + 1\}$ the equation has two roots in $\mathbf{F}$ and the trace condition from Lemma 1 includes a permutation polynomial by Lemma 2.

When we substitute these roots to the next equation this second equation has two roots for exactly one of the previous roots by Lemma 2. And so on; we can always "drop down" one of the two roots. The last equation has two solutions U_1 but exactly one of them satisfies the condition $\text{Tr}_m\left(U_1/s^2\right) = 0$.   □

To prove the next theorem we need one lemma.

Lemma 4. *For any $b \in \mathbf{F}_{2^k}$ we have the following identities*

$$\overline{W_k(s, b^2s^2 + bs + b^2)} = F_k(s, bs^2 + bs + b^2)$$

$$= (f_{k-1}(b) + 1)(f_k(s) + 1) + f_{k-1}(b^2) + 1$$

$$= \text{Tr}_k(b)(f_k(s) + 1) + \text{Tr}_k(b^2), \quad \text{if} \quad k = 2^l \ .$$

Proof. The last equality is clear by Remark 1. The first two identities can be proved by induction with the following steps.

$$\overline{W_{k+1}(s, b^2 s^2 + bs + b^2)}$$
$$= \overline{(T^2 + w_k(s)T) \circ (W_k(s, b^2 s^2 + bs + b^2)}$$
$$= \overline{W_k(s, b^2 s^2 + bs + b^2)^2 + w_k(s)W_k(s, b^2 s^2 + bs + b^2)}$$
$$= \overline{W_k(s, b^2 s^2 + bs + b^2)}^2 + f_k(s)\overline{W_k(s, b^2 s^2 + bs + b^2)}$$
$$= F_{k+1}(s, bs^2 + bs + b^2)$$
$$= (T^2 + f_k(s)T) \circ [(f_{k-1}(b) + 1)(f_k(s) + 1) + f_{k-1}(b^2) + 1]$$
$$= (f_{k-1}(b)^2 + 1)(f_k(s)^2 + 1) + f_{k-1}(b^2)^2 + 1$$
$$\quad + (f_{k-1}(b) + 1)(f_k(s)^2 + f_k(s)) + f_k(s)f_{k-1}(b^2) + f_k(s)$$
$$= (f_{k-1}(b)^2 + f_{k-1}(b))(f_k(s)^2 + f_k(s)) + f_{k-1}(b^2)^2 + f_{k-1}(b^2)$$
$$= (f_k(b) + 1)(f_{k+1}(s) + 1) + f_k(b^2) + 1 \ . \qquad \square$$

Theorem 5. *For every $l \geq 1$ the equation $P_{2^l}(s, T) = (s + 1)W_{2^l}(s, T)$ holds.*

Proof. By (1) we have

$$P_k(s, T) = D_{2^k+1}(s, T) + D_{2^k+1}(1, T) + s^{2^k+1} + 1$$

and Theorem 3 gives us a factorization

$$P_k(s, T) = (s + 1)\left(\prod_{i=1}^{2^{k-1}} (s^2 + \beta_i s + 1 + \beta_i^2 T) + \left(s^{2^k} + s^{2^k-1} + \cdots + s + 1\right)\right)$$

$$= (s + 1)\left(\prod_{i=1}^{2^{k-1}} \left(\frac{1}{\beta_i^2}s^2 + \frac{1}{\beta_i}s + \frac{1}{\beta_i^2} + T\right) + \prod_{i=1}^{2^{k-1}} (s^2 + \beta_i s + 1)\right)$$

since $\prod \beta_i$ is the coefficient of $s^{2^{k-1}}$ in the rightmost product and it is equal to 1. It is also clear, that $\mathrm{Tr}_k(\beta_i) = 1$ since $s^2 + \beta_i s + 1$ do not have roots in $\mathbf{F}_{2^k}$.

$P_k(s, T)$ is a linearized polynomial with respect to the variable T and above we have divided it to an affine polynomial plus a constant term. The roots of the affine part are easily seen and the roots of $P_k(s, T)$ are differences of them, i.e.

$$\left(\frac{1}{\beta_1^2} + \frac{1}{\beta_i^2}\right)s^2 + \left(\frac{1}{\beta_1} + \frac{1}{\beta_i}\right)s + \left(\frac{1}{\beta_1^2} + \frac{1}{\beta_i^2}\right), \qquad i = 1, \ldots, 2^{k-1}$$

are some of them. Actually, with an identity

$$A := \left\{\frac{1}{\beta_1} + \frac{1}{\beta_i} \,\middle|\, i = 1, \ldots, 2^{k-1}\right\} = \{\alpha \in \mathbf{F}_{2^k} \mid \mathrm{Tr}_k(\alpha) = 0\}$$

it is clear that the roots form a suitable size subspace and we have all the roots. All in all, the decomposition can be presented in a simpler form

$$P_k(s,T) = (s+1) \prod_{\alpha \in A} \left(T + \alpha^2 s^2 + \alpha s + \alpha^2\right) .$$

If we can show that every root of $P_k(s,T)$ is a root of $W_k(s,T)$ we are done since the polynomials have the same degree. By Lemma 4 and $k = 2^l$ we have

$$\overline{W_k(s, \alpha^2 s^2 + \alpha s + \alpha^2)} = \mathrm{Tr}_k(\alpha)(f_k(s) + 1) + \mathrm{Tr}_k(\alpha^2) = 0$$

for every $\alpha \in A$ which by Lemma 3 finishes the proof.   □

Now we explain the phrase "variants of cyclotomic polynomials".

Corollary 4. *As polynomials in $\mathbf{F}_2[s]$ we have identities*

$$(s+1) \prod_{i=0}^{2^l-1} w_i(s) = s \prod_{i=0}^{2^l-1} f_i(s) = s \prod_{i|2^{2^l}-1} Q_i(s) = s + s^{2^{2^l}} .$$

Proof. For cyclotomic polynomials $Q_i(s)$ the identity is well known. Compare the coefficient of T in polynomial $P_{2^l}(s,T)$ by definition and by previous theorem

$$s^{2^{2^l}} - 1 = (s+1)w_1(s)w_2(s)\ldots w_{2^l-1}(s)$$

and multiplying both sides with $w_0(s) = s$ gives the result for $w_n(s)$'s. With Lemma 3 we get the relation for $f_n(s)$'s.   □

In addition, the degrees of polynomials $w_n(s)$, $f_n(s)$, and $Q_n(s)$ and the degrees of their irreducible factors over $\mathbf{F}_2$ seem to coincide.

4 Coding Theory Preliminaries

Below we have an example of extended binary cyclic codes of length q. Parity-check matrices of these codes can be described with a primitive element α of $\mathbf{F}$.

Definition 4. *Let $m \geq 3$ be odd. The extended binary two-error-correcting BCH-like code $\mathcal{B}_k$ of length $q = 2^m$ is defined by a parity-check matrix*

$$\begin{pmatrix} 1 & 1 & 1 & 1 & \cdots & 1 \\ 0 & 1 & \alpha & \alpha^2 & \cdots & \alpha^{q-2} \\ 0 & 1 & \alpha^{2^k+1} & \alpha^{(2^k+1)2} & \cdots & \alpha^{(2^k+1)(q-2)} \end{pmatrix}$$

where $1 \leq k \leq (m-1)/2$ and $\gcd(m,k) = 1$.

The code $\mathcal{B}_1$ is the usual two-error-correcting extended BCH code. The codes $\mathcal{B}_k$ are pairwise nonequivalent [1, Sect. 4.3] and have parameters $[q, q - 2m - 1, 6]$. The dual code $\mathcal{B}_k^\perp$ has only 3 weights in the range $1 \leq i \leq q - 3$ and by the famous Assmus–Mattson theorem the supports of codewords of Hamming weight 6 in $\mathcal{B}_k$ form a 3-$(q, 6, (q - 8)/6)$ design.

We consider linear $\mathbf{Z}_4$-codes of length q which are subgroups of $\mathbf{Z}_4^q$ with componentwise addition. Let $R = \mathrm{GR}(4, m)$ be a Galois ring of characteristic 4 with $q^2 = 4^m$ elements. The multiplicative group of units R^* contains a unique cyclic subgroup $\langle \beta \rangle$ of order $q - 1$. Every element of R can be expressed uniquely as $A + 2B$, where $A, B \in \mathcal{T}$ and

$$\mathcal{T} = \{0, 1, \beta, \ldots, \beta^{q-2}\} \ .$$

Let $\mu : \mathbf{Z}_4 \to \mathbf{F}_2$ denote the modulo 2 reduction map. We extend μ to R and $\mathbf{Z}_4^q$ in a natural way, and then $\mu(\mathcal{T}) = \mathbf{F}$ and $\mu(\mathbf{Z}_4^q) = \mathbf{F}_2^q$.

Definition 5. *Let $m \geq 3$ be odd. The $\mathbf{Z}_4$-Goethals-like code $\mathcal{G}_k$ of length $q = 2^m$ is defined by a parity-check matrix*

$$\begin{pmatrix} 1 & 1 & 1 & 1 & \cdots & 1 \\ 0 & 1 & \beta & \beta^2 & \cdots & \beta^{q-2} \\ 0 & 2 & 2\beta^{2^k+1} & 2\beta^{(2^k+1)2} & \cdots & 2\beta^{(2^k+1)(q-2)} \end{pmatrix}$$

where $1 \leq k \leq (m - 1)/2$ and $\gcd(m, k) = 1$.

The codes $\mathcal{G}_1$ and $\mathcal{G}_k$ were introduced in [4] and [6], respectively, with the results stated in the next theorem (see also [7] for the fact 1.). These codes are pairwise nonequivalent [11] and have $2^{2q-3m-2}$ codewords which means that their binary Gray images have four times as many codewords as BCH codes of the same length and minimum distance.

Remark 2. The restriction $1 \leq k \leq (m - 1)/2$ comes from the fact $\mathcal{G}_k = \mathcal{G}_{m-k}$. There are 90 pairs (m, k) of suitable parameters for $\mathbf{Z}_4$-Goethals code $\mathcal{G}_k$ with $5 \leq m \leq 29$. 86 of these pairs can be presented in the form (m, k) where $2^l \equiv \pm k \pmod{m}$ for some $l \geq 1$. In theorem 1 and its corollaries we have the restriction $k = 2^l$ which means that we can prove the existence of the designs in almost all codes $\mathcal{G}_k$ with $5 \leq m \leq 29$. The smallest parameters with which we can not get the designs in this paper are $(17, \{3, 5, 6, 7\})$ and the next ones are $(31, \{3, 5, 6, \ldots\})$.

By [5, Lemma 2] a word $(c_X)_{X \in \mathcal{T}} \in \mathbf{Z}_4^q$, with $C_j = \{\mu(X) \mid c_X = j\}$ for $j \in \mathbf{Z}_4$, is a codeword of $\mathcal{G}_k$ if and only if it satisfies the following equations over $\mathbf{F}$:

$$\sum_{x \in \mathbf{F}} c_x = 0 \quad (\text{in } \mathbf{Z}_4) \qquad\qquad \sum_{x \in C_1 \cup C_3} x = 0$$

$$\sum_{\substack{x, y \in C_1 \cup C_3 \\ x < y}} x y = \sum_{x \in C_2 \cup C_3} x^2 \qquad\qquad \sum_{x \in C_1 \cup C_3} x^{2^k+1} = 0 \ . \tag{3}$$

where $\leq$ is some total order on $\mathbf{F}$. As we have equations over $\mathbf{F}$ we think from now on that the codewords are indexed with the elements of $\mathbf{F}$.

Theorem 6.   *1.* $d_L(\mathcal{G}_k) = 8$;
2. $\mu(\mathcal{G}_k) = \{\mu(\mathbf{c}) \mid \mathbf{c} \in \mathcal{G}_k\} = \mathcal{B}_k$;
3. $\mathcal{G}_k \cap 2\mathbf{Z}_4^q = \{2\mathbf{d} \mid \mu(\mathbf{d}) \in \mathcal{H}\}$ *where* $\mathcal{H}$ *is the extended Hamming code;*
4. *the automorphism group of* $\mathcal{G}_k$ *contains the doubly transitive group of affine permutations*
$$x \mapsto ax + b, \qquad a \in \mathbf{F}^*, \ b \in \mathbf{F} \ .$$

5   Proof of the Main Theorem

The supports of codewords of Hamming weight 8 in all $\mathbf{Z}_4$-Goethals-like codes $\mathcal{G}_k$ have the same structure as was found in the codes $\mathcal{G}_1$ by Ranto [12] . In particular, we have in every $\mathcal{G}_k$ codewords of cwe-type (complete weight enumerator) $X^6 Z^2$ which means codewords with six 1's and two 3's. Among the supports of these codewords we have a special class of supports considered in the main theorem: they are disjoint unions of two nonparallel 2-flats when we think them as subsets of $\mathbf{F}$ equipped with the m-dimensional affine geometry. In addition, it is easy to see [13] that the two 3's occur in different 2-flats. This set of supports is nonempty whenever $m \geq 5$ and explains this restriction in the results. The interested reader is referred to [12, 13] for more details.

   To prove the main theorem we have to show that any three distinct coordinate positions are included in equally many supports of the special type described above. By the automorphisms described in Theorem 6 we can assume that these positions are 0, 1, and an arbitrary element $a \in \mathbf{F} \setminus \{0, 1\}$. We divide the supports into different groups according to the way the fixed positions 0, 1, and a are divided among the 2-flats and also according to the positions of the two 3's. Altogether, there are 22 such groups.

Table 2. Needed combinations of three fixed positions

| | 1 | 1 | 1 | 3 | 1 | 1 | 1 | 3 | |
Case	x_1	x_2	x_3	x_4	y_1	y_2	y_3	y_4	Frequency
(0a)	0	1	a						$(q-8)/6$
(1a)	0	1			a				$\frac{2(q-8)}{3}$
(1b)	0	1						a	
(2a)	0			1	a				$\frac{q-8}{2}$
(2b)	1			0	a				
(3a)	0			1				a	$\frac{q-8}{6}$
(3b)	1			0				a	

   We can use automorphisms $x \mapsto x+1$ and $x \mapsto x/a$ to reduce the case analysis and the 7 cases which we really have to count are shown in Table 2. The number of supports belonging to each combination with a fixed a is also shown. The cases left out are: 3 cases similar to (0a) with the same frequency and 2 copies of block (1a)-(3b) with the same frequencies.

Next we verify the different frequencies and by summing them up we claim that λ is equal to $14(q-8)/3$ and the supports, indeed, form a 3-design.

5.1   Syndrome Equations

Next we consider the equations which the support $\{x_1, x_2, x_3, x_4, y_1, y_2, y_3, y_4\}$ from Table 2 should satisfy. The sets $\{x_1, x_2, x_3, x_4\}$ and $\{y_1, y_2, y_3, y_4\}$ form the two 2-flats and 3's are thought to be in the positions x_4 and y_4. By (3) and the 2-flat structure the following equations should hold:

$$\sigma_1(x_1, x_2, x_3, x_4) = \sigma_1(y_1, y_2, y_3, y_4) = 0$$
$$\sigma_2(x_1, x_2, x_3) = \sigma_2(y_1, y_2, y_3) \tag{4}$$
$$S_{2^k+1}(x_1, x_2, x_3, x_4) = S_{2^k+1}(y_1, y_2, y_3, y_4).$$

where $\sigma_k(a_1, \ldots, a_n) = \sum_{1 \le i_1 < \cdots < i_k \le n} a_{i_1} \cdots a_{i_k}$ is the kth elementary symmetric polynomial and $S_k(a_1, \ldots, a_n) = \sum_{i=1}^n a_i^k$ is the sum of kth powers.

If the variables x_i and y_i are distinct, the corresponding support is of the desired type. The only possible overlapping of the variables x_i and y_i satisfying (4) is the case where the 2-flats are equal and $x_4 = y_4$; that is, they form the support $\{0, 1, a, a+1\}$ of cwe-type Y^4 (four 2's). This kind of supports correspond to codewords of the extended binary Hamming code $\mathcal{H}$, see Theorem 6, and they form a 3-$(q, 4, 1)$ design by Assmus–Mattson theorem. Hence the solutions of (4) have one extra codeword which must be excluded.

5.2   Case (0a)

As mentioned after Definition 4 there are $(q-8)/6$ codewords of Hamming weight 6 in $\mathcal{B}_k$ that contain the three fixed coordinates. These codewords can be uniquely lifted to codewords in $\mathcal{G}_k$ of cwe-type X^6Y: the codeword of $\mathcal{B}_k$ satisfies two of the four equations in (3) and suitably positioning a single 2 makes the remaining two equations hold, too. This 2-symbol can not be within the original support of size 6 as $d_L(\mathcal{G}_k) = 8$.

We can lift the same codeword as above in three different ways to a codeword of cwe-type X^4YZ^2 in $\mathcal{G}_k$ such that the three fixed coordinates are all 1's: choose two 3-positions from the three positions which are not fixed and find the unique position for 2. We have all in all $(1+3)(q-8)/6 = 2(q-8)/3$ such codewords of cwe-type X^6Y and X^4YZ^2. The geometric connection between the supports of size 7 and the supports in the main theorem is described in [12, 13] and in the case (0a) this connection is now recalled in Table 3.

We counted the frequency in the left hand side of Table 3, i.e. verified Corollary 1 in relevant cases first. In the table some 3-flat codewords from $\mathcal{G}_k$ are shown to be differences of codewords connected to Theorem 1 and Corollary 1. If a support of a codeword includes a 2-flat, elements of one 2-flat are underlined.

Table 3. Structural dependence of blocks in Theorem 1 and Corollary 1 in Case (0a)

Corollary 1 → Theorem 1			Theorem 1 → Corollary 1		
Fix	Comb	Freq	Fix	Comb	Freq
111 1112			111 31113		
33313111	1		13333111	1	
111 33111		$\frac{2(q-8)}{3}$	111 2111		$\frac{q-8}{6}$
111 1332			111 31113		
31133113	1		13311133	3	
111 13113			111 2 133		

From one codeword in Corollary 1 we get one codeword in the main theorem; to the other direction we get 4 codewords from one codeword. Therefore the frequency in the right hand side must be $(q - 8)/6$ which concludes this case.

5.3   Cases (1a) and (1b)

Next we study the case (1a), so $x_1 = 0$, $x_2 = 1$, $x_4 = x_3 + 1$, $y_1 = a$, and $y_4 = a + y_2 + y_3$.

The syndrome equations (3) imply that $x_3 = a(y_2 + y_3) + y_2 y_3$ and

$$1 + x_3^{2^k+1} + (x_3 + 1)^{2^k+1} = a^{2^k+1} + y_2^{2^k+1} + y_3^{2^k+1} + (a + y_2 + y_3)^{2^k+1} .$$

Substituting the first equation to the second we get

$$W(U + V) = UV$$

where $W = a + a^{2^k}$, $U = y_2 + y_2^{2^k}$, and $V = y_3 + y_3^{2^k}$ for any k.

It is well known that the mapping $u \mapsto u + u^{2^k}$ is two-to-one and its image is $T_0 = \{u \in \mathbf{F} \mid \mathrm{Tr}_m(u) = 0\}$. Now we have for all k the following equation

$$W(U + V) = UV \qquad W, U, V \in T_0 . \tag{5}$$

In [12] it was noticed that the number of solutions does not depend on a when $k = 1$. Therefore the number of solutions of (5) does not depend on W and this holds now for all k. One value of $W = a + a^{2^k}$ corresponds to a and $a+1$ simultaneously but this is not a problem since there are equally many codewords for the values a and $a+1$ as can be seen via the automorphism $x \mapsto x+1$. Hence the number of solutions of (5) does not depend on a and by [12] it is equal to $2(q - 8)/3$.

In the case (1b) we have $x_1 = 0$, $x_2 = 1$, $x_4 = x_3 + 1$, $y_4 = a$, and $y_1 = a + y_2 + y_3$. By (4) we derive $x_3 = a(y_2 + y_3) + y_2 y_3 + (y_2 + y_3)^2$ and

$$W(U + V) = UV + (U + V)^2 \qquad W, U, V \in T_0 .$$

With the same argument as above the number of solutions depends neither on k nor on a. Considering U and V as roots of a quadratic equation $T^2 + (U+V)T = UV$ we see also that the value of

$$\mathrm{Tr}_m\left(\frac{W}{U+V}\right)$$

determines whether the solution (U,V) belongs to the case (1a) or (1b).

5.4   Cases (2a) and (2b)

In the case (2a) we have $x_1 = 0$ and $x_4 = 1$, and in the case (2b) $x_1 = 1$ and $x_4 = 0$, and in both cases $y_1 = a$. We denote $x = x_2$, $\sigma_1 = y_2 + y_3$, and $\sigma_2 = y_2 y_3$ which implies $x_3 = x + 1$ and $y_4 = a + \sigma_1$ by (4). One of the equations

$$\begin{aligned} x + x^2 &= a\sigma_1 + \sigma_2 & \text{(2a)} \\ 1 + x + x^2 &= a\sigma_1 + \sigma_2 & \text{(2b)} \end{aligned} \tag{6}$$

holds and now the third equation in (4) transforms to

$$x + x^{2^k} = a^{2^k+1} + y_2^{2^k+1} + y_3^{2^k+1} + (a+\sigma_1)^{2^k+1} \; . \tag{7}$$

Suppose now that k is even. With a telescopic identity

$$x + x^{2^k} = \sum_{i=0}^{k-1} \left(x+x^2\right)^{2^i} = \sum_{i=0}^{k-1} \left(1 + x + x^2\right)^{2^i}$$

we can consider the cases (2a) and (2b) simultaneously and writing (7) down with a Dickson polynomial we derive

$$\sum_{i=0}^{k-1} (a\sigma_1 + \sigma_2)^{2^i} = a^{2^k+1} + D_{2^k+1}(\sigma_1, \sigma_2) + (a+\sigma_1)^{2^k+1} \; .$$

We substitute $\sigma_2 = T + a\sigma_1 + a^2$ and get

$$\sum_{i=0}^{k-1} \left(T + a^2\right)^{2^i} = a^{2^k+1} + D_{2^k+1}\left(\sigma_1, T + a\sigma_1 + a^2\right) + (a+\sigma_1)^{2^k+1} \; . \tag{8}$$

By (2) and $a^{2^k+1} + (a+\sigma_1)^{2^k+1} = D_{2^k+1}\left(\sigma_1, a\sigma_1 + a^2\right)$ we have

$$\sum_{i=0}^{k-1} \left(T + a^2\right)^{2^i} = D_{2^k+1}(\sigma_1, T) + \sigma_1^{2^k+1} \; .$$

By regrouping the terms we arrive at the equation

$$P_k(\sigma_1, T) = \sum_{i=0}^{k-1} \left(\sigma_1^{2^k+1-2^{i+1}} + 1\right) T^{2^i} = \sum_{i=1}^{k} a^{2^i} \; . \tag{9}$$

Now we suppose that $k = 2^l$ with some $l \geq 1$. By Theorems 4 and 5 the above equation has $(q-8)/2$ solutions $(\sigma_1, T) \in \mathbf{F}^a \times \mathbf{F}$ with $\mathrm{Tr}_m(T/\sigma_1^2) = \mathrm{Tr}_m(\sigma_2/\sigma_1^2) = 0$ which gives us the variables y_2 and y_3. The restriction $\sigma_1 \in \mathbf{F}^a$ is for avoiding the extra codeword mentioned in Subsection 5.1. The variable x can be solved from exactly one of the equations (6). The other solution $x + 1$ refers to the same codeword and all in all we have $(q-8)/2$ codewords containing the three coordinates 0, 1, and a.

5.5   Cases (3a) and (3b)

The setting in these cases differs from the previous subsection such that (6) is replaced by

$$x + x^2 = a\sigma_1 + \sigma_2 + \sigma_1^2 \qquad \text{(3a)}$$
$$1 + x + x^2 = a\sigma_1 + \sigma_2 + \sigma_1^2 \qquad \text{(3b)}$$

(10)

so there is one additional term σ_1^2 in both equations. The ideas are exactly the same as above. By substituting $\sigma_2 = T + a\sigma_1 + a^2 + \sigma_1^2$ we replace (8) by

$$\sum_{i=0}^{k-1} \left(T + a^2\right)^{2^i} = a^{2^k+1} + D_{2^k+1}\left(\sigma_1, T + a\sigma_1 + a^2 + \sigma_1^2\right) + (a + \sigma_1)^{2^k+1}$$

and using twice equality (2) we have

$$\sum_{i=0}^{k-1} \left(T + a^2\right)^{2^i} = D_{2^k+1}(\sigma_1, T) + (k+1)\sigma_1^{2^k+1} .$$

When k is even we get the same equation (9) as above and when k is a power of 2 we can also calculate the number of roots.

Theorem 4 holds also in these cases with the difference that $\text{Tr}_m(\sigma_2/\sigma_1^2) = \text{Tr}_m(T/\sigma_1^2) + 1$ and hence the solutions here are exactly those which were ruled out in the last step of that theorem. Again the variable x can be solved from one of the equations (10) but this time every codeword is counted three times: we can choose two positions referring to σ_1 and σ_2 in three ways. All in all we have $(q-8)/6$ codewords containing the three coordinates 0, 1, and a.

6   Conclusions and Further Research

We have shown that for all the designs which have been found from the code $\mathcal{G}_1$ so far we can find a design with the same parameters from the code $\mathcal{G}_k$ when $k = 2^l$. The decomposition technique described in this paper does not work for parameters $k \neq 2^l$: for example, one can easily check that $P_6(s,T) \neq (s+1)W_6(s,T)$. Recently [9], the authors managed to prove the existence of the designs for every k with a different analysis of the polynomial P_k.

It should be possible to prove more about the nonequivalence of these designs. This is related to the question of BCH-like codes $\mathcal{B}_k$ and whether or not their minimum weight codewords generate the code, see [13, Sect. 5.3]. In addition, several other open problems can be found from [13].

Acknowledgments

We thank Ph.D. Petri Rosendahl for valuable discussions and anonymous referees for helpful comments.

References

1. P. Charpin. Open problems on cyclic codes. In V. S. Pless and W. C. Huffman, editors, *Handbook of Coding Theory*, volume I, chapter 11, pages 963–1063. Elsevier, 1998.
2. S. D. Cohen and R. W. Matthews. Exceptional polynomials over finite fields. *Finite Fields Appl.*, 1(3):261–277, 1995.
3. D. Goss. *Basic Structures of Function Field Arithmetic*. Springer, second edition, 1998.
4. A. R. Hammons, Jr., P. V. Kumar, A. R. Calderbank, N. J. A. Sloane, and P. Solé. The $\mathbb{Z}_4$-linearity of Kerdock, Preparata, Goethals, and related codes. *IEEE Trans. Inform. Theory*, 40(2):301–319, 1994.
5. T. Helleseth and P. V. Kumar. The algebraic decoding of the Z_4-linear Goethals code. *IEEE Trans. Inform. Theory*, 41(6):2040–2048, 1995.
6. T. Helleseth, P. V. Kumar, and A. Shanbhag. Codes with the same weight distributions as the Goethals codes and the Delsarte-Goethals codes. *Des. Codes Cryptogr.*, 9(3):257–266, 1996.
7. T. Helleseth, J. Lahtonen, and K. Ranto. A simple proof to the minimum distance of Z_4-linear Goethals-like codes. *J. Complexity*, 20(2–3):297–304, 2004.
8. T. Helleseth, C. Rong, and K. Yang. On t-designs from codes over $\mathbf{Z}_4$. *Discrete Math.*, 238:67–80, 2001.
9. J. Lahtonen, K. Ranto, and R. Vehkalahti. 3-Designs from all $\mathbf{Z}_4$-Goethals-like codes with block size 7 and 8. *Finite Fields Appl.*, 2005. To appear.
10. R. Lidl, G. L. Mullen, and G. Turnwald. *Dickson Polynomials*, volume 65 of *Pitman Monographs and Surveys in Pure and Applied Mathematics*. Longman Scientific & Technical, Harlow, 1993.
11. K. Ranto. On algebraic decoding of the $\mathbf{Z}_4$-linear Goethals-like codes. *IEEE Trans. Inform. Theory*, 46(6):2193–2197, 2000.
12. K. Ranto. Infinite families of 3-designs from $\mathbf{Z}_4$-Goethals codes with block size 8. *SIAM J. Discrete Math.*, 15(3):289–304, 2002.
13. K. Ranto. $\mathbf{Z}_4$-*Goethals Codes, Decoding and Designs*. PhD thesis, University of Turku, Oct. 2002. http://www.tucs.fi/publications/insight.php?id=phdRanto02a.
14. D.-J. Shin, P. V. Kumar, and T. Helleseth. 3-Designs from the Z_4-Goethals codes via a new Kloosterman sum identity. *Des. Codes Cryptogr.*, 28(3):247–263, 2003.

Space-Time Code Designs Based on the Generalized Binary Rank Criterion with Applications to Cooperative Diversity

A. Roger Hammons Jr.*

The Johns Hopkins University Applied Physics Laboratory,
Laurel, MD 20723, USA
`roger.hammons@jhuapl.edu`

Abstract. Li and Xia have recently investigated the design of space-time codes that achieve full spatial diversity for quasi-synchronous cooperative communications. They show that certain of the binary space-time trellis codes derived from the Hammons-El Gamal stacking construction are delay tolerant and can be used in the multilevel code constructions by Lu and Kumar to produce delay tolerant space-time codes for PSK and QAM signaling. In this paper, we present a generalized stacking criterion for maximal rank-d binary codes and develop new explicit constructions. We also present several multilevel space-time code constructions for certain AM-PSK constellations that generalize the recent Lu-Kumar unified construction. Following the approach by Li and Xia, we show that, if the binary constituent codes used in these AM-PSK constructions are delay tolerant, so are the multilevel codes, making them well-suited for quasi-synchronous cooperative diversity applications.

1 Introduction

It is well-known that wireless communications over Rayleigh fading channels can benefit from the simultaneous use of multiple antennas at both the transmitter and receiver to convey information either more reliably or at higher rates than would be possible for single antenna systems. In certain applications, however, it may be infeasible or not cost effective to equip terminals with the additional hardware. Therefore, there has been significant recent research interest in applying multiple-input multiple-output (MIMO) techniques to cooperative networks. In such networks, the individual terminals may be poorly equipped, but they can overcome their limitations by pooling resources with other terminals.

Since the cooperative terminals do not necessarily share a common reference, Li and Xia [12] argue that cooperative diversity schemes are fundamentally asynchronous and therefore the design of space-time codes should address the case of asynchronicity explicitly. Li and Xia investigate the design of space-time trellis codes that yield full spatial diversity for any number of quasi-synchronous

---

* This material is based upon work supported by the National Science Foundation under Grant No. CCR-0325781.

Ø. Ytrehus (Ed.): WCC 2005, LNCS 3969, pp. 69–84, 2006.

cooperating relays. They have shown that certain of the trellis codes derived from the Hammons-El Gamal stacking construction are delay tolerant—in the sense that full diversity is preserved despite random delays among the various transmissions—and thus suitable for cooperative diversity schemes. They provide necessary (but not sufficient) conditions for the trellis codes to be delay tolerant. They have also shown that, when these codes are used in the multilevel Lu-Kumar construction for PSK and QAM modulation, the resulting space-time codes also achieve full spatial diversity in quasi-synchronous cooperative operations.

This paper extends the previous work. First, from the generalized binary rank criterion, we develop a generalized stacking construction that applies to maximal rank-d binary codes. These include, as special cases, the family of Gabidulin codes described by Lu and Kumar. We also present various multi-level space-time code constructions for certain AM-PSK constellations that generalize the Lu-Kumar multi-level construction. These are intimately related to various constructions first developed by the author for noncooperative multiple-input multiple-output (MIMO) communication systems [7] [8]. Following the approach by Li and Xia, we show that, if the binary constituent codes used in these AM-PSK constructions are delay tolerant and therefore suitable for use in cooperative diversity schemes, so are the multilevel codes.

2   Background

Let $\mathcal{C}$ be a code of length MT, with $M \leq T$, over the discrete alphabet Ω. The codewords of $\mathcal{C}$ are presented as $M \times T$ matrices in which the (m,t)-th entry $a_{m,t} \in \Omega$ represents the information symbol that is modulated and transmitted from the m-th transmit antenna at transmission interval t. If all of the pairwise differences between distinct modulated code word matrices have rank at least d over $\mathbb{C}$, then the space-time code is called an $M \times T$ rank-d code. In the special case that all of the nontrivial pairwise differences between modulated code words are of full rank M, the space-time code is called an $M \times T$ full-rank code.

There is a tradeoff [2] between achievable transmission rate and achievable transmit diversity level for space-time codes. Full-rank space-time codes can achieve transmission rates no greater than one symbol per transmission interval. For rank d space-time codes, the maximum transmission rate is $M-d+1$ symbols per transmission interval. Equivalently, the size of an $M \times T$ rank-d space-time code cannot exceed $q^{T(M-d+1)}$, where q is the size of the signaling constellation Ω. Codes meeting this upper limit are referred to as maximal.

In [3], Hammons and El Gamal developed a method of algebraic space-time code design for BPSK and QSPK modulation in which the rank of modulated code words over the field $\mathbb{C}$ is inferred from the rank of their projections as matrices over the binary field $\mathbb{F}$. This work was extended and further refined by Liu *et al.* [4] and Lu and Kumar [6].

The connection between modulated space-time codes, with entries from $\mathbb{Z}[\theta] \subset \mathbb{C}$, where θ be a complex root of unity, and binary codes over $\mathbb{F} = GF(2)$ is through the isomorphism $\mathbb{Z}[\theta]/(1-\theta) \cong \mathbb{F}$. We will let $\mu : \mathbb{Z}[\theta] \to \mathbb{F}$ denote the corresponding projection modulo $1 - \theta$. It is straightforward to show [6] that, if C is a complex matrix with entries from $\mathbb{Z}[\theta]$ whose binary projection $\mu(C)$ is of rank d over $\mathbb{F}$, then C is also of rank at least d over $\mathbb{C}$. We refer to this as the *generalized binary rank criterion*.

3 Stacking Construction for Maximal Rank-d Binary Codes

We first introduce a generalization of the Hammons-El Gamal stacking construction [3] applicable to the design of binary maximal rank-d codes. We will use the following notation. Given a set of linear transformations $T_1, T_2, \ldots, T_M : \mathbb{F}^K \to \mathbb{F}^T$ and a binary matrix $\mathbf{A} \in \mathbb{F}^{M \times M}$, we form a new set of linear transformations $T_1^{\mathbf{A}}, T_2^{\mathbf{A}}, \ldots, T_M^{\mathbf{A}}$ according to the relation

$$\begin{bmatrix} T_1^{\mathbf{A}} \\ T_2^{\mathbf{A}} \\ \vdots \\ T_M^{\mathbf{A}} \end{bmatrix} = \mathbf{A} \begin{bmatrix} T_1 \\ T_2 \\ \vdots \\ T_M \end{bmatrix}.$$

These are called the $\mathbf{A}$-modified linear transformations.

Theorem 1 (Generalized Stacking Construction). *Let K, M, R, and T be positive integers with $K = RT$ and $M \leq T$. Let $T_1, T_2, \ldots, T_M$ be linear transformations from $\mathbb{F}^K$ to $\mathbb{F}^T$. The code $\mathcal{S}$ consists of all binary matrices of the form*

$$\mathbf{c} = \begin{bmatrix} \bar{x}T_1 \\ \bar{x}T_2 \\ \vdots \\ \bar{x}T_M \end{bmatrix},$$

where $\bar{x} \in \mathbb{F}^K$.

Suppose that, for all nonsingular $\mathbf{A} \in \mathbb{F}^{M \times M}$, the intersection of the null spaces of any set of R or more of the $\mathbf{A}$-modified transformations $T_1^{\mathbf{A}}, T_2^{\mathbf{A}}, \ldots, T_M^{\mathbf{A}}$ is trivial. Then $\mathcal{S}$ is a binary maximal rank $d = M - R + 1$ code.

Proof. The rate of the code is $\log(|\mathcal{S}|)/T = R$. Hence, by the rate-diversity tradeoff, it must be shown that the difference between any two BPSK-modulated code word matrices is of rank at least d over the complex numbers. The binary rank criterion states that it is sufficient to show that every non-zero code word of $\mathcal{S}$ has rank at least d over $\mathbb{F}$. Suppose to the contrary that, for some $\bar{x} \neq \bar{0}$, we have

$$\mathbf{c} = \begin{bmatrix} \bar{x}T_1 \\ \bar{x}T_2 \\ \vdots \\ \bar{x}T_M \end{bmatrix}$$

has rank $s < d$ over $\mathbb{F}$. Then, there is a nonsingular matrix $\mathbf{A}$ such that $\mathbf{Ac}$ has exactly $M - s$ rows that are identically zero. Equivalently, $\bar{x}$ is a nontrivial member of the null spaces of $M - s$ of the $\mathbf{A}$-modified transformations. Since $M - s \geq R$, this contradicts the assumption that the intersection of every set of R or more of the $\mathbf{A}$-modified transforms is trivial. □

Explicit examples of the generalized stacking construction are easily found. For example, the companion matrix construction of [3] [5] produces an $M \times M$ maximal full-rank binary code using transformations that correspond to multiplication by linearly independent field elements in a Galois field. Our objective is to generalize that construction to provide a family of maximal rank-d binary codes. We first need the following lemma.

Lemma 1. *Let* $\beta_1, \beta_2, \ldots, \beta_T$ *be a basis for* $GF(2^T)$, *and let* σ *denote the Frobenius automorphism of* $GF(2^T)$. *Then the matrix*

$$\Lambda = \begin{bmatrix} \beta_1 & \sigma(\beta_1) & \sigma^2(\beta_1) & \cdots & \sigma^{T-1}(\beta_1) \\ \beta_2 & \sigma(\beta_2) & \sigma^2(\beta_2) & \cdots & \sigma^{T-1}(\beta_2) \\ \vdots & \vdots & \vdots & \ddots & \vdots \\ \beta_T & \sigma(\beta_T) & \sigma^2(\beta_T) & \cdots & \sigma^{T-1}(\beta_T) \end{bmatrix}$$

is nonsingular over $GF(2^T)$ *and has determinant 1. Furthermore, every* $\ell \times \ell$ *contiguous submatrix of* Λ *is nonsingular over* $GF(2^T)$.

Proof. The first statement is Lemma 18 of Chapter 4 in [1]. Consider the $\ell \times \ell$ submatrix

$$\Lambda_{m,n} = \begin{bmatrix} y_1 & \sigma(y_1) & \cdots & \sigma^{\ell-1}(y_1) \\ y_2 & \sigma(y_2) & \cdots & \sigma^{\ell-1}(y_2) \\ \vdots & \vdots & \ddots & \vdots \\ y_\ell & \sigma(y_\ell) & \cdots & \sigma^{\ell-1}(y_\ell) \end{bmatrix},$$

where $y_i = \sigma^n(\beta_{m+i-1})$ for $i = 1, 2, \ldots, \ell$. Suppose that $\bar{a} = (a_1, a_2, \ldots, a_\ell) \in (GF(2^T))^\ell$ satisfies $\Lambda_{m,n}\bar{a}^{\mathrm{T}} = \bar{0}$. Then $y_1, y_2, \ldots, y_\ell$ are roots of the polynomial

$$f(x) = a_1 x + a_2 x^2 + \cdots + a_\ell x^{2^{\ell-1}}.$$

We note that the sum of any two roots of $f(x)$ is also a root. Since $y_1, y_2, \ldots, y_\ell$ are linearly independent over $\mathbb{F}$, every binary linear combination of them produces a distinct root of $f(x)$. Therefore, $f(x)$ is a polynomial of degree at most $2^{\ell-1}$ with at least 2^ℓ distinct roots in $GF(2^T)$. We conclude that $f(x)$ is the zero polynomial, $\bar{a}$ is the zero vector, and $\Lambda_{m,n}$ is nonsingular as claimed. □

For the generalized companion matrix construction, we need the following notation. Let $\mathcal{B} = \{\beta_1, \beta_2, \cdots, \beta_T\}$ be a basis for the finite field $GF(2^T)$ over $\mathbb{F}$. For any $\gamma \in GF(2^T)$, let $\bar{\gamma}$ denote its vector of coordinates with respect to the basis $\mathcal{B}$. Furthermore, let $T_\gamma : GF(2^T) \longrightarrow GF(2^T)$ denote the mapping $x \mapsto x\gamma$, and let $\mathbf{M}_\gamma$ denote the $T \times T$ matrix representation of T_γ having the property that $\bar{x}\mathbf{M}_\gamma = \bar{y}$ iff $T_\gamma(x) = y$.

Theorem 2 (Companion Matrix Construction). *Let M, R, and T be positive integers with $R \leq M \leq T$, and let $d = M - R + 1$. Choose $\beta_1, \beta_2, \ldots, \beta_M$ to be linearly independent elements in $GF(2^T)$. Let S denote the code given by the generalized stacking construction applied to the $RT \times T$ binary matrices $\mathbf{G}_1, \mathbf{G}_2, \ldots, \mathbf{G}_M$, where*

$$\mathbf{G}_i = \begin{bmatrix} \mathbf{M}_{\beta_i} \\ \mathbf{M}_{\sigma(\beta_i)} \\ \vdots \\ \mathbf{M}_{\sigma^{R-1}(\beta_i)} \end{bmatrix}.$$

Then S is an $M \times T$ maximal rank-d binary code.

Proof. Suppose that

$$\mathbf{c} = \begin{bmatrix} \bar{x}\mathbf{G}_1 \\ \bar{x}\mathbf{G}_2 \\ \vdots \\ \bar{x}\mathbf{G}_M \end{bmatrix}$$

is a code word in S having rank less than d. Then there is a nonsingular binary matrix $\mathbf{A}$ for which $\mathbf{A}\mathbf{c}$ has R rows that are identically zero. Without loss of generality, we may assume that the first R rows are zero—that is,

$$\bar{x}\mathbf{G}_1^{\mathbf{A}} = \bar{x}\mathbf{G}_2^{\mathbf{A}} = \cdots = \bar{x}\mathbf{G}_R^{\mathbf{A}} = 0, \tag{1}$$

where

$$\mathbf{G}_i^{\mathbf{A}} = \sum_{j=1}^{M} a_{i,j}\mathbf{G}_j$$

and $a_{i,j}$ is the (i,j)-th element of $\mathbf{A}$. Letting $\bar{x} = \begin{pmatrix} \bar{x}_1 & \bar{x}_2 & \cdots & \bar{x}_R \end{pmatrix}$, where the $\bar{x}_i \in \mathbb{F}^T$, we have that (1) is equivalent to

$$\bar{x}_1\mathbf{M}_{y_1} + \bar{x}_2\mathbf{M}_{\sigma(y_1)} + \cdots + \bar{x}_R\mathbf{M}_{\sigma^{R-1}(y_1)} = 0$$
$$\bar{x}_1\mathbf{M}_{y_2} + \bar{x}_2\mathbf{M}_{\sigma(y_2)} + \cdots + \bar{x}_R\mathbf{M}_{\sigma^{R-1}(y_2)} = 0$$

$$\vdots$$

$$\bar{x}_1\mathbf{M}_{y_R} + \bar{x}_2\mathbf{M}_{\sigma(y_R)} + \cdots + \bar{x}_R\mathbf{M}_{\sigma^{R-1}(y_R)} = 0,$$

where $y_i = \sum_{j=1}^{M} a_{i,j}\beta_j \in GF(2^T)$. Identifying T-tuple $\bar{x}_i$ as the coordinates of $x_i \in GF(2^T)$, we therefore have $\begin{pmatrix} x_1 & x_2 & \cdots & x_R \end{pmatrix}$ as a solution to the following matrix equation over $GF(2^T)$:

$$\begin{bmatrix} y_1 & \sigma(y_1) & \cdots & \sigma^{R-1}(y_1) \\ y_2 & \sigma(y_2) & \cdots & \sigma^{R-1}(y_2) \\ \vdots & \vdots & \ddots & \vdots \\ y_R & \sigma(y_R) & \cdots & \sigma^{R-1}(y_R) \end{bmatrix} \begin{bmatrix} x_1 \\ x_2 \\ \vdots \\ x_R \end{bmatrix} = \bar{0}.$$

By Lemma 1, this matrix is non-singular, implying $x_1 = x_2 = \cdots = x_R = 0$. Hence, $\mathbf{c} = 0$.   □

Remark 1. The choice of a polynomial basis, $\beta_i = \alpha^{i-1}$ where α is primitive in $GF(2^T)$, gives the Gabidulin maximal rank-d binary codes discussed in Lu and Kumar [6]. The Lu-Kumar representation is different, however, and the connection with the stacking construction is not obvious in their approach. Bases other than polynomial bases (e.g. normal bases) may also be used, so the new construction of Theorem 2 is more general than the Gabidulin construction.

Remark 2. In [3], full-rank binary convolutional codes are derived as examples of the Hammons-El Gamal stacking construction. In [6], Lu and Kumar generalize the construction to produce maximal rank-d binary convolutional codes with generator matrix

$$\mathbf{G} = \begin{bmatrix} g_{0,0}(D) & g_{1,0}(D) & \cdots & g_{M-1,0}(D) \\ g_{0,1}(D) & g_{1,1}(D) & \cdots & g_{M-1,1}(D) \\ \vdots & \vdots & \ddots & \vdots \\ g_{0,R-1}(D) & g_{1,R-1}(D) & \cdots & g_{M-1,R-1}(D) \end{bmatrix},$$

where $g_{m,r} = (D^m)^{2^r} \pmod{f(D)}$ and $f(D)$ is a primitive polynomial of degree $T \geq M$. Identifying D with the primitive element α in $GF(2^T)$, we see that the columns of the matrix $\mathbf{G}$ correspond precisely to the transformations used in the general companion matrix construction for the special case of $\beta_i = \alpha^{i-1}$. Hence, the general form of the companion matrix construction in Theorem 2 leads to the following class of maximal rank-d convolutional codes.

Corollary 1. *Let M, R, and T be positive integers with $R \leq M \leq T$, and let $d = M - R + 1$. Let α be primitive in $GF(2^T)$ with minimal polynomial $f(D)$. Choose $\beta_1, \beta_2, \ldots, \beta_M$ to be linearly independent elements in $GF(2^T)$ over $\mathbb{F}$, with binary expansions*

$$\beta_i = \sum_{j=0}^{T-1} b_{i,j}\alpha^j.$$

Let S denote the convolutional code with generator matrix

$$\mathbf{G} = \begin{bmatrix} g_{0,0}(D) & g_{1,0}(D) & \cdots & g_{M-1,0}(D) \\ g_{0,1}(D) & g_{1,1}(D) & \cdots & g_{M-1,1}(D) \\ \vdots & \vdots & \ddots & \vdots \\ g_{0,R-1}(D) & g_{1,R-1}(D) & \cdots & g_{M-1,R-1}(D) \end{bmatrix},$$

where $g_{m,r} = (g_m(D))^{2^r} \pmod{f(D)}$ and $g_m(D) = b_{m,0} + b_{m,1}D + \cdots + b_{m,T-1}D^{T-1}$. Then S is an $M \times T$ maximal rank-d binary code.

Remark 3. The corollary is a restatement of Theorem 22 in [6].

4   Diversity-Preserving Multilevel Constructions

The generalized binary rank criterion and general stacking construction lead to natural multi-level space-time code designs for traditional constellations, including the following signaling alphabets:

- 2^K-PAM, consisting of the points

$$s = \sum_{k=0}^{K-1} 2^k (-1)^{a_k},$$

  for $\bar{a} = (a_0, a_1, \ldots, a_{K-1}) \in \mathbb{F}^K$.
- 4^K-QAM, consisting of the points

$$s = (1 + i) \sum_{k=0}^{K-1} 2^k i^{a_k + 2b_k},$$

  for $\bar{a} = (a_0, a_1, \ldots, a_{K-1})$ and $\bar{b} = (b_0, b_1, \ldots, b_{K-1})$ in $\mathbb{F}^K$.
- 2^K-PSK, consisting of the points

$$s = \theta^a,$$

  where $a \in \mathbb{Z}_{2^K} = \{0, 1, 2, \ldots, 2^K - 1\}$ and θ is a complex, primitive 2^K-th root of unity.
- 2^{K+1}-AM-PSK, consisting of the points

$$s = r^a \theta^b,$$

  where $a \in \mathbb{F}$, $b \in \mathbb{Z}_{2^K} = \{0, 1, 2, \ldots, 2^K - 1\}$, θ is a complex, primitive 2^K-th root of unity, and $r > 1$.

The set of $M \times T$ matrices over an alphabet Ω will be denoted by $\Omega^{M \times T}$. When $A = [a_{i,j}]$ is a matrix with entries in $\mathbb{Z}_{2^K}$, we write θ^A for the matrix whose (i, j)-th entry is $\theta^{a_{i,j}}$. For matrices A and B, the matrix $A \odot B$ is their Hermitian (*i.e.*, componentwise) product.

4.1   Unified AM-PSK Construction

Theorem 3. *Let $\mathcal{A}_1, \mathcal{A}_2, \ldots, \mathcal{A}_L$ be $M \times T$ binary codes with $M \leq T$ of ranks $d_1, d_2, \ldots, d_L$ respectively. Let K and U be positive integers, and let*

$$\{ \mathcal{C}_{u,k} \; : \; 0 \leq u < U, \; 0 \leq k < K \}$$

be a collection of $M \times T$ binary codes of ranks $d_{u,k}$ respectively. From these, form the following set of 2^K-ary codes:

$$\mathcal{C}_u = \left\{ C_u = \sum_{k=0}^{K-1} 2^k C_{u,k} \; : \; C_{u,k} \in \mathcal{C}_{u,k} \right\} \quad (0 \leq u < U).$$

Let κ be a non-zero complex number, $\eta \in 2\mathbb{Z}[\theta]$, and θ be a complex primitive 2^K-th root of unity. Choose $\nu \in \mathbb{Z}[\theta]$ such that $\nu \equiv 0 \ (mod \ 1 - \theta)$, $\eta^{U-1} \mid \nu$ in $\mathbb{Z}[\theta]$, and $\nu / \eta^{U-1} \equiv 0 \ (mod \ 1 - \theta)$. Set $r_i = 2\nu^i + 1$ for $i = 1, 2, \ldots, L$.
  Then the modulated space-time code defined by

$$\mathcal{S} = \left\{ S = \left(\bigodot_{i=1}^{L} r_i^{A_i} \right) \odot \kappa \sum_{u=0}^{U-1} \eta^u \theta^{C_u} \ : \ A_i \in \mathcal{A}_i \text{ and } C_u \in \mathcal{C}_u \right\}$$

achieves transmit diversity at least d, where

$$d = \min\{d_i, d_{u,k} : 1 \le i \le L, 0 \le u < U, 0 \le k < K\}.$$

Proof. One must show that, whenever S and S' are distinct code word matrices in $\mathcal{S}$, the difference $\Delta S = S - S'$ is of rank at least d over $\mathbb{C}$.
  Consider the partial products of S defined by

$$\sigma_0 = \kappa \sum_{u=0}^{U-1} \eta^u \theta^{C_u},$$

$$\sigma_\ell = \left(\bigodot_{i=1}^{\ell} r_i^{A_i} \right) \odot \sigma_0, \qquad (1 \le \ell \le L).$$

Then

$$\sigma_\ell = \sigma_{\ell-1} + (r_\ell - 1) A_\ell \odot \sigma_{\ell-1}$$
$$= \sigma_0 + (r_1 - 1) A_1 \odot \sigma_0 + (r_2 - 1) A_2 \odot \sigma_1 + \cdots + (r_\ell - 1) A_\ell \odot \sigma_{\ell-1}.$$

Similarly, $\sigma'_0, \sigma'_1, \ldots, \sigma'_L$ denote the partial products of S'.
  We now have

$$\Delta S = (\sigma_0 - \sigma'_0) + 2\nu D, \tag{2}$$

where

$$D = \sum_{i=1}^{L} \nu^{i-1} \left(A_i \odot \sigma_{i-1} - A'_i \odot \sigma'_{i-1} \right). \tag{3}$$

To show that ΔS is of rank at least d over $\mathbb{C}$, there are two cases to consider.

  Case 1. $\sigma_0 = \sigma'_0$
  Let ℓ be the smallest index i for which $A_i \ne A'_i$. Then $\sigma_i = \sigma'_i$ for $i < \ell$. From (2) and (3), we have

$$\frac{\Delta S}{2\nu^\ell} = \sum_{i=0}^{L-\ell} \nu^i \left(A_{\ell+i} \odot \sigma_{\ell+i-1} - A'_{\ell+i} \odot \sigma'_{\ell+i-1} \right).$$

Since all but the first term are multiples of ν, we see

$$\frac{\Delta S}{2\nu^\ell} \equiv A_\ell \oplus A'_\ell \qquad (mod \ 1 - \theta).$$

By the binary rank criterion and choice of code $\mathcal{A}_\ell$, ΔS is of rank at least $d_\ell \ge d$ over $\mathbb{C}$.

Case 2. $\sigma_0 \neq \sigma_0'$

Let (u^*, k^*) denote the lexicographically first index pair (u, k) for which $C_{u,k} \neq C_{u,k}'$. Then

$$\Delta S = \kappa \eta^{u^*} \theta P \odot \left(\theta^{2^{k^*}} Q - \theta^{2^{k^*}} Q' \right) + 2\nu D + \kappa \eta^{u^*+1} E \,,$$

where

$$E = \sum_{i=0}^{U-u^*-2} \eta^i \left(\theta C_{u^*+i+1} - \theta C_{u^*+i+1}' \right),$$

$$P = \sum_{i=0}^{k^*-1} 2^i C_{u^*,i}, \quad Q = \sum_{i=0}^{K-k^*-1} 2^i C_{u^*,k^*+i}, \quad \text{and} \quad Q' = \sum_{i=0}^{K-k^*-1} 2^i C_{u^*,k^*+i}' \,.$$

Then

$$\frac{\Delta S}{\kappa \eta^{u^*}(1 - \theta^{2^{k^*}})} = \theta P \odot \left(\frac{\theta^{2^{k^*}} Q - \theta^{2^{k^*}} Q'}{1 - \theta^{2^{k^*}}} \right) + \left(\frac{2}{1 - \theta^{2^{k^*}}} \right) \left(\frac{\nu}{\eta^{u^*}} \right) D$$

$$+ \left(\frac{\eta}{1 - \theta^{2^{k^*}}} \right) E \,.$$

The terms in parentheses on the right hand side are either scalars in $\mathbb{Z}[\theta]$ or matrices with entries in $\mathbb{Z}[\theta]$. The two rightmost summands are congruent to 0 (mod $1 - \theta$). It is straightforward to show that $(1 - \theta^{2^\ell}) \mid (\theta^{2^\ell m} - \theta^{2^\ell n})$ in $\mathbb{Z}[\theta]$ and that

$$\frac{\theta^{2^\ell m} - \theta^{2^\ell n}}{1 - \theta^{2^\ell}} \equiv \bar{m} \oplus \bar{n} \pmod{1 - \theta},$$

where $\bar{m}$ and $\bar{n}$ denote the modulo 2 projections of m and n, respectively. Hence,

$$\frac{\Delta S}{\kappa \eta^{u^*}(1 - \theta^{2^{k^*}})} \equiv C_{u^*,k^*} \oplus C_{u^*,k^*}' \pmod{1 - \theta}.$$

By the binary rank criterion, ΔS is of rank at least $d_{u^*,k^*} \geq d$ over $\mathbb{C}$, which completes the proof. $\qquad \square$

Remark 4. The radii r_i may be specified more generally [7] [9] as follows. First choose non-zero $\eta \in 2\mathbb{Z}[\theta]$. Then choose $\epsilon_1, \epsilon_2, \ldots, \epsilon_L \in \mathbb{Z}[\theta]$ satisfying both

$$\epsilon_i \equiv 0 \pmod{1 - \theta} \text{ for all } i = 1, 2, \ldots, L$$

and

$$\eta^{U-1} \mid \epsilon_1 \text{ in } \mathbb{Z}[\theta] \text{ with } \epsilon_1/\eta^{U-1} \equiv 0 \pmod{1 - \theta}.$$

Finally, define $\nu_i = \prod_{\ell=1}^{i} \epsilon_\ell$, and set $r_i = 2\nu_i + 1$ for $i = 1, 2, \ldots, L$. The proof carries through as before.

Remark 5. When $L = 0$, by proper choice of parameters, one produces space-time codes that achieve the rate-diversity tradeoff for 2^m-PAM, 4^m-QAM, and 2^m-PSK constellations. For PAM, one chooses $\eta = 2$, $K = 1$, $U = m$, $\kappa = 1$, $\theta = -1$; for QAM, one chooses $\eta = 2$, $K = 2$, $U = m$, $\kappa = 1 + i$, $\theta = i$; and for PSK, one chooses $\eta = 2$, $K = m$, $U = 1$, $\kappa = 1$, $\theta = e^{2\pi i / 2^m}$. When all of the constituent codes are maximal rank-d binary codes, the resulting construction is the Lu-Kumar unified construction [6]. When $L > 0$ and the PSK parameters are selected, one gets space-time codes for AM-PSK modulations consisting of multiple rings of PSK. These are discussed in more detail in [7] [8] [9] [10].

4.2   Special a Constructions

In [7] [9], the author introduced the so-called "Special A" construction for rate-diversity optimal space-time codes in which one or more of the binary matrices A_i in the Unified AM-PSK Construction are derived as functions of the nonbinary matrices C_u. The canonical examples of which are the binary-component projection mappings. (This construction was further explored by Lu [11].) Generalizing Theorem 3 in the same way, we have the following construction.

Theorem 4. *Let $\mathcal{A}_1, \mathcal{A}_2, \ldots, \mathcal{A}_L$ be $M \times T$ binary codes of rank $d_1, d_2, \ldots, d_L$, respectively, with $M \leq T$. Let K and U be positive integers, and let*

$$\{\, \mathcal{C}_{u,k} \; : \; 0 \leq u < U, \; 0 \leq k < K \,\}$$

be a collection of $M \times T$ binary codes of rank $d_{u,k}$, respectively. From these, form the following set of 2^K-ary codes:

$$\mathcal{C}_u = \left\{\, C_u = \sum_{k=0}^{K-1} 2^k \, C_{u,k} \; : \; C_{u,k} \in \mathcal{C}_{u,k} \right\} \quad (0 \leq u < U).$$

Let κ be a non-zero complex number, and θ be a complex primitive 2^K-th root of unity. Choose non-zero $\eta \in 2\mathbb{Z}[\theta]$ and $\epsilon_1, \epsilon_2, \ldots, \epsilon_{L+\ell} \in \mathbb{Z}[\theta]$ such that $\epsilon_i \equiv 0$ (mod $1 - \theta$) for all $i = 1, 2, \ldots, L + \ell$. Furthermore, we require that $\eta^{U-1} \mid \epsilon_1$ in $\mathbb{Z}[\theta]$ and $\epsilon_1 / \eta^{U-1} \equiv 0$ (mod $1 - \theta$). Set $\nu_i = \prod_{k=1}^{i} \epsilon_k$ and $r_i = 2\nu_i + 1$ for $i = 1, 2, \ldots, L + \ell$.

For $i = 1, 2, \ldots, \ell$, let $\Psi_i : \mathcal{C}_0 \times \mathcal{C}_1 \times \cdots \times \mathcal{C}_{U-1} \to \mathcal{B}_i$ be functions that map U-tuples of non-binary code word matrices to binary $M \times T$ matrices, the ranges $\mathcal{B}_i$ being arbitrary.

Then the modulated space-time code defined by

$$\mathcal{S} = \left\{\; S = \left(\bigodot_{i=1}^{L} r_i^{A_i} \right) \odot \left(\bigodot_{i=1}^{L} r_{L+i}^{B_i} \right) \odot \kappa \sum_{u=0}^{U-1} \eta^u \theta^{C_u} \right.$$

$$: \; A_i \in \mathcal{A}_i, \; B_i = \Psi_i(C_0, C_1, \ldots, C_{U-1}) \in \mathcal{B}_i, \text{ and } C_u \in \mathcal{C}_u\}$$

achieves transmit diversity at least d, where

$$d = \min\{d_i, d_{u,k} : 1 \leq i \leq L, 0 \leq u < U, 0 \leq k < K\}.$$

4.3   Linear Transformations

Let $\tau : \mathbb{F}^\nu \to \mathbb{F}^\nu$ be a linear transformation. For $\nu = KU$, we define the induced vector component mappings via the action

$$\tau : \bar{x} \in \mathbb{F}^{KU} \mapsto (\tau_0(\bar{x}), \tau_1(\bar{x}), \ldots, \tau_{U-1}(\bar{x})) \in \mathbb{F}^K \times \cdots \times \mathbb{F}^K.$$

In this notation, the scalar components of τ are then indexed by the pairs (u, k) so that

$$\tau_u(\bar{x}) = (\tau_{u,0}(\bar{x}), \tau_{u,1}(\bar{x}), \ldots, \tau_{u,K-1}(\bar{x})).$$

We may extend this map to a linear transformation $\tau : (\mathbb{F}^{M \times T})^\nu \to (\mathbb{F}^{M \times T})^\nu$, acting on ν-tuples of $M \times T$ binary matrices by applying it componentwise. Let $\bar{A} = (A_0, A_1, \ldots, A_{\nu-1}) \in (\mathbb{F}^{M \times T})^\nu$, where A_i is an $M \times T$ matrix whose (m,t)-th entry is $a_{m,t}^{(i)}$. Then $\tau(\bar{A}) = (\tau_0(\bar{A}), \tau_1(\bar{A}), \ldots, \tau_{\nu-1}(\bar{A}))$, where $\tau_i(\bar{A})$ is the matrix whose (m,t)-th entry is the i-th component of $\tau(a_{m,t}^{(0)}, a_{m,t}^{(1)}, \ldots, a_{m,t}^{(\nu-1)})$.

Theorem 5 (Hammons [10]). *Let $\mathcal{A}$ be an $M \times T$ rank-d binary code. Let K and U be positive integers. Let κ be a non-zero complex number, θ be a complex primitive 2^K-th root of unity, and η be a non-zero element of $2\mathbb{Z}[\theta]$. Let $\tau : \mathbb{F}^{KU} \to \mathbb{F}^{KU}$ be a nonsingular linear transformation.*

Then the modulated space-time code defined by

$$\mathcal{S}^\tau = \left\{ S = \kappa \sum_{u=0}^{U-1} \eta^u \theta(\tau_u(\tilde{A})) : \ \tilde{A} = (\tilde{A}_0, \tilde{A}_1, \ldots, \tilde{A}_{U-1}) \in \mathcal{A}^{KU} \right\}$$

achieves transmit diversity d.

Proof. Let $S = \kappa \sum_{u=0}^{U-1} \eta^u \theta(\tau_u(\tilde{A}))$ and $S' = \kappa \sum_{u=0}^{U-1} \eta^u \theta(\tau_u(\tilde{A}'))$ be distinct code words in $\mathcal{S}$. Let (u^*, k^*) denote the lexicographically first index pair (u, k) for which $\tau_{u,k}(\tilde{A}) \neq \tau_{u,k}(\tilde{A}')$. Then

$$\Delta S = S - S' = \kappa \eta^{u^*} \theta^P \odot \left(\theta^{2^{k^*}} Q - \theta^{2^{k^*}} Q' \right) + \kappa \eta^{u^*+1} E,$$

where

$$P = \sum_{i=0}^{k^*-1} 2^i \tau_{u^*,i}(\tilde{A}), \quad Q = \sum_{i=0}^{K-k^*-1} 2^i \tau_{u^*,k^*+i}(\tilde{A}), \quad Q' = \sum_{i=0}^{K-k^*-1} 2^i \tau_{u^*,k^*+i}(\tilde{A}'),$$

$$\text{and} \quad E = \sum_{i=0}^{U-u^*-2} \eta^i \left[\theta(\tau_{u^*+i+1}(\tilde{A})) - \theta(\tau_{u^*+i+1}(\tilde{A}')) \right].$$

Then

$$\frac{\Delta S}{\kappa \eta^{u^*}(1 - \theta^{2^{k^*}})} = \theta^P \odot \left(\frac{\theta^{2^{k^*}} Q - \theta^{2^{k^*}} Q'}{1 - \theta^{2^{k^*}}} \right) + \left(\frac{\eta}{1 - \theta^{2^{k^*}}} \right) E.$$

The terms in parentheses on the right hand side are either scalars in $\mathbb{Z}[\theta]$ or matrices with entries in $\mathbb{Z}[\theta]$. The rightmost summand is congruent to 0 (mod $1 - \theta$). Hence,

$$\frac{\Delta S}{\kappa \eta^{u^*}(1 - \theta^{2^{k^*}})} \equiv \tau_{u^*,k^*}(\tilde{A}) \oplus \tau_{u^*,k^*}(\tilde{A}') \quad (\text{mod } 1 - \theta).$$

Both $\tau_{u^*,k^*}(\tilde{A})$ and $\tau_{u^*,k^*}(\tilde{A}')$ are code words in $\mathcal{A}$; hence, their sum has rank at least d over $\mathbb{F}$. By the binary rank criterion, ΔS is of rank at least d over $\mathbb{C}$, which completes the proof. $\qquad \square$

4.4   Non-binary Extensions

The constructions all generalize in a straightforward manner to the case of p^K-PAM, p^K-PSK, p^{2K}-QAM, and related constellations, when $p \geq 2$ is prime.

Theorem 6. *Let $p \geq 2$ be prime. Let $\mathcal{A}$ be an $M \times T$ rank-d code over the alphabet $\mathbb{F}_p$. Let K and U be positive integers. Let κ be a non-zero complex number, θ be a complex primitive p^K-th root of unity, and η be a non-zero element of $p\mathbb{Z}[\theta]$. Let $\sigma : \mathbb{F}_p^{KU} \to \mathbb{F}_p^{KU}$ be a nonsingular linear transformation.*
   Then the modulated space-time code defined by

$$\mathcal{S}^\sigma = \left\{ S = \kappa \sum_{u=0}^{U-1} \eta^u \theta^{\sum_{k=0}^{K-1} p^k \sigma_{u,k}(\tilde{A})} : \ \tilde{A} = (\bar{A}_0, \bar{A}_1, \ldots, \bar{A}_{U-1}) \in \mathcal{A}^{KU} \right\}$$

achieves transmit diversity d.

Proof. If θ is a complex primitive p^ν-th root of unity, then $\mathbb{Z}[\theta]/(1 - \theta) \cong \mathbb{F}_p$; and, for $0 \leq \ell < \nu$,

$$\frac{\theta^{p^\ell m} - \theta^{p^\ell n}}{1 - \theta^{p^\ell}} \equiv \bar{m} \oplus_p \bar{n} \quad (\text{mod } 1 - \theta),$$

where $\bar{m}$ and $\bar{n}$ denote the modulo p projections of m and n, respectively, and $\oplus_p$ denotes modulo p addition. (See [6] for details.) Thus, the underlying algebra and technical details of the proof are the same as in the $p = 2$ case. $\qquad \square$

4.5   Rate-Diversity Optimal Space-Time Codes

The multilevel space-time code constructions presented in sections IV.A through IV.C are diversity preserving. In particular, per Theorems 3 through 6, when the constituent binary codes are rank-d codes, so are the multilevel codes. When the constituent binary codes are maximal rank-d codes, it turns out that the multilevel codes in fact achieve the rate-diversity trade-off. See [9] and [10] for further details.

5   Applications to Cooperative Diversity

We consider the cooperative communications model in which a source node communicates to a destination node in a two-step process. In the first step, the source broadcasts its message conventionally to both the destination node and any potential relay nodes. In the second step, the relay nodes use a decode-and-forward strategy and retransmit the message simultaneously to the destination node. Since the retransmissions are intended to overlap in both time and frequency, the relay nodes cooperatively implement a space-time channel code.

Unlike conventional space-time coding, however, not all of the relays may successfully decode the original transmission, so the number of transmitters is a random variable. Fortunately, this does not substantially affect the design of the cooperative space-time code [12]. In addition, the relays can be dispersed geographically; so propagation delays and timing uncertainties may result in the destination node receiving certain of the transmissions later than others. If the relative timing delays exceed a symbol interval, the performance of the space-time code can be adversely impacted. In particular, the space-time code may no longer achieve full spatial diversity [12].

Following the Li-Xia approach [12] to the design of cooperative space-time codes, we assume that each relay terminal knows an upper bound on the worst-case differential delay and transmits fill symbols corresponding to binary 0's at the beginning and end of each code word to cover the potential mismatch. The problem then becomes to design the space-time code so that full diversity is achieved irrespective of the transmission delays—that is, the code word matrices must be of full rank even when the rows of the matrix are slipped out of alignment by arbitrary amounts (up to the maximum specified delay).

These ideas are illustrated by the following examples.

Example 1. Consider the full-rank companion matrix construction using the standard basis $\mathcal{B} = \{1, \alpha, \alpha^2\}$ for $GF(8)$, where α is primitive and $\alpha^3 + \alpha + 1 = 0$. The corresponding maximal rank-3 code $\mathcal{S}$ consists of the binary matrices:

$$\mathbf{c} = \begin{bmatrix} a & b & c \\ c & a+c & b \\ b & b+c & a+c \end{bmatrix}, \quad \forall a, b, c \in \mathbb{F}.$$

One notes that det $\mathbf{c} = a + b + c + ab + ac + bc + abc$, which is equal to zero if and only if $a = b = c = 0$, confirming that $\mathcal{S}$ achieves full rank (3-level spatial diversity) in accordance with the stacking construction.

Suppose that, in the cooperative scenario, the first transmission (first row) is delayed by one symbol compared to the second and third transmissions. In this case, we say that the relative delay profile for these transmissions is $\Delta = (1, 0, 0)$. Then the binary code word matrices effectively become

$$\mathbf{c}^\Delta = \begin{bmatrix} 0 & a & b & c \\ c & a+c & b & 0 \\ b & b+c & a+c & 0 \end{bmatrix}.$$

We denote the set of all 3×4 matrices derived from $\mathcal{S}$ in accordance with delay profile Δ as the code $\mathcal{S}^\Delta$.

For $a = 1$ and $b = c = 0$, we have

$$\mathbf{c}^\Delta = \begin{bmatrix} 0 & 1 & 0 & 0 \\ 0 & 1 & 0 & 0 \\ 0 & 0 & 1 & 0 \end{bmatrix} \in \mathcal{S}^\Delta,$$

which has rank 2 instead of 3. Thus, the code $\mathcal{S}^\Delta$ is not delay tolerant.

Example 2. Consider the binary code $\mathcal{S}$ produced by the stacking construction using the transformations described by the matrices

$$\mathbf{M}_1 = \begin{bmatrix} 1 & 0 & 0 & 0 \\ 0 & 1 & 0 & 0 \\ 0 & 0 & 1 & 0 \\ 0 & 0 & 0 & 1 \end{bmatrix} \quad \text{and} \quad \mathbf{M}_2 = \begin{bmatrix} 0 & 1 & 1 & 0 \\ 0 & 0 & 1 & 1 \\ 1 & 1 & 0 & 1 \\ 1 & 0 & 1 & 0 \end{bmatrix},$$

It is easy to see that $\mathcal{S}$ consists of the 16 code word matrices of the form

$$\mathbf{c} = \begin{bmatrix} a & b & c & d \\ c+d & a+c & a+b+d & b+c \end{bmatrix},$$

where $a, b, c, d \in \mathbb{F}$. One first notes that neither of the rows of $\mathbf{c}$ is identically zero unless $a = b = c = d = 0$. Thus, to show that any delayed variant of $\mathbf{c}$ has full rank, it suffices to check that the sum of its two rows is also nonzero whenever one of $a, b, c,$ or d is nonzero. A relative delay of two symbols results in the code word matrices of the form

$$\begin{bmatrix} 0 & 0 & a & b & c & d \\ c+d & a+c & a+b+d & b+c & 0 & 0 \end{bmatrix}$$

or

$$\begin{bmatrix} a & b & c & d & 0 & 0 \\ 0 & 0 & c+d & a+c & a+b+d & b+c \end{bmatrix}.$$

It is easy to check that, in either case, they all have rank 2 unless $a = b = c = d = 0$. By enumerating all the delayed variants of $\mathbf{c}$, one discovers that, for all delay profiles Δ, the space-time code $\mathcal{S}^\Delta$ achieves full spatial diversity.

Li and Xia [12] have recently shown that certain of the full-rank binary trellis codes derived from the stacking construction are delay tolerant—in the sense that full diversity is preserved despite random delays among the various transmissions—and thus suitable for cooperative diversity schemes. They provide necessary (but not sufficient) conditions for the trellis codes to be delay tolerant. They have also shown that, when these codes are used in the multilevel Lu-Kumar construction for PSK and QAM modulation (special cases of the Unified AM-PSK Construction of Theorem 3), the resulting space-time codes also achieve full spatial diversity in quasi-synchronous cooperative operations.

In sections IV.A through IV.C, we have presented several generalizations of the Lu-Kumar multilevel construction. We may summarize the results of Theorems 3 through 6 as follows: If $\mathcal{F}$ is a family of $M \times T$ binary codes achieving at least d-level spatial diversity, then the corresponding multilevel code $\mathcal{S}_{\mathcal{F}}$ built from $\mathcal{F}$ in accordance with the theorem also achieves spatial diversity at least d.

This implies that the constructions preserve preserve delay tolerance—that is, the delay tolerance of the multilevel code is at least as great as that of the constituent binary codes. Thus, we have the following general result.

Theorem 7 (Preservation of Delay Tolerance). *The multilevel Unified AM-PSK Construction and its Special $\mathcal{A}$ and Linear Transformation Variations are suitable for asynchronous cooperative diversity schemes if the binary constituent codes used in these constructions are so suitable. Specifically, if the binary constituent codes provide d-level spatial diversity under BPSK modulation for a given delay profile, then the multilevel AM-PSK space-time code also achieves d-level spatial diversity for the same delay profile.*

Proof. Consider a delay profile $\Delta = (\delta_1, \delta_2, \ldots, \delta_M)$ in which δ_i denotes the relative delay of the signal received from the i-th transmit antenna. Let $\delta_{\min} (= 0)$ and $\delta_{\max}$ respectively denote the minimum and maximum of the relative delays. From the receiver's perspective, the space-time code being used is effectively $\mathcal{S}_{\mathcal{F}}^{\Delta}$, the set of matrices of dimension $M \times (T + \delta_{\max})$ produced by transforming the $M \times T$ code word matrices of $\mathcal{S}_{\mathcal{F}}$ in accordance with the delay profile Δ as indicated in Examples 1 and 2. Let $\mathcal{F}^{\Delta}$ denote the family of binary codes similarly transformed in accordance with Δ.

By Theorems 3 through 6, if the binary codes of $\mathcal{F}^{\Delta}$ achieve spatial diversity at least d, so does the multilevel space-time code $\mathcal{S}_{\mathcal{F}^{\Delta}}$. But $\mathcal{S}_{\mathcal{F}}^{\Delta} = \mathcal{S}_{\mathcal{F}^{\Delta}}$, so the result is proven. □

Remark 6. The binary code of Example 1 is not delay tolerant, whereas the binary code of Example 2 is fully delay tolerant. By Theorem 7, the multilevel codes built in accordance with Theorems 3 through 6, using the binary code of Example 2 as the constituent codes, will also be fully delay tolerant.

6   Conclusion

In this paper, we have presented a generalized stacking construction for maximal rank-d binary codes and have developed explicit examples including a generalized companion matrix construction. The generalized companion matrix construction includes as special cases the family of Gabidulin codes described by Lu and Kumar [6]. We have also presented several multilevel space-time code constructions for certain AM-PSK constellations that generalize the Lu-Kumar unified construction [6]. Following the approach by Li and Xia [12], we have shown that, if the binary constituent codes used in these AM-PSK constructions are delay tolerant, so are the multilevel codes. Together these results provide a flexible architecture for space-time coding that is well-suited for quasi-synchronous cooperative communications.

References

1. F.J. MacWilliams and N.J.A. Sloane, *The Theory of Error-Correcting Codes*, North-Holland Mathematical Library, Elsevier Science Publishers, 1977.
2. V. Tarokh, N. Seshadri, and A.R. Calderbank, "Space-Time Codes for High Data Rate Wireless Communication: Performance Criterion and Code Construction," *IEEE Transactions on Information Theory*, vol. 44, March 1998, pp. 744–765.
3. A.R. Hammons Jr. and H. El Gamal, "On the Theory of Space-Time Codes for PSK Modulation," *IEEE Transactions on Information Theory*, vol. 46, no. 2, March 2000, pp. 524–542.
4. Y. Liu, M.P. Fitz, and O.Y. Takeshita, "A Rank Criterion for QAM Space-Time Codes," *IEEE Transactions on Information Theory*, vol. 48, no. 12, December 2002, pp. 3062–3079.
5. H. El Gamal and A. R. Hammons Jr, "On the design and performance of algebraic space-time codes for BPSK and QPSK modulation," *IEEE Transactions on Communications*, June 2002.
6. H.F. Lu and P.V. Kumar, "A Unified Construction of Space-Time Codes with Optimal Rate-Diversity Tradeoff," *IEEE Transactions on Information Theory*, vol. 51, no. 5, May 2005, pp. 1709–1730.
7. A.R. Hammons Jr., "New AM-PSK Space-Time Codes Achieving the Rate-Diversity Tradeoff," in *International Symposium on Information Theory and Its Applications*, Parma, Italy, 2004.
8. A.R. Hammons Jr., "On Space-Time Codes and Constellation Labelling," in *IEEE Information Theory Workshop*, San Antonio, Texas, 2004.
9. A.R. Hammons Jr., "Space-Time Codes that Achieve the Rate-Diversity Tradeoff for Certain AM-PSK Modulations," submitted to *IEEE Transactions on Information Theory*, March 2004.
10. A.R. Hammons Jr., "Space-Time Codes for Linearly Labelled PAM, PSK, QAM, and Related Constellations," submitted to *IEEE Information Theory Transactions*, Aug. 2004.
11. H.-F. Lu, "Generalized Super-Unified Constructions for Space-Time Codes and their Subset-Subcodes that Achieve Rate-Diversity Tradeoff," preprint.
12. Y. Li and X.-G. Xia, "Full Diversity Distributed Space-Time Trellis Codes for Asynchronous Cooperative Communications," preprint.

Geometric Conditions for the Extendability of Ternary Linear Codes

Tatsuya Maruta[*] and Kei Okamoto

Department of Mathematics and Information Sciences,
Osaka Prefecture University
Sakai, Osaka 599-8531, Japan

Abstract. We give the necessary and sufficient conditions for the extendability of ternary linear codes of dimension k, $4 \leq k \leq 6$, with minimum distance $d \equiv 1$ or $2 \pmod 3$ from a geometrical point of view. We also give the necessary and sufficient conditions for the extendability of ternary linear codes with diversity $(\theta_{k-2}, 3^{k-2})$, $(\theta_{k-2}+3^{k-3}, 4 \cdot 3^{k-3})$, $(\theta_{k-2} - 3^{k-3}, 5 \cdot 3^{k-3})$ for $k \geq 6$, where $\theta_j = (3^{j+1} - 1)/2$.

1 Introduction

Let $V(n,q)$ denote the vector space of n-tuples over $\mathrm{GF}(q)$, the field of q elements. A linear code C of length n, dimension k and minimum (Hamming) distance d over $\mathrm{GF}(q)$ is referred to as an $[n, k, d]_q$ code. The *weight* of a vector $\boldsymbol{x} \in V(n,q)$, denoted by $wt(\boldsymbol{x})$, is the number of nonzero coordinate positions in $\boldsymbol{x}$. Let A_i be the number of codewords of C with weight i. We only consider *non-degenerate* codes having no coordinate which is identically zero.

The code obtained by deleting the same coordinate from each codeword of C is called a *punctured code* of C. If there exists an $[n+1, k, d+1]_q$ code C' which gives C as a punctured code, C is called *extendable* (to C') and C' is an *extension* of C. See [1-4,8] for the known results about the extendability of q-ary linear codes.

Let C be an $[n, k, d]_3$ code with $k \geq 3$, $\gcd(3, d) = 1$. The *diversity* (Φ_0, Φ_1) of C is given as the pair of integers:

$$\Phi_0 = \frac{1}{2} \sum_{3|i, i \neq 0} A_i, \quad \Phi_1 = \frac{1}{2} \sum_{i \neq 0, d \pmod 3} A_i,$$

where the notation $x|y$ means that x is a divisor of y. Let $\mathcal{D}_k$ be the set of all possible diversities of C. $\mathcal{D}_k$ has been determined in [5] for $k \leq 6$ and in [6] for $k \geq 7$. For $k \geq 3$, let $\mathcal{D}_k^*$ and $\mathcal{D}_k^+$ be as follows:

$$\mathcal{D}_k^* = \{(\theta_{k-2}, 0), (\theta_{k-3}, 2 \cdot 3^{k-2}), (\theta_{k-2}, 2 \cdot 3^{k-2}), (\theta_{k-2} + 3^{k-2}, 3^{k-2})\},$$
$$\mathcal{D}_k^+ = \mathcal{D}_k \setminus \mathcal{D}_k^*,$$

---

[*] This research has been partially supported by Grant-in-Aid for Scientific Research of Japan Society for the Promotion of Science under Contract Number 17540129.

Ø. Ytrehus (Ed.): WCC 2005, LNCS 3969, pp. 85–99, 2006.
© Springer-Verlag Berlin Heidelberg 2006

where $\theta_j = (3^{j+1} - 1)/2$. It is known that $\mathcal{D}_k^*$ is included in $\mathcal{D}_k$ and that C is extendable if $(\Phi_0, \Phi_1) \in \mathcal{D}_k^*$ ([5]). We define Φ_e as follows:

$$\Phi_e = \frac{1}{2} \sum_{d < i \equiv d (\mathrm{mod}\ 3)} A_i.$$

Since C is extendable when $(\Phi_0, \Phi_1) \in \mathcal{D}_k^*$, it suffices to investigate the extendability of C when $(\Phi_0, \Phi_1) \in \mathcal{D}_k^+$. It is also known that $\mathcal{D}_3^+ = \{(4,3)\}$ and that an $[n, 3, d]_3$ code with diversity $(4,3)$ is extendable if and only if $\Phi_e > 0$ ([5]). So, we consider the following problem:

Problem. Find the necessary and sufficient conditions for the extendability of an $[n, k, d]_3$ code with $\gcd(3, d) = 1$, $k \geq 4$, whose diversity is in $\mathcal{D}_k^+$.

2   Geometric Preliminaries

We denote by $\mathrm{PG}(r, q)$ the projective geometry of dimension r over $\mathrm{GF}(q)$. A *j-flat* is a projective subspace of dimension j in $\mathrm{PG}(r, q)$. 0-flats, 1-flats, 2-flats, 3-flats and $(r - 1)$-flats are called *points, lines, planes, solids* and *hyperplanes* respectively as usual. We denote by $\mathcal{F}_j$ the set of j-flats of $\mathrm{PG}(r, q)$ and denote by θ_j the number of points in a j-flat, i.e. $\theta_j = |\mathrm{PG}(j, q)| = (q^{j+1} - 1)/(q - 1)$, where $|T|$ denotes the number of elements in T for a given set T.

For an $[n, k, d]_q$ code C with a generator matrix G, the columns of G can be considered as a multiset of n points in $\Sigma = \mathrm{PG}(k - 1, q)$ denoted by $\bar{G}$. An *i-point* is a point of Σ which has multiplicity i in $\bar{G}$. Let Σ_i be the set of i-points in Σ. For any subset S of Σ we define *the multiplicity of S with respect to C* as

$$m_C(S) = \sum_{i=1}^{\gamma_0} i \cdot |S \cap \Sigma_i|,$$

where $\gamma_0 = \max\{i \mid \text{an } i\text{-point exists}\}$.

Then we obtain the partition $\Sigma = \Sigma_0 \cup \Sigma_1 \cup \cdots \cup \Sigma_{\gamma_0}$ such that

$$n = m_C(\Sigma),$$
$$n - d = \max\{m_C(\pi) \mid \pi \in \mathcal{F}_{k-2}\}.$$

Conversely such a partition of Σ as above gives an $[n, k, d]_q$ code in the natural manner. Since $(n + 1) - (d + 1) = n - d$, we get the following.

Lemma 1. *C is extendable if and only if there exists a point $P \in \Sigma$ such that $m_C(\pi) < n - d$ for all hyperplanes π through P.*

Let Σ^* be the dual space of Σ (considering $\mathcal{F}_{k-2}$ as the set of points of Σ^*). Then Lemma 1 is equivalent to the following:

Lemma 2. *C is extendable if and only if there exists a hyperplane Π of Σ^* such that*

$$\Pi \subset \{\pi \in \mathcal{F}_{k-2} \mid m_C(\pi) < n - d\}.$$

Now, let C be an $[n, k, d]_3$ code with diversity (Φ_0, Φ_1), $\gcd(3, d) = 1$, $k \geq 3$, and let $\mathcal{F}_j^*$ be the set of j-flats of Σ^*, i.e., $\mathcal{F}_j^* = \mathcal{F}_{k-2-j}, 0 \leq j \leq k - 2$. We define F_0, F_1, F_e, F and $\bar{F}$ as follows:

$$F_0 = \{\pi \in \mathcal{F}_0^* \mid m_C(\pi) \equiv n \pmod{3}\},$$
$$F_1 = \{\pi \in \mathcal{F}_0^* \mid m_C(\pi) \not\equiv n, \ n - d \pmod{3}\},$$
$$F_e = \{\pi \in \mathcal{F}_0^* \mid m_C(\pi) < n - d, \ m_C(\pi) \equiv n - d \pmod{3}\},$$
$$F = F_0 \cup F_1, \quad \bar{F} = F \cup F_e.$$

Then we have $\Phi_0 = |F_0|, \Phi_1 = |F_1|, \Phi_e = |F_e|$ since $|\{\pi \in \mathcal{F}_{k-2} \mid m_C(\pi) = i\}| = A_{n-i}/(q - 1)$. Lemma 2 implies the following:

Lemma 3. *C is extendable if and only if $\bar{F}$ contains a hyperplane of Σ^*.*

We consider the extendability of C from this geometrical point of view. A t-flat Π of Σ^* with $|\Pi \cap F_0| = i$, $|\Pi \cap F_1| = j$ is called an $(i, j)_t$ flat. A $(1, 0)_0$ flat is just a point of F_0. An $(i, j)_1$ flat, an $(i, j)_2$ flat and an $(i, j)_3$ flat are called an (i, j)-*line*, an (i, j)-*plane* and an (i, j)-*solid* respectively.

Remark. We defined F_0, F_1, F_e as subsets of Σ^*. Alternatively, one can define F_0, F_1, F_e as subsets of $\mathrm{PG}(k - 1, 3)$ as follows. Let $G = [g_0, g_1, \cdots, g_{k-1}]^\mathrm{T}$ be a generator matrix of C, $g_j \in V(n, 3)$, where M^T stands for the transpose of a matrix M. Then any codeword $c \in C$ can be written as $v \cdot G = \sum_{i=0}^{k-1} v_i g_i$ for some $v = (v_0, v_1, \cdots, v_{k-1}) \in V(k, 3)$. Since there are $2\Phi_e$ vectors $v = (v_0, v_1, \cdots, v_{k-1}) \in V(k, 3)$ such that

$$wt(v \cdot G) \equiv d \pmod{3}, \quad wt(v \cdot G) > d \tag{2.1}$$

and since $2 \cdot v$ also satisfies (2.1), one can select Φ_e vectors $v \in V(k, 3)$ satisfying (2.1) any two of which are linearly independent, say $R_1, \cdots, R_{\Phi_e}$. Similarly, one can find Φ_0 vectors $v \in V(k, 3)$ with $wt(v \cdot G) \equiv 0 \pmod{3}$, say $P_1, \cdots, P_{\Phi_0}$, and Φ_1 vectors $v \in V(k, 3)$ with $wt(v \cdot G) \not\equiv 0, d \pmod{3}$, say $Q_1, \cdots, Q_{\Phi_1}$, so that any two of $P_1, \cdots, P_{\Phi_0}, Q_1, \cdots, Q_{\Phi_1}$ are linearly independent. Then the vectors $P_1, \cdots, P_{\Phi_0}, Q_1, \cdots, Q_{\Phi_1}, R_1, \cdots, R_{\Phi_e}$ are considered as distinct points of $\mathrm{PG}(k - 1, 3)$, and F_0, F_1, F_e are defined as:

$$F_0 = \{P_1, \cdots, P_{\Phi_0}\}, \quad F_1 = \{Q_1, \cdots, Q_{\Phi_1}\}, \quad F_e = \{R_1, \cdots, R_{\Phi_e}\}.$$

Let Λ_1 be the set of all possible (i, j) for which an (i, j)-line exists in $\mathcal{F}_1^*$. Then we have

$$\Lambda_1 = \{(1, 0), (0, 2), (2, 1), (1, 3), (4, 0)\},$$

see [5]. Assume $2 \leq t \leq k - 1$ and let $\Pi \in \mathcal{F}_t^*$. Denote by $c_{i,j}^{(t)}$ the number of $(i, j)_{t-1}$ flats in Π and let $\varphi_s^{(t)} = |\Pi \cap F_s|$, $s = 0, 1$. The pair $(\varphi_0^{(t)}, \varphi_1^{(t)})$ is called the *diversity of* Π and the list of $c_{i,j}^{(t)}$'s is called its *spectrum*. Let Λ_t be the set of all possible $(\varphi_0^{(t)}, \varphi_1^{(t)})$. Λ_t and the corresponding spectra are determined as in Table 1 for $t = 2$ and as in Table 2 for $t = 3$. For $t \geq 2$ we set Λ_t^- as

$$\Lambda_t^- = \{(\theta_{t-1}, 0), (\theta_{t-2}, 2 \cdot 3^{t-1}), (\theta_{t-1}, 2 \cdot 3^{t-1}), (\theta_{t-1} + 3^{t-1}, 3^{t-1}), (\theta_{t-1}, 3^t), (\theta_t, 0)\}.$$

It is known that Λ_t^- is included in Λ_t for all $t \geq 2$ ([5]).

Table 1.

$\varphi_0^{(2)}$	$\varphi_1^{(2)}$	$c_{1,0}^{(2)}$	$c_{0,2}^{(2)}$	$c_{2,1}^{(2)}$	$c_{1,3}^{(2)}$	$c_{4,0}^{(2)}$
4	0	12	0	0	0	1
1	6	2	9	0	2	0
4	3	4	3	6	0	0
4	6	0	3	6	4	0
7	3	1	0	9	1	2
4	9	0	0	0	12	1
13	0	0	0	0	0	13

Table 2.

$\varphi_0^{(3)}$	$\varphi_1^{(3)}$	$c_{4,0}^{(3)}$	$c_{1,6}^{(3)}$	$c_{2,1}^{(3)}$	$c_{4,3}^{(3)}$	$c_{4,6}^{(3)}$	$c_{7,3}^{(3)}$	$c_{4,9}^{(3)}$	$c_{13,0}^{(3)}$
13	0	39	0	0	0	0	0	0	1
4	18	2	36	0	0	0	0	2	0
13	9	4	3	27	0	0	6	0	0
10	15	0	10	15	15	0	0	0	0
16	12	0	0	12	12	16	0	0	0
13	18	0	3	0	27	6	4	0	0
22	9	1	0	0	0	0	36	1	2
13	27	0	0	0	0	0	0	39	1
40	0	0	0	0	0	0	0	0	40

Lemma 4 ([5]). *For $t \geq 2$, the spectrum corresponding to each diversity in Λ_t^- is uniquely determined as follows:*

(1) $(c_{\theta_{t-2},0}^{(t)}, c_{\theta_{t-1},0}^{(t)}) = (\theta_t - 1, 1)$ for $(\varphi_0^{(t)}, \varphi_1^{(t)}) = (\theta_{t-1}, 0)$;

(2) $(c_{\theta_{t-2},0}^{(t)}, c_{\theta_{t-3},2\cdot3^{t-2}}^{(t)}, c_{\theta_{t-2},3^{t-1}}^{(t)}) = (2, \theta_t - \theta_1, 2)$ for $(\varphi_0^{(t)}, \varphi_1^{(t)}) = (\theta_{t-2}, 2\cdot3^{t-1})$;

(3) $(c_{\theta_{t-3},2\cdot3^{t-2}}^{(t)}, c_{\theta_{t-2},2\cdot3^{t-2}}^{(t)}, c_{\theta_{t-2}+3^{t-2},3^{t-2}}^{(t)}, c_{\theta_{t-2},3^{t-1}}^{(t)}) = (3, \theta_t - \theta_2, 6, 4)$ for $(\varphi_0^{(t)},$ $\varphi_1^{(t)}) = (\theta_{t-1}, 2\cdot3^{t-1})$;

(4) $(c_{\theta_{t-2},0}^{(t)}, c_{\theta_{t-2}+3^{t-2},3^{t-2}}^{(t)}, c_{\theta_{t-2},3^{t-1}}^{(t)}, c_{\theta_{t-1},0}^{(t)}) = (1, \theta_t - \theta_1, 1, 2)$ for $(\varphi_0^{(t)}, \varphi_1^{(t)}) =$ $(\theta_{t-1}+3^{t-1}, 3^{t-1})$;

(5) $(c_{\theta_{t-2},3^{t-1}}^{(t)}, c_{\theta_{t-1},0}^{(t)}) = (\theta_t - 1, 1)$ for $(\varphi_0^{(t)}, \varphi_1^{(t)}) = (\theta_{t-1}, 3^t)$;

(6) $c_{\theta_{t-1},0}^{(t)} = \theta_t$ for $(\varphi_0^{(t)}, \varphi_1^{(t)}) = (\theta_t, 0)$.

An s-flat S in Π is called the *axis of Π* (*of type (a,b)*) if every hyperplane of Π not containing S has the same diversity (a,b) and if there is no hyperplane of Π through S whose diversity is (a,b). Then the spectrum of Π satisfies $c_{a,b}^{(t)} = \theta_t - \theta_{t-1-s}$ and the axis is unique if it exists. The axis is helpful to characterize the geometrical structure of Π. For example, (2) of Lemma 4 yields that there exist two $(\theta_{t-2}, 0)_{t-1}$ flats and two $(\theta_{t-2}, 3^{t-1})_{t-1}$ flats through a fixed $(\theta_{t-2}, 0)_{t-2}$ flat which is the axis of Π. The following lemma is obtained from Lemma 4.

Lemma 5. *Let Π be a t-flat in Σ^*, $t \geq 2$.*

(1) *Π is a $(\theta_{t-1}, 0)_t$ flat if and only if Π contains a $(\theta_{t-1}, 0)_{t-1}$ flat which is the axis of type $(\theta_{t-2}, 0)$.*

(2) *Π is a $(\theta_{t-1}, 3^t)_t$ flat if and only if Π contains a $(\theta_{t-1}, 0)_{t-1}$ flat which is the axis of type $(\theta_{t-2}, 3^{t-1})$.*

(3) *Π is a $(\theta_{t-2}, 2\cdot3^{t-1})_t$ flat if and only if Π contains a $(\theta_{t-2}, 0)_{t-2}$ flat which is the axis of type $(\theta_{t-3}, 2\cdot3^{t-2})$.*

(4) *Π is a $(\theta_{t-1} + 3^{t-1}, 3^{t-1})_t$ flat if and only if Π contains a $(\theta_{t-2}, 0)_{t-2}$ flat which is the axis of type $(\theta_{t-2} + 3^{t-2}, 3^{t-2})$.*

It is easy to see the geometrical structure of Π whose diversity is in Λ_t^- except for the type of (3) in Lemma 4. As for the type (3) of Lemma 4, see [5] for $t = 2$ and Section 4 for $t \geq 3$.

Set $\Lambda_t^+ = \Lambda_t \setminus \Lambda_t^-$. The diversities in Λ_t^+ and the corresponding spectra for $t \geq 4$ are determined as follows.

Lemma 6 ([6]).

(1) *When t is odd (≥ 5):*

$$\Lambda_t^+ = \{(\theta_{t-1}, 3^{t-1})\} \cup \{(\theta_{t-1} - 3^{T+1+s}, \theta_{t-1} + \theta_{T+s} + 1), (\theta_{t-1} + 3^{T+1+s}, \theta_{t-1} - \theta_{T+s}) \mid 0 \leq s \leq T\} \cup \{(\theta_{t-1}, \theta_{t-1} - \theta_{T+s}), (\theta_{t-1}, \theta_{t-1} + \theta_{T+s} + 1) \mid 1 \leq s \leq T\},$$

where $T = (t-3)/2$. The spectrum corresponding to each diversity is uniquely determined as follows:

(A-1) $c_{\theta_{t-2}-3^{T+1}, \theta_{t-2}+\theta_T+1}^{(t)} = \theta_{t-1} - 3^{T+1}$, $c_{\theta_{t-2}, \theta_{t-2}-\theta_T}^{(t)} = c_{\theta_{t-2}, \theta_{t-2}+\theta_T+1}^{(t)} = \theta_{t-1} + \theta_T + 1$ for $(\varphi_0^{(t)}, \varphi_1^{(t)}) = (\theta_{t-1} - 3^{T+1}, \theta_{t-1} + \theta_T + 1)$;

(A-2) $c_{\theta_{t-2}, \theta_{t-2}-\theta_T}^{(t)} = c_{\theta_{t-2}, \theta_{t-2}+\theta_T+1}^{(t)} = \theta_{t-1} - \theta_T$, $c_{\theta_{t-2}+3^{T+1}, \theta_{t-2}-\theta_T}^{(t)} = \theta_{t-1} + 3^{T+1}$ for $(\varphi_0^{(t)}, \varphi_1^{(t)}) = (\theta_{t-1} + 3^{T+1}, \theta_{t-1} - \theta_T)$;

(A-3) $(c_{\theta_{t-2}, 0}^{(t)}, c_{\theta_{t-3}, 2\cdot3^{t-2}}^{(t)}, c_{\theta_{t-2}, 3^{t-2}}^{(t)}, c_{\theta_{t-2}+3^{t-2}, 3^{t-2}}^{(t)}) = (4, 3, \theta_t - \theta_2, 6)$ for $(\varphi_0^{(t)}, \varphi_1^{(t)}) = (\theta_{t-1}, 3^{t-1})$;

(A-4) $c_{\theta_{t-2}-3^{T+1+s}, \theta_{t-2}+\theta_{T+s}+1}^{(t)} = \theta_{t-1-2s} - 3^{T+1-s}$, $c_{\theta_{t-2}, \theta_{t-2}-\theta_{T+s}}^{(t)} = c_{\theta_{t-2}, \theta_{t-2}+\theta_{T+s}+1}^{(t)} = \theta_{t-1-2s} + \theta_{T-s} + 1$, $c_{\theta_{t-2}-3^{T+s}, \theta_{t-2}+\theta_{T-1+s}+1}^{(t)} = \theta_t - \theta_{t-2s}$ for $(\varphi_0^{(t)}, \varphi_1^{(t)}) = (\theta_{t-1} - 3^{T+1+s}, \theta_{t-1} + \theta_{T+s} + 1)$, $1 \leq s \leq T$;

(A-5) $c_{\theta_{t-2}, \theta_{t-2}-\theta_{T+s}}^{(t)} = c_{\theta_{t-2}, \theta_{t-2}+\theta_{T+s}+1}^{(t)} = \theta_{t-1-2s} - \theta_{T-s}$, $c_{\theta_{t-2}+3^{T+1+s}, \theta_{t-2}-\theta_{T+s}}^{(t)} = \theta_{t-1-2s} + 3^{T+1-s}$, $c_{\theta_{t-2}+3^{T+s}, \theta_{t-2}-\theta_{T-1+s}}^{(t)} = \theta_t - \theta_{t-2s}$ for $(\varphi_0^{(t)}, \varphi_1^{(t)}) = (\theta_{t-1} + 3^{T+1+s}, \theta_{t-1} - \theta_{T+s})$, $1 \leq s \leq T$;

(A-6) $c_{\theta_{t-2}, \theta_{t-2}-\theta_{T+s}}^{(t)} = \theta_{t-2s}$, $c_{\theta_{t-2}-3^{T+s}, \theta_{t-2}+\theta_{T-1+s}+1}^{(t)} = \theta_{t-2s} - \theta_{T+1-s}$, $c_{\theta_{t-2}+3^{T+s}, \theta_{t-2}-\theta_{T-1+s}}^{(t)} = \theta_{t-2s} + \theta_{T+1-s} + 1$, $c_{\theta_{t-2}, \theta_{t-2}-\theta_{T-1+s}}^{(t)} = \theta_t - \theta_{t+1-2s}$ for $(\varphi_0^{(t)}, \varphi_1^{(t)}) = (\theta_{t-1}, \theta_{t-1} - \theta_{T+s})$, $1 \leq s \leq T$;

(A-7) $c_{\theta_{t-2}-3^{T+s}, \theta_{t-2}+\theta_{T-1+s}+1}^{(t)} = \theta_{t-2s} - \theta_{T+1-s}$, $c_{\theta_{t-2}+3^{T+s}, \theta_{t-2}-\theta_{T-1+s}}^{(t)} = \theta_{t-2s} + \theta_{T+1-s} + 1$, $c_{\theta_{t-2}, \theta_{t-2}+\theta_{T+s}+1}^{(t)} = \theta_{t-2s}$, $c_{\theta_{t-2}, \theta_{t-2}+\theta_{T-1+s}+1}^{(t)} = \theta_t - \theta_{t+1-2s}$ for $(\varphi_0^{(t)}, \varphi_1^{(t)}) = (\theta_{t-1}, \theta_{t-1} + \theta_{T+s} + 1)$, $1 \leq s \leq T$.

(2) *When t is even (≥ 4):*

$$\Lambda_t^+ = \{(\theta_{t-1}, 3^{t-1})\} \cup \{(\theta_{t-1}, \theta_{t-1} - \theta_{U+1+s}), (\theta_{t-1}, \theta_{t-1} + \theta_{U+1+s} + 1) \mid 0 \leq s \leq U\} \cup \{(\theta_{t-1} - 3^{U+1+s}, \theta_{t-1} + \theta_{U+s} + 1), (\theta_{t-1} + 3^{U+1+s}, \theta_{t-1} - \theta_{U+s}) \mid 1 \leq s \leq U+1\},$$

where $U = (t-4)/2$. The spectrum corresponding to each diversity is uniquely determined as follows:

(B-1) $c^{(t)}_{\theta_{t-2},\theta_{t-2}-\theta_{U+1}} = \theta_{t-1}$, $c^{(t)}_{\theta_{t-2}-3^{U+1},\theta_{t-2}+\theta_U+1} = \theta_{t-1} - \theta_{U+1}$,

$c^{(t)}_{\theta_{t-2}+3^{U+1},\theta_{t-2}-\theta_U} = \theta_{t-1} + \theta_{U+1} + 1$ for $(\varphi_0^{(t)}, \varphi_1^{(t)}) = (\theta_{t-1}, \theta_{t-1} - \theta_{U+1})$;

(B-2) $c^{(t)}_{\theta_{t-2}-3^{U+1},\theta_{t-2}+\theta_U+1} = \theta_{t-1} - \theta_{U+1}$, $c^{(t)}_{\theta_{t-2}+3^{U+1},\theta_{t-2}-\theta_U} = \theta_{t-1} + \theta_{U+1} + 1$,

$c^{(t)}_{\theta_{t-2},\theta_{t-2}+\theta_{U+1}+1} = \theta_{t-1}$ for $(\varphi_0^{(t)}, \varphi_1^{(t)}) = (\theta_{t-1}, \theta_{t-1} + \theta_{U+1} + 1)$;

(B-3) $(c^{(t)}_{\theta_{t-2},0}, c^{(t)}_{\theta_{t-3},2\cdot 3^{t-2}}, c^{(t)}_{\theta_{t-2},3^{t-2}}, c^{(t)}_{\theta_{t-2}+3^{t-2},3^{t-2}}) = (4, 3, \theta_t - \theta_2, 6)$

for $(\varphi_0^{(t)}, \varphi_1^{(t)}) = (\theta_{t-1}, 3^{t-1})$;

(B-4) $c^{(t)}_{\theta_{t-2}-3^{U+1+s},\theta_{t-2}+\theta_{U+s}+1} = \theta_{t-2s} - 3^{U+2-s}$, $c^{(t)}_{\theta_{t-2},\theta_{t-2}-\theta_{U+s}} =$

$c^{(t)}_{\theta_{t-2},\theta_{t-2}+\theta_{U+s}+1} = \theta_{t-2s} + \theta_{U+1-s} + 1$, $c^{(t)}_{\theta_{t-2}-3^{U+s},\theta_{t-2}+\theta_{U-1+s}+1} = \theta_t - \theta_{t+1-2s}$

for $(\varphi_0^{(t)}, \varphi_1^{(t)}) = (\theta_{t-1} - 3^{U+1+s}, \theta_{t-1} + \theta_{U+s} + 1)$, $1 \leq s \leq U + 1$;

(B-5) $c^{(t)}_{\theta_{t-2},\theta_{t-2}-\theta_{U+s}} = c^{(t)}_{\theta_{t-2},\theta_{t-2}+\theta_{U+s}+1} = \theta_{t-2s} - \theta_{U+1-s}$,

$c^{(t)}_{\theta_{t-2}+3^{U+1+s},\theta_{t-2}-\theta_{U+s}} = \theta_{t-2s} + 3^{U+2-s}$, $c^{(t)}_{\theta_{t-2}+3^{U+s},\theta_{t-2}-\theta_{U-1+s}} = \theta_t - \theta_{t+1-2s}$

for $(\varphi_0^{(t)}, \varphi_1^{(t)}) = (\theta_{t-1} + 3^{U+1+s}, \theta_{t-1} - \theta_{U+s})$, $1 \leq s \leq U + 1$;

(B-6) $c^{(t)}_{\theta_{t-2},\theta_{t-2}-\theta_{U+1+s}} = \theta_{t-1-2s}$, $c^{(t)}_{\theta_{t-2}-3^{U+1+s},\theta_{t-2}+\theta_{U+s}+1} = \theta_{t-1-2s} - \theta_{U+1-s}$,

$c^{(t)}_{\theta_{t-2}+3^{U+1+s},\theta_{t-2}-\theta_{U+s}} = \theta_{t-1-2s} + \theta_{U+1-s} + 1$, $c^{(t)}_{\theta_{t-2},\theta_{t-2}-\theta_{U+s}} = \theta_t - \theta_{t-2s}$ for

$(\varphi_0^{(t)}, \varphi_1^{(t)}) = (\theta_{t-1}, \theta_{t-1} - \theta_{U+1+s})$, $1 \leq s \leq U$;

(B-7) $c^{(t)}_{\theta_{t-2}-3^{U+1+s},\theta_{t-2}+\theta_{U+s}+1} = \theta_{t-1-2s} - \theta_{U+1-s}$, $c^{(t)}_{\theta_{t-2}+3^{U+1+s},\theta_{t-2}-\theta_{U+s}} =$

$\theta_{t-1-2s} + \theta_{U+1-s} + 1$, $c^{(t)}_{\theta_{t-2},\theta_{t-2}+\theta_{U+1+s}+1} = \theta_{t-1-2s}$, $c^{(t)}_{\theta_{t-2},\theta_{t-2}+\theta_{U+s}+1} = \theta_t -$

θ_{t-2s} for $(\varphi_0^{(t)}, \varphi_1^{(t)}) = (\theta_{t-1}, \theta_{t-1} + \theta_{U+1+s} + 1)$, $1 \leq s \leq U$.

3   Main Results

In this section, we give the geometric conditions and the main theorems on the extendability of ternary linear codes. For $k \geq 4$, let $(C_k\text{-}0)$, $(C_k\text{-}1)$ and $(C_k\text{-}2)$ be the following conditions:

(C_k-0) there exists a $(\theta_{k-4}, 0)_{k-3}$ flat δ_1 in Σ^* satisfying $\delta_1 \setminus F_0 \subset F_e$,

(C_k-1) $(C_k\text{-}0)$ holds and there exists a $(\theta_{k-4}, 3^{k-3})_{k-3}$ flat δ_2 in Σ^* such that $\delta_1 \cap \delta_2$ is a $(\theta_{k-4}, 0)_{k-4}$ flat,

(C_k-2) there exist two $(\theta_{k-4}, 0)_{k-3}$ flats δ_1, δ_2 in Σ^* such that $\delta_1 \cap \delta_2$ is a $(\theta_{k-4}, 0)_{k-4}$ flat with $(\delta_1 \cup \delta_2) \setminus (\delta_1 \cap \delta_2) \subset F_e$.

We denote by $\langle \chi_1, \chi_2, \cdots \rangle$ the smallest flat containing subsets $\chi_1, \chi_2, \cdots$ of Σ^*. For $k = 4$ we consider two more conditions:

(C_4-3) there are three non-collinear points $R_1, R_2, R_3 \in F_e$ such that the three lines $\langle R_1, R_2 \rangle$, $\langle R_2, R_3 \rangle$, $\langle R_3, R_1 \rangle$ are $(0, 2)$-lines,

(C$_4$-4) there are three non-collinear points $Q_1, Q_2, Q_3 \in F_1$ such that the three lines $\langle Q_1, Q_2 \rangle$, $\langle Q_2, Q_3 \rangle$, $\langle Q_3, Q_1 \rangle$ are $(0, 2)$-lines each of which contains two points of F_e.

For $k \geq 5$, let (C$_k$-3) and (C$_k$-4) be the following conditions:

(C$_k$-3) there exist three $(\theta_{k-5}, 0)_{k-4}$ flats $\delta_1, \delta_2, \delta_3$ through a fixed $(\theta_{k-5}, 0)_{k-5}$ flat L such that $\langle \delta_1, \delta_2 \rangle$, $\langle \delta_2, \delta_3 \rangle$, $\langle \delta_3, \delta_1 \rangle$ form distinct $(\theta_{k-5}, 2 \cdot 3^{k-4})_{k-3}$ flats and that $(\delta_1 \cup \delta_2 \cup \delta_3) \setminus L \subset F_e$ holds,
(C$_k$-4) there exist a $(\theta_{k-5}, 0)_{k-5}$ flat L, three $(\theta_{k-5}, 3^{k-4})_{k-4}$ flats $\delta_1', \delta_2', \delta_3'$ through L, and six $(\theta_{k-5}, 0)_{k-4}$ flats $\delta_1, \cdots, \delta_6$ through L such that $\langle \delta_i', \delta_j' \rangle$ forms a $(\theta_{k-5}, 2 \cdot 3^{k-4})_{k-3}$ flat containing two of $\delta_1, \cdots, \delta_6$ for $1 \leq i < j \leq 3$ and that $(\cup_{i=1}^6 \delta_i) \setminus L \subset F_e$ holds.

For $k = 5$ we consider two more conditions:

(C$_5$-5) there exist a $(4,0)$-line l and four skew $(1,0)$-lines l_1, l_2, l_3, l_4 such that each of $l_1, ..., l_4$ meets l and that $\langle l_1, l_2, l_3, l_4 \rangle \in \mathcal{F}_3^*$ and $(\cup_{i=1}^4 l_i) \setminus l \subset F_e$ hold,
(C$_5$-6) there exist a $(2,1)$-line l_0 containing two points $P_1, P_2 \in F_0$ and two $(1,0)$-lines l_1, l_2 (resp. l_1', l_2') through P_1 (resp. P_2) such that $l = \langle l_1, l_2 \rangle \cap \langle l_1', l_2' \rangle$ and $m_i = \langle Q_0, Q_i \rangle$ are $(0, 2)$-lines for $i = 1, 2$, where $l_0 \cap F_1 = \{Q_0\}$, $l \cap F_1 = \{Q_1, Q_2\}$ and that $(\cup_{i=1}^2 (l_i \cup l_i' \cup m_i)) \setminus F \subset F_e$ holds.

We define the conditions (C$_k$-5) and (C$_k$-6) for $k \geq 6$ as follows:

(C$_k$-5) there exist a $(\theta_{k-4}, 0)_{k-4}$ flat δ, a $(\theta_{k-6}, 0)_{k-6}$ flat H in δ and four $(\theta_{k-5}, 0)_{k-4}$ flats $\delta_1, \cdots, \delta_4$ such that $\delta_1 \setminus \delta, \cdots, \delta_4 \setminus \delta$ are mutually disjoint and that $\delta_1 \cap \delta, \cdots, \delta_4 \cap \delta$ are distinct $(k-5)$-flats through H and that $\langle \delta_1, \cdots, \delta_4 \rangle \in \mathcal{F}_{k-2}^*$ and $(\cup_{i=1}^4 \delta_i) \setminus \delta \subset F_e$ hold,
(C$_k$-6) there exist a $(\theta_{k-5} + 3^{k-5}, 3^{k-5})_{k-4}$ flat δ_0 (containing the $(\theta_{k-5}, 0)_{k-5}$ flats L_1, L_2 and the $(\theta_{k-6}, 3^{k-5})_{k-5}$ flat L_0), two $(\theta_{k-5}, 0)_{k-4}$ flats δ_1, δ_2 (resp. δ_1', δ_2') through L_1 (resp. L_2) such that $\delta = \langle \delta_1, \delta_2 \rangle \cap \langle \delta_1', \delta_2' \rangle$, $\delta_3 = \langle L_0, M_1 \rangle$ and $\delta_4 = \langle L_0, M_2 \rangle$ form $(\theta_{k-6}, 2 \cdot 3^{k-5})_{k-4}$ flats and that $(\delta_1' \cup \delta_2' \cup (\cup_{i=0}^4 \delta_i)) \setminus F \subset F_e$ holds, where M_1 and M_2 are the $(\theta_{k-6}, 3^{k-5})_{k-5}$ flats in δ.

For $k = 6$ we consider extra two conditions:

(C$_6$-7) there exist a $(4,9)$-plane δ, a $(4,0)$-line $l = \{P_1, \cdots, P_4\}$ in δ, and three non-coplanar $(1,0)$-lines l_{i1}, l_{i2}, l_{i3} through P_i with $\Delta_i = \langle l_{i1}, l_{i2}, l_{i3} \rangle$ for each i $(1 \leq i \leq 4)$ such that $\Delta_1, \cdots, \Delta_4$ are distinct solids through δ and that $(\cup_{i=1}^4 \cup_{j=1}^3 l_{ij}) \setminus l \subset F_e$ holds,
(C$_6$-8) there exist a $(4,0)$-plane δ, a $(4,0)$-line $l = \{P_1, \cdots, P_4\}$ in δ, and three non-coplanar $(1,0)$-lines l_{i1}, l_{i2}, l_{i3} through P_i none of which lie on δ with $\Delta_i = \langle l_{i1}, l_{i2}, l_{i3} \rangle$ for each i $(1 \leq i \leq 4)$ such that $\Delta_1, \cdots, \Delta_4$ are distinct solids through δ and that $(\delta \cup (\cup_{i=1}^4 \cup_{j=1}^3 l_{ij})) \setminus l \subset F_e$ holds.

$(C_5$-6)

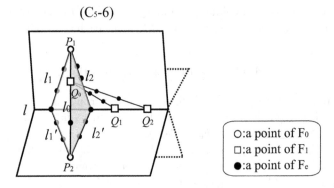

Let C be an $[n, k, d]_3$ code with diversity $(\Phi_0, \Phi_1) \in \mathcal{D}_k^+$, $d \equiv 1$ or $2 \pmod 3$, $k \geq 3$. Since $\mathcal{D}_3^+ = \{(4, 3)\}$, $\mathcal{D}_4^+ = \{(13, 9), (10, 15), (16, 12)\}$ and $\mathcal{D}_k^+ = \Lambda_{k-1}^+$ for $k \geq 5$ ([5],[6]), we have $|\mathcal{D}_k| = 2k - 1$ for all $k \geq 3$. It is known that an $[n, 4, d]_3$ code with diversity $(\Phi_0, \Phi_1) \in \mathcal{D}_4^+$ is not extendable if $\Phi_e < 3$ for $k = 4$ ([5]). The conditions $(C_4$-0$)$-$(C_4$-4$)$ are used to check the extendability of $[n, 4, d]_3$ codes.

Theorem 1 ([6]). *Let C be an $[n, 4, d]_3$ code with diversity $(\Phi_0, \Phi_1) \in \mathcal{D}_4^+$, $\gcd(3, d) = 1$. Then C is extendable if and only if one of the conditions indicated in Table 3 holds.*

For the case when $k = 5$, C is not extendable if $\Phi_e < 9$ when $(\Phi_0, \Phi_1) \neq (40, 36)$ or if $\Phi_e < 12$ when $(\Phi_0, \Phi_1) = (40, 36)$ ([5]). Otherwise, we need to check whether one of the conditions $(C_5$-0$)$-$(C_5$-6$)$ holds or not according to the diversity of C.

Theorem 2 ([7]). *Let C be an $[n, 5, d]_3$ code with diversity $(\Phi_0, \Phi_1) \in \mathcal{D}_5^+$, $\gcd(3, d) = 1$. Then C is extendable if and only if one of the conditions indicated in Table 4 holds.*

	Table 3.
(Φ_0, Φ_1)	conditions
(13,9)	$(C_4$-1$)$, $(C_4$-4$)$
(10,15)	$(C_4$-2$)$, $(C_4$-3$)$, $(C_4$-4$)$
(16,12)	$(C_4$-0$)$, $(C_4$-3$)$

	Table 4.
(Φ_0, Φ_1)	conditions
(40,27)	$(C_5$-1$)$, $(C_5$-4$)$
(31,45)	$(C_5$-2$)$, $(C_5$-3$)$, $(C_5$-4$)$, $(C_5$-6$)$
(40,36)	$(C_5$-4$)$, $(C_5$-5$)$, $(C_5$-6$)$
(40,45)	$(C_5$-3$)$, $(C_5$-5$)$, $(C_5$-6$)$
(49,36)	$(C_5$-0$)$, $(C_5$-3$)$, $(C_5$-5$)$

For $k = 6$, C is not extendable if $\Phi_e < 27$ when $(\Phi_0, \Phi_1) \in \{(121,81), (94,135), (121,135), (148,108)\}$ or if $\Phi_e < 36$ when $(\Phi_0, \Phi_1) \in \{(121,108), (112,126), (130,117)\}$ ([5]). Otherwise, we need to check whether one of the conditions $(C_6$-0$)$-$(C_5$-8$)$ holds or not according to the diversity of C.

Theorem 3. *Let C be an $[n, 6, d]_3$ code with diversity $(\Phi_0, \Phi_1) \in \mathcal{D}_6^+$, $\gcd(3, d) = 1$. Then C is extendable if and only if one of the conditions indicated in Table 5 holds.*

Table 5.

(Φ_0, Φ_1)	conditions
(121, 81)	$(C_6\text{-}1)$, $(C_6\text{-}4)$
(94,135)	$(C_6\text{-}2)$, $(C_6\text{-}3)$, $(C_6\text{-}4)$, $(C_6\text{-}6)$
(121,108)	$(C_6\text{-}4)$, $(C_6\text{-}5)$, $(C_6\text{-}6)$, $(C_6\text{-}8)$
(112,126)	$(C_6\text{-}6)$, $(C_6\text{-}7)$, $(C_6\text{-}8)$
(130,117)	$(C_6\text{-}5)$, $(C_6\text{-}7)$, $(C_6\text{-}8)$
(121,135)	$(C_6\text{-}3)$, $(C_6\text{-}5)$, $(C_6\text{-}6)$, $(C_6\text{-}7)$
(148,108)	$(C_6\text{-}0)$, $(C_6\text{-}3)$, $(C_6\text{-}5)$

The result for $(\Phi_0, \Phi_1) = (121, 81), (94, 135), (148, 108)$ in Theorem 3 can be generalized to the following Theorems 4-6 respectively.

Theorem 4. *Let C be an $[n, k, d]_3$ code with diversity $(\theta_{k-2}, 3^{k-2})$, $gcd(3, d) = 1$, $k \geq 5$. Then C is extendable if and only if either the conditions $(C_k\text{-}1)$ or $(C_k\text{-}4)$ holds.*

Theorem 5. *Let C be an $[n, k, d]_3$ code with diversity $(\theta_{k-2} - 3^{k-3}, 5 \cdot 3^{k-3})$, $gcd(3, d) = 1$, $k \geq 6$. Then C is extendable if and only if one of the conditions $(C_k\text{-}2)$, $(C_k\text{-}3)$, $(C_k\text{-}4)$, $(C_k\text{-}6)$ holds.*

Theorem 6. *Let C be an $[n, k, d]_3$ code with diversity $(\theta_{k-2} + 3^{k-3}, 4 \cdot 3^{k-3})$, $gcd(3, d) = 1$, $k \geq 6$. Then C is extendable if and only if one of the conditions $(C_k\text{-}0)$, $(C_k\text{-}3)$, $(C_k\text{-}5)$ holds.*

4   Proof of Theorems 3 – 6

Theorem 2 can be proved using the following lemma.

Lemma 7 ([7]). *Let Δ be a solid in Σ^*.*

(1) *Δ is a $(13, 9)$-solid with $\Delta \setminus F \subset F_e$ if and only if Δ satisfies $(C_5\text{-}4)$.*
(2) *Δ is a $(10, 15)$-solid with $\Delta \setminus F \subset F_e$ if and only if Δ satisfies $(C_5\text{-}6)$.*
(3) *Δ is a $(16, 12)$-solid with $\Delta \setminus F \subset F_e$ if and only if Δ satisfies $(C_5\text{-}5)$.*
(4) *Δ is a $(13, 18)$-solid with $\Delta \setminus F \subset F_e$ if and only if Δ satisfies $(C_5\text{-}3)$.*

It is easy to see that the point L in $(C_5\text{-}3)$ (resp. in $(C_5\text{-}4)$) is the axis of a $(13, 18)$-solid (resp. a $(13, 9)$-solid). The following lemma is obtained from the proof of Lemma 7(2), see [7].

Lemma 8. *Let Δ be a $(10, 15)$-solid and assume $(C_5\text{-}6)$ holds in Δ except for the condition that $m_i = \langle Q_0, Q_i \rangle$ is a $(0, 2)$-line for $i = 1, 2$. Then m_1 and m_2 are necessarily $(0, 2)$-lines in Δ.*

We give the geometric characterizations of some 4-flats in Σ^* before proving Theorem 3.

Lemma 9. *For a 4-flat Π in Σ^*, Π is a $(40, 54)_4$ flat with $\Pi \setminus F \subset F_e$ if and only if $(C_6\text{-}3)$ holds in Π.*

Lemma 10. *For a 4-flat Π in Σ^*, Π is a $(40, 27)_4$ flat with $\Pi \setminus F \subset F_e$ if and only if $(C_6\text{-}4)$ holds in Π.*

Lemma 11. *For a 4-flat Π in Σ^*, Π is a $(49, 36)_4$ flat with $\Pi \setminus F \subset F_e$ if and only if $(C_6\text{-}5)$ holds in Π.*

See Lemmas 15,16,18 for the proofs of Lemmas 9,10,11, respectively.

Lemma 12. *For a 4-flat Π in Σ^*, Π is a $(31, 45)_4$ flat with $\Pi \setminus F \subset F_e$ if and only if $(C_6\text{-}6)$ holds in Π.*

Proof. ("only if" part:) Assume that Π is a $(31, 45)_4$ flat with $\Pi \setminus F \subset F_e$. Then the spectrum of Π is $(c_{4,18}^{(4)}, c_{13,9}^{(4)}, c_{10,15}^{(4)}, c_{13,18}^{(4)}) = (10, 15, 81, 15)$ by Lemma 6, see (B4) with $t = 4, s = 1$. Take two $(4,18)$-solids Δ_1, Δ_2 in Π meeting in a $(1,6)$-plane δ, where δ contains exactly two $(1,3)$-lines, say M_1, M_2. Let L_i be the axis of Δ_i, which is a $(4,0)$-line in Δ_i for $i = 1, 2$ by Lemma 5. Then there are exactly two $(4,0)$-planes through L_i in Δ_i, say δ_1, δ_2 for $i = 1$ and δ'_1, δ'_2 for $i = 2$. Put $H = L_1 \cap L_2$. Then H is the point of F_0 in δ. Since Π has no $(13,0)$-plane, $\delta_0 = \langle L_1, L_2 \rangle$ is a $(7,3)$-plane by Table 1. Let L_0 be the $(1,3)$-line in δ_0. It suffices to show that $\delta_3 = \langle L_0, M_1 \rangle$ and $\delta_4 = \langle L_0, M_2 \rangle$ are $(1,6)$-planes. Take a point $P_i(\neq P)$ on L_i for $i = 1, 2$ and let l be a $(0,2)$-line in δ which consists of $Q_1, Q_2 \in F_1$ and $R_1, R_2 \in F_e$. Then $\langle P_i, l \rangle$ is a $(1,6)$-plane, for there are two $(4,9)$-planes and two $(4,0)$-planes through L_i in Δ_i. Note that the line $\langle P_1, P_2 \rangle$ in δ_0 is a $(2,1)$-line. Containing two $(1,3)$-lines $\langle P_1, Q_1 \rangle$, $\langle P_2, Q_1 \rangle$ and a $(2,1)$-line $\langle P_1, P_2 \rangle$, $\langle P_1, Q_1, P_2 \rangle$ is a $(4,6)$-plane. Meanwhile, containing two $(1,0)$-lines $\langle P_1, R_1 \rangle$, $\langle P_2, R_1 \rangle$ and a $(2,1)$-line $\langle P_1, P_2 \rangle$, $\langle P_1, R_1, P_2 \rangle$ is a $(4,3)$-plane. Hence, containing three type of planes $\langle P_1, l \rangle$, $\langle P_1, Q_1, P_2 \rangle$, and $\langle P_1, R_1, P_2 \rangle$, it follows from Table 2 that $\langle l, P_1, P_2 \rangle$ is a $(10,15)$-solid. Then, by Lemma 8, $\langle Q_0, Q_i \rangle$ is a $(0,2)$-line, where $Q_0 = L_0 \cap \langle P_1, P_2 \rangle$. Hence δ_3 and δ_4 are $(1,6)$-planes.

("if" part:) Assume $(C_6\text{-}6)$ holds in Π. Put $H = L_1 \cap L_2$ and let Δ be a solid in Π not containing H. Setting $l_0 = \Delta \cap \delta_0$, $l = \Delta \cap \delta$, $Q_0 = \Delta \cap L_0$ and $P_i = \Delta \cap L_i$, $l_i = \Delta \cap \delta_i$, $l'_i = \Delta \cap \delta'_i$, $Q_i = \Delta \cap M_i$ for $i = 1, 2$, the condition $(C_5\text{-}6)$ holds in Δ. Hence Δ is a $(10,15)$-solid by Lemma 7(2) and Π satisfies $c_{10,15}^{(4)} \geq \theta_4 - \theta_3 = 81$. This implies that Π is a $(31, 45)_4$ flat. $\square$

From the proof of Lemma 12, the point $H = L_1 \cap L_2$ in the condition $(C_6\text{-}6)$ is the axis of a $(31, 45)_4$ flat Π. Now, let Δ_0 be a $(13,18)$-solid in Π with the axis H_0 and let δ' be a $(1,6)$-plane in Δ_0. Then the point of F_0 in δ' is H_0 since every $(1,6)$-plane in Δ_0 contains the axis of Δ_0. Note that there are exactly two $(4,18)$-solids through δ_0. It follows from the "only if" part of the previous proof that H_0 coincides with H. Hence we have the following.

Corollary 1. *The axis of any $(13, 18)$-solid in a $(31, 45)_4$ flat Π coincides with the axis of Π.*

Lemma 13. *For a 4-flat Π in Σ^*, Π is a $(40, 45)_4$ flat with $\Pi \setminus F \subset F_e$ if and only if $(C_6\text{-}7)$ holds in Π.*

Proof. ("only if" part:) Assume that Π is a $(40,45)_4$ flat with $\Pi \setminus F \subset F_e$. Then the spectrum of Π is $(c_{10,15}^{(4)}, c_{16,12}^{(4)}, c_{13,18}^{(4)}) = (36, 45, 40)$ by Lemma 6, see (B2). For a $(4,9)$-plane δ in Π, there are exactly four $(13,18)$-solids, say $\Delta_1, \cdots, \Delta_4$, through δ in Π. Let l be the axis of δ, which is the $(4,0)$-line in δ. For $1 \leq i \leq 4$, let P_i be the axis of Δ_i and let l_{i1}, l_{i2}, l_{i3} be the three $(1,0)$-lines through P_i in Δ_i. Then P_i is a point of l since every $(4,9)$-plane in Δ_i contains the axis of Δ_i. Suppose $P_1 = P_2$. Since $\langle l_{11}, l_{12} \rangle$ is a $(1,6)$-plane in Δ_1 and since there are no $(1,0)$-lines three of which are meeting in a point of F_0 in a $(10,15)$-solid from Lemma 7(2), $\langle l_{11}, l_{12}, l_{21} \rangle$ is a $(13,18)$-solid in Π. On the other hand, there is only one $(13,18)$-solid through a fixed $(1,6)$-plane in Π, a contradiction. Hence we have $l = \{P_1, P_2, P_3, P_4\}$, so that (C$_6$-7) holds.

("if" part:) Assume (C$_6$-7) holds in Π. Then, Δ_i is a $(4,18)$-solid or a $(13,18)$-solid, for there are three non-coplanar $(1,0)$-lines through a fixed point of F_0 in Δ_i and there is a $(4,9)$-plane δ in Δ_i. Since Π has exactly $45(= (18-9) \times 4 + 9)$ points of F_1, Π is a $(31,45)_4$ flat or a $(40,45)_4$ flat. Suppose Π is a $(31,45)_4$ flat. Then we may assume that $\Delta_1, \Delta_2, \Delta_3$ are $(13,18)$-solids and that Δ_4 is a $(4,18)$-solid. Since all of the planes through l in Δ_i other than δ are $(7,3)$-planes, the three $(1,0)$-lines contained in these three $(7,3)$-planes are just l_{i1}, l_{i2}, l_{i3}, and P_i is the axis of Δ_i for $1 \leq i \leq 3$. This is contradictory to Corollary 1. Hence Π is a $(40,45)_4$ flat. $\quad \square$

The following Lemma can be proved similarly to the proof of Lemma 13.

Lemma 14. *Let Π be an $(i_0, i_1)_4$ flat in Σ^* with $(i_0, i_1) \neq (40, 0)$. Then Π is a $(40, 36)_4$ flat with $\Pi \setminus F \subset F_e$ if and only if (C$_6$-8) holds in Π.*

Proof of Theorem 3. We prove Theorem 3 only for the case $(\Phi_0, \Phi_1) = (94, 135)$. Other cases can be proved by similar arguments using Lemmas 9-14.

("only if" part:) Assume that C is extendable. Then there is an $(i, j)_4$ flat Π in Σ^* satisfying $\Pi \setminus F \subset F_e$. We have $(i, j) \in \{(13, 54), (40, 27), (31, 45), (40, 54)\}$ by Lemma 6. If Π is a $(13,54)_4$ flat, then (C$_6$-2) holds since there are exactly two $(13,0)$-solids and two $(13,27)$-solids through a fixed $(13,0)$-plane (see Lemma 4(2)). If Π is a $(40,54)_4$ flat, a $(40,27)_4$ flat or a $(31,45)_4$ flat, then (C$_6$-3), (C$_6$-4) or (C$_6$-6) holds respectively by Lemmas 9,10,12.

("if" part:) Recall that an $(i, j)_4$ flat in Σ^* satisfies $(i, j) \in \{(13, 54), (40, 27), (31, 45), (40, 54)\}$ when $(\Phi_0, \Phi_1) = (94, 135)$. We first assume that (C$_6$-3) holds and let Π_1 be the 4-flat containing $\delta_1, \delta_2, \delta_3$. Then Π_1 is a $(40,54)_4$ flat containing 27 points of F_e by Lemma 9. Hence C is extendable by Lemma 3. It can be also proved similarly that C is extendable if either (C$_6$-4) or (C$_6$-10) holds by Lemmas 10,12. Now, assume that (C$_6$-2) holds and let Π_2 be the 4-flat containing δ_1, δ_2. Then Π_2 is a $(40,27)_4$ flat or a $(13,54)_4$ flat by Lemmas 4,6 since Π_2 contains at least two $(13,0)$-solids. If Π_2 is a $(40,27)_4$ flat, then for any $(13,0)$-plane δ there are exactly one $(13,0)$-solid and three $(22,9)$-solids through δ in Π_2, contradicting (C$_6$-2). So Π_2 is a $(13,54)_4$ flat containing 54 points of F_e. Hence C is extendable by Lemma 3. $\quad \square$

Lemma 15. *Let Π be a t-flat in Σ^* with $t \geq 3$. Then Π is a $(\theta_{t-1}, 2 \cdot 3^{t-1})_t$ flat with $\Pi \setminus F \subset F_e$ if and only if (C$_{t+2}$-3) holds in Π.*

Proof. ("only if" part:) Assume that Π is a $(\theta_{t-1}, 2 \cdot 3^{t-1})_t$ flat. Then the spectrum of Π is given as (3) of Lemma 4:

$$(c_{\theta_{t-3}, 2\cdot 3^{t-2}}^{(t)}, c_{\theta_{t-2}, 2\cdot 3^{t-2}}^{(t)}, c_{\theta_{t-2}+3^{t-2}, 3^{t-2}}^{(t)}, c_{\theta_{t-2}, 3^{t-1}}^{(t)}) = (3, \theta_t - \theta_2, 6, 4).$$

For a $(\theta_{t-3}, 0)_{t-2}$ flat δ_1, it follows from (2)-(5) of Lemma 4 that there are exactly two $(\theta_{t-3}, 2\cdot 3^{t-2})_{t-1}$ flats, say Δ_1, Δ_2, and two $(\theta_{t-2}+3^{t-2}, 3^{t-2})_{t-1}$ flats through δ_1 in Π. Let L be the axis of δ_1, which is a $(\theta_{t-3}, 0)_{t-3}$ flat by Lemma 5(1). Then there is a $(\theta_{t-3}, 0)_{t-2}$ flat, say δ_2 (resp. δ_3) through L other than δ_1 in Δ_1 (resp. Δ_2). A $(t-1)$-flat containing at least two $(\theta_{t-3}, 0)_{t-2}$ flats in Π is only a $(\theta_{t-3}, 2\cdot 3^{t-2})_{t-1}$ flat by Lemma 4, whence $\langle \delta_2, \delta_3 \rangle$ is a $(\theta_{t-3}, 2\cdot 3^{t-2})_{t-1}$ flat. Since $|(\delta_1 \cup \delta_2 \cup \delta_3) \setminus L| = 3(\theta_{t-2} - \theta_{t-3}) = \theta_t - (\theta_{t-1} + 2\cdot 3^{t-1}) = |\Pi \setminus F|$, we have $(\delta_1 \cup \delta_2 \cup \delta_3) \setminus L \subset F_e$. Hence $(C_{t+2}\text{-3})$ holds.

("if" part:) We proceed by induction on t. The conclusion holds for $t = 3$ by Lemma 7(4). Now, assume that $t \geq 4$ and that $(C_{t+2}\text{-3})$ holds in Π. Take a hyperplane π of Π not containing L and put $L_i = \pi \cap \delta_i$, $i = 1, 2, 3$, $H = \pi \cap L$. Since $\langle L_i, L_j \rangle$ contains two $(\theta_{t-4}, 0)_{t-3}$ flats meeting in a $(\theta_{t-4}, 0)_{t-4}$ flat, $\langle L_i, L_j \rangle$ is a $(\theta_{t-4}, 2\cdot 3^{t-3})_{t-2}$ flat or a $(\theta_{t-3}, 0)_{t-2}$ flat for $1 \leq i < j \leq 3$. On the other hand, $\langle \delta_i, \delta_j \rangle$ contains no $(\theta_{t-3}, 0)_{t-2}$ flat other than δ_i, δ_j by Lemma 4(2). So, each $\langle L_i, L_j \rangle$ is a $(\theta_{t-4}, 2\cdot 3^{t-3})_{t-2}$ flat. Applying the inductive assumption to $\pi = \langle L_1, L_2, L_3 \rangle$, π is a $(\theta_{t-2}, 2\cdot 3^{t-2})_{t-1}$ flat. Hence, counting the points of F_0 and F_1 in the four $(t-1)$-flats through $\langle L_1, L_2 \rangle$, we have $|\Pi \cap F_0| = (\theta_{t-2} - \theta_{t-4})3 + \theta_{t-3} = \theta_{t-1}$, $|\Pi \cap F_1| = (2\cdot 3^{t-2} - 2\cdot 3^{t-3})3 + 2\cdot 3^{t-2} = 2\cdot 3^{t-1}$. This completes the proof. $\square$

Lemma 16. *Let Π be a t-flat in Σ^* with $t \geq 3$. Then Π is a $(\theta_{t-1}, 3^{t-1})_t$ flat with $\Pi \setminus F \subset F_e$ if and only if $(C_{t+2}\text{-4})$ holds in Π.*

Proof. ("only if" part:) Assume that Π is a $(\theta_{t-1}, 3^{t-1})_t$ flat. Then the spectrum of Π is given as (A3) or (B3) of Lemma 6:

$$(c_{\theta_{t-2}, 0}^{(t)}, c_{\theta_{t-3}, 2\cdot 3^{t-2}}^{(t)}, c_{\theta_{t-2}, 3^{t-2}}^{(t)}, c_{\theta_{t-2}+3^{t-2}, 3^{t-2}}^{(t)}) = (4, 3, \theta_t - \theta_2, 6).$$

For a $(\theta_{t-3}, 3^{t-2})_{t-2}$ flat δ'_1 in Π there are exactly two $(\theta_{t-3}, 2\cdot 3^{t-2})_{t-1}$ flats, say Δ_2, Δ_3, and two $(\theta_{t-2}+3^{t-2}, 3^{t-2})_{t-1}$ flats through δ'_1 in Π by Lemmas 4,6. Let L be the axis of δ'_1, which is a $(\theta_{t-3}, 0)_{t-3}$ flat in δ'_1 by Lemma 5(2). Then there are two $(\theta_{t-3}, 3^{t-2})_{t-2}$ flats through L other than δ'_1, say δ'_2 in Δ_2 and δ'_3 in Δ_3, and four $(\theta_{t-3}, 0)_{t-2}$ flats $\delta_1, \cdots, \delta_4$ through L with $\delta_1, \delta_2 \subset \Delta_2$, $\delta_3, \delta_4 \subset \Delta_3$. It follows from Lemmas 4,6 and the spectrum of Π that a $(t-1)$-flat containing two $(\theta_{t-3}, 3^{t-2})_{t-2}$ flats in Π is a $(\theta_{t-3}, 2\cdot 3^{t-2})_{t-1}$ flat. Hence $\langle \delta'_2, \delta'_3 \rangle$ is a $(\theta_{t-3}, 2\cdot 3^{t-2})_{t-1}$ flat containing two $(\theta_{t-3}, 0)_{t-2}$ flats, say δ_5, δ_6. Thus $(C_{t+2}\text{-4})$ holds.

("if" part:) The proof is by induction on t. For $t = 3$, our assertion follows by Lemma 7(1). Now, assume that $t \geq 4$ and that $(C_{t+2}\text{-4})$ holds in Π. Take a hyperplane π of Π not containing L and put $L_i = \pi \cap \delta'_i$, $i = 1, 2, 3$, $H = \pi \cap L$. By a similar argument to the proof of the previous lemma, we can deduce that $\langle L_i, L_j \rangle$ $(1 \leq i < j \leq 3)$ is a $(\theta_{t-4}, 2\cdot 3^{t-3})_{t-2}$ flat. Applying the inductive assumption to $\pi = \langle L_1, L_2, L_3 \rangle$, π is a $(\theta_{t-2}, 3^{t-2})_{t-1}$ flat. A t-flat containing $(\theta_{t-3}, 2 \cdot 3^{t-2})_{t-1}$ flats and $(\theta_{t-2}, 3^{t-2})_{t-1}$ flats is a $(\theta_{t-1}, 3^{t-1})_t$ flat or a $(\theta_{t-1} - 3^{t-2}, \theta_{t-1} + \theta_{t-3} + 1)_t$ flat. Since Π contains $2 \cdot 3^{t-1}$ points of F_e, Π is a $(\theta_{t-1}, 3^{t-1})_t$ flat. $\square$

It follows from the proofs of Lemmas 15,16 that the $(\theta_{t-3}, 0)_{t-3}$ flat L in (C$_{t+2}$-3) (resp. in (C$_{t+2}$-4)) is the axis of a $(\theta_{t-1}, 2 \cdot 3^{t-1})_t$ flat (resp. a $(\theta_{t-1}, 3^{t-1})_t$ flat). Hence we get the following.

Corollary 2. Let Π be a t-flat in Σ^*, $t \geq 3$.
(1) Π is a $(\theta_{t-1}, 3^{t-1})_t$ flat if and only if Π contains a $(\theta_{t-3}, 0)_{t-3}$ flat which is the axis of type $(\theta_{t-2}, 3^{t-2})$.
(2) Π is a $(\theta_{t-1}, 2 \cdot 3^{t-1})_t$ flat if and only if Π contains a $(\theta_{t-3}, 0)_{t-3}$ flat which is the axis of type $(\theta_{t-2}, 2 \cdot 3^{t-2})$.

As for the axis of a $(\theta_{t-1}, 3^{t-1})_t$ flat, the following lemma also holds.

Lemma 17. For $t \geq 3$, let Π be a $(\theta_{t-1}, 3^{t-1})_t$ flat in Σ^* with the axis L.
(1) The axis of any $(\theta_{t-3}, 2 \cdot 3^{t-2})_{t-1}$ flat in Π coincides with L.
(2) Let δ be a $(\theta_{t-2}, 0)_{t-2}$ flat and let Δ be a $(\theta_{t-2} + 3^{t-2}, 3^{t-2})_{t-1}$ flat through δ in Π. Then the $(\theta_{t-3}, 0)_{t-2}$ flat δ_0 in Δ contains L.
(3) Let $\Delta_1, \Delta_2, \Delta_3$ be the $(\theta_{t-2}, 3^{t-2})_{t-1}$ flats in Π through a fixed $(\theta_{t-3}, 0)_{t-2}$ flat δ. Then the axes of $\Delta_1, \Delta_2, \Delta_3$ are the same $(\theta_{t-4}, 0)_{t-4}$ flat in L.
(4) Every $(\theta_{t-3}, 3^{t-2})_{t-2}$ flat in Π contains L.

Proof. (1) and (4) are obtained from Corollary 2(1) and the proof of Lemma 16.
(2) Let H be the axis of Δ. Then δ_0 contains H by Lemma 5(4). Since there is a $(\theta_{t-3}, 2 \cdot 3^{t-2})_{t-1}$ flat through δ_0 in Π, we have $H = L$ by (1) of this lemma.
(3) Let L' be the axis of δ and put $S = L \cap L'$. For a $(\theta_{t-3}, 2 \cdot 3^{t-2})_{t-1}$ flat Δ, $\Delta \cap \Delta_i$ is a $(\theta_{t-4}, 2 \cdot 3^{t-3})_{t-2}$ flat through S. Hence S is the axis of Δ_i for $i = 1, 2, 3$. $\square$

Lemma 18. Let Π be a t-flat in Σ^* with $t \geq 4$. Then Π is a $(\theta_{t-1} + 3^{t-2}, 4 \cdot 3^{t-2})_t$ flat with $\Pi \setminus F \subset F_e$ if and only if (C$_{t+2}$-5) holds in Π.

Proof. ("only if" part:) Assume that Π is a $(\theta_{t-1} + 3^{t-2}, 4 \cdot 3^{t-2})_t$ flat. The spectrum of Π is given as

$$(c^{(t)}_{\theta_{t-2}, 3^{t-2}}, c^{(t)}_{\theta_{t-2} + 3^{t-3}, 4 \cdot 3^{t-3}}, c^{(t)}_{\theta_{t-2}, 2 \cdot 3^{t-2}}, c^{(t)}_{\theta_{t-2} + 3^{t-2}, 3^{t-2}}) = (12, \theta_t - \theta_3, 12, 16),$$

see (A-5) with $s = T$ and (B-5) with $s = U + 1$ of Lemma 6. For a $(\theta_{t-2}, 0)_{t-2}$ flat δ in Π, there are exactly four $(\theta_{t-2} + 3^{t-2}, 3^{t-2})_{t-1}$ flats, say $\Delta_1, \cdots, \Delta_4$, through δ in Π. Let δ_i be the $(\theta_{t-3}, 0)_{t-2}$ flat in Δ_i and let L_i be the $(\theta_{t-3}, 0)_{t-3}$ flat $\delta_i \cap \delta$ which is the axis of δ_i for $1 \leq i \leq 4$. If $L_i = L_j$, then $\langle \delta_i, \delta_j \rangle$ is a $(\theta_{t-2}, 3^{t-2})_{t-1}$ flat since it contains at least two $(\theta_{t-3}, 0)_{t-2}$ flats in Π, $1 \leq i < j \leq 4$. This contradicts that there is exactly one $(\theta_{t-3}, 0)_{t-2}$ flat through a fixed $(\theta_{t-3}, 0)_{t-3}$ flat in a $(\theta_{t-2}, 3^{t-2})_{t-1}$ flat. Hence $L_i \neq L_j$ for $1 \leq i < j \leq 4$. Put $L_1 \cap L_j = H$, where $j > 1$. Let Δ be one of the $(t-1)$-flats in Π through δ_1 other than Δ_1.

Then Δ is a $(\theta_{t-2}, 3^{t-2})_{t-1}$ flat. Let δ_j' be the $(\theta_{t-3}, 3^{t-2})_{t-2}$ flat through L_j in Δ_j. Then $\Delta \cap \delta_j' = (\Delta \cap \Delta_j) \cap \delta_j'$ is a $(\theta_{t-4}, 3^{t-3})_{t-3}$ flat containing H since $\Delta \cap \Delta_j$ is a $(t-2)$-flat through L_1. Hence H is the axis of Δ by Lemma 17(4). This yields that $L_1 \cap L_2 = L_1 \cap L_3 = L_1 \cap L_4$. Hence $(C_{t+2}\text{-}5)$ holds.

("if" part:) Assume $(C_{t+2}\text{-}5)$ holds and let Π be the t-flat containing $\delta_1, \cdots, \delta_4$. Then $\Delta_i = \langle \delta_i, \delta \rangle$ is a $(\theta_{t-2}, 0)_{t-1}$ flat or a $(\theta_{t-2} + 3^{t-2}, 3^{t-2})_{t-1}$ flat for $1 \leq i \leq 4$. If x of $\Delta_1, \cdots, \Delta_4$ are $(\theta_{t-2} + 3^{t-2}, 3^{t-2})_{t-1}$ flats, then Π is a $(\theta_{t-2} + 3^{t-2}x, 3^{t-2}x)_t$ flat with $x = 3$ or 4 since $(\theta_{t-2} + 3^{t-2}x, 3^{t-2}x) \notin \Lambda_t$ for $x \leq 2$. We also have $x \neq 3$ by Lemma 17(2). Hence Π is a $(\theta_{t-1} + 3^{t-2}, 4 \cdot 3^{t-2})_t$ flat.   □

Now, assume that Π is a $(\theta_{t-1} - 3^{t-2}, 5 \cdot 3^{t-2})_t$ flat. Then the spectrum of Π is given by Lemma 6 as

$$(c_{\theta_{t-3}, 2 \cdot 3^{t-2}}^{(t)}, c_{\theta_{t-2}, 3^{t-2}}^{(t)}, c_{\theta_{t-2} - 3^{t-3}, 5 \cdot 3^{t-3}}^{(t)}, c_{\theta_{t-2}, 2 \cdot 3^{t-2}}^{(t)}) = (10, 15, \theta_t - \theta_3, 15),$$

see (A-4) with $s = T$ and (B-4) with $s = U + 1$. Take two $(\theta_{t-3}, 2 \cdot 3^{t-2})_{t-1}$ flats Δ_1, Δ_2 in Π meeting in a $(\theta_{t-4}, 2 \cdot 3^{t-3})_{t-2}$ flat. Let L_i be the axis of Δ_i, which is a $(\theta_{t-3}, 0)_{t-3}$ flat by Lemma 5 for $i = 1, 2$. Then it can be proved that the $(\theta_{t-4}, 0)_{t-4}$ flat $H = L_1 \cap L_2$ forms the axis of Π of type $(\theta_{t-2} - 3^{t-3}, 5 \cdot 3^{t-3})$. A similar argument to the proof of Lemma 12 yields the following lemma by induction on t.

Lemma 19. *Let Π be a t-flat in Σ^* with $t \geq 4$. Then Π is a $(\theta_{t-1} - 3^{t-2}, 5 \cdot 3^{t-2})_t$ flat with $\Pi \setminus F \subset F_e$ if and only if $(C_{t+2}\text{-}6)$ holds in Π.*

The following result for $t = 4$ is obtained from the proof of Lemma 12.

Corollary 3. *For $t \geq 4$, let Π be a $(\theta_{t-1} - 3^{t-2}, 5 \cdot 3^{t-2})_t$ flat and assume $(C_{t+2}\text{-}6)$ holds in Π except for the condition that $\delta_3 = \langle L_0, M_1 \rangle$ and $\delta_4 = \langle L_0, M_2 \rangle$ are $(\theta_{t-4}, 2 \cdot 3^{t-3})_{t-2}$ flats. Then δ_3 and δ_4 are necessarily $(\theta_{t-4}, 2 \cdot 3^{t-3})_{t-2}$ flats in Π.*

Finally we give the proof of Theorem 5. Theorems 4 and 6 can be proved similarly using Lemmas 15-19.

Proof of Theorem 5

("only if" part:) Assume that C is extendable. Then there is an $(i, j)_{k-2}$ flat Π in Σ^* satisfying $\Pi \setminus F \subset F_e$ by Lemma 3. From Lemma 6 we have

$$(i, j) \in \{(\theta_{k-4}, 2 \cdot 3^{k-3}), (\theta_{k-3}, 3^{k-3}), (\theta_{k-3} - 3^{k-4}, 5 \cdot 3^{k-4}), (\theta_{k-3}, 2 \cdot 3^{k-3})\},$$

see (A-4) with $s = T$ and (B-4) with $s = U + 1$. If Π is a $(\theta_{k-4}, 2 \cdot 3^{k-3})_{k-2}$ flat, then $(C_k\text{-}2)$ holds since there are exactly two $(\theta_{k-4}, 0)_{k-3}$ flats and two $(\theta_{k-4}, 3^{k-3})_{k-3}$ flats through a fixed $(\theta_{k-4}, 0)_{k-4}$ flat in Π. If Π is a $(\theta_{k-3}, 2 \cdot 3^{k-3})_{k-2}$ flat, a $(\theta_{k-3}, 3^{k-3})_{k-2}$ flat or a $(\theta_{k-3} - 3^{k-4}, 5 \cdot 3^{k-4})_{k-2}$ flat, then $(C_k\text{-}3)$, $(C_k\text{-}4)$ or $(C_k\text{-}6)$ holds respectively by Lemmas 15,16,19.

("if" part:) Assume that $(C_k\text{-}2)$ holds and let Π_1 be the $(k-2)$-flat containing δ_1, δ_2. Then Π_1 is a $(\theta_{k-3}, 3^{k-3})_{k-2}$ flat or a $(\theta_{k-4}, 2 \cdot 3^{k-3})_{k-2}$ flat since Π_1

contains at least two $(\theta_{k-4},0)_{k-3}$ flats. If Π_1 is a $(\theta_{k-3},3^{k-3})_{k-2}$ flat, then for any $(\theta_{k-4},0)_{k-4}$ flat δ there are exactly one $(\theta_{k-4},0)_{k-3}$ flat and three $(\theta_{k-4}+3^{k-4},3^{k-4})_{k-3}$ flats through δ in Π_1, contradicting (C$_k$-2). So Π_1 is a $(\theta_{k-4},2\cdot 3^{k-3})_{k-2}$ flat containing $2\cdot 3^{k-3}$ points of F_e. Hence C is extendable by Lemma 3. Next, assume that one of the conditions (C$_k$-3), (C$_k$-4), (C$_k$-6) holds. We take a $(k-2)$-flat Π as $\Pi = \langle \delta_1, \delta_2, \delta_3 \rangle$ for (C$_k$-3), $\Pi = \langle \delta_1', \delta_2', \delta_3' \rangle$ for (C$_k$-4) and $\Pi = \langle \delta_1, \delta_2, \delta_3, \delta_4 \rangle$ for (C$_k$-6). Then $\Pi \setminus F \subset F_e$ holds by Lemmas 15,16,19, and C is extendable by Lemma 3.   □

References

1. Hill, R.: An extension theorem for linear codes, Des. Codes Cryptogr. 17 (1999) 151–157.
2. Maruta, T.: On the extendability of linear codes, Finite Fields Appl. 7 (2001) 350–354.
3. Maruta, T.: Extendability of linear codes over GF(q) with minimum distance d, gcd(d,q) = 1, Discrete Math. 266 (2003) 377–385.
4. Maruta, T.: A new extension theorem for linear codes, Finite Fields Appl., 10 (2004) 674–685.
5. Maruta, T.: Extendability of ternary linear codes, Des. Codes Cryptogr., 35 (2005) 175-190.
6. Maruta, T., Okamoto, K.: Some improvements to the extendability of ternary linear codes, Finite Fields Appl., to appear.
7. Okamoto, K., Maruta, T.: Extendability of ternary linear codes of dimension five, Proc. 9th International Workshop in Algebraic and Combinatorial Coding Theory (ACCT), Kranevo, Bulgaria, 2004, pp. 312-318.
8. Simonis, J.: Adding a parity check bit, IEEE Trans. Inform. Theory 46 (2000) 1544–1545.

On the Design of Codes for DNA Computing*

Olgica Milenkovic[1] and Navin Kashyap[2]

[1] University of Colorado, Boulder, CO 80309, USA
olgica.milenkovic@colorado.edu
[2] Queen's University, Kingston, ON, K7L 3N6, Canada
nkashyap@mast.queensu.ca

Abstract. In this paper, we describe a broad class of problems arising in the context of designing codes for DNA computing. We primarily focus on design considerations pertaining to the phenomena of secondary structure formation in single-stranded DNA molecules and non-selective cross-hybridization. Secondary structure formation refers to the tendency of single-stranded DNA sequences to fold back upon themselves, thus becoming inactive in the computation process, while non-selective cross-hybridization refers to unwanted pairing between DNA sequences involved in the computation process. We use the Nussinov-Jacobson algorithm for secondary structure prediction to identify some design criteria that reduce the possibility of secondary structure formation in a codeword. These design criteria can be formulated in terms of constraints on the number of complementary pair matches between a DNA codeword and some of its shifts. We provide a sampling of simple techniques for enumerating and constructing sets of DNA sequences with properties that inhibit non-selective hybridization and secondary structure formation. Novel constructions of such codes include using cyclic reversible extended Goppa codes, generalized Hadamard matrices, and a binary mapping approach. Cyclic code constructions are particularly useful in light of the fact we prove that the presence of a cyclic structure reduces the complexity of testing DNA codes for secondary structure formation.

1 Introduction

The field of DNA-based computation was established in a seminal paper by Adleman [2], in which he described an experiment involving the use of DNA molecules to solve a specific instance of the directed travelling salesman problem. DNA sequences within living cells of eukaryotic species appear in double helices (alternatively, *duplexes*), in which one strand of nucleotides is chemically attached to its complementary strand. However, in DNA-based computation, only relatively short *single-stranded* DNA sequences, referred to as *oligonucleotides*, are used. The computing process simply consists of allowing these oligonucleotide strands

---

* This work was supported in part by a research grant from the Natural Sciences and Engineering Research Council (NSERC) of Canada. Portions of this work were presented at the 2005 IEEE International Symposium on Information Theory (ISIT'05) held in Adelaide, Australia.

Ø. Ytrehus (Ed.): WCC 2005, LNCS 3969, pp. 100–119, 2006.

to self-assemble to form long DNA molecules via the process of *hybridization*.
Hybridization is the process in which oligonucleotides with long regions of complementarity bond with each other. The astounding parallelism of biochemical
reactions makes a DNA computer capable of parallel-processing information on
an enormously large scale. However, despite its enormous potential, DNA-based
computing is unlikely to completely replace electronic computing, due to the
inherent unreliability of biochemical reactions, as well as the sheer speed and
flexibility of silicon-based devices [30]. Nevertheless, there exist special applications for which they may represent an attractive alternative or the only available
option for future development. These include cell-based computation systems for
cancer diagnostics and treatment [3], and ultra-high density storage media [17].
Such applications require the design of oligonucleotide sequences that allow for
operations to be performed on them with a high degree of reliability.

The process of self-assembly in DNA computing requires the oligonucleotide
strands (codewords) participating in the computation to selectively hybridize in
a manner compatible with the goals of the computation. If the codewords are
not chosen appropriately, unwanted (non-selective) hybridization may occur. For
many applications, even more detrimental is the fact that an oligonucleotide sequence may *self-hybridize*, *i.e.*, fold back onto itself, forming a secondary structure which prevents the sequence from participating in the computation process
altogether[1]. For example, a large number of read-out failures in the DNA storage
system described in [17] was attributed to the formation of *hairpins*, a special
secondary structure formed by oligonucleotide sequences. The number of computational errors in a DNA system designed for solving an instance of a 3-SAT
problem [5] were reduced by generating DNA sequences that avoid folding and
undesired hybridization phenomena. Similar issues were reported in [4], where a
DNA-based computer was used for breaking the Digital Encryption Standard.

Even if hybridization can be made error-free and no detrimental folding of
sequences occurs, there remain other reliability issues to be dealt with. One
such issue is DNA duplex stability [6],[18]: here, a hybridized pair of sequences
has to remain in a duplex formation for a sufficiently long period of time in order
for the extraction and sequence "sifting" processes to be performed accurately.
It was observed in [6] that the stability of duplexes depends on the combinatorial
structure of the sequences, more precisely, on the combination of adjacent pairs
of bases present in the oligonucleotide strands.

It must be pointed out that the problem of designing sets of codewords that
have properties suitable for DNA computing purposes can be considered to be
partially solved from the computational point of view. There exist many software
packages, such as the Vienna package [29] and the mfold web server [32], that can
predict the secondary structure of a single-stranded DNA (or RNA) sequence.
But such procedures can often be computationally expensive when large numbers
of sequences are sought, or if the sequences are long. Furthermore, they do not

---

[1] This is not a problem with all DNA-based systems; there exist DNA-based computer
logic circuits for which specific folding patterns are actually required by the system
architecture itself [25].

provide any insight into the combinatorial nature of the problems at hand. Such insight is extremely valuable from the perspective of functional genomics, for which one of the outstanding principles is that the folding structure of a sequence is closely related to its biological function [7].

Until now, the focus of coding for DNA computing [1],[8],[10],[14],[18],[23] was on constructing large sets of DNA codewords with fixed base frequencies (constant GC-content) and prescribed minimum distance properties. When used in DNA computing experiments, such sets of codewords are expected to lead to very rare hybridization errors. The largest families of linear codes avoiding hybridization errors were described in [10], while bounds on the size of such codes were derived in [18] and [14]. As an example, it was shown in [10] that there exist 94595072 codewords of length 20 with minimum Hamming distance $d = 5$ and with exactly 10 **G/C** bases. In comparison, without disclosing their design methods, Shoemaker *et al.* reported [24] the existence of only 9105 DNA sequences of length 20, at Hamming distance at least 5, free of secondary structure at temperatures of $61 \pm 5\ ^oC$. Since ambient temperature and chemical composition have a significant influence on the secondary structure of oligonucleotides, it is possible that this number is even smaller for other environmental parameters.

The aim of this paper is to provide a broad description of the kinds of problems that arise in coding for DNA computing, and in particular, to stress the fact that DNA code design must take secondary structure considerations into account. We provide the necessary biological background and terminology in Section 2 of the paper. Section 3 contains a detailed description of the secondary structure considerations that must go into the design of DNA codes. By studying the well-known Nussinov-Jacobson algorithm for secondary structure prediction, we show how the presence of a cyclic structure in a DNA code reduces the complexity of the problem of testing the codewords for secondary structure. We also use the algorithm to argue that imposing constraints on the number of complementary base pair matches between a DNA sequence and some of its shifts could inhibit the occurrence of sequence folding. In Section 4, consider the enumeration of sequences satisfying some of these shift constraints. Finally, in Section 5, we provide a sampling of techniques for constructing cyclic DNA codes with properties that are believed to limit non-selective hybridization and/or self-hybridization. Among the many possible approaches for code design, those resulting in large families with simple descriptions are pursued.

2   Background and Notation

We start by introducing some basic definitions and concepts relating to DNA sequences. The oligonucleotide[2] sequences used for DNA computing are *oriented* words over a four-letter alphabet, consisting of four *bases* — two *purines*, adenine (**A**) and guanine (**G**), and two *pyrimidines*, thymine (**T**) and cytosine (**C**). A

---

[2] Usually, the word 'oligonucleotide' refers to single-stranded nucleotide chains consisting of a few dozen bases; we will however use the same word to refer to single-stranded DNA sequences composed of any number of bases.

DNA strand is oriented due to the asymmetric structure of the sugar-phosphate backbone. It is standard to designate one end of a strand as $3'$ and the other as $5'$, according to the number of the free carbon molecule. Only strands of opposite orientation can hybridize to form a stable duplex. A *DNA code* is simply a set of (oriented) sequences over the alphabet $Q = \{\mathbf{A}, \mathbf{C}, \mathbf{G}, \mathbf{T}\}$.

Each purine base is the *Watson-Crick complement* of a unique pyrimidine base (and *vice versa*) — adenine and thymine form a complementary pair, as do guanine and cytosine. We describe this using the notation $\overline{\mathbf{A}} = \mathbf{T}$, $\overline{\mathbf{T}} = \mathbf{A}$, $\overline{\mathbf{C}} = \mathbf{G}$, $\overline{\mathbf{G}} = \mathbf{C}$. The chemical ties between the two WC pairs are different — $\mathbf{C}$ and $\mathbf{G}$ pair through three hydrogen bonds, while $\mathbf{A}$ and $\mathbf{T}$ pair through two hydrogen bonds. We will assume that hybridization only occurs between complementary base pairs, although certain semi-stable bonds between mismatched pairs form relatively frequently due to biological mutations.

Let $\mathbf{q} = q_1 q_2 \ldots q_n$ be a word of length n over the alphabet Q. For $1 \leq i \leq j \leq n$, we will use the notation $\mathbf{q}_{[i,j]}$ to denote the subsequence $q_i q_{i+1} \ldots q_j$. Furthermore, the sequence obtained by reversing $\mathbf{q}$, *i.e.*, the sequence $q_n q_{n-1} \ldots q_1$, will be denoted by $\mathbf{q}^R$. The *Watson-Crick complement*, or *reverse-complement*, of $\mathbf{q}$ is defined to be $\mathbf{q}^{RC} = \overline{q_n}\, \overline{q_{n-1}} \ldots \overline{q_1}$, where $\overline{q_i}$ denotes the Watson-Crick complement of q_i. For any pair of length-n words $\mathbf{p} = p_1 p_2 \ldots p_n$ and $\mathbf{q} = q_1 q_2 \ldots q_n$ over the alphabet Q, the Hamming distance $d_H(p, q)$ is defined as usual to be the number of positions i at which $p_i \neq q_i$. We further define the *reverse Hamming distance* between the words $\mathbf{p}$ and $\mathbf{q}$ to be $d_H^R(\mathbf{p}, \mathbf{q}) = d_H(\mathbf{p}, \mathbf{q}^R)$. Similarly, their *reverse-complement Hamming distance* is defined to be $d_H^{RC}(\mathbf{p}, \mathbf{q}) = d_H(\mathbf{p}, \mathbf{q}^{RC})$. For a DNA code $\mathcal{C}$, we define its minimum (Hamming) distance, minimum reverse (Hamming) distance, and minimum reverse-complement (Hamming) distance in the obvious manner:

$$d_H(\mathcal{C}) = \min_{\mathbf{p}, \mathbf{q} \in \mathcal{C}, \mathbf{p} \neq \mathbf{q}} d_H(\mathbf{p}, \mathbf{q}), \qquad d_H^R(\mathcal{C}) = \min_{\mathbf{p}, \mathbf{q} \in \mathcal{C}} d_H^R(\mathbf{p}, \mathbf{q})$$

$$d_H^{RC}(\mathcal{C}) = \min_{\mathbf{p}, \mathbf{q} \in \mathcal{C}} d_H^{RC}(\mathbf{p}, \mathbf{q})$$

We also extend the above definitions of sequence complements, reversals, d_H, d_H^R and d_H^{RC} to sequences and codes over an arbitrary alphabet $\mathcal{A}$, for an appropriately defined complementation map from $\mathcal{A}$ onto $\mathcal{A}$. For example, for $\mathcal{A} = \{0, 1\}$, we define complementation as usual via $\overline{0} = 1$ and $\overline{1} = 0$.

Hybridization between a pair of distinct DNA sequences is referred to as *cross-hybridization*, to distinguish it from self-hybridization or sequence folding. The distance measures defined above come into play when evaluating cross-hybridization properties of DNA words under the assumption of a perfectly rigid DNA backbone. As an example, consider two DNA codewords $3' - \mathbf{AAGCTA} - 5'$ and $3' - \mathbf{ATGCTA} - 5'$ at Hamming distance one from each other. For such a pair of codewords, the reverse complement of the first codeword, namely $3' - \mathbf{TAGCTT} - 5'$, will show a very large affinity to hybridize with the second codeword. In order to prevent such a possibility, one could impose a minimum Hamming distance constraint, $d_H(\mathcal{C}) \geq d_{\min}$, for some sufficiently large value of $d_{\min}$. On the other hand, in order to prevent unwanted hybridization between two

DNA codewords, one could try to ensure that the reverse-complement distance between all codewords is larger then a prescribed threshold, i.e. $d^{RC}(\mathcal{C}) \geq d_{min}^{RC}$. Indeed, if the reverse-complement distance between two codewords is small, as for example in the case of the DNA strands $3' - \textbf{AAGCTA} - 5'$ and $3' - \textbf{TACCTT} - 5'$, then there is a good chance that the two strands will hybridize.

Hamming distance is not the only measure that can be used to assess DNA cross-hybridization patterns. For example, if the DNA sugar-phosphate backbone is taken to be a perfectly elastic structure, then it is possible for bases not necessarily at the same position in two strands to pair with each other. Here, it is assumed that bases not necessarily at the same position in two strands can pair with each other. For example, consider the two sequences $3' - \textbf{A}_1^{(1)}\textbf{A}_2^{(1)}\textbf{C}_1^{(1)}\textbf{C}_2^{(1)}\textbf{A}_3^{(1)}\textbf{G}_1^{(1)}\textbf{A}_4^{(1)}\textbf{A}_5^{(1)} - 5'$ and $3' - \textbf{G}_3^{(2)}\textbf{G}_2^{(2)}\textbf{T}_3^{(2)}\textbf{T}_2^{(2)}\textbf{A}_1^{(2)}\textbf{G}_2^{(2)}\textbf{G}_1^{(2)}\textbf{T}_1^{(2)} - 5'$. Under the "perfectly elastic backbone" model, hybridization between the subsequences of not necessarily consecutive bases, $3' - \textbf{A}_1^{(1)}\textbf{C}_1^{(1)}\textbf{C}_2^{(1)}\textbf{A}_3^{(1)}\textbf{A}_4^{(1)} - 5'$ and $5' - \textbf{T}_1^{(2)}\textbf{G}_1^{(2)}\textbf{G}_2^{(2)}\textbf{T}_2^{(2)}\textbf{T}_3^{(2)} - 3'$, is plausible. The relevant distance measure for this model is the *Levenshtein distance* [15], which for a pair of sequences $\textbf{p}$ and $\textbf{q}$, is defined to be smallest number, $d_L(\textbf{p}, \textbf{q})$, of insertions and deletions needed to convert $\textbf{p}$ to $\textbf{q}$. A study of DNA codes with respect to this metric can be found in [8]. The recent work of D'yachkov *et al.* [9] considers a distance measure that is a slight variation on the Levenshtein metric, and seems to fit better in the DNA coding context than the Hamming or Levenshtein metrics.

Another important code design consideration linked to the process of oligonucleotide hybridization pertains to the GC-content of sequences in a DNA code. The *GC-content*, $w_{GC}(\textbf{q})$, of a DNA sequence $\textbf{q} = q_1q_2 \ldots q_n$ is defined to be the number of indices i such that $q_i \in \{\textbf{G}, \textbf{C}\}$. A DNA code in which all codewords have the same GC-content, w, is called a *constant GC-content code*. The constant GC-content requirement assures similar thermodynamic characteristics for all codewords, and is introduced in order to ensure that all hybridization operations take place in parallel, *i.e.*, roughly at the same time. The GC-content is usually required to be in the range of 30–50% of the length of the code.

One other issue associated with hybridization that we will mention is that of the stability of the resultant DNA duplexes. The duplexes formed during the hybridization phase of the computation process must remain paired for the entire duration of the long "post-processing" phase in which the sequences are extracted and sifted through to determine the result of the computation. As observed in [6], the stability of DNA duplexes depends closely on the sequence of bases in the individual strands; thus, it should be possible to take duplex stability into account while designing DNA codes. We will, however, not touch upon this topic further in this paper.

3   Secondary Structure Considerations

Probably the most important criterion in designing codewords for DNA computing purposes is that the codewords should not form secondary structures that

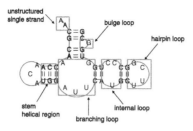

Fig. 1. DNA/RNA secondary structure model (reprinted from [19])

cause them to become computationally inactive. A secondary structure is formed by a chemically active oligonucleotide sequence folding back onto itself by complementary base pair hybridization. As a consequence of the folding, elaborate spatial structures are formed, the most important components of which are loops (including branching, internal, hairpin and bulge loops), stem helical regions, as well as unstructured single strands[3]. Figure 1 illustrates these structures for an RNA strand[4]. It has been shown experimentally that the most important factors influencing the secondary structure of a DNA sequence are the number of base pairs in stem regions, the number of base pairs in a hairpin loop region as well as the number of unpaired bases.

For a collection of interacting entities, one measure commonly used for assessing the system's property is the free energy. The stability and form of a secondary configuration is usually governed by this energy, the general rule-of-thumb being that a secondary structure minimizes the free energy associated with a DNA sequence. The free energy of a secondary structure is determined by the energy of its constituent pairings, and consequently, its loops. Now, the energy of a pairing depends on the bases involved in the pairing as well as all bases adjacent to it. Adding complication is the fact that in the presence of other neighboring pairings, these energies change according to some nontrivial rules.

Nevertheless, some simple dynamic programming techniques can be used to *approximately* determine base pairings in a secondary structure of a oligonucleotide DNA sequence. Among these techniques, the Nussinov-Jacobson (NJ) folding algorithm [22] is one of the simplest and most widely used schemes.

3.1 The Nussinov-Jacobson Algorithm

The NJ algorithm is based on the assumption that in a DNA sequence $q_1 q_2 \ldots q_n$, the energy of interaction, $\alpha(q_i, q_j)$, between the pair of bases (q_i, q_j) is independent

---

[3] We do not consider more complicated structures such as the so-called "pseudoknots"; the general problem of determining secondary structure including pseudoknots is known to be NP-complete.

[4] Oligonucleotide DNA sequences are structurally very similar to RNA sequences, which are by their very nature single-stranded, and consist of the same bases as DNA strands, except for thymine being replaced by uracil (**U**).

of all other base pairs. The interaction energies $\alpha(q_i, q_j)$ are negative quantities whose values usually depend on the actual choice of the base pair (q_i, q_j). One frequently used set of values for RNA sequences is [7]

$$\alpha(q_i, q_j) = \begin{cases} -5 & \text{if } (q_i, q_j) \in \{(\mathbf{G}, \mathbf{C}), (\mathbf{C}, \mathbf{G})\} \\ -4 & \text{if } (q_i, q_j) \in \{(\mathbf{A}, \mathbf{T}), (\mathbf{T}, \mathbf{A})\} \\ -1 & \text{if } (q_i, q_j) \in \{(\mathbf{G}, \mathbf{T}), (\mathbf{T}, \mathbf{G})\}. \end{cases}$$

The value of -1 used for the pairs $(\mathbf{G}, \mathbf{T})$ and $(\mathbf{T}, \mathbf{G})$ indicates a certain frequency of bonding between these mismatched pairs. We will, however, focus our attention only on pairings between Watson-Crick complements. In addition, in order to simplify the discussion, we will restrict our attention to a uniform interaction energy model with $\alpha(q_i, q_j) = -1$ whenever q_i and q_j are Watson-Crick complements and $\alpha(q_i, q_j) = 0$ otherwise.

Let $E_{i,j}$ denote the minimum free energy of the subsequence $q_i \ldots q_j$. The independence assumption allows us to compute the minimum free energy of the sequence $q_1 q_2 \ldots q_n$ through the recursion

$$E_{i,j} = \min \begin{cases} E_{i+1,j-1} + \alpha(q_i, q_j), \\ E_{i,k-1} + E_{k,j}, & i < k \leq j, \end{cases} \tag{1}$$

where $E_{i,i} = E_{i,i-1} = 0$ for $i = 1, 2, ..., n$. The value of $E_{1,n}$ is the minimum free energy of a secondary structure of $q_1 q_2 \ldots q_n$. Note that $E_{1,n} \leq 0$. A large negative value for the free energy, $E_{1,n}$, of a sequence is a good indicator of the presence of a secondary structure in the physical DNA sequence.

The NJ algorithm can be described in terms of free-energy tables, an example of which is shown in Figure 2. In a free-energy table, the entry at position (i, j) (the top left position being $(1, 1)$), contains the value of $E_{i,j}$. The table is filled out by initializing the entries on the main diagonal and on the first lower sub-diagonal of the matrix to zero, and calculating the energy levels according to the recursion in (1). The calculations proceed successively through the upper diagonals: entries at positions $(1, 2), (2, 3), ..., (n - 1, n)$ are calculated first, followed by entries at positions $(1, 3), (2, 4), ..., (n - 2, n)$, and so on. Note that the entry at $(i, j), j > i$, depends on $\alpha(i, j)$ and the entries at $(i, l), l = i, \ldots, j-1, (l, j), l = i+1, \ldots, n-1$, and $(i + 1, j - 1)$. The complexity of the NJ algorithm is $O(n^3)$, since each of the $O(n^2)$ entries requires $O(n)$ computations [19].

The minimum-energy secondary structure itself can be found by the *backtracking algorithm* [22] which retraces the steps of the NJ algorithm (for a description of the backtracking algorithm, the reader is referred to [19]). Figure 2 shows the minimum-energy structure of the sequence **GGGAAATCC**, as determined by the backtracking algorithm. The trace-back path through the free-energy table is indicated by the boldface entries in the table.

From a DNA code design point of view, it would be of considerable interest to determine a set of amenable properties that oligonucleotide sequences should possess so as to either facilitate testing for secondary structure, or exhibit a very low probability for forming such a structure. We next make some straightforward, yet important, observations about the NJ algorithm that provide us with some guidelines for DNA code design.

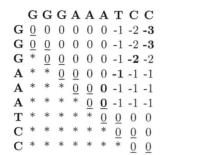

	G	G	G	A	A	A	T	C	C
G	0	0	0	0	0	0	-1	-2	-3
G	0	0	0	0	0	0	-1	-2	-3
G	*	0	0	0	0	0	-1	-2	-2
A	*	*	0	0	0	0	-1	-1	-1
A	*	*	*	0	0	0	-1	-1	-1
A	*	*	*	*	0	0	-1	-1	-1
T	*	*	*	*	*	0	0	0	0
C	*	*	*	*	*	*	0	0	0
C	*	*	*	*	*	*	*	0	0

Fig. 2. Free-energy table for the sequence **GGGAAATCC**, along with its secondary structure as obtained by backtracking through the table

3.2   Testing for Secondary Structure

One design principle that arises out of a study of the NJ algorithm is that DNA codes should contain a cyclic structure. The key idea behind this principle is based on the observation that once the free-energy table, and consequently, the minimum free energy of a DNA sequence $\mathbf{q}$ has been computed, the corresponding computation for any cyclic shift of $\mathbf{q}$ becomes easy. This idea is summarized in the following proposition.

Proposition 1. *The overall complexity of computing the free-energy tables of a DNA codeword $q_1 q_2 \ldots q_n$ and* all *of its cyclic shifts is $O(n^3)$.*

Sketch of Proof. It is enough to show that the free-energy table of the cyclic shift $\mathbf{q}^* = q_n q_1 \ldots q_{n-1}$ can be obtained from the table of $\mathbf{q} = q_1 \ldots q_n$ in $O(n^2)$ steps. The sets of subsequences contained within the positions $1, \ldots, n-1$ of $\mathbf{q}$ and within the positions $2, \ldots, n$ of $\mathbf{q}^*$ are the same. This implies that only entries in the first row of the energy table of $\mathbf{q}^*$ have to be computed. Computing each entry in the first row involves $O(n)$ operations, resulting in a total complexity of $O(n^2)$. □

The above result shows that the complexity of testing a DNA code with M length-n codewords for secondary structure is reduced from $O(Mn^3)$ to $O(Mn^2)$, if the code is cyclic. It is also worth pointing out that a cyclic code structure can also simplify the actual production of the DNA sequences that form the code.

Example 1. The minimal free energies of the sequence shown in Figure 3(a) and all its cyclic shifts lie in the range -0.24 to -0.41 kcal/mol. None of these sequences has a secondary structure. On the other hand, for the sequence in Figure 3(b), all its cyclic shifts have a secondary structure, and the minimal free energies are in the range -1.05 to -1.0 kcal/mol. The actual construction of these sequences is described in Example 3 in Section 5.2. Their secondary structures have been determined using the Vienna RNA/DNA secondary structure package [29], which is based on the NJ algorithm, but which uses more accurate values for the parameters $\alpha(q_i, q_j)$, as well as sophisticated prediction methods for base pairing probabilities.

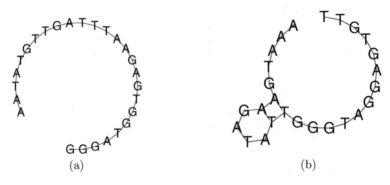

Fig. 3. Secondary structures of two DNA codewords at a temperature of $37^\circ C$

3.3   Avoiding Formation of Secondary Structure

While testing DNA sequences for secondary structure is one aspect of the code design process, it is equally important to know how to design codewords that have a low tendency to form secondary structures. The obvious approach here would be to identify properties of the sequence of bases in an oligonucleotide that would encourage secondary structure formation, so that we could then try to construct codewords which do not have those properties. For example, it seems intuitively clear that if a sequence $\mathbf{q}$ has long, non-overlapping segments $\mathbf{s}_1$ and $\mathbf{s}_2$ such that $\mathbf{s}_1 = \mathbf{s}_2^{RC}$, then there is a good chance that $\mathbf{q}$ will fold to enable $\mathbf{s}_1$ to bind with $\mathbf{s}_2^R$ thus forming a stable structure. Actually, we can slightly strengthen the above condition for folding by requiring that $\mathbf{s}_1$ and $\mathbf{s}_2$ be spaced sufficiently far apart, since a DNA oligonucleotide usually does not make sharp turns, *i.e.*, does not bend over small regions. In any case, the logic is that a sequence that avoids such a scenario should not fold. Unfortunately, this is not quite true: it is not necessarily the longest regions of reverse-complementarity in a sequence that cause a secondary structure to form, as demonstrated by the example in Figure 4. The longest regions of reverse-complementarity in the sequence in the figure are actually the segments of length 7 at either end, which do not actually hybridize with each other within the secondary structure.

A subtler approach to finding properties that inhibit folding consists of identifying components of secondary structures that have a destabilizing effect on the structure. Since the DNA sugar-phosphate backbone is a semi-rigid

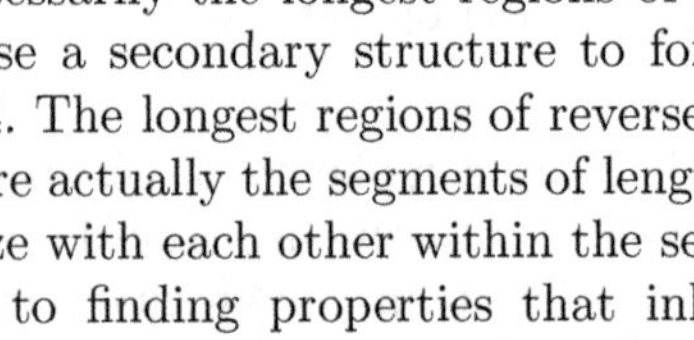

Fig. 4. Secondary structure of the sequence **CGTAA...TTACG**

structure, it is reasonable to expect that long loops (especially hairpin loops) tend to destabilize a secondary structure, unless they are held together by an even longer string of stacked base-pairs, which is an unlikely occurrence.

To identify what could induce a hairpin loop to form in a DNA sequence, we enlist the help of the free-energy tables from the NJ algorithm. As an illustrative example, consider the table and corresponding secondary structure in Figure 2. The secondary structure consists of three stacked base-pairs, and a hairpin loop involving two **A**'s. The three stacked base-pairs correspond to the three diagonal steps ($-3 \to -2$, $-2 \to -1$ and $-1 \to 0$) made in the trace-back path indicated by boldface entries in the table; the hairpin loop corresponds to the vertical segment formed by the two 0's in the trace-back path. In general, a vertical segment involving m '0' entries from the first m upper diagonals indicates the presence of a hairpin loop of length m. For sufficiently large m, such a loop would have a destabilizing effect on any nearby stacked base-pairs, leading to an unravelling of the overall structure.

Thus, if the first m upper diagonals of the free-energy table of a DNA sequence $\mathbf{q} = q_1 q_2 \ldots q_n$ contain only zero-valued entries, then a hairpin loop of size m is necessarily present in the secondary structure. Consequently, it is very likely that even if base pairing is possible, the overall structure will be unstable[5]. It is easy to verify that the first m upper diagonals in the free-energy table contain only zeros if and only if $\mathbf{q}$ and any of its first $m - 1$ shifts contain no complementary base pairs at the same positions, i.e., $q_i \neq \overline{q_{i+j}}$ for $1 \leq j \leq m - 1$ and $1 \leq i \leq n - j$.

Relaxing the above argument a little, we see that from the stand-point of designing DNA codewords without secondary structure, it is desirable to have codewords for which the sums of the elements on each of the first few diagonals in their free-energy tables are either all zero or of some very small absolute value. This requirement can be rephrased in terms of requiring a DNA sequence to satisfy a "shift property", in which a sequence and its first few shifts have few or no complementary base pairs at the same positions.

In the following section, we define a shift property of a sequence more rigorously, and provide some results on the enumeration of DNA sequences satisfying certain shift properties.

4 Enumerating DNA Sequences Satisfying a Shift Property

Recall that for $q \in Q = \{\mathbf{A}, \mathbf{C}, \mathbf{G}, \mathbf{T}\}$, $\bar{q}$ denotes the Watson-Crick complement of q.

Definition 1. *Given a DNA sequence* $\mathbf{q} = q_1 q_2 \ldots q_n$, *we define for* $0 \leq i \leq n - 1$, *the* ith *matching number,* $\mu_i(\mathbf{q})$, *of* $\mathbf{q}$ *to be the number of indices* $\ell \in \{1, 2, \ldots, n - i\}$ *such that* $q_\ell = \overline{q_{i+\ell}}$.

---

[5] The no sharp turn constraint implies that one can restrict its attention only to the fifth, sixth, ..., m-th upper diagonals, but for reasons of simplicity, we will consider only the previously described scenario.

A shift property of $\mathbf{q}$ *is any sort of restriction imposed on the matching numbers* $\mu_i(\mathbf{q})$.

Enumerating sequences having various types of shift properties is useful because doing so yields upper bounds on the size of DNA codes whose codewords satisfy such properties. We present a few such combinatorial results here.

Given $s \geq 1$, let $g_s(n)$ denote the number of sequences, $\mathbf{q}$, of length n for which $\mu_i(\mathbf{q}) = 0$, $i = 1, ..., s$. For $n \leq s$, we take $g_s(n)$ to be $g_{n-1}(n)$.

Lemma 2. *For all* $n > 1$, $g_{n-1}(n) = 4(2^n - 1)$.

Proof. It is clear that a DNA sequence is counted by $g_{n-1}(n)$ iff it contains no pair of complementary bases. Such a sequence must be over one of the alphabets $\{\mathbf{A}, \mathbf{G}\}$, $\{\mathbf{A}, \mathbf{C}\}$, $\{\mathbf{T}, \mathbf{G}\}$ and $\{\mathbf{T}, \mathbf{C}\}$. There are $4(2^n - 1)$ such sequences, since there are 2^n sequences over each of these alphabets, of which $\mathbf{A}^n$, $\mathbf{T}^n$, $\mathbf{G}^n$ and $\mathbf{C}^n$ are each counted twice. $\qquad\square$

Lemma 3. *For all* $n > s$,

$$g_s(n) = 2g_s(n - 1) + g_s(n - s).$$

Proof. Let $\mathcal{G}_s(n)$ denote the set of all sequences $\mathbf{q}$ of length n for which $\mu_i(\mathbf{q}) = 0$, $i = 1, ..., s$. Thus, $|\mathcal{G}_s(n)| = g_s(n)$. Note that for any $\mathbf{q} \in \mathcal{G}_s(n)$, $\mathbf{q}_{[n-s,n]}$ cannot contain a complementary pair of bases, and hence cannot contain three distinct bases. Let $\mathcal{E}(n)$ denote the set of sequences $q_1 q_2 \dots q_n \in \mathcal{G}_s(n)$ such that $q_{n-s+1} = q_{n-s+2} = \cdots = q_n$, and let $\mathcal{U}(n) = \mathcal{G}_s(n) \setminus \mathcal{E}(n)$. We thus have $|\mathcal{E}(n)| + |\mathcal{U}(n)| = g_s(n)$. Each sequence in $\mathcal{E}(n)$ is obtained from some sequence $q_1 q_2 \dots q_{n-s+1} \in \mathcal{G}_s(n - s + 1)$ by appending $s - 1$ bases, $q_{n-s+2}, \dots, q_n$, all equal to q_{n-s+1}. Hence, $|\mathcal{E}(n)| = |\mathcal{G}_s(n - s + 1)| = g_s(n - s + 1)$, and therefore, $|\mathcal{U}(n)| = g_s(n) - g_s(n - s + 1)$.

Now, observe that each sequence $q_1 q_2 \dots q_n \in \mathcal{G}_s(n)$ is obtained by appending a single base, q_n, to some sequence $q_1 q_2 \dots q_{n-1} \in \mathcal{G}_s(n - 1)$. If $q_1 q_2 \dots q_{n-1}$ is in fact in $\mathcal{E}(n - 1)$, then there are three choices for q_n. Otherwise, if $q_1 q_2 \dots q_{n-1} \in \mathcal{U}(n - 1)$, there are only two possible choices for q_n. Hence,

$$\begin{aligned} g_s(n) &= 3\,|\mathcal{E}(n - 1)| + 2\,|\mathcal{U}(n - 1)| \\ &= 3\,g_s(n - s) + 2\,(g_s(n - 1) - g_s(n - s)) \end{aligned}$$

This proves the claimed result. $\qquad\square$

From Lemmas 2 and 3, we obtain the following result.

Theorem 4. *The generating function* $G_s(z) = \sum_{z=1}^{\infty} g_s(n)z^{-n}$ *is given by*

$$G_s(z) = 4 \cdot \frac{z^{s-1} + z^{z-2} + \cdots + z + 1}{z^s - 2z^{s-1} - 1}.$$

It can be shown that for $s > 1$, the polynomial $\psi_s(z) = z^s - 2z^{s-1} - 1$ in the denominator of $G_s(z)$ has a real root, ρ_s, in the interval $(2,3)$, and $s - 1$ other roots within the unit circle. It follows that $g_s(n) \sim \beta_s(\rho_s)^n$ for some constant $\beta_s > 0$. It is easily seen that ρ_s decreases as s increases, and that $\lim_{s \to \infty} \rho_s = 2$.

Theorem 5. *Given an $s \in \{1, 2, \ldots, n-1\}$, the number of length-$n$ DNA sequences q such that $\mu_s(q) = m$, is $\binom{n-s}{m} 4^s 3^{n-s-m}$.*

Proof. Let $B_s(n, m)$ be the set of length-n DNA sequences q such that $\mu_s(q) = m$. A sequence $q = q_1 q_2 \ldots q_n$ is in $B_s(n, m)$ iff the set $I = \{i : q_i = \overline{q_{i-s}}\}$ has cardinality m. So, to construct such a sequence, we first arbitrarily pick $q_1, q_2, \ldots, q_s$ and an $I \subset \{s+1, s+2, \ldots, n\}$, $|I| = m$, which can be done in $4^s \binom{n-s}{m}$ ways. The rest of q is constructed recursively: for $i \geq s+1$, set $q_i = \overline{q_{i-s}}$ if $i \in I$, and pick a $q_i \neq \overline{q_{i-1}}$ if $i \notin I$. Thus, there are 3 choices for each $i \geq s+1$, $i \notin I$, and hence a total of $\binom{n-s}{m} 4^s 3^{n-s-m}$ sequences q in $B_s(n, m)$.

The enumeration of DNA sequences satisfying any sort of shift property becomes considerably more difficult if we bring in the additional requirement of constant GC-content. The following result can be proved by applying the powerful Goulden-Jackson method of combinatorial enumeration [11, Section 2.8]. The result is a direct application of Theorem 2.8.6 and Lemma 2.8.10 in [11], and the details of the algebraic manipulations involved are omitted.

Theorem 6. *The number of DNA sequences q of length n and GC-content w, such that $\mu_1(q) = 0$, is given by the coefficient of $x^n y^w$ in the (formal) power series expansion of*

$$\Phi(x, y) = \left(1 - \frac{2x}{1+x} - \frac{2xy}{1+xy}\right)^{-1}.$$

5 Some DNA Code Constructions

Having in previous sections described some of the code design problems in the context of DNA computing, we present some sample solutions in this section. We mainly focus on constructions of cyclic codes, since as mentioned earlier, the presence of a cyclic structure reduces the complexity of testing DNA codes for secondary structure formation, and also simplifies the DNA sequence fabrication procedure. We have seen that other properties desirable in DNA codes include large minimum Hamming distance, large minimum reverse-complement distance, constant GC-content, and the shift properties introduced in Sections 3.3 and 4. The codes presented in this section are constructed in such a way as to possess some subset of these properties. There are many such code constructions possible, so we pick some that are easy to describe and result in sufficiently large codes. Due to the restrictions imposed on the code design methods with respect to testing for secondary structure, the resulting codes are sub-optimal with respect to the codeword cardinality criteria [10].

5.1   DNA Codes from Cyclic Reversible Extended Goppa Codes

The use of reversible cyclic codes for the construction of DNA sequences was previously proposed in [1] and [23]. Here, we will follow a more general approach that allows for the construction of large families of DNA codes with a certain guaranteed minimum distance and minimum reverse-complement distance, based on extended Goppa codes over $GF(2^2)$ [28].

Recall that a code $\mathcal{C}$ is said to be reversible if $\mathbf{c} \in \mathcal{C}$ implies that $\mathbf{c}^R \in \mathcal{C}$ [16, p. 206]. It is a well-known fact that a cyclic code is reversible if and only if its generator polynomial $g(z)$ is self-reciprocal, i.e., $z^{deg(g(z))}g(z^{-1}) = \pm g(z)$. Given an $[n, k, d]$ reversible cyclic code, $\mathcal{C}$, over $GF(2^2)$ with minimum distance d, consider the code $\widehat{\mathcal{C}}$ obtained by first eliminating all the self-reversible codewords (i.e., codewords $\mathbf{c}$ such that $\mathbf{c}^R = \mathbf{c}$), and then choosing one half of the remaining codewords such that no codeword and its reverse are selected simultaneously. If r is the number of self-reversible codewords in $\mathcal{C}$, then $\widehat{\mathcal{C}}$ is a nonlinear code with $(4^k - r)/2$ codewords of length n, and furthermore, $d_H(\widehat{\mathcal{C}}) \geq d$ and $d_H^R(\widehat{\mathcal{C}}) \geq d$. The value of r can be determined easily, as shown below.

Proposition 7. *A reversible cyclic code of dimension k over $GF(q)$ contains $q^{\lceil k/2 \rceil}$ self-reversible codewords.*

Proof. If $\mathbf{a} = a_0 a_1 \ldots a_{n-1}$ is a self-reversible codeword, then the polynomial $a(z) = a_0 + a_1 z + \ldots a_{n-1} z^{n-1}$ is self-reciprocal. Let $g(z)$ be the generator polynomial for the code, so that $a(z) = i_a(z)g(z)$ for some polynomial $i_a(z)$ of degree at most $k - 1$. Since $g(z)$ and $a(z)$ are self-reciprocal, so is $i_a(z)$. Hence, $i_a(z)$ is uniquely determined by the coefficients of its $\lceil k/2 \rceil$ least-order terms z^i, $i = 0, 1, \ldots, \lceil k/2 \rceil - 1$, and there are exactly $q^{\lceil k/2 \rceil}$ choices for these coefficients. $\square$

The code $\widehat{\mathcal{C}}$ defined above can be thought of as a DNA code by identifying $GF(2^2)$ with the DNA alphabet $Q = \{\mathbf{A}, \mathbf{C}, \mathbf{G}, \mathbf{T}\}$. Let $\mathcal{D}$ be the code obtained from $\widehat{\mathcal{C}}$ by means of the following simple modification: for each $\mathbf{c} \in \mathcal{C}$, replace each of the first $\lfloor n/2 \rfloor$ symbols of $\mathbf{c}$ by its Watson-Crick complement. It is clear that $\mathcal{D}$ has the same number of codewords as $\widehat{\mathcal{C}}$, and that $d_H(\mathcal{D}) \geq d$ as well. It can also readily be seen that if n is even, then $d_H^{RC}(\mathcal{D}) = d_H^R(\widehat{\mathcal{C}})$, and if n is odd, then $d_H^{RC}(\mathcal{D})$ may be one less than $d_H^R(\widehat{\mathcal{C}})$. In any case, we have $d_H^{RC}(\mathcal{D}) \geq d - 1$.

We apply the above construction to a class of extended Goppa codes that are known to be reversible and cyclic. We first recall the definition of a Goppa code.

Definition 2. *[16, p. 338] Let $\mathcal{L} = \{\alpha_1, ..., \alpha_n\} \subseteq GF(q^m)$, for q a power of a prime and $m, n \in \mathbb{Z}^+$. Let $g(z)$ be a polynomial of degree $\delta < n$ over $GF(q^m)$ such that $g(z)$ has no root in $\mathcal{L}$. The Goppa code, $\Gamma(\mathcal{L})$, consists of all words $(c_1, ..., c_n), c_i \in GF(q)$ such that $\sum_{i=1}^{n} \frac{c_i}{z - \alpha_i} \equiv 0 \mod g(z)$. $\Gamma(\mathcal{L})$ is a code of length n, dimension $k \geq n - m\delta$ and minimum distance $d \geq \delta + 1$.*

The polynomial $g(z)$ in the definition above is referred to as the *Goppa polynomial*. We shall consider Goppa codes derived from Goppa polynomials of the form $g(z) = [(z - \beta_1)(z - \beta_2)]^a$, for some integer a. Two choices for the roots β_1, β_2 and the corresponding *location sets* $\mathcal{L}$ are of interest: (i) $\beta_1, \beta_2 \in GF(q^m)$, $\mathcal{L} = GF(q^m) - \{\beta_1, \beta_2\}$, $n = q^m - 2$; (ii) $\mathcal{L} = GF(q^m)$, with $\beta_1, \beta_2 \in GF(q^{2m})$, such that $\beta_2 = \beta_1^{q^m}$, $\beta_1 = \beta_2^{q^m}$, and $n = q^m$.

It was shown in [28] that for such a choice of $g(z)$ and for an ordering of the location set $\mathcal{L}$ satisfying $\alpha_i + \alpha_{n+1-i} = \beta_1 + \beta_2$, the extended Goppa codes obtained by adding an overall parity check to $\Gamma(\mathcal{L})$ in the above cases are reversible and cyclic. The extended code has the same dimension as $\Gamma(\mathcal{L})$, but the minimum distance is now at least $2a + 2$. Applying the DNA code construction described earlier to such a family of extended Goppa codes over $GF(4)$, we obtain the following theorem.

Theorem 8. *For arbitrary positive integers a, m, there exist cyclic DNA codes $\mathcal{D}$ such that $d_H(\mathcal{D}) \geq 2a + 2$ and $d_H^{RC}(\mathcal{D}) \geq 2a + 1$, having the following parameters:*

(i) length $n = 4^m + 1$, and number of codewords $M \geq \frac{1}{2}(4^{2^{2m} - 2ma} - 4^{2^{2m-1} - ma})$;

(ii) length $n = 4^m - 1$, and number of codewords $M \geq \frac{1}{2}(4^{2^{2m} - 2(ma+1)} - 4^{2^{2m-1} - (ma+1)})$.

Example 2. Let $\mathcal{L} = GF(2^2)$, with $q = 2^2, m = 1$, and let $\beta_1 = \alpha$, $\beta_2 = \alpha^4$, for a primitive element α of $GF(2^4)$. We take the Goppa polynomial to be $g(z) = (z - \beta_1)(z - \beta_2)$, so that $a = 1$. The extended Goppa code over $GF(2^2)$ obtained from these parameters is a code of length 5, dimension 2 and minimum distance 4.

We list out the elements of $GF(2^2)$ as $\{0, 1, \theta, 1 + \theta\}$, and make the identification $0 \leftrightarrow \mathbf{G}$, $1 \leftrightarrow \mathbf{C}$, $\theta \leftrightarrow \mathbf{T}$, $1 + \theta \leftrightarrow \mathbf{A}$, The DNA code $\mathcal{D}$ constructed as outlined in this section has $d_H(\mathcal{D}) = d_H^R(\mathcal{D}) = 4$ and $d_H^{RC}(\mathcal{D}) = 3$, and consists of the following six codewords: **CGTTC, CAAAT, CTCCA, GCCTT, GGAGA, ACTAA**.

5.2   DNA Codes from Generalized Hadamard Matrices

Hadamard matrices have long been used to construct constant-weight [16, Chap. 2] and constant-composition codes [26]. We continue this tradition by providing constructions of cyclic codes with constant GC-content, and good minimum Hamming and reverse-complement distance properties.

A *generalized Hadamard matrix* $H \equiv H(n, \mathbb{C}_m)$ is an $n \times n$ square matrix with entries taken from the set of mth roots of unity, $\mathbb{C}_m = \{e^{-2\pi i \ell/m}, \ell = 0, ..., m-1\}$, that satisfies $HH^* = nI$. Here, I denotes the identity matrix of order n, while $*$ stands for complex-conjugation. We will only concern ourselves with the case $m = p$ for some prime p. A necessary condition for the existence of generalized Hadamard matrices $H(n, \mathbb{C}_p)$ is that $p|n$. The *exponent matrix*, $E(n, \mathbb{Z}_p)$, of $H(n, \mathbb{C}_p)$ is the $n \times n$ matrix with entries in $\mathbb{Z}_p = \{0, 1, 2, \ldots, p - 1\}$, obtained by replacing each entry $(e^{-2\pi i})^\ell$ in $H(n, \mathbb{C}_p)$ by the exponent ℓ.

A generalized Hadamard matrix H is said to be in *standard form* if its first row and column consist of ones only. The $(n-1) \times (n-1)$ square matrix formed by the remaining entries of H is called the *core* of H, and the corresponding submatrix of the exponent matrix E is called the core of E. Clearly, the first row and column of the exponent matrix of a generalized Hadamard matrix in standard form consist of zeros only. It can readily be shown (see *e.g.*, [13]) that the rows of such an exponent matrix must satisfy the following two properties: (i) in each of the nonzero rows of the exponent matrix, each element of $\mathbb{Z}_p$ appears a constant number, n/p, of times; and (ii) the Hamming distance between any two rows is $n(p-1)/p$. We will only consider generalized Hadamard matrices that are in standard form.

Several constructions of generalized Hadamard matrices are known (see [13] and the references therein). A particularly nice general construction is given by the following result from [13].

Theorem 9. [13, Theorem II] *Let* $N = p^k - 1$ *for* p *prime and* $k \in \mathbb{Z}^+$. *Let* $g(x) = c_0 + c_1 x + c_2 x^2 + \dots + c_{N-k} x^{N-k}$ *be a monic polynomial over* $\mathbb{Z}_p$, *of degree* $N - k$, *such that* $g(x)h(x) = x^N - 1$ *over* $\mathbb{Z}_p$, *for some monic irreducible polynomial* $h(x) \in \mathbb{Z}_p[x]$. *Suppose that the vector* $(0, c_0, c_1, \dots, c_{N-k}, c_{N-k+1}, \dots, c_{N-1})$, *with* $c_i = 0$ *for* $N - k < i < N$, *has the property that it contains each element of* $\mathbb{Z}_p$ *the same number of times. Then the* N *cyclic shifts of the vector* $\mathbf{g} = (c_0, c_1, \dots, c_{N-1})$ *form the core of the exponent matrix of some Hadamard matrix* $H(p^k, \mathbb{C}_p)$.

Thus, the core of $E \equiv E(p^k, \mathbb{Z}_p)$ (and hence, $H(p^k, \mathbb{C}_p)$) guaranteed by the above theorem is a circulant matrix consisting of all the $N = p^k - 1$ cyclic shifts of its first row. We refer to such a core as a *cyclic core*. Each element of $\mathbb{Z}_p$ appears in each row of E exactly $(N+1)/p = p^{k-1}$ times, and the Hamming distance between any two rows is exactly $(N+1)(p-1)/p = (p-1)p^{k-1}$. Thus, the N rows of the core of E form a constant-composition code consisting of the N cyclic shifts of some word of length N over the alphabet $\mathbb{Z}_p$, with the Hamming distance between any two codewords being $(p-1)p^{k-1}$.

DNA codes with constant GC-content can obviously be constructed from constant-composition codes over $\mathbb{Z}_p$ by mapping the symbols of $\mathbb{Z}_p$ to the symbols of the DNA alphabet, $Q = \{\mathbf{A}, \mathbf{C}, \mathbf{G}, \mathbf{T}\}$. For example, using the cyclic constant-composition code of length $3^k - 1$ over $\mathbb{Z}_3$ guaranteed by Theorem 9, and using the mapping that takes 0 to $\mathbf{A}$, 1 to $\mathbf{T}$ and 2 to $\mathbf{G}$, we obtain a DNA code $\mathcal{D}$ with $3^k - 1$ codewords and a GC-content of 3^{k-1}. Clearly, $d_H(\mathcal{D}) = 2 \cdot 3^{k-1}$, and in fact, since $\overline{\mathbf{G}} = \mathbf{C}$ and no codeword in $\mathcal{D}$ contains the symbol $\mathbf{C}$, we also have $d_H^{RC}(\mathcal{D}) \geq 3^{k-1}$. We summarize this in the following corollary to Theorem 9.

Corollary 10. *For any* $k \in \mathbb{Z}^+$, *there exist DNA codes* $\mathcal{D}$ *with* $3^k - 1$ *codewords of length* $3^k - 1$, *with constant GC-content equal to* 3^{k-1}, $d_H(\mathcal{D}) = 2 \cdot 3^{k-1}$, $d_H^{RC}(\mathcal{D}) \geq 3^{k-1}$. *and in which each codeword is a cyclic shift of a fixed generator codeword* $\mathbf{g}$.

Example 3. Each of the following vectors generates a cyclic core of a Hadamard matrix [13]:

$$\mathbf{g}^{(1)} = (2220122120200111021121210200),$$

$$\mathbf{g}^{(2)} = (2021221022200101211201110 0).$$

DNA codes can be obtained from such generators by mapping $\{0,1,2\}$ onto $\{\mathbf{A}, \mathbf{T}, \mathbf{G}\}$. Although all such mappings yield codes with (essentially) the same parameters, the actual choice of mapping has a strong influence on the secondary structure of the codewords. For example, the codeword in Figure 3(a) was obtained from $\mathbf{g}^{(1)}$ via the mapping $0 \rightarrow \mathbf{A}$, $1 \rightarrow \mathbf{T}$, $2 \rightarrow \mathbf{G}$, while the codeword in Figure 3(b) was obtained from the same generator $\mathbf{g}^{(1)}$ via the mapping $0 \rightarrow \mathbf{G}$, $1 \rightarrow \mathbf{T}$, $2 \rightarrow \mathbf{A}$.

5.3   Code Constructions Via a Binary Mapping

The problem of constructing DNA codes with some of the properties desirable for DNA computing can be made into a binary code design problem by mapping the DNA alphabet onto the set of length-two binary words as follows:

$$\mathbf{A} \rightarrow 00, \quad \mathbf{T} \rightarrow 01, \quad \mathbf{C} \rightarrow 10, \quad \mathbf{G} \rightarrow 11. \tag{2}$$

The mapping is chosen so that the first bit of the binary image of a base uniquely determines the complementary pair to which it belongs.

Let $\mathbf{q}$ be a DNA sequence. The sequence $b(\mathbf{q})$ obtained by applying coordinatewise to $\mathbf{q}$ the mapping given in (2), will be called the *binary image* of $\mathbf{q}$. If $b(\mathbf{q}) = b_0 b_1 b_2 \ldots b_{2n-1}$, then the subsequence $e(\mathbf{q}) = b_0 b_2 \ldots b_{2n-2}$ will be referred to as the *even subsequence* of $b(\mathbf{q})$, and $o(\mathbf{q}) = b_1 b_3 \ldots b_{2n-1}$ will be called the *odd subsequence* of $b(\mathbf{q})$. Thus, for example, for $\mathbf{q} = \mathbf{ACGTCC}$, we have $b(\mathbf{q}) = 001011011010$, $e(\mathbf{q}) = 011011$ and $o(\mathbf{q}) = 001100$. Given a DNA code $\mathcal{C}$, we define its *even component* $\mathcal{E}(\mathcal{C}) = \{e(\mathbf{p}) : \mathbf{p} \in \mathcal{C}\}$, and its *odd component* $\mathcal{O}(\mathcal{C}) = \{o(\mathbf{p}) : \mathbf{p} \in \mathcal{C}\}$.

It is clear from the choice of the binary mapping that the GC-content of a DNA sequence $\mathbf{q}$ is equal to the Hamming weight of the binary sequence $e(\mathbf{q})$. Consequently, a DNA code $\mathcal{C}$ is a constant GC-content code if and only if its even component, $\mathcal{E}(\mathcal{C})$, is a constant-weight code. Other properties of a DNA code can also be expressed in terms of properties of its even and code components (for example, see Lemma 11 below). Thus if we have binary codes $\mathcal{B}_1$ and $\mathcal{B}_2$ with suitable properties, then we can construct a good DNA code, whose binary image is equivalent to $\mathcal{B}_1 \times \mathcal{B}_2$, that has $\mathcal{B}_1$ and $\mathcal{B}_2$ as its even and odd components. We present two such constructions here.

Construction B1.   Let $\mathcal{B}$ be a binary code consisting of M codewords of length n and minimum distance $d_{\min}$, such that $\mathbf{c} \in \mathcal{B}$ implies that $\bar{\mathbf{c}} \in \mathcal{B}$. For $w > 0$, consider the constant-weight subcode $\mathcal{B}_w = \{\mathbf{u} \in \mathcal{B} : w_H(\mathbf{u}) = w\}$, where $w_H(\cdot)$ denotes Hamming weight. Choose $w > 0$ such that $n \geq 2w + \lceil d_{\min}/2 \rceil$, and consider a DNA code, $\mathcal{C}_w$, with the following choice for its even and odd components:

$$\mathcal{E}_w = \{\mathbf{a}\overline{\mathbf{b}} : \mathbf{a}, \mathbf{b} \in \mathcal{B}_w\}, \qquad \mathcal{O} = \{\mathbf{a}\mathbf{b}^{RC} : \mathbf{a}, \mathbf{b} \in \mathcal{B}, \mathbf{a} <_{\text{lex}} \mathbf{b}\},$$

where $<_{\text{lex}}$ denotes lexicographic ordering. The $\mathbf{a} <_{\text{lex}} \mathbf{b}$ in the definition of $\mathcal{O}$ ensures that if $\mathbf{a}\mathbf{b}^{RC} \in \mathcal{O}$, then $\mathbf{b}\mathbf{a}^{RC} \notin \mathcal{O}$, so that distinct codewords in $\mathcal{O}$ cannot be reverse-complements of each other.

The code $\mathcal{E}_w$ has $|\mathcal{B}_w|^2$ codewords of length $2n$ and constant weight n. Furthermore, $d_H(\mathcal{E}_w) \geq d_{\min}$ and $d_H^R(\mathcal{E}_w) \geq d_{\min}$, the first of these inequalities following from the fact that $\mathcal{B}_w$ is a subset of codewords in $\mathcal{B}$. To prove the second inequality, note that for any two distinct codewords $\mathbf{a}\overline{\mathbf{b}}$ and $\mathbf{c}\overline{\mathbf{d}}$, we have

$$d_H(\mathbf{a}\overline{\mathbf{b}}, \mathbf{d}^{RC}\mathbf{c}^R) = d_H(\mathbf{a}, \mathbf{d}^{RC}) + d_H(\overline{\mathbf{b}}, \mathbf{c}^R) = d_H(\mathbf{a}, \mathbf{d}^{RC}) + d_H(\mathbf{c}, \mathbf{b}^{RC}).$$

Since $\mathbf{b}$ and $\mathbf{d}$ both have weight w, it follows that $\mathbf{b}^{RC}$ and $\mathbf{d}^{RC}$ have weight $n - w$. Due to the constraint on the weight w, we have $d_H(\mathbf{a}, \mathbf{d}^{RC}) \geq \lceil d_{\min}/2 \rceil$, and similarly, $d_H(\mathbf{c}, \mathbf{b}^{RC}) \geq \lceil d_{\min}/2 \rceil$. Therefore, for all $\mathbf{a}, \mathbf{b}, \mathbf{c}, \mathbf{d} \in \mathcal{B}_w$, we must have $d_H(\mathbf{a}\overline{\mathbf{b}}, \mathbf{d}^{RC}\mathbf{c}^R) \geq 2\lceil d_{\min}/2 \rceil \geq d_{\min}$.

The code $\mathcal{O}$ has $M(M-1)/2$ codewords of length $2n$. Clearly, $d_H(\mathcal{O}) \geq d_{\min}$, since the component codewords of $\mathcal{O}$ are taken from $\mathcal{B}$. Similarly, $d_H^{RC}(\mathcal{O}) \geq d_{\min}$, to prove which we only have to observe that for any pair of codewords $\mathbf{a}\mathbf{b}^{RC}$ and $\mathbf{c}\mathbf{d}^{RC}$, $d_H(\mathbf{a}\mathbf{b}^{RC}, \mathbf{d}\mathbf{c}^{RC}) = d_H(\mathbf{a}, \mathbf{d}) + d_H(\mathbf{c}, \mathbf{b}) \geq d_{\min}$.

Therefore, the DNA code

$$\mathcal{C} = \bigcup_{w=d_{\min}}^{w_{\max}} \mathcal{C}_w,$$

with $w_{\max} = (n - \lceil d_{\min}/2 \rceil)/2$, has $\frac{1}{2} M(M-1) \sum_{w=d_{\min}}^{w_{\max}} |A_w|^2$ codewords of length $2n$, and satisfies $d_H(\mathcal{B}) \geq d_{\min}$ and $d_H^{RC}(\mathcal{B}) \geq d_{\min}$.

The following lemma (whose simple proof we omit) records a trivial result that is useful for our next construction. For notational ease, given binary words $\mathbf{x} = (x_i)$ and $\mathbf{y} = (y_i)$, we define $\mathbf{x} \oplus \mathbf{y} = (x_i + y_i)$, the sum being taken modulo-2, and $\mathbf{x} * \mathbf{y} = (x_i y_i)$.

Lemma 11. *Let $\mathbf{q}$ be a length-n sequence over the DNA alphabet Q. For $i \in \{1, 2, \ldots, n-1\}$, defining $\sigma_i = e(\mathbf{q}_{[1,n-i]}) \oplus e(\mathbf{q}_{[i+1,n]})$, and $\tau_i = o(\mathbf{q}_{[1,n-i]}) \oplus o(\mathbf{q}_{[i+1,n]})$, we have*

$$\mu_i(\mathbf{q}) = w_H(\overline{\sigma_i} * \tau_i)$$

where $\overline{\sigma_i}$ denotes the complement of the binary sequence σ_i.

Construction B2. Let $\mathcal{C}$ be the DNA code obtained by choosing the set of non-zero codewords of a cyclic simplex code of length $n = 2^m - 1$ for both the even and odd code components. Recall that a cyclic simplex code of dimension m is a constant-weight code of length $n = 2^m - 1$ and minimum-distance 2^{m-1}, composed of the all-zeros codeword and the n distinct cyclic shifts of any non-zero codeword [12, Chapter 8]. It is clear that the DNA code $\mathcal{C}$ is a cyclic code contains $(2^m - 1)^2$ codewords of length $2^m - 1$ and GC-content 2^{m-1}.

We claim that $\mathcal{C}$ has the property that for all $i \in \{1, 2, \ldots, n-1\}$ and $\mathbf{q} \in \mathcal{C}$, $\mu_i(\mathbf{q}) \leq 2^{m-2}$. To see this, observe first that for any $\mathbf{q} \in \mathcal{C}$, σ_i and τ_i, defined as in Lemma 11, are just truncations of codewords from the simplex code. Since the simplex code is a constant-weight code, with minimum distance 2^{m-1}, each pair of codewords shares exactly 2^{m-2} positions containing 1's. This implies that for each pair of simplex codewords, there are exactly 2^{m-2} positions in which one codeword contains all 1's, while the other contains all 0's. Such positions are precisely what is counted by $w_H(\overline{\sigma}_i * \tau_i)$ in Lemma 11 which proves our claim.

Example 4. Consider the DNA code resulting from Construction B2 using the cyclic simplex code generated by the codeword 1110100. The DNA code contains 49 codewords of length 7. The minimum Hamming distance of the code is 4, and the codewords all have GC-content equal to 4. A selected subset of codewords from this code is listed below:

TGGCTCA, TCCGTGA, CACGGTC, TAGCCTG,

CATGGCT, GATCCGT, GGGAGAA, GGAGAAG.

The last two codewords consist of the bases **G** and **A** only, and clearly satisfy $\mu_i = 0$ for all $i \leq 7$. On the other hand, for the first three codewords we have $\mu_1 = 1$, while for the next three codewords we see that $\mu_1 = 2$ (meeting the upper bound claimed in the construction). Evaluation of this code using the Vienna secondary structure package [29] shows that none of the 49 codewords exhibits a secondary structure.

As a final remark, we note that the problem of constructing a DNA code that can be efficiently tested for secondary structure using the NJ algorithm can also be reformulated in terms of specifications for the even code component. If the even component code is cyclic, and each codeword in the even component is combined with codewords from the odd component, then "approximate" testing can be performed in the following manner. The codeword from the even component code $x_1, \ldots, x_n$, $x_i \in \{0, 1\}$ is tested by the NJ algorithm following the steps outlined in Section 3.1, except that the pairing energies are found according to

$$\alpha(x_i, x_j) = \begin{cases} -1 & \text{if } x_i\, x_j \in \{00, 11\} \\ 0 & \text{if } x_i\, x_j \in \{01, 10\}. \end{cases}$$

The result of the NJ algorithm for the even component codeword represents the worst-case scenario for DNA sequence folding. If the free energy of some even component codeword, $\mathbf{b}$, exceeds a certain threshold (which can be determined by a combination of probabilistic and experimental results), all the DNA sequences $\mathbf{q}$ such that $e(\mathbf{q}) = \mathbf{b}$ are subjected to an additional test by the algorithm. If the free energy of $\mathbf{b}$ is below a given threshold, then one can be reasonably sure that none of the DNA sequences $\mathbf{q}$ that have $\mathbf{b}$ as their even subsequence will form a secondary structure.

References

1. T. Abualrub and A. Ghrayeb, "On the construction of cyclic codes for DNA computing," preprint.
2. L.M. Adleman, "Molecular computation of solutions to combinatorial problems," *Science*, vol. 266, pp. 1021–1024, Nov. 1994.
3. Y. Benenson, B. Gil, U. Ben-Dor, R. Adar and E. Shapiro, "An autonomous molecular computer for logical control of gene expression," *Nature*, vol. 429, pp. 423–429, May 2004.
4. D. Boneh, C. Dunworth, and R. Lipton, "Breaking DES using a molecular computer," *Technical Report CS-TR-489-95*, Department of Computer Science, Princeton University, USA, 1995.
5. R.S. Braich, N. Chelyapov, C. Johnson, P.W.K. Rothemund and L. Adleman, "Solution of a 20-variable 3-SAT problem on a DNA computer," *Science*, vol. 296, pp. 492–502, April 2002.
6. K. Breslauer, R. Frank, H. Blocker, and L. Marky, "Predicting DNA duplex stability from the base sequence," *Proc. Natl. Acad. Sci. USA*, vol. 83, pp. 3746–3750, 1986.
7. P. Clote and R. Backofen, *Computational Molecular Biology – An Introduction*, Wiley Series in Mathematical and Computational Biology, New York, 2000.
8. A. D'yachkov, P.L. Erdös, A. Macula, V. Rykov, D. Torney, C-S. Tung, P. Vilenkin and S. White, "Exordium for DNA codes," *J. Comb. Optim.*, vol. 7, no. 4, pp. 369–379, 2003.
9. A. D'yachkov, A. Macula, T. Renz, P. Vilenkin and I. Ismagilov, "New results on DNA codes," *Proc. IEEE Int. Symp. Inform. Theory (ISIT'05)*, Adelaide, Australia, pp. 283–287, Sept. 2005.
10. P. Gaborit and O.D. King, "Linear constructions for DNA codes," *Theoretical Computer Science*, vol. 334, no. 1-3, pp. 99–113, April 2005.
11. I.P. Goulden and D.M. Jackson, *Combinatorial Enumeration*, Dover, 2004.
12. J.I. Hall, Lecture notes on error-control coding, available online at http://www.mth.msu.edu/~jhall/.
13. I. Heng and C.H. Cooke, "Polynomial construction of complex Hadamard matrices with cyclic core," *Applied Mathematics Letters*, vol. 12, pp. 87–93, 1999.
14. O.D. King, "Bounds for DNA codes with constant GC-content," *The Electronic Journal of Combinatorics*, vol. 10, no. 1, #R33, 2003.
15. V.I. Levenshtein, "Binary codes capable of correcting deletions, insertions, and reversals," *Dokl. Akad. Nauk SSSR*, vol. 163, no. 4, pp. 845–848, 1965 (Russian). English translation in *Soviet Physics Doklady*, vol. 10, no. 8, pp. 707–710, 1966.
16. F.J. MacWilliams and N.J.A. Sloane, *The Theory of Error-Correcting Codes*, North-Holland, Amsterdam, 1977.
17. M. Mansuripur, P.K. Khulbe, S.M. Kuebler, J.W. Perry, M.S. Giridhar and N. Peyghambarian, "Information storage and retrieval using macromolecules as storage media," *University of Arizona Technical Report*, 2003.
18. A. Marathe, A. E. Condon and R. M. Corn, "On combinatorial DNA word design," *J. Comput. Biol.*, vol. 8, pp. 201–219, 2001.
19. S. Mneimneh, "Computational Biology Lecture 20: RNA secondary structures," available online at engr.smu.edu/~saad/courses/cse8354/lectures/lecture20.pdf.
20. O. Milenkovic, "Generalized Hamming and coset weight enumerators of isodual codes," accepted for publication in *Designs, Codes and Cryptography*.

21. O. Milenkovic and N. Kashyap, "DNA codes that avoid secondary structures," *Proc. IEEE Int. Symp. Inform. Theory (ISIT'05)*, Adelaide, Australia, pp. 288–292, Sept. 2005.

22. R. Nussinov and A.B. Jacobson, "Fast algorithms for predicting the secondary structure of single stranded RNA ," *Proc. Natl. Acad. Sci. USA*, vol. 77, no. 11, pp. 6309–6313, 1980.

23. V. Rykov, A.J. Macula, D. Torney and P. White, "DNA sequences and quaternary cyclic codes," in *Proc. IEEE Int. Symp. Inform. Theory (ISIT'01)*, Washington DC, p. 248, June 2001.

24. D.D. Shoemaker, D.A. Lashkari, D. Morris, M. Mittman and R.W. David, "Quantitative phenotye analysis of yeast deletion mutants using a highly parallel molecular bar-coding strategy," *Nature Genetics*, vol. 16, pp. 450–456, Dec. 1996.

25. M.N. Stojanovic, D. Stefanovic, "A deoxyribozyme-based molecular automaton," *Nature Biotechnology* vol. 21, pp. 1069–1074, 2003.

26. M. Svanström, P.R.J. Östergard and G.T. Bogdanova, "Bounds and constructions for ternary constant-composition codes," *IEEE Trans. Inform. Theory*, vol. 48, no. 1, pp. 101–111, Jan. 2002.

27. S. Tsaftaris, A. Katsaggelos, T. Pappas and E. Papoutsakis, "DNA computing from a signal processing viewpoint," *IEEE Signal Processing Magazine*, pp. 100–106, Sept. 2004.

28. K.K. Tzeng and K.P. Zimmermann, "On extending Goppa codes to cyclic codes," *IEEE Trans. Inform. Theory*, vol. IT-21, pp. 712–716, Nov. 1975.

29. The Vienna RNA Secondary Structure Package, `http://rna.tbi.univie.ac.at/cgi-bin/RNAfold.cgi`.

30. E. Winfree, "DNA computing by self-assembly," *The Bridge*, vol. 33, no. 4, pp. 31–38, 2003. Also available online at `http://www.dna.caltech.edu/Papers/FOE_2003_final.pdf`.

31. D.H. Wood, "Applying error correcting codes to DNA computing," in *Proc. 4th Int. Meeting on DNA Based Computers*, 1998, pp. 109–110.

32. M. Zuker, "Mfold web server for nucleic acid folding and hybridization prediction," *Nucleic Acids Res.*, vol. 31, no. 13, pp. 3406–15, 2003. Web access at `http://www.bioinfo.rpi.edu/~zukerm/rna/`.

Open Problems Related to Algebraic Attacks on Stream Ciphers

Anne Canteaut*

INRIA - projet CODES
B.P. 105 - 78153 Le Chesnay cedex, France
Anne.Canteaut@inria.fr

Abstract. The recently developed algebraic attacks apply to all key-stream generators whose internal state is updated by a linear transition function, including LFSR-based generators. Here, we describe this type of attacks and we present some open problems related to their complexity. We also investigate the design criteria which may guarantee a high resistance to algebraic attacks for keystream generators based on a linear transition function.

1 Introduction

In an additive stream cipher, the ciphertext is obtained by adding bitwise the plaintext to a pseudo-random sequence called the keystream. The keystream generator is a finite state automaton whose initial internal state is derived from the secret key and from a public initial value by a key-loading algorithm. At each time unit, the keystream digit produced by the generator is obtained by applying a *filtering function* to the current internal state. The internal state is then updated by a *transition function*. Both filtering function and transition function must be chosen carefully in order to make the underlying cipher resistant to known-plaintext attacks. In particular, the filtering function must not leak too much information on the internal state and the transition function must guarantee that the sequence formed by the successive internal states has a high period.

Stream ciphers are mainly devoted to applications which require either an exceptional encryption rate or an extremely low implementation cost in hardware. Therefore, a linear transition function seems to be a relevant choice as soon as the filtering function breaks the inherent linearity. Amongst all possible linear transition functions, those based on LFSRs are very popular because they are appropriated for low-cost hardware implementations, produce sequences with good statistical properties and can be easily analyzed. LFSR-based generators have been extensively studied. It is known that the involved filtering function must

---
* This work was supported in part by the European Commission through the IST Programme under Contract IST-2002-507932 ECRYPT. The information in this document reflects only the author's views, is provided as is and no warranty is given that the information is fit for any particular purpose. The user thereof uses the information at its sole risk and liability.

Ø. Ytrehus (Ed.): WCC 2005, LNCS 3969, pp. 120–134, 2006.

satisfy some well-defined criteria (such as a high nonlinearity, a high correlation-immunity order,...), and the designers of such generators now provide evidence that their ciphers cannot be broken by the classical attacks.

However, the recent progress in research related to algebraic attacks, introduced by Courtois and Meier [11], seems to threaten all keystream generators based on a linear transition function. In this context, it is important to determine whether such ciphers are still secure or not. Here, we investigate some related open problems, concerning the complexity of algebraic attacks (and of their variants) and concerning the design criteria of LFSR-based stream ciphers which guarantee a high resistance to these cryptanalytic techniques.

2 Basic Principle of Algebraic Attacks

Here, we focus on binary keystream generators based on a linear transition function, which can be described as follows. We denote by $\mathbf{x}_t$ the n-bit internal state of the generator at time t. The filtering function f is first assumed to be a Boolean function of n variables, i.e., at time t the generator outputs only one bit, $s_t = f(\mathbf{x}_t)$. The transition function is supposed to be *linear* and is denoted by $L : \mathbf{F}_2^n \to \mathbf{F}_2^n$. Therefore, we have

$$s_t = f(L^t(\mathbf{x}_0)) ,$$

where $\mathbf{x}_0$ is the initial state. We only consider the case where both the filtering function and the transition function are publicly known, i.e., independent from the secret key. Two popular constructions known as nonlinear filter generators and combination generators fit the previous model.

The basic principle of algebraic attacks goes back to Shannon's work [26, Page 711]: these techniques consist in expressing the whole cipher as a large system of multivariate algebraic equations, which can be solved to recover the secret key. A major parameter which influences the complexity of such an attack is then the degree of the underlying algebraic system. When the transition is linear, any keystream bit can obviously be expressed as a function of degree $\deg(f)$ in the initial state bits. Therefore, it is known for a long time that the filtering function involved in such a stream cipher must have a high degree.

However, as pointed out by Courtois and Meier [11], the keystream generator may be vulnerable to algebraic attacks even if the degree of the algebraic function is high. Actually, the attack applies as soon as there exist relations of low degree between the output and the inputs of the filtering function f. Such relations correspond to low degree multiples of f, i.e., to relations $g(x)f(x) = h(x)$ for some g where h has a low degree. But, it was proved in [21, 24] that, in the case of algebraic attacks over $\mathbf{F}_2$, the existence of any such relation is equivalent to the existence of a low degree *annihilator* of f or of $(1 + f)$, in the sense of Definition 1. Indeed, if $g(x)f(x) = h(x)$ with $\deg(h) \leq d$, we obtain, by multiplying this equation by $f(x)$, that

$$g(x) [f(x)]^2 = h(x)f(x) = g(x)f(x) = h(x) ,$$

leading to $h(x) [1 + f(x)] = 0$.

Definition 1. *Let f be a Boolean function of n variables. The annihilator ideal of f, denoted by $AN(f)$, is the set of all Boolean functions g of n variables such that*

$$g(x)f(x) = 0, \quad \forall x \in \mathbf{F}_2^n .$$

Moreover, for any degree d, we denote by $AN_d(f)$ the set of all annihilators of f with degree at most d:

$$AN_d(f) = \{g \in AN(f), \ \deg(g) \le d\} .$$

Since the keystream bit at time t is defined by $s_t = f \circ L^t(\mathbf{x_0})$, we deduce that:

- if $s_t = 1$, any function g in $AN(f)$ leads to $g \circ L^t(\mathbf{x_0}) = 0$;
- if $s_t = 0$, any function h in $AN(1 + f)$ leads to $h \circ L^t(\mathbf{x_0}) = 0$.

Therefore, if we collect the relations associated to all functions of degree at most d in $AN(f) \cup AN(f + 1)$ for N known keystream bits, we obtain a system of equations of degree d depending on n variables, $x_1, \ldots, x_n$, which correspond to the bits of the initial state:

$$\begin{cases} g \circ L^t(x_1, \ldots, x_n) \ \forall g \in AN_d(f), & \forall \, 0 \le t < N \text{ such that } s_t = 1 \\ h \circ L^t(x_1, \ldots, x_n) \ \forall h \in AN_d(1 + f), \ \forall \, 0 \le t < N \text{ such that } s_t = 0 \end{cases} \quad (1)$$

The n-bit initial state can then be recovered by solving this multivariate polynomial system.

3   Complexity of Algebraic Attacks

Solving a multivariate polynomial system such as (1) is a typical problem studied in computer algebra. In order to get a rough estimate of the complexity of algebraic attacks for determining the suitable parameters for the keystream generator, we only focus on the simplest technique, called *linearization*. It consists in identifying the system with a linear system of $\sum_{i=1}^{d} \binom{n}{i}$ variables, where each product of i bits of the initial state ($1 \le i \le d$) is seen as a new variable. The entire initial state is then recovered by a Gaussian reduction (or by more sophisticated techniques) whose time complexity is roughly

$$\left(\sum_{i=1}^{d} \binom{n}{i} \right)^{\omega} \simeq n^{\omega d} ,$$

where ω is the exponent of the matrix inversion algorithm, i.e., $\omega \simeq 2.37$ [9].

However, the previous estimation of the attack complexity is based on two hypotheses. It is first assumed that almost all monomials of degree d appear in System (1). This clearly corresponds to the worst situation for the attacker, but we can wonder whether some weak choices for the transition function L and for the filtering function f can provide a system involving a small proportion of all possible monomials only, leading to a faster attack.

Open problem 1. *Determine the number of monomials in $x_1, \ldots, x_n$ involved in System (1), depending on the choice of L and f.*

A probably much stronger assumption in the usual complexity estimation is that the system can always be solved: it is usually supposed that the knowledge of

$$N \simeq \frac{2n^d}{d! \, (\dim A_d(f) + \dim A_d(1+f))}$$

keystream bits lead to a system with $\sum_{i=1}^{d} \binom{n}{i}$ linearly independent equations. It then raises the following open issue.

Open problem 2. *Determine the rank of System (1) depending on the choice of functions L and f.*

Obviously, this question has an influence on the number of keystream bits re-quired for the attack. But, a more crucial point is that the attack using equations of degree d may be infeasible even if a huge keystream segment is available. This situation occurs when the system generated by N keystream bits is underdeter-mined for any value of N. A natural related question is to determine whether the equations corresponding to a given annihilator g are different for all keystream bits, i.e., whether there exists some T less than the period of $\{L^t, t \geq 0\}$ such that $g \circ L^T(x) = g(x)$ for all $x \in \mathbf{F}_2^n$. It is clear that such an integer T divides the period of $\{L^t, t \geq 0\}$. This observation leads to the following result when L corresponds to the next-state function of an LFSR.

Proposition 1. *Let L be the next-state function of an LFSR of length n with primitive feedback polynomial. Let g be a Boolean function of n variables. If $2^n - 1$ is a prime, then all functions $g \circ L^t$, for $0 \leq t \leq 2^n - 1$, are distinct.*

But, when $(2^n - 1)$ is not a prime, there always exist filtering functions f such that some of their annihilators $g \in AN(f)$, $g \neq 0$, lead to a sequence $\{g \circ L^t, 0 \leq t \leq 2^n - 1\}$ with a small period, as pointed out in the following toy example.

Example 1. Let us consider the LFSR of length 4 with primitive feedback poly-nomial $P(x) = x^4 + x + 1$ and the 4-variable filtering function f defined by

$$f(x_1, \ldots, x_4) = x_3 + x_4 + x_1 x_2 + x_2 x_3 + x_1 x_2 x_3 + x_1 x_2 x_4 + x_1 x_3 x_4 + x_2 x_3 x_4 \ .$$

Then, the function $g(x_1, \ldots, x_4) = 1 + x_2 + x_3 + x_4 + x_2 x_4 + x_3 x_4$ belongs to $AN(f)$ and it satisfies

$$g \circ L^t(x_1, \ldots, x_4) = g \circ L^{t \bmod 5}(x_1, \ldots, x_4)$$

for all t. Actually, when $\mathbf{F}_2^4$ is identified with the finite field with 16 elements defined by the primitive polynomial P, we have $g(x) = g(x\alpha^5)$, where α is a root of P.

However, when a function g in $AN(f)$ has such a strong periodic structure, this also holds for the filtering function, implying that the keystream can be easily distinguished from a random sequence.

Proposition 2. *Let f be a Boolean function of n variables and let g be a nonzero function in $AN(f) \cup AN(1 + f)$. If $g \circ L^T = g$ for some integer T, then there exists $t_0 < T$ such that all keystream bits $s_{t_0+iT}, i \geq 0$ are equal for at least one initial state. Moreover, if L corresponds to the next-state function of an LFSR with primitive feedback polynomial, then all $s_{t_0+iT}, i \geq 0$ are equal for some $t_0 < T$ for all nonzero initial states when $\deg(g) \neq n$.*

Proof. Since g is not the zero function, there exists some $a \in \mathbf{F}_2^n$ such that $g(a) = 1$, implying $g \circ L^{iT}(a) = 1$ for all $i \geq 0$. Because g belongs to $AN(f)$ (resp. $AN(1 + f)$), we deduce that f (resp. $(1 + f)$) vanishes at points $L^{iT}(a)$, for all $i \geq 0$. Therefore, the keystream generated from initial state $\mathbf{x}_0$ is such that $s_{t_0+iT}, i \geq 0$ are equal for some $t_0 < T$ as soon as an internal state a with $g(a) = 1$ can be reached from $\mathbf{x}_0$. For an LFSR with maximum period, all internal states are generated for each nonzero $\mathbf{x}_0$, except the all-zero state. Thus, the property holds unless g is the function of degree n which vanishes at all points except 0.

However, the previous propositions only investigate the possibility that all equations derived from a given annihilator may be equal. The question of their linear dependency is still open. We can nevertheless conjecture from the previous discussion that, if the rank of the system involved in an algebraic attack highly differs from the rank of a random system, the corresponding keystream generator is probably vulnerable to a distinguishing attack.

If we assume that System (1) behaves like a random system with respect to both previously discussed properties, it clearly appears that the relevant parameter in the context of algebraic attacks against such stream ciphers is the so-called *algebraic immunity* of the filtering function.

Definition 2. *The* algebraic immunity *of a Boolean function f, denoted by $AI(f)$, is the lowest degree achieved by a nonzero function in $AN(f) \cup AN(1+f)$.*

It is worth noticing that the previous definition may be inappropriate when we consider algebraic attacks against other families of ciphers, for instance against block ciphers or combiners with memory. In such cases, the annihilator ideals of f and of $(1 + f)$ may play very different roles [3].

In our case, the time-complexity of algebraic attacks based on linearization is roughly

$$\mathcal{O}\left(n^{\omega AI(f)}\right) \text{ where } \omega \simeq 2.37$$

and the associated data-complexity, i.e., the required number of keystream bits, is $\mathcal{O}\left(n^{AI(f)}\right)$, but it is probably reduced when the number of functions of degree $AI(f)$ in $AN(f) \cup AN(1 + f)$ increases. Thus, we can derive from this approximation a lower bound on the algebraic immunity of the filtering function which must be satisfied in order to resist algebraic attacks. If we suppose that the size of the internal state is minimal with respect to key-size k, i.e., that $n = 2k$ (it is known that the size of the internal state must be at least twice the key size in order to resist time-memory-data trade-off attacks), the complexity of the attack is greater than the complexity of an exhaustive search on the key when

$$AI(f) \geq 0.42 \left\lceil \frac{k}{1 + \log_2 k} \right\rceil .$$

For instance, in a filter generator with a 128-bit key and a 256-bit internal state, the algebraic immunity of the filtering function must be at least 7.

But, the secure minimum value for the algebraic immunity is probably higher since more efficient techniques than linearization can be used for solving the algebraic system. Actually, this problem has been extensively studied in computer algebra and it is well-known that some methods based on Gröbner basis algorithms efficiently apply. The most recent and powerful algorithms, F4 and F5, are due to Faugère [19, 27, 20]. It was recently proved [18, 5] that F4 is more efficient than the extended linearization algorithm (XL) proposed by Courtois, Klimov, Patarin and Shamir [12]; XL actually computes a Gröbner basis in the particular context of algebraic attacks. And Algorithm F5 is strictly more efficient than all previous ones. Another technique, called XSL, has also been presented by Courtois and Pieprzyk [14] but its complexity and its implementation feasibility are still controversial.

Some recent results on the complexities of F4 and F5 can be found in [6, 7]. However, it is worth noticing that all these results only hold in the so-called semi-regular case. Therefore, the major problem is to determine whether the system involved in algebraic attacks behaves like a random system or not with respect to the previously mentioned algorithms. We would like to emphasize that it does not make sense to use some complexity results for the semi-regular case if we do not have any hint on the behaviour of the system. For instance, the public challenge on the asymmetric cryptosystem Hidden Field Equations (HFE) was broken by Faugère with F5 whereas the attack was infeasible according to its complexity in the generic case [22].

Open problem 3. *Does System (1) behave like a semi-regular system in the sense of [6]?*

4   Algebraic Immunity of Filtering Functions

Obviously, the algebraic immunity of the filtering function highly influences the complexity of the attack even if the estimation of the time complexity for solving the underlying system is still an open problem.

4.1   General Properties of the Algebraic Immunity

The set $AN(f)$ of all annihilating functions of f is obviously an ideal in the ring of all Boolean functions, and it is generated by $(1 + f)$. It consists of the $2^{2^n - wt(f)}$ functions of n variables which vanish on the support of f, i.e., on all x such that $f(x) = 1$, where $wt(f)$ denotes the size of the support of f. The number of functions of degree at most d in $AN(f)$ is equal to 2^κ where κ is the dimension of the kernel of the matrix obtained by restricting the Reed-Muller code of length 2^n and order d to the support of f. In other words, the rows of this matrix correspond to the evaluations of the monomials of degree at most d on $\{x, f(x) = 1\}$. Since this matrix has $\sum_{i=0}^{d} \binom{n}{i}$ rows and $wt(f)$ columns, its kernel is non-trivial when

$$\sum_{i=0}^{d} \binom{n}{i} > wt(f) .$$

Similarly, $AN(1 + f)$ contains some functions of degree d or less if

$$\sum_{i=0}^{d} \binom{n}{i} > 2^n - wt(f) .$$

Thus, as pointed out in [15], the algebraic immunity of an n-variable function is related to its Hamming weight. Most notably, for odd n, only balanced functions can have optimal algebraic immunity. A trivial corollary is also that, for any n-variable Boolean function, we have $AI(f) \leq \lceil n/2 \rceil$.

   Another interesting property is that the highest possible algebraic immunity for a function is related to the number of its 0-linear structures. Let $\mathcal{S}_0(f)$ be the set of all 0-linear structures for f, i.e., $\mathcal{S}_0(f) = \{a \in \mathbf{F}_2^n, f(x + a) = f(x), \forall x\}$. Then,

$$AI(f) \leq \left\lceil \frac{n - \dim(\mathcal{S}_0(f))}{2} \right\rceil .$$

This bound is important for instance in the case of filtered LFSRs, since the filtering function usually depends only on a small subset of the internal state bits. We deduce from the previous discussion that if an m-variable Boolean function is used for filtering the n-bit internal state of the generator, the complexity of the algebraic attack will be at most $n^{\frac{wm}{2}}$. Therefore, the cipher resists algebraic attacks only if the number m of variables of the filtering function satisfies

$$m \geq 0.84 \left[\frac{k}{1 + \log_2(k)} \right] ,$$

where k is the key-size and where the initial state is supposed to be twice longer than the key. For instance, a filter generator with a 128-bit key and a 256-bit internal state must use a filtering function of at least 16 variables. Here again, the secure number of variables is probably higher than the previous bound which is based on the complexity of linearization.

4.2   Algebraic Immunity of Random Balanced Functions

For 5-variable functions, it is possible to compute the algebraic immunity of all Boolean functions using the classification due to Berlekamp and Welch (because algebraic immunity is invariant under composition by a linear permutation). We here focus on balanced functions because they are the only ones that may have optimal algebraic immunity for n odd. We can compute the algebraic immunity of all $601, 080, 390$ balanced functions of 5 variables:

$AI(f)$	1	2	3
nb. of balanced f	62	403,315,208	197,765,120
proportion of balanced f	10^{-7}	0.671	0.329

Another interesting quantity is the number of linearly independent annihilators of degree at most 2 for all balanced functions of 5 variables:

$\dim(AN_2(f))$	0	1	2	3	4	5
proportion of balanced f	0.329	0.574	0.094	0.002	$2 \cdot 10^{-5}$	10^{-7}

An important observation is that both sets $AN_2(f)$ and $AN_2(1+f)$ have the same dimension for all balanced functions except for one function and its complement (up to linear equivalence). This raises the following open problem.

Open problem 4. *For balanced Boolean functions f, is there a general relationship between $AN(f)$ and $AN(1+f)$?*

Similar simulations can be performed as far as the functions of n variables are classified into equivalence classes under composition by a linear permutation. But, such a classification only exist for $n = 6$ and for cubic functions up to 8 variables.

Even if some well-known constructions of cryptographic Boolean functions have been proved to have a low algebraic immunity, probabilistic arguments tend to show that the proportion of balanced functions with low algebraic immunity is very small. It has been proved in [24] that the probability that a balanced function of n variables has algebraic immunity less than $0.22n$ tends to zero when n tends to infinity. An upper bound on the probability that a balanced function has an annihilator of degree less than d is also given. This bound involves a part of the weight enumerator of $RM(d, n)$ and any new information on its complete weight distribution can clearly improve the result. However, both following problems are still open.

Open problem 5. *Determine the average value of the algebraic immunity for a balanced function of n variables.*

Open problem 6. *Determine the proportion of balanced Boolean functions of n variables with optimal algebraic immunity.*

4.3 Boolean Functions with Optimal Algebraic Immunity

A first relationship between the annihilators of f and of $1 + f$ can be exhibited for functions with optimal algebraic immunity. Actually, all annihilators of a balanced n-variable function f have maximal degree $\lfloor \frac{n+1}{2} \rfloor$ if and only if the support of f corresponds to a subset of 2^{n-1} columns of the Reed-Muller code of length 2^n and order $\lfloor \frac{n-1}{2} \rfloor$ with maximal rank. When n is odd, such a set is an information set for the Reed-Muller code of order $\frac{n-1}{2}$ which has dimension 2^{n-1}. Then, a relationship between $\deg(AN(f))$ and $\deg(AN(1 + f))$ can be derived from the fact that this code is a self-dual code.

Proposition 3. *Let C be a linear self-dual code. If I is an information set for C, then its complement is an information set too.*

Proof. Let I be an information set for C. Then, there exists a generator matrix for C which can be decomposed into $G = (Id, M)_I$ where the first part corresponds to the positions in I. Let us now assume that the complement of I is not an information set for C. This means that there exists a nonzero codeword of the form $c = (c', 0)_I$ in C. Since C is self-dual, c belongs to the dual code. Therefore, $Gc = 0$, implying that some columns of the identity matrix sum up to zero, a contradiction.

We can immediately derive the following result.

Theorem 1. *Let n be an odd integer and f be a balanced Boolean function of n variables. Then, f has optimal algebraic immunity $\frac{n+1}{2}$ if and only if $AN(f)$ does not contain any nonzero function of degree strictly less than $\frac{n+1}{2}$.*

A few classes of Boolean functions with optimal algebraic immunity have been recently exhibited. An iterative construction which provides an infinite family of balanced Boolean functions with optimal algebraic immunity is presented in [16]. Another example of functions with optimal algebraic immunity is the majority symmetric function depending on an odd number of variables, i.e., the function which outputs 1 if and only if the Hamming weight of its input vector is greater than or equal to $\frac{n+1}{2}$. This property was first proved in [23, Theorem 1] in terms of information sets for the self-dual Reed-Muller code, and it is also mentioned in [17].

4.4   Algebraic Immunity and Other Cryptographic Criteria

Besides the Hamming weight of the function, its nonlinearity is also related to its algebraic immunity [15]. It can be proved that, for any linear function φ, the algebraic immunity of $f + \varphi$ is at most $AI(f) + 1$. Therefore, any function f of n variables with algebraic immunity at least d satisfies

$$\mathcal{NL}(f) \geq \sum_{i=0}^{d-2} \binom{n}{i} .$$

It follows that any function with optimal algebraic immunity has a high nonlinearity, more precisely

$$\mathcal{NL}(f) \geq \begin{cases} 2^{n-1} - \binom{n}{\frac{n-1}{2}} & \text{if } n \text{ is odd} \\ 2^{n-1} - \frac{1}{2}\binom{n}{\frac{n}{2}} - \binom{n}{\frac{n}{2}-1} & \text{if } n \text{ is even} \end{cases}$$

A high nonlinearity and a high algebraic immunity are then compatible criteria. Another important consequence is that the nonlinearity of a function may be a sufficient criterion to decide whether it has low algebraic immunity (but the converse is not true).

Another cryptographic property that implies that a function does not have a maximal algebraic immunity is the notion of *normality*. A function is said to be k-normal (resp. k-weakly normal) if there exists an affine subspace of dimension k on which the function is constant (resp. affine). Since the minimum weight codewords of $RM(r, n)$ are those whose support is an affine subspace of dimension $n - r$, we deduce that any k-normal function f of n variables has algebraic immunity at most $n - k$. Similarly, any k-weakly normal function has algebraic immunity at most $n - k + 1$. Non-normal (and non-weakly normal) functions may be good candidates if we want to construct functions with optimal nonlinearity.

The existence of links between algebraic immunity and other cryptographic criteria remains unknown. For instance, the relation between the distance of a function to all low-degree functions (i.e., its distance to $RM(d, n)$) and its algebraic immunity is still unclear. Correlation-immunity does not seem to be a priori incompatible with optimal algebraic immunity: there exists a 1-resilient function of 5 variables with optimal algebraic immunity. However, the link with all known criteria must be investigated further.

4.5   Algebraic Immunity of Known Constructions

Some bounds have been established on the algebraic immunity of the cryptographic functions obtained by applying classical constructions. First, the algebraic immunity of a function can be derived from the algebraic immunities of its restrictions to a given hyperplane and to its complement [15]. For instance, if

$$f(x_1, \ldots, x_n) = (1 + x_n)f_1(x_1, \ldots, x_{n-1}) + x_n f_2(x_1, \ldots, x_{n-1}) ,$$

we have:

- if $AI(f_1) \neq AI(f_2)$, then $AI(f) = \min(AI(f_1), AI(f_2)) + 1$;
- if $AI(f_1) = AI(f_2)$, then $AI(f) \in \{AI(f_1), AI(f_1) + 1\}$.

Therefore, it is obvious how to construct a function of $2t$ variables with optimal algebraic immunity from two functions of $(2t - 1)$ variables with respective algebraic immunities equal to t and to $(t - 1)$. But, constructing a function of $(2t + 1)$ variables with optimal algebraic immunity from two functions of $2t$ variables is much more difficult since both restrictions must have optimal algebraic immunity and they must also satisfy some additional conditions.

Some bounds on the algebraic immunities of some classical constructions, such as the Maiorana-McFarland family, can be found in [24, 15, 25].

4.6   Computing the Algebraic Immunity of a Boolean Function

The basic algorithm for computing the algebraic immunity of an n-variable function consists in performing a Gaussian elimination on the generator matrix of the punctured $RM(\lfloor \frac{n-1}{2} \rfloor, n)$ restricted to the support of f. This matrix has $wt(f)$ columns and $k(\lfloor \frac{n-1}{2} \rfloor, n) = \sum_{i=0}^{\lfloor \frac{n-1}{2} \rfloor} \binom{n}{i}$ rows. Therefore, the algorithm requires $k^2(\lfloor \frac{n-1}{2} \rfloor, n)wt(f)$ operations, which is close to 2^{3n-3} when f is balanced. As noted in [24], the complexity can be significantly reduced if we only want to check whether a function has annihilators of small degree d, since we do not need to consider all positions in the support of f. Indeed, considering a number of columns which is only slightly higher that the code dimension $k(d, n)$

is usually sufficient for proving that a function does not admit any annihilator of degree d. A technique for reducing the size of the matrix over which the Gaussian elimination is performed is presented in [24]. The idea is that the elements in the support of f with low Hamming weight provide simple equations that can be removed from the matrix by a substitution step. However, due to the lack of simulation results, it is very hard to evaluate the time complexity of the substitution step in practice.

Gröbner bases algorithms such as F5 provide other techniques for computing the size of the annihilator ideal. But they need to be compared with the basic techniques in this particular context.

5   Resistance to Fast Algebraic Attacks

At CRYPTO 2003, Courtois presented some important improvements on algebraic attacks, called *fast algebraic attacks* [10]. The refinement first relies on the existence of some low degree relations between the bits of the initial state and not only one but several consecutive keystream bits. In other words, the attacker wants to find some low degree relations g between the inputs and outputs of

$$F_m \colon \mathbf{F}_2^n \to \mathbf{F}_2^m$$
$$x \mapsto (f(x), f(L(x)), \dots, f(L^{m-1}(x)))$$

where L is the linear transition function. This function is very similar to the so-called *augmented function* defined in [1]. The fact that the augmented function may be much weaker than the filtering function, i.e., than F_0 with the previous notation, has been pointed out by Anderson [1] in the context of correlation attacks. However, finding the low degree relations between the n inputs and m outputs of F_m becomes infeasible when m increases. The direct algorithm used for a function S with n inputs and m outputs consists in finding the low degree annihilators for the characteristic function Φ_S of S, which is the Boolean function of $(n + m)$ variables defined by

$$\Phi_S(x_1, \dots, x_n, y_1, \dots, y_m) = 1 \text{ if and only if } y_i = S_i(x_1, \dots, x_n), \forall i .$$

Due to its high complexity, it can only be used for small values of m. For instance, if we consider a Boolean function of 20 variables, it may have algebraic immunity 10. But, there always exist relations of degree at most 7 involving 4 consecutive keystream bits together. The problem is that determining whether relations of degree less than or equal to 6 exist in this case requires the computation of the kernel of a matrix of 120 GBytes. And even checking whether relations of degree 3 exist involves a 2.7 GByte-matrix. Mounting algebraic attacks based on the augmented function is then related to the following problem.

Open problem 7. *Find an algorithm which determines the low-degree relations for the augmented function.*

More generally, we can wonder whether the particular form of the augmented function has an influence on the degree of the annihilator ideal of its characteristic function. For instance, the existence of a general relationship between the algebraic immunity of a Boolean function and the algebraic immunity of the associated augmented function is still unclear. The fact that the augmented function is a very special case of multi-output functions may lead to new theoretical results or to dedicated algorithms in that case. For instance, a very particular property of the augmented function is that all its Boolean components are linearly equivalent. This raises the following open question, which is clearly related to algebraic attacks against block ciphers which use power functions as S-Boxes, like the AES.

Open problem 8. *Does the linear equivalence between all output components of a multi-output function influence its algebraic immunity?*

Since the computation of low degree relations involving several keystream bits is usually infeasible, Courtois proposed to focus on particular subclasses of relations that can be obtained much faster. The relations considered in the attack are given by linear combinations of relations of the form

$$g(x_0, \ldots, x_{\ell-1}, s_t, \ldots, s_{t+m})$$

where the terms of highest degree do not involve any keystream bits. Then, an additional precomputation step consists in determining the linear combinations of the previous relations which cancel out the highest degree monomials. Some algorithms for this step have been proposed in [10, 2]. This technique helps to decrease the degree of the relations used in the attack for different practical examples. But, here again, we do not have any theoretical result connecting the algebraic immunity of the function and the existence of such low degree linear combinations.

6 Using More Sophisticated Filtering Functions

Many stream ciphers do not use a simple Boolean filtering function; they prefer more sophisticated mappings in order to render the attacks more difficult or in order to increase the throughput of the generator.

Multi-output Boolean functions. A basic technique for increasing the speed of the generator consists in using a filtering function with several outputs. Such functions are called vectorial Boolean functions, or S-boxes by analogy with block ciphers. But, as pointed out in [28], the resistance of the generator to fast correlation attacks usually decreases with the number of output bits of the function. For a single output function, the attack exploits the fact that the output may be approximated by an affine function of the input variables. But, for a function S with m outputs, the attacker can apply any Boolean function g of m variables to the output vector $(y_1, \ldots, y_m)$ and he or she can perform the attack on the resulting sequence $z = g(y_1, \ldots, y_m)$. Therefore, the relevant parameter is not the nonlinearity of the vectorial function, which is the lowest Hamming distance between any linear combination of the components of S and the affine functions, but the so-called *unrestricted nonlinearity* [8], which is the lowest distance between any function $g \circ S$ and the affine functions, where g varies in the set of all nonzero Boolean functions of m variables.

For similar reasons, the algebraic immunity of a vectorial function tends to decrease with the number of output bits. For an S-box with n inputs and m outputs, there exists a relation of degree at most d in the input variables (and of any degree in the output variables) if

$$\sum_{i=0}^{d} \binom{n}{i} > 2^{n-m} .$$

A particular case of generators based on multi-output Boolean functions are the word-oriented ciphers. In order to increase the performance of software implementations, many ciphers use LFSRs over an extension field $\mathbf{F}_{2^m}$ and the associated filtering function is usually a mapping from $\mathbf{F}_{2^m}^n$ into $\mathbf{F}_{2^m}$. This technique is used in many recent stream ciphers, e.g. in SNOW 2.0. The associated filtering function can obviously be seen as a vectorial Boolean function with mn inputs and m outputs. Consequently, all results previously mentioned apply, but the major open issue here is to determine whether word-oriented attacks can be mounted which exploit the particular structure of the function defined as a polynomial over $\mathbf{F}_{2^m}$.

Functions with memory. In some keystream generators, the filtering function is replaced by a finite automaton with some memory bits. An example is the E_0 keystream generator used in the Bluetooth wireless LAN system, which uses a combining function with 4 inputs and 4 memory bits. However, (fast) algebraic attacks [4] can still be applied on such systems. Armknecht and Krause proved that, for any filtering function of n variables with M memory bits, there always exists a relation of degree at most $\lceil \frac{n(M+1)}{2} \rceil$ between $(M+1)$ consecutive output bits and the bits of the initial state, for a given initial assignment of the memory bits. Obviously, relations of lower degree may exist. For instance, the function used in E_0 provides a relation of degree 4 involving 4 consecutive output bits, which leads to an algebraic attack of running-time around 2^{67} [4]. General results on algebraic attacks against combiners with memory can be found in [3, 13].

The main open issue related to the use of such sophisticated functions is to improve the efficiency of the algorithms for computing their algebraic immunity for a large number of input variables. Another related open problem is to find some general constructions which guarantee a high resistance to all these attacks.

Acknowledgements

Many thanks to Daniel Augot and Matthew Parker for their contributions to this work.

References

1. R. J. Anderson. Searching for the optimum correlation attack. In *Fast Software Encryption - FSE'94*, volume 1008 of *Lecture Notes in Computer Science*, pages 137–143. Springer-Verlag, 1995.
2. F. Armknecht. Improving fast algebraic attacks. In *Fast Software Encryption - FSE 2004*, volume 3017 of *Lecture Notes in Computer Science*, pages 65–82. Springer-Verlag, 2004.
3. F. Armknecht. Algebraic attacks and annihilators. In *Proceedings of the Western European Workshop on Research in Cryptology (WEWoRC 2005)*, Lecture Notes in Informatics. Springer-Verlag, 2005. To appear.
4. F. Armknecht and M. Krause. Algebraic attacks on combiners with memory. In *Advances in Cryptology - CRYPTO 2003*, volume 2729 of *Lecture Notes in Computer Science*, pages 162–176. Springer-Verlag, 2003.
5. G. Ars, J.-C. Faugère, H. Imai, M. Kawazoe, and M. Sugita. Comparison between XL and Gröbner basis algorithms. In *Advances in Cryptology - ASIACRYPT 2004*, volume 3329 of *Lecture Notes in Computer Science*, pages 338–353. Springer-Verlag, 2004.
6. M. Bardet, J-C. Faugère, B. Salvy, and B-Y. Yang. On the complexity of Gröbner basis computation of semi-regular overdetermined algebraic equations. In *Proc. International Conference on Polynomial System Solving (ICPSS'2004)*, 2004.
7. M. Bardet, J-C. Faugère, B. Salvy, and B-Y. Yang. Asymptotic behaviour of the degree of regularity of semi-regular polynomial systems. In *MEGA 2005*, Porto Conte, Italy, May 2005.
8. C. Carlet and E. Prouff. On a new notion of nonlinearity relevant to multi-output pseudo-random generators. In *Selected Areas in Cryptography - SAC 2003*, volume 3006 of *Lecture Notes in Computer Science*, pages 291–305. Springer-Verlag, 2004.
9. D. Coppersmith and S. Winograd. Matrix multiplication via arithmetic programming. *Journal of Symbolic Computation*, (9):251–280, 1990.
10. N. Courtois. Fast algebraic attacks on stream ciphers with linear feedback. In *Advances in Cryptology - CRYPTO 2003*, volume 2729 of *Lecture Notes in Computer Science*, pages 176–194. Springer-Verlag, 2003.
11. N. Courtois and W. Meier. Algebraic attacks on stream ciphers with linear feedback. In *Advances in Cryptology - EUROCRYPT 2003*, volume 2656 of *Lecture Notes in Computer Science*, pages 345–359. Springer-Verlag, 2003.
12. N. T. Courtois, A. Klimov, J. Patarin, and A. Shamir. Efficient algorithms for solving overdefined systems of multivariate polynomial equations. In *Advances in Cryptology - EUROCRYPT 2000*, volume 1807 of *Lecture Notes in Computer Science*, pages 392–407. Springer-Verlag, 2000.
13. N.T. Courtois. Algebraic attacks on combiners with memory and several outputs. In *ICISC 2004*, volume 3506 of *Lecture Notes in Computer Science*, pages 3–20. Springer-Verlag, 2005.
14. N.T. Courtois and J. Pieprzyk. Cryptanalysis of block ciphers with overdefined systems of equations. In *Advances in Cryptology - ASIACRYPT 2002*, volume 2502 of *Lecture Notes in Computer Science*, pages 267–287. Springer-Verlag, 2002.
15. D.K. Dalai, K.C. Gupta, and S. Maitra. Results on algebraic immunity for cryptographically significant Boolean functions. In *Progress in Cryptology - Indocrypt 2004*, volume 1880 of *Lecture Notes in Computer Science*, pages 92–106. Springer-Verlag, 2004.

16. D.K. Dalai, K.C. Gupta, and S. Maitra. Cryptographically significant Boolean functions: Construction and analysis in terms of algebraic immunity. In *Fast Software Encryption - FSE 2005*, volume 3357 of *Lecture Notes in Computer Science*, pages 98–111. Springer-Verlag, 2005.

17. D.K. Dalai, S. Sarkar, and S. Maitra. Balanced Boolean functions with maximum possible algebraic immunity. Preprint, April 2005.

18. C. Diem. The XL algorithm and a conjecture from commutative algebra. In *Advances in Cryptology - ASIACRYPT 2004*, volume 3329 of *Lecture Notes in Computer Science*, pages 323–337. Springer-Verlag, 2004.

19. J.-C. Faugère. A new efficient algorithm for computing Gröbner bases (F_4). *Journal of Pure and Applied Algebra*, 139(1-3):61–88, 1999.

20. J.-C. Faugère. A new efficient algorithm for computing Gröbner bases without reduction to zero (F_5). In *Proceedings of the 2002 international symposium on Symbolic and algebraic computation*. ACM, 2002.

21. J.-C. Faugère and G. Ars. An algebraic cryptanalysis of nonlinear filter generators using Gröbner bases. Technical Report 4739, INRIA, 2003. Available at ftp://ftp.inria.fr/INRIA/publication/publi-pdf/RR/RR-4739.pdf.

22. J.-C. Faugère and A. Joux. Algebraic cryptanalysis of hidden field equation (HFE) cryptosystems using Gröbner bases. In *Advances in Cryptology - CRYPTO 2003*, volume 2729 of *Lecture Notes in Computer Science*. Springer-Verlag, 2003.

23. J.D. Key, T.P. McDonough, and V.C. Mavron. Information sets and partial permutation decoding for codes from finite geometries. *Finite Fields and Their Applications*, 2005. To appear.

24. W. Meier, E. Pasalic, and C. Carlet. Algebraic attacks and decomposition of Boolean functions. In *Advances in Cryptology - EUROCRYPT 2004*, volume 3027 of *Lecture Notes in Computer Science*, pages 474–491. Springer-Verlag, 2004.

25. E. Pasalic. On algebraic immunity of Maiorana-McFarland like functions and applications of algebraic attack. In *Proceedings of the ECRYPT Symmetric Key Encryption Workshop (SKEW)*, Aarhus, Danemark, May 2005.

26. C.E. Shannon. Communication theory of secrecy systems. *Bell system technical journal*, 28:656–715, 1949.

27. A. Steel. Allan Steel's Gröbner basis timings page, 2004. http://magma.maths.usyd.edu.au/users/allan/gb/.

28. M. Zhang and A. Chan. Maximum correlation analysis of nonlinear S-boxes in stream ciphers. In *Advances in Cryptology - CRYPTO 2000*, volume 1880 of *Lecture Notes in Computer Science*, pages 501–514. Springer-Verlag, 2000.

On the Non-linearity and Sparsity of Boolean Functions Related to the Discrete Logarithm in Finite Fields of Characteristic Two

Nina Brandstätter[1], Tanja Lange[2,*], and Arne Winterhof[1]

[1] Johann Radon Institute for Computational and Applied Mathematics
Austrian Academy of Sciences
Altenbergerstraße 69, A-4040 Linz, Austria
{nina.brandstaetter, arne.winterhof}@oeaw.ac.at
[2] Department of Mathematics
Technical University of Denmark, Matematiktorvet – Bygning 303
DK-2800 Kongens Lyngby, Denmark
t.lange@mat.dtu.dk

In memory of Klaus Burde

Abstract. In public-key cryptography the discrete logarithm has gained increasing interest as a one-way function. This paper deals with the particularly interesting case of the discrete logarithm in finite fields of characteristic two.

We obtain bounds on the maximal Fourier coefficient, i.e., on the non-linearity, on the degree and the sparsity of Boolean functions interpolating the discrete logarithm in finite fields of characteristic two. These bounds complement earlier results for finite fields of odd characteristic.

The proofs of the results for odd characteristic involve quadratic character sums and are not directly extendable to characteristic two. Here we use a compensation for dealing with the quadratic character.

Keywords: Discrete logarithm, Boolean functions, non-linearity, degree, sparsity, maximal Fourier coefficient.

1 Introduction

Let $\mathbb{F}_q$ be a finite field with $q = 2^r$ elements and let γ be a generator of the multiplicative group $\mathbb{F}_q^*$. The *discrete logarithm* (or *index*) $\operatorname{ind}_\gamma(\xi)$ of $\xi \in \mathbb{F}_q^*$ to the base γ is the integer $0 \leq l \leq q-2$ such that $\xi = \gamma^l$. In this paper we

* The work described in this paper has been supported in part by the European Commission through the IST Programme under Contract IST-2002-507932 ECRYPT. The information in this document reflects only the author's views, is provided as is and no guarantee or warranty is given that the information is fit for any particular purpose. The user thereof uses the information at its sole risk and liability.

Ø. Ytrehus (Ed.): WCC 2005, LNCS 3969, pp. 135–143, 2006.
© Springer-Verlag Berlin Heidelberg 2006

study properties of Boolean functions which provide information on the discrete logarithm. Namely, let $B(U_0, U_1, \ldots, U_{r-1})$ be a Boolean function satisfying

$$B(k_0, k_1, \ldots, k_{r-1}) = \begin{cases} 0, & \text{if } \operatorname{ind}_\gamma(\xi_k) \text{ is even,} \\ 1, & \text{if } \operatorname{ind}_\gamma(\xi_k) \text{ is odd,} \end{cases} \tag{1}$$

where $\xi_k = k_0\beta_0 + k_1\beta_1 + \cdots + k_{r-1}\beta_{r-1} \in \mathbb{F}_q^*$ and $k_i \in \{0,1\}$ for some fixed ordered basis $\{\beta_0, \beta_1, \ldots, \beta_{r-1}\}$ of $\mathbb{F}_q$ over $\mathbb{F}_2$. Hence, B provides the least significant bit of $\operatorname{ind}_\gamma(\xi)$ for $\xi \in \mathbb{F}_q^*$.

For finite fields of odd characteristic such Boolean functions were studied in [8, 9, 13]. The proofs involve sums over the quadratic character. Since there is no quadratic character in fields of even characteristic the proofs are not directly extendable to characteristic two. However, this case is particularly interesting for cryptographic applications (see e.g. [10]). Here we use a compensation for dealing with quadratic characters to extend some selected results of [8, 9, 13] to characteristic two.

In Section 3 we estimate the maximal *Fourier coefficient* $\max\limits_{a \in \mathbb{F}_2^r} \hat{B}(a)$ of B, where

$$\hat{B}(a) := \sum_{u \in \mathbb{F}_2^r} (-1)^{B(u)+<a,u>} \tag{2}$$

and $< a, u >$ denotes the standard inner product. This provides a lower bound for the *non-linearity* $\mathcal{NL}(B)$ of B, i.e., the minimum Hamming distance to affine functions, because of the relation

$$\mathcal{NL}(B) = 2^{r-1} - \frac{1}{2} \max_{a \in \mathbb{F}_2^r} |\hat{B}(a)|.$$

For the significance of this notion see [2, 3, 4, 5, 6, 14, 16].

In Section 4 we prove bounds on the *sparsity*, i.e., the number of nonzero coefficients, and *degree* of B.

Interestingly, our upper bound on the maximal Fourier coefficient is better than the analog result for finite fields of odd characteristic ($O(q^{1/2} \log q)$ vs. $O(q^{7/8} \log q)$). In contrast, our lower bound on the sparsity is weaker, which seems to be unnatural.

2   Preliminary Results

Let χ be a primitive (multiplicative) character of $\mathbb{F}_q$ and put $\eta := \chi^{-1}(\gamma)$. Define the function $\psi : \mathbb{F}_q^* \to \mathbb{C}$ (cf. [1]) by

$$\psi(X) := \frac{2}{q-1} \left(\sum_{j=1}^{q-2} \frac{1}{\eta^j + 1} \chi^j(X) + \frac{1}{2} \right).$$

The following Lemma shows that ψ characterizes the least significant bit of the discrete logarithm.

Lemma 1. *For $\xi \in \mathbb{F}_q^*$ we have*

$$\psi(\xi) = \begin{cases} 1, & \text{ind}_\gamma(\xi) \text{ is even,} \\ -1, & \text{otherwise.} \end{cases}$$

Proof. We start with the well-known relation

$$\frac{1}{q-1} \sum_{j=0}^{q-2} \chi^j(\xi) = \begin{cases} 1, \xi = 1, \\ 0, \text{otherwise,} \end{cases} \quad \xi \in \mathbb{F}_q^*. \tag{3}$$

Then (3) implies for $0 \leq m \leq q-2$,

$$\frac{1}{q-1} \sum_{j=0}^{q-2} \eta^{jm} \chi^j(\xi) = \begin{cases} 1, \text{ind}_\gamma(\xi) = m, \\ 0, \text{otherwise.} \end{cases} \tag{4}$$

Now we get

$$\psi(\xi) = \frac{2}{q-1} \left(\sum_{j=1}^{q-2} \frac{1}{\eta^j + 1} \chi^j(\xi) + \frac{1}{2} \right)$$

$$= \frac{2}{q-1} \left(\frac{q}{2} + \sum_{j=1}^{q-2} \chi^j(\xi) \frac{\eta^{2jq/2} - 1}{\eta^{2j} - 1} \right) - 1$$

$$= \frac{2}{q-1} \sum_{j=0}^{q-2} \chi^j(\xi) \sum_{l=0}^{q/2-1} \eta^{2jl} - 1$$

$$= \sum_{l=0}^{q/2-1} \frac{2}{q-1} \sum_{j=0}^{q-2} \eta^{2jl} \chi^j(\xi) - 1$$

and the result follows by (4).   □

For the following bound on Gaussian sums see [11, Theorem 2G] and [11, Theorem 2C'].

Lemma 2. *Let χ be a nontrivial multiplicative character and ψ be a nontrivial additive character of $\mathbb{F}_q$. Let $f(X) \in \mathbb{F}_q[X]$ be a polynomial which is not an $\text{ord}(\chi)$-th power with m distinct roots in its splitting field. Then for $y \in \mathbb{F}_q$ we have*

$$\left| \sum_{k=0}^{q-1} \chi(f(\xi_k)) \psi(y\xi_k) \right| \leq \begin{cases} mq^{1/2}, & y \neq 0, \\ (m-1)q^{1/2}, & y = 0. \end{cases}$$

For the following bound on incomplete character sums see [15, Section 3, p. 469].

Lemma 3. *Let χ be a nontrivial multiplicative character of $\mathbb{F}_q$ and $f(X) \in \mathbb{F}_q[X]$ a monic polynomial which is not an $\text{ord}(\chi)$-th power and has m distinct* *zeros in its splitting field over* $\mathbf{F}_q$. *Then we have for any additive subgroup* V *of* $\mathbf{F}_q$ *and* $a \in \mathbf{F}_q^*$,

$$\left| \sum_{\xi \in V} \chi(af(\xi)) \right| \le mq^{1/2}.$$

The following bound will be useful for the proofs.

Lemma 4. *Let* η *be a complex primitive* $(q-1)$-*th root of unity. For* $q \ge 4$ *we have*

$$\sum_{j=1}^{q-2} \left| \frac{1}{\eta^j + 1} \right| < 0.785(q-1)\log_2(q-1).$$

Proof. We have

$$\left| \frac{1}{\eta^j + 1} \right| = \left| \frac{1}{\eta^{-j} + 1} \right| = \left| \frac{\eta^{-j} - 1}{\eta^{-2j} - 1} \right| = \left| \frac{\eta^{-2jq/2} - 1}{\eta^{-2j} - 1} \right| = \left| \sum_{l=0}^{q/2-1} \eta^{-2jl} \right|.$$

Using [7, Theorem 1] we have

$$\sum_{j=1}^{q-2} \left| \sum_{l=0}^{q/2-1} \eta^{-2jl} \right| = \sum_{j=1}^{q-2} \left| \frac{\sin(\pi jq/(2(q-1)))}{\sin(\pi j/(q-1))} \right|$$

$$\le \frac{4}{\pi^2}(q-1)\ln(q-1) + 0.38(q-1) + 0.608$$

$$+ 0.116 \frac{\gcd(q/2, q-1)^2}{q-1}$$

from which the result follows as $\gcd(q/2, q-1) = 1$.   □

3   A Bound for the Maximum Fourier Coefficient

In this section we prove an upper bound for the maximal Fourier coefficient $\hat{B}$ of B given by (2).

Theorem 1. *Let* B *be defined as in* (1). *Then we have for* $q \ge 4$

$$\max_{a \in \mathbf{F}_2^r} |\hat{B}(a)| < 2q^{1/2}\log_2(q-1).$$

Proof. Since $(-1)^{B(k_1,\dots,k_r)} = \psi(\xi_k)$ for $\xi_k \neq 0$ we have for any $a \in \mathbf{F}_2^r$

$$\hat{B}(a) = \sum_{k=1}^{2^r-1} \psi(\xi_k)(-1)^{\langle k,a \rangle} + (-1)^{B(0,\dots,0)},$$

where $\langle k, a \rangle = \langle (k_0, \ldots, k_{r-1}), (a_0, \ldots, a_{r-1}) \rangle$. Put

$$S(a) = \sum_{k=0}^{2^r-1} \psi(\xi_k)(-1)^{\langle k,a \rangle}$$

where we additionally define $\psi(0) = 0$. Then

$$|\hat{B}(a)| \leq |S(a)| + 1.$$

Note that the mapping $\psi_a : \xi_k \in \mathbb{F}_{2^r} \mapsto (-1)^{\langle k,a \rangle}$ is an additive character of $\mathbb{F}_{2^r}$. We get

$$|S(a)| = \left| \sum_{k=0}^{2^r-1} \psi(\xi_k)\psi_a(\xi_k) \right|$$

$$= \left| \sum_{k=0}^{2^r-1} \frac{2}{q-1} \left(\sum_{j=1}^{q-2} \frac{1}{\eta^j+1} \psi^j(\xi_k) + \frac{1}{2} \right) \psi_a(\xi_k) \right|$$

$$= \left| \sum_{k=0}^{2^r-1} \frac{1}{q-1} \psi_a(\xi_k) + \sum_{k=0}^{2^r-1} \frac{2}{q-1} \sum_{j=1}^{q-2} \frac{1}{\eta^j+1} \psi^j(\xi_k)\psi_a(\xi_k) \right|$$

$$\leq \left| \sum_{k=0}^{2^r-1} \frac{1}{q-1} \psi_a(\xi_k) \right| + \frac{2}{q-1} \sum_{j=1}^{q-2} \left| \frac{1}{\eta^j+1} \right| \left| \sum_{k=0}^{2^r-1} \psi^j(\xi_k)\psi_a(\xi_k) \right|.$$

So it follows from Lemma 2 and Lemma 4 that

$$|S(a)| \leq \begin{cases} 1.57 \log_2(q-1)q^{1/2}, & a \neq 0, \\ 2^r/(q-1), & a = 0, \end{cases}$$

which yields the result. $\qquad\qquad\qquad\qquad\qquad\qquad\qquad\qquad\qquad\square$

Corollary 1. *Let B be defined as in (1). Then we have for $q \geq 4$*

$$\mathcal{NL}(B) \geq 2^{r-1} - q^{1/2} \log_2(q-1).$$

4   Lower Bounds on Sparsity and Degree of B

The aim of this section is to provide a lower bound on the sparsity of Boolean functions satisfying (1). This bound holds for an arbitrary basis $\{\beta_0, \beta_1, \ldots, \beta_{r-1}\}$ of the finite field showing that there is no Boolean function of extremely low sparsity providing the least significant bit of the discrete logarithm. However, the bound is much lower than one would expect. We also show that for some special basis a much larger bound can be proven.

Theorem 2. *Let $B(U_0, U_1, \ldots, U_{r-1})$ be a Boolean function satisfying (1). For $q \geq 4$ we have*

$$\mathrm{spr}(B) > \frac{\log_2 q}{6 \log_2 \log_2 q} - 2.$$

Proof. Define the integer a by $2^a > \mathrm{spr}(B) + 1 \geq 2^{a-1}$. For each $1 \leq m < 2^a$ with binary expansion $m = \sum_{i=0}^{a-1} m_i 2^i$, $m_0, \ldots, m_{a-1} \in \{0,1\}$, we consider the function

$$B_m(U_0, U_1, \ldots, U_{r-a-1}) := B(U_0, U_1, \ldots, U_{r-a-1}, m_0, m_1, \ldots, m_{a-1}).$$

The number of distinct monomials in $U_0, \ldots, U_{r-a-1}$ occurring in all the B_m cannot exceed $\mathrm{spr}(B)$. Since $2^a - 1 > \mathrm{spr}(B)$ we find a non-trivial linear combination

$$\sum_{m=1}^{2^a-1} c_m B_m(U_0, U_1, \ldots, U_{r-a-1}), \quad c_1, \ldots, c_{2^a-1} \in \mathbb{F}_2$$

which vanishes identically.

The function ψ introduced in Section 2 satisfies

$$\psi(\xi_k) = (-1)^{B(k_0, k_1, \ldots, k_{r-1})}, \quad 1 \leq k < 2^r,$$

by Lemma 1. Now we vary the first $r - a$ variables to obtain bounds on 2^a. For $0 \leq k < 2^{r-a}$ we have

$$\prod_{m=1}^{2^a-1} \psi(\xi_{k+2^{r-a}m})^{c_m} = (-1)^{\sum_{m=1}^{2^a-1} c_m B_m(k_0, k_1, \ldots, k_{r-a-1})} = 1.$$

Let H be the Hamming weight of $(c_1, c_2, \ldots, c_{2^a-1})$ and denote by N the set of integers $m \in [1, 2^a - 1]$ for which $c_m = 1$. We sum over all possible k and use the definition of ψ to get

$$2^{r-a} = \sum_{k=0}^{2^{r-a}-1} \prod_{m=1}^{2^a-1} \psi(\xi_{k+2^{r-a}m})^{c_m}$$

$$= \sum_{k=0}^{2^{r-a}-1} \prod_{m=1}^{2^a-1} \left(\frac{2}{q-1} \left(\sum_{j=1}^{q-2} \frac{1}{\eta^j+1} \chi^j(\xi_{k+2^{r-a}m}) + \frac{1}{2} \right) \right)^{c_m}$$

$$= \left(\frac{2}{q-1} \right)^H \sum_{s=0}^{H} \left(\frac{1}{2} \right)^{H-s} \sum_{m_1,\ldots,m_s \in N} \sum_{j_1,\ldots,j_s=1}^{q-2} \frac{1}{\prod_{i=1}^{s}(\eta^{j_i}+1)}$$

$$\sum_{k=0}^{2^{r-a}-1} \prod_{i=1}^{s} \chi^{j_i}(\xi_{k+2^{r-a}m}).$$

By Lemmas 3 and 4 we obtain

$$2^{r-a} \leq \left(\frac{2}{q-1} \right)^H \sum_{s=0}^{H} \left(\frac{1}{2} \right)^{H-s} \sum_{m_1,\ldots,m_s \in N} \sum_{j_1,\ldots,j_s=1}^{q-2} \frac{1}{\prod_{i=1}^{s}(\eta^{j_i}+1)} sq^{1/2}$$

$$< \left(\frac{2}{q-1} \right)^H \sum_{s=0}^{H} \left(\frac{1}{2} \right)^{H-s} \binom{H}{s} (0.785(q-1))^s \log_2^s(q-1) sq^{1/2}$$

$$\leq q^{1/2} \left(\frac{2}{q-1}\right)^H H\left(0.785(q-1)\log_2(q-1) + 1/2\right)^H$$

$$\leq 2^a q^{1/2} \left(1.57 \log_2(q-1) + \frac{1}{q-1}\right)^{2^a}.$$

Since otherwise the bound is trivial we may assume $2^a \leq r$. Thus we get

$$\frac{q^{1/2}}{r^2} \leq \frac{q^{1/2}}{2^{2a}} \leq \left(1.57 \log_2(q-1) + 1/3\right)^{2^a},$$

which allows to obtain

$$\mathrm{spr}(B) \geq 2^{a-1} - 1$$
$$> 0.5 \left(\log_2(q^{1/2}/r^2) / \log_2(1.57 \log_2(q-1) + 1/3)\right) - 1$$

and by observing that

$$\log_2(1.57 \log_2(q-1) + 1/3) \leq 1.5 \log_2(\log_2 q) \quad \text{for } q \geq 4$$

the result follows.   □

Remark 1. Our estimate is much weaker than the comparable result for odd characteristic (see [8, 13]). The main reason is of technical nature, namely ψ, the compensation for the quadratic character in the proof, is not multiplicative and the product introduces 2^a factors which results in an exponent of 2^a instead of a factor of 2^a.

Much stronger bounds can be shown if one uses a special basis. However, a possible attacker using a Boolean function is not restricted to use a special basis. We now consider a fixed basis given by $\beta_i = \gamma^i$, $0 \leq i \leq r-1$. To distinguish the Boolean function working for this special basis we denote it by B_γ. We point out that obviously one can always use a linear change of variables between the bases $\{1, \gamma, \ldots, \gamma^{r-1}\}$ and $\{\beta_0, \ldots, \beta_{r-1}\}$ to represent the field elements with respect to a different basis. However, there is no reason that for a different basis the degree of B cannot be smaller.

We use a proof technique introduced in [8, 12, 13]. Up to now no application in even characteristic was possible.

Proposition 1. *Let B_γ be a Boolean function satisfying (1). We have*

$$\deg(B_\gamma) \geq r - 1 \text{ and } \mathrm{spr}(B_\gamma) \geq q/4.$$

Proof. Define the Boolean function F by

$$F(U_0, \ldots, U_{r-2}) := B_\gamma(U_0, \ldots, U_{r-2}, 0) + B_\gamma(0, U_0, \ldots, U_{r-2}).$$

As $(0, k_0, \ldots, k_{r-2})$ represents $\sum_{i=0}^{r-2} k_i \gamma^{i+1} = \gamma \sum_{i=0}^{r-2} k_i \gamma^i$, i.e., γ times the first input, we have that exactly one of $\sum_{i=0}^{r-2} k_i \gamma^i$ and $\gamma \sum_{i=0}^{r-2} k_i \gamma^i$ has even discrete logarithm. Hence, for each nonzero vector $(k_0, \ldots, k_{r-2})$ one has $F(k_0, \ldots, k_{r-2}) = 1$. Evaluating $F(0, 0, \ldots, 0)$ gives 0 independent of the ambiguous value of $B_\gamma(0, 0, \ldots, 0)$. Hence, F is non-constant and we get

$$F(U_0, \ldots, U_{r-2}) = \prod_{i=0}^{r-2}(1 + U_i) + 1.$$

From the definition of F we have

$$\deg(B_\gamma) \geq \deg(F) = r - 1$$

and

$$\operatorname{spr}(B_\gamma) \geq \lceil 0.5\operatorname{spr}(F) \rceil = \lceil 0.5(2^{r-1} - 1) \rceil = 2^{r-2} = q/4,$$

which completes the proof. $\square$

Acknowledgment

Part of the research was done during a visit of the second author to the Johann Radon Institute for Computational and Applied Mathematics, Linz, Austria. She wishes to thank the institute for hospitality and financial support. The first and the third author are supported by the Austrian Science Fund (FWF) grant S8313. The third author is also supported by the Austrian Academy of Sciences.

We would like to thank Claude Carlet and Philippe Langevin for valuable comments.

References

1. N. Brandstätter and A. Winterhof, Approximation of the discrete logarithm in finite fields of even characteristic by real polynomials, *Arch. Math. (Brno)*, to appear.
2. C. Carlet, On Cryptographic Complexity of Boolean Functions, *Finite Fields with Applications to Coding Theory, Cryptography and Related Areas*, Springer-Verlag, Berlin, 2002, 53–69.
3. C. Carlet, On the degree, nonlinearity, algebraic thickness, and nonnormality of Boolean functions with developments on symmetric functions, *IEEE Trans. Inform. Theory*, **50** (2004), 2178–2185.
4. C. Carlet and C. Ding, Highly nonlinear mappings, *J. Compl.*, **20** (2004), 205–244.
5. C. Carlet and P. Sarkar, Spectral domain analysis of correlation immune and resilient Boolean functions, *Finite Fields and Their Appl.*, **8** (2002), 120–130.
6. C. Carlet and Y. Tarannikov, Covering sequences of Boolean functions and their cryptographic significance, *Designs, Codes and Cryptography*, **25** (2002), 263–279.
7. T. Cochrane, On a trigonometric inequality of Vinogradov, *J. Number Theory*, **27** (1987), 9–16.
8. T. Lange and A. Winterhof, Incomplete character sums over finite fields and their application to the interpolation of the discrete logarithm by Boolean functions, *Acta Arith.*, **101** (2002), 223–229.

9. T. Lange and A. Winterhof, Interpolation of the discrete logarithm in F_q by Boolean functions and by polynomials in several variables modulo a divisor of $q - 1$, *Discrete Appl. Math.*, **128** (2003), 193–206.
10. A. Menezes, P. van Oorschot, S. Vanstone, *Handbook of Applied Cryptography*, CRC Press, Boca Raton, 1996.
11. W.M. Schmidt, *Equations over finite fields*, Lect. Notes Math. 536, Springer, Berlin, 1976.
12. I.E. Shparlinski, *Number Theoretic Methods in Cryptography*, Birkhäuser, Basel, 1999.
13. I.E. Shparlinski, *Cryptographic Applications of Analytic Number Theory. Complexity lower bounds and pseudorandomness*, Birkhäuser, Basel, 2003.
14. P. Štanikă, Nonlinearity, local and global avalanche characteristics of balanced Boolean functions, *Discr. Math.*, **248** (2002), 181–193.
15. A. Winterhof, Incomplete additive character sums and applications, in: Jungnickel, D. and Niederreiter, H. (eds.): Finite fields and applications, 462–474, Springer, Heidelberg, 2001.
16. Y. Zheng and X. M. Zhang, Connections among nonlinearity, avalanche and correlation immunity, *Theor. Comp. Sci.*, **292** (2003), 697–710.

Interpolation of Functions Related to the Integer Factoring Problem

Clemens Adelmann[1] and Arne Winterhof[2]

[1] Institut für Analysis und Algebra,
Technische Universität Braunschweig,
Pockelsstraße 14, D-38106 Braunschweig, Germany
c.adelmann@tu-bs.de
[2] Johann Radon Institute for Computational and Applied Mathematics,
Altenberger Straße 69, A-4040 Linz, Austria
arne.winterhof@oeaw.ac.at

Dedicated to Henning Stichtenoth on the occasion of his 60th birthday

Abstract. The security of the RSA public key cryptosystem depends on the intractability of the integer factoring problem. This paper shall give some theoretical support to the assumption of hardness of this number theoretic problem.

We obtain lower bounds on degree, weight, and additive complexity of polynomials interpolating functions related to the integer factoring problem, including Euler's totient function, the divisor sum functions, Carmichael's function, and the RSA-function.

These investigations are motivated by earlier results of the same flavour on the interpolation of discrete logarithm and Diffie-Hellman mapping.

Keywords: polynomials, degree, weight, additive complexity, factoring problem, RSA-problem, Euler's totient function, divisor sum function, Carmichael's function.

1 Introduction

Computationally difficult number theoretic problems like the discrete logarithm problem or the integer factoring problem play a fundamental role in public key cryptography. The Diffie-Hellman key exchange depends on the intractability of the discrete logarithm problem and the RSA cryptosystem is based on the hardness of the integer factoring problem (see e. g. [27, Chapter 3]).

In the monograph [40] (or its predecessor [38]) and the series of papers [2, 3, 4, 8, 10, 14, 15, 16, 17, 18, 19, 20, 21, 22, 23, 24, 25, 26, 30, 31, 32, 37, 43, 44, 45] several results on discrete logarithm problem and Diffie-Hellman problem supporting the assumption of their hardness were proven. In particular, it was shown that there are no low degree or sparse interpolation polynomials of discrete logarithm and Diffie-Hellman mapping for a large set of given data. In the present paper we prove analog results for functions related to the integer

Ø. Ytrehus (Ed.): WCC 2005, LNCS 3969, pp. 144–154, 2006.
© Springer-Verlag Berlin Heidelberg 2006

factoring problem. We restrict ourselves to the case of factoring 'RSA-integers' $N = pq$ with two odd primes $p < q$.

In Section 3 we investigate real and integer interpolation polynomials of mappings allowing to factor N, including *Euler's totient function*

$$\varphi(pq) = (p-1)(q-1),$$

Carmichael's function

$$\lambda(pq) = \frac{\varphi(pq)}{\gcd(p-1, q-1)},$$

and *divisor sum functions*

$$\sigma_n(pq) = (p^n + 1)(q^n + 1)$$

with a small positive integer n, and '*factoring functions*'

$$\psi_{n,m}(pq) = p^n q^m$$

with small different nonnegative integers n and m.

In Section 4 we prove a lower bound on degree and weight of an integer polynomial representing the *RSA-function*

$$f(x) \equiv x^d \bmod pq, \quad x \in S,$$

for a subset S of $\mathbb{Z}^*_{pq} = \{1 \le x < pq : \gcd(x, pq) = 1\}$ and some integer d with $\gcd(d, (p-1)(q-1)) = 1$.

We collect some auxiliary results on polynomials in the next section.

2   Preliminaries

A proof of the following useful relation between the number of zeros and the degree of a multivariate polynomial, which extends the well-known relation for univariate polynomials, can be found in [11, Lemma 6.44.].

Lemma 1. *Let D be an integral domain, $n \in \mathbb{N}$, $S \subseteq D$, and $f \in D[X_1, \ldots, X_n]$ be a polynomial of total degree d, with at least N zeros in S^n. If f is not the zero polynomial, then we have*

$$d \ge \frac{N}{|S|^{n-1}}.$$

The *additive complexity* $C_\pm(f)$ of a polynomial $f(X)$ is the smallest number of '+' and '−' signs necessary to write down this polynomial. In [33, 34] the number of different zeros of a real polynomial was estimated in terms of its additive complexity.

Lemma 2. *For a nonzero polynomial $f(X) \in \mathbb{R}[X]$ having N different real zeros we have*

$$C_\pm(f) \geq \left(\frac{1}{5}\log(N)\right)^{1/2},$$

where $\log(N)$ is the binary logarithm.

In [35, 36] the following improvement was obtained for integer polynomials.

Lemma 3. *For a nonzero polynomial $f(X) \in \mathbb{Z}[X]$ having N different rational zeros we have*

$$\log(N) = O(C_\pm(f)\log(C_\pm(f))).$$

The *weight* $w(f)$ of a polynomial f is the number of its nonzero coefficients. For polynomials over a finite field $\mathbb{F}_q$ of q elements we have the following lower bound on the weight (see [40, Lemma 2.5]).

Lemma 4. *Let $f(X) \in \mathbb{F}_q[X]$ be a nonzero polynomial of degree at most $q - 2$ with N different zeros in $\mathbb{F}_q^*$. Then we have*

$$w(f) \geq \frac{q-1}{q-1-N}.$$

Obviously, for any univariate polynomial f we have

$$C_\pm(f) \leq w(f) - 1 \leq \deg(f).$$

3   Interpolation of Factoring Functions

For example, the knowledge of the value

$$\varphi(N) = (p-1)(q-1)$$

of Euler's totient function at an integer $N = pq$ with unknown primes p and q is sufficient to determine p and q by solving the quadratic equation

$$X^2 + (\varphi(N) - N - 1)X + N = 0. \tag{1}$$

In general, let $g(X)$ and $h(X)$ be (known) real rational functions, such that the product $g(X)h(N/X)$ is not constant. Then from the knowledge of the values in $N = pq$ of a function f with the property

$$f(N) = g(p)h(q) = g(p)h(N/p)$$

we can determine the unknown factors p and q of N by solving an algebraic equation which is derived from

$$g(X)h(N/X) = f(N) \tag{2}$$

by clearing denominators and negative powers of X.

If we could interpolate the function f by a polynomial of low degree or low additive complexity and the degree of the algebraic equation derived from (2) were small, then we could efficiently factorize N. Hence, it becomes important to prove lower bounds on degree and additive complexity of such interpolation polynomials.

First we prove lower bounds on degree and additive complexity of a real polynomial with some special prescribed values.

Proposition 1. *For $M \geq 3$ let*

$$0 < a_1 < a_2 < \ldots < a_M$$

be a set of ordered reals,

$$g : \{a_1, a_2, \ldots, a_{M-1}\} \to \mathbb{R},$$

$$h : \{a_2, a_3, \ldots, a_M\} \to \mathbb{R},$$

real valued functions, and G the unique interpolation polynomial of g of degree at most $M - 2$. Let $f \in \mathbb{R}[X]$ be a polynomial satisfying

$$f(a_i a_j) = g(a_i)h(a_j), \quad 1 \leq i < j \leq M.$$

If there exist $1 \leq i < j \leq M - 1$ such that

$$G\left(\frac{a_i a_j}{a_M}\right) h(a_M) \neq g(a_i)h(a_j) \tag{3}$$

then we have

$$\deg(f) \geq M - 1,$$

$$C_{\pm}(f) \geq \left(\frac{1}{5}\log(M - 1)\right)^{1/2} - C_{\pm}(G) - 1,$$

and if $f(a_M X) - h(a_M)G(X) \in \mathbb{Q}[X]$ and $a_1, \ldots, a_{M-1} \in \mathbb{Q}$ then we have

$$C_{\pm}(f) + C_{\pm}(G) = \Omega\left(\frac{\log(M)}{\log\log(M)}\right).$$

Proof. The polynomial

$$F(X) = f(a_M X) - G(X)h(a_M) \tag{4}$$

is not identically zero by (3) and has zeros at $a_1, \ldots, a_{M-1}$. So we have

$$\max(\deg(f), M - 2) \geq \max(\deg(f), \deg(G)) \geq \deg(F) \geq M - 1$$

by Lemma 1 and thus $\deg(f) \geq M - 1$. By Lemma 2 and observing that

$$C_{\pm}(F) \leq C_{\pm}(f) + C_{\pm}(G) + 1$$

we obtain our second assertion. The third assertion follows by Lemma 3 if we multiply (4) with the least common denominator of the coefficients of F.   □

Condition (3) in Proposition 1 is necessary and natural. For example, if the given values are

$$g(a_i) = h(a_i) = a_i^n, \quad i = 1, \ldots, M,$$

with $M \geq n + 2$, they determine the interpolation polynomial $f(X) = X^n$ of degree $n \leq M - 2$ having additive complexity 0. However, the interpolation polynomial of g is $G(X) = X^n$ and we have

$$G\left(\frac{a_i a_j}{a_M}\right) h(a_M) = a_i^n a_j^n = g(a_i) h(a_j), \quad 1 \leq i < j \leq M - 1,$$

contradicting (3).

On the other hand, if g and h are polynomials of small degree with respect to M, then (3) being not valid implies that $g(X)h(Y) = g(XY/a_M)h(a_M)$ by Lemma 1. Hence, for each fixed curve $Y = N/X$ the polynomial $g(X)h(N/X)$ is constant and (2) cannot be used to determine the factorization of N.

Proposition 1 provides lower bounds on degree and additive complexity of real polynomials f interpolating several well-known functions, as generalizations of Euler's totient function

$$\varphi_n(pq) = (p^n - 1)(q^n - 1), \quad n \neq 0, \tag{5}$$

and generalized divisor sums

$$\sigma_n(pq) = (p^n + 1)(q^n + 1), \quad n \neq 0, \tag{6}$$

but also 'factoring functions' $\psi_{n,m}$ of the form

$$\psi_{n,m}(pq) = p^n q^m, \quad n \neq m, \tag{7}$$

where n and m are nonnegative integers and p and q are primes with $p < q$.

Theorem 1. *For $M \geq 3$ let $p_1 < p_2 < \ldots < p_M$ be a set of primes and F a function of the form (5), (6), or (7). Let $f \in \mathbb{R}[X]$ be a polynomial satisfying*

$$f(p_i p_j) = F(p_i p_j), \quad 1 \leq i < j \leq M.$$

Then we have

$$\deg(f) \geq M - 1$$

and

$$C_\pm(f) \geq \left(\frac{1}{5} \log(M - 1)\right)^{1/2} - 2.$$

Proof. Since the functions

$$f_n(X) = \left(\left(\frac{a}{X}\right)^n - 1\right)(X^n - 1), \quad a > 0, \ n = 1, 2, \ldots,$$

are decreasing for $x > \sqrt{a}$ we have for all $1 \leq i < j < k \leq M$,

$$\left(\left(\frac{p_i p_j}{p_k} \right)^n - 1 \right) (p_k^n - 1) < (p_i^n - 1)(p_j^n - 1)$$

and (3) is satisfied in case of generalizations of Euler's totient function. Since

$$f_n(X) = \left(\left(\frac{a}{X} \right)^n + 1 \right) (X^n + 1), \quad a > 0, \ n = 1, 2, \ldots,$$

are increasing for $x > \sqrt{a}$ we have

$$\left(\left(\frac{p_i p_j}{p_k} \right)^n + 1 \right) (p_k^n + 1) > (p_i^n + 1)(p_j^n + 1)$$

and (3) is satisfied in case of generalized divisor sums. Trivially, we have

$$\left(\frac{p_i p_j}{p_k} \right)^n p_k^m \neq p_i^n p_j^m$$

for all $n \neq m$ and (3) is satisfied in case of 'factoring functions'. Now the Theorem follows by Proposition 1. $\qquad\square$

Proposition 1 does not apply to the Carmichael function

$$\lambda(N) = \frac{\varphi(N)}{\gcd(p - 1, q - 1)}, \quad N = pq,$$

with two odd primes $p \neq q$, which can also be used to factorize N.

We first study how λ can be used to factor N.

Proposition 2. *Let* $N = pq$ *be a product of two unknown odd primes* $p < q$ *and put* $\Delta = \lfloor N/\lambda(N) \rfloor$. *Then either* $\Delta = p$ *or* p *and* q *are the solutions of the quadratic equation*

$$X^2 + (\Delta\lambda(N) - N - 1)X + N = 0.$$

Proof. Put $g = \gcd(p - 1, q - 1)$. Then we have

$$\frac{N}{\lambda(N)} - \frac{2g}{p - 1} < g < \frac{N}{\lambda(N)} - \frac{2g}{q - 1}.$$

If $g = p - 1$, then we have $N/\lambda(N) = p + p/(q - 1)$, such that $\Delta = p$. If $g \leq (p - 1)/2$, then the above inequalities give $N/\lambda(N) - 1 < g < N/\lambda(N)$ and thus $\Delta = g$. Hence in this case we have

$$\varphi(N) = \Delta\lambda(N)$$

and can determine p and q from the quadratic equation (1). $\qquad\square$

Next we prove an analog of Theorem 1 for the Carmichael function. Let $\tau(x)$ denote the number of positive divisors of an integer x.

Theorem 2. *For $M \geq 3$ let $p_1 < p_2 < \ldots < p_M$ be a set of primes and $f \in \mathbb{R}[X]$ be a polynomial satisfying*

$$f(p_i p_j) = \frac{(p_i - 1)(p_j - 1)}{\gcd(p_i - 1, p_j - 1)}, 1 \leq i < j \leq M.$$

Put $T = \min_{1 \leq i \leq M} \tau(p_i - 1)$. Then we have

$$\deg(f) \geq \frac{M - 1}{T}$$

and

$$C_{\pm}(f) \geq \left(\frac{1}{5} \log \left(\frac{M - 1}{T}\right)\right)^{1/2} - 2.$$

Proof. Choose $1 \leq k \leq M$ with

$$\tau(p_k - 1) = \min_{1 \leq i \leq M} \tau(p_i - 1).$$

For each divisor d of $p_k - 1$ we define a polynomial

$$F_d(X) = f(p_k X) - \frac{(X - 1)(p_k - 1)}{d}.$$

Then each p_i with $1 \leq i \leq M$ and $i \neq k$ is a zero of at least one F_d. These polynomials are not identically zero. Otherwise, for three different primes p_i, p_j, p_k, $F_d(p_i p_j / p_k) = 0$ yields a monic quadratic equation in p_k with constant term $p_i p_j$, and the only possible integral solutions p_k have to be divisors of $p_i p_j$, which is impossible by assumption. Now the result follows analogously to the proof of Proposition 1 by the pigeon hole principle. □

Remark. The dependence of the result on T may indicate that factoring integers $N = pq$ is easier if $p - 1$ and $q - 1$ are smooth which fits to the expected running time of Pollard's $p - 1$ factoring algorithm. On the other hand the expected running time of the (in general faster) number field sieve does not depend on the factorization of $p - 1$ and $q - 1$.

4   Interpolation of the RSA-Function

The *RSA problem* is the following: Given a positive integer N that is a product of two distinct odd primes p and q, a positive integer e such that $\gcd(e, (p - 1)(q - 1)) = 1$, and an integer c, find an integer m such that $m^e \equiv c \bmod N$. In other words, if d is an (unknown) integer with $ed \equiv 1 \bmod (p-1)(q-1)$ then we have to evaluate the mapping $f(x) = x^d$ in c. The following result excludes the existence of very simple interpolation polynomials of this mapping in the case of low public exponent e.

Theorem 3. *Let $N = pq$ be the product of two odd primes with $p < q$. Choose integers $d, e > 1$ such that $ed \equiv 1 \bmod (p-1)(q-1)$. Let $S \subseteq \mathbb{Z}_N^*$ be a set of size $s \geq 2$. If $f(X) = \sum_{i=0}^m a_i X^i \in \mathbb{Z}[X]$ is a polynomial with degree $m < (q-1)/e$ and $\gcd(a_0, \ldots, a_m, N) = 1$ which satisfies*

$$f(x) \equiv x^d \bmod N \quad \text{for all } x \in S,$$

then we have

$$\deg(f) \geq \max\left(\frac{s}{e(p-1)}, \frac{s^{1/2}}{e}\right) \quad \text{and} \quad w(f) \geq \left(\frac{s}{(p-1)(q-1)-s}\right)^{1/e}.$$

Proof. Put $F(X) = f(X)^e - X$. Since $s \geq 2$ and $e > 1$ the interpolation polynomial $f(X)$ is not constant and we have

$$\deg(F) = e \deg(f).$$

For $n \geq 1$ let $Z_n(F)$ denote the number of different zeros of $F \bmod n$ lying in $\mathbb{Z}_n^*$. We have $Z_{pq}(F) = Z_p(F)Z_q(F)$ by the Chinese Remainder Theorem. ¿From our conditions on f we infer that $\deg(F) < q - 1$. Thus

$$s \leq Z_p(F)Z_q(F) \leq (p-1)Z_q(F) \leq (p-1)\deg(F) = e(p-1)\deg(f).$$

If $s < (p-1)^2$ then we may assume $\deg(F) = e\deg(f) < p - 1$ and get

$$s \leq Z_p(F)Z_q(F) \leq (\deg(F))^2 = (e\deg(f))^2.$$

By Lemma 4 and the same arguments we get

$$w(F) \geq \frac{q-1}{q-1-Z_q(F)} \geq \frac{q-1}{q-1-s/(p-1)} = \frac{(p-1)(q-1)}{(p-1)(q-1)-s},$$

and the last statement is a consequence of $w(F) \leq (w(f))^e + 1$. □

If d is small then e has to be large and the lower bounds become very weak. In this case the attack of [42] for small d (see also [5, Section 3]) solves the RSA-problem. It should be also mentioned that for low public exponents e attacks on RSA are known [6, 7, 13].

5 Some Related Results

In [1] it was shown that if the discrete logarithm problem in $\mathbb{Z}_N^*$ can be solved in polynomial time, then N can be factored in polynomial time, and the Diffie-Hellman problem in $\mathbb{Z}_N^*$ is at least as difficult as the problem of factoring N. Most of the results on the discrete logarithm and the Diffie-Hellman mapping modulo a prime in [40] can be extended to composite moduli. Such results can also be regarded as complexity lower bounds on functions related to the factoring problem of the same flavour as in this paper.

The linear complexity of several sequences related to the factoring problem including RSA-generator, Blum-Blum-Shub-generator, and two prime generator was investigated in [4, 9, 12, 39].

Finally, we mention that an analog of Theorem 3 for the LUC cryptosystem can be easily proven, where instead of monomial X^d Dickson polynomials are used (see [28, 29, 41]).

Acknowledgments

Parts of this paper were written during a visit of the first author to RICAM. He wishes to thank the Austrian Academy of Sciences for hospitality and financial support. The second author is supported by the Austrian Academy of Sciences and by the Austrian Science Fund (FWF) grant S8313.

We wish to thank Tanja Lange for helpful discussions.

References

1. E. Bach, *Discrete logarithms and factoring*, Report No. UCB/CSD-84-186, Computer Science Division (EECS), University of California, Berkeley, California, 1984.
2. N. Brandstätter, T. Lange, and A. Winterhof, On the Non-Linearity and Sparsity of Boolean Functions Related to the Discrete Logarithm in finite fields of characteristic two, Proc. Workshop on Coding and Cryptography WCC 2005, Lect. Notes Comput. Sci., to appear.
3. N. Brandstätter and A. Winterhof, Approximation of the discrete logarithm in finite fields of even characteristic by real polynomials, Arch. Math. (Brno), to appear.
4. N. Brandstätter and A. Winterhof, Some notes on the two-prime generator, IEEE Trans. Inform. Theory **51** (2005), 3654–3657.
5. D. Boneh, Twenty years of attacks on the RSA cryptosystem, *Notices Amer. Math. Soc.* **46** (1999), 203–213.
6. D. Boneh and R. Venkatesan, Breaking RSA may not be equivalent to factoring (extended abstract), *Advances in cryptology—EUROCRYPT '98 (Espoo), Lecture Notes in Comput. Sci.* **1403**, Springer, Berlin, 1998, 59–71.
7. D. Coppersmith, Finding a small root of univariate modular equation, *Advances in cryptology—EUROCRYPT '96 (Saragossa, 1996), Lecture Notes in Comput. Sci.* **1070**, Springer, Berlin, 1996, 155–165,.
8. D. Coppersmith and I. Shparlinski, On polynomial approximation of the discrete logarithm and the Diffie-Hellman mapping, *J. Cryptology* **13** (2000), 339–360.
9. C. Ding, Linear complexity of generalized cyclotomic binary sequences of order 2, *Finite Fields Appl.* **3** (1997), 159–174.
10. C. Ding and T. Helleseth, On cyclotomic generator of order r, *Inform. Process. Lett.* **66** (1998), 21–25.
11. J. von zur Gathen and J. Gerhard, *Modern Computer Algebra*, Cambridge University Press, New York, 1999.
12. F. Griffin and I. Shparlinski, On the linear complexity profile of the power generator, *IEEE Trans. Inform. Theory* **46** (2000), 2159–2162.
13. J. Hastad, Solving simultaneous modular equations of low degree, *SIAM J. Comput.* **17** (1988), 336–341.

14. E. Kiltz and A. Winterhof, Lower bounds on weight and degree of bivariate polynomials related to the Diffie-Hellman mapping, *Bull. Austral. Math. Soc.* **69** (2004), 305–315.
15. E. Kiltz and A. Winterhof, Polynomial interpolation of cryptographic functions related to Diffie-Hellman and discrete logarithm problem, *Discrete Appl. Math.* **154** (2006), 326–336..
16. S. Konyagin, T. Lange, and I. Shparlinski, Linear complexity of the discrete logarithm, *Des. Codes Cryptogr.* **28** (2003), 135–146.
17. T. Lange and A. Winterhof, Polynomial interpolation of the elliptic curve and XTR discrete logarithm, *Proceedings of the 8th Annual International Computing and Combinatorics Conference (COCOON'02) (Singapore, 2002)*, Springer, 2002, 137-143.
18. T. Lange and A. Winterhof, Incomplete character sums over finite fields and their application to the interpolation of the discrete logarithm by Boolean functions, *Acta Arith.* **101** (2002), 223–229.
19. T. Lange and A. Winterhof, Interpolation of the discrete logarithm in F_q by Boolean functions and by polynomials in several variables modulo a divisor of $q - 1$, International Workshop on Coding and Cryptography (WCC 2001) (Paris), *Discrete Appl. Math.* **128** (2003), 193–206.
20. T. Lange and A. Winterhof, Interpolation of the elliptic curve Diffie-Hellman mapping, *Lecture Notes in Comput. Sci.* **2643**, Springer, Berlin, 2003, 51–60.
21. E. El Mahassni and I. Shparlinski, Polynomial representations of the Diffie-Hellman mapping, *Bull. Austral. Math. Soc.* **63** (2001), 467–473.
22. W. Meidl and A. Winterhof, Lower bounds on the linear complexity of the discrete logarithm in finite fields, *IEEE Trans. Inform. Theory* **47** (2001), 2807–2811.
23. W. Meidl and A. Winterhof, A polynomial representation of the Diffie-Hellman mapping, *Appl. Algebra Engrg. Comm. Comput.* **13** (2002), 313–318.
24. G.C. Meletiou, Explicit form for the discrete logarithm over the field GF(p, k), *Arch. Math. (Brno)* **29** (1993), 25–28.
25. G.C. Meletiou, Explicit form for the discrete logarithm over the field GF(p, k), *Bul. Inst. Politeh. Iaşi. Secţ. I. Mat. Mec. Teor. Fiz.* **41(45)** (1995), 1–4.
26. G. Meletiou and G.L. Mullen, A note on discrete logarithms in finite fields, *Appl. Algebra Engrg. Comm. Comput.* **3** (1992), 75–78.
27. A.J. Menezes, P.C. van Oorschot, and S.A. Vanstone, *Handbook of applied cryptography. With a foreword by Ronald L. Rivest*, CRC Press Series on Discrete Mathematics and its Applications, CRC Press, Boca Raton, FL, 1997.
28. W.B. Müller and W. Nöbauer, Some remarks on public-key cryptosystems, Studia Sci. Math. Hungar. 16 (1981) 71–76.
29. W.B. Müller and R. Nöbauer, Cryptanalysis of the Dickson-scheme, Lecture Notes in Comput. Sci. 219 (1985) 50–61.
30. G.L. Mullen and D. White, A polynomial representation for logarithms in GF(q), *Acta Arith.* **47** (1986), 255–261.
31. H. Niederreiter, A short proof for explicit formulas for discrete logarithms in finite fields, *Appl. Algebra Engrg. Comm. Comput.* **1** (1990), 55–57.
32. H. Niederreiter and A. Winterhof, Incomplete character sums and polynomial interpolation of the discrete logarithm, *Finite Fields Appl.* **8** (2002), 184–192.
33. J.-J. Risler, Hovansky's theorem and complexity theory. Ordered fields and real algebraic geometry (Boulder, Colo., 1983), *Rocky Mountain J. Math.* **14** (1984), 851–853.
34. J.-J. Risler, Additive complexity and zeros of real polynomials, *SIAM J. Comput.* **14** (1985), 178–183.

35. J.M. Rojas, Additive complexity and p-adic roots of polynomials, *Lecture Notes in Comput. Sci.* **2369**, Springer, Berlin, 2002, 506–516.

36. J.M. Rojas, Arithmetic multivariate Descartes' rule, *Amer. J. Math.* **126** (2004), 1–30.

37. T. Satoh, On degrees of polynomial interpolations related to elliptic curve cryptography (Extended abstract), Workshop on Coding and Cryptography (WCC) 2005, 55–61.

38. I. Shparlinski, *Number theoretic methods in cryptography. Complexity lower bounds*, Progress in Computer Science and Applied Logic, 17, Birkhäuser, Basel, 1999.

39. I. Shparlinski, On the linear complexity of the power generator, *Des. Codes Cryptogr.* **23** (2001), 5–10.

40. I. Shparlinski, *Cryptographic applications of analytic number theory. Complexity lower bounds and pseudorandomness*, Progress in Computer Science and Applied Logic, 22, Birkhäuser, Basel, 2003.

41. P. Smith and M. Lennon, LUC: a new public key system, in: Proceedings of the Ninth IFIP Int. Symp. on Computer Security, North Holland, 1993, 103–117.

42. M. Wiener, Cryptanalysis of short RSA secret exponents, *IEEE Trans. Inform. Theory* **36** (1990), 553–558.

43. A. Winterhof, A note on the interpolation of the Diffie-Hellman mapping, *Bull. Austral. Math. Soc.* **64** (2001), 475–477.

44. A. Winterhof, Polynomial interpolation of the discrete logarithm, *Des. Codes Cryptogr.* **25** (2002), 63–72.

45. A. Winterhof, A note on the linear complexity profile of the discrete logarithm in finite fields, *Progress Comp. Sci. Appl. Logic* **23** (2004), 359–367.

On Degrees of Polynomial Interpolations Related to Elliptic Curve Cryptography

Takakazu Satoh

Department of Mathematics,
Tokyo Institute of Technology, Tokyo, 152-8551, Japan
satowc5bg@mathpc-satoh.math.titech.ac.jp

Abstract. We study two topics on degrees of polynomials which interpolate cryptographic functions. The one is concerned with elliptic curve discrete logarithm (ECDL) on curves with an endomorphism of degree 2 or 3. For such curves, we obtain a better lower bound of degrees for polynomial interpolation of ECDL. The other deals with degrees of polynomial interpolations of embeddings of a subgroup of the multiplicative group of a finite field to an elliptic curve.

Keywords: Elliptic curves, polynomial interpolation, division polynomials.

1 Introduction

Lange and Winterhof gave many results on degrees of polynomial interpolations of functions related to discrete logarithm based cryptosystems including elliptic curve cryptography in their series of papers [3], [4], [5]. We give two more results by using their technique. The one is concerned with their result on polynomial interpolation for elliptic curve discrete logarithms (ECDL). Let $E/\mathbf{F}_p$ be an elliptic curve and let P be a point of order l. We denote the X-coordinate of nP by x_n for $n \notin l\mathbf{Z}$. Then the following inequality holds for a degree of a polynomial which interpolates ECDL with the base point P.

Theorem *(Lange and Winterhof[5, Proposition 2]). Let $F(X) \in \mathbf{F}_p[X]$, $p \geq 7$, satisfy*

$$F(x_n) = n \quad and \quad F(x_{2n}) = 2n, \quad n \in S$$

for a subset $S \subset \{1, \ldots, [l/4]\}$. Then we have

$$\deg F \geq \frac{\#S}{4}.$$

The constant $\frac{1}{4}$ comes from the degree of the multiplication by 2 map, which is 4. Some elliptic curves have endomorphisms of smaller degree. In Section 2, we replace the multiplication by 2 map by endomorphisms of degree two or three and obtain a better lower bound (but smaller range of S) for such curves.

The other result deals with so-called pairing inversion. Let q be a power of p and consider an arbitrary elliptic curve $E/\mathbf{F}_q$. Let $B \in E(\mathbf{F}_q)$ and put its

Ø. Ytrehus (Ed.): WCC 2005, LNCS 3969, pp. 155–163, 2006.

order l. Assume that l is odd and that $\mathbf{F}_q$ contains a primitive l-th root ζ of unity. (Hence l must be prime to p.) Using the Weil paring or the Tate pairing, we can compute a group isomorphism $\langle B \rangle \to \langle \zeta \rangle$. As is shown by Verheul[9, Sect. 3], the computational complexity for evaluating its inverse isomorphism $t : \langle \zeta \rangle \to \langle B \rangle$ is related to the computational complexity of elliptic curve discrete log problems. Here $\xi(P)$ stands for the X-coordinate of $P \in E - \{\mathcal{O}\}$. Let $S \subset \mathbf{Z}/l\mathbf{Z} - \{0\}$. Assume that $\#S > 2$ and that a univariate polynomial f interpolates the map t over S in the sense that

$$f(\zeta^n) = \xi(nB) \ \ and \ \ f(\zeta^{2n}) = \xi(2nB)$$

for all $n \in S$. We show in Section 3 that $\deg f \geq \frac{1}{5}\#S$ for odd p and $\deg f \geq \frac{1}{2}\#S$ for $p = 2$. The idea of proof is flipping the method of the proof for Lange and Winterhof[5, Proposition 2] inside out.

Notation. Let E be an elliptic curve defined over some perfect field. We assume E is given by the Weierstrass model. We denote the X-coordinate function and the Y-coordinate function by ξ and η, respectively. Let $\tau := -\xi/\eta$ be its local parameter at the point $\mathcal{O}$ at infinity. The order of zero of a rational function f at $P \in E$ is denoted by $\mathrm{ord}_P f$. As usual, we understand that $-\mathrm{ord}_P f$ is the order of pole at P when $\mathrm{ord}_P f < 0$. For a prime p and its power q, we let $\mathrm{Gal}(\mathbf{F}_q/\mathbf{F}_p)$ act on $\mathbf{F}_q[X]$ coefficient wise. We denote the p-th power Frobenius automorphism by σ.

2   Endomorphisms of Small Degree

We begin with a simple (perhaps well known) lemma which is used later. After we have done with it, we list elliptic curves with an endomorphism of small degree and give a new bound for degree of polynomial interpolation for such curves.

Lemma 1. *Let $\varphi \in \mathrm{End}(E)$ be separable. Let $P \in \mathrm{Ker}\varphi$. Then, $\mathrm{ord}_P \xi \circ \varphi = -2$.*

Proof. Let V_{-P} be the translation by $-P$ map, i.e., $V_{-P}(A) = A - P$. Then $\tau_P := \tau \circ V_{-P}$ is a local parameter at P. Since φ is separable, we have an expansion $\tau \circ \varphi = c\tau + O(\tau^2)$ with a non-zero constant c. Recall $\xi = \tau^{-2} + O(1)$ (see e.g. Silverman[7, Chap. 4]). So, $\xi \circ \varphi = c^{-2}\tau^{-2} + O(1)$. Applying V_{-P} on the right, we see $\xi \circ \varphi \circ V_{-P} = c^{-2}\tau^{-2} \circ V_{-P} + O(1)$. Since $P \in \mathrm{Ker}\varphi$, it holds that

$$\xi \circ \varphi = \xi \circ \varphi \circ V_{-P} = c^{-2}\tau_P^{-2} + O(1). \qquad \square$$

Now, there are 7 pairs consisting of an imaginary quadratic algebraic integer β_i (up to multiplication by units) whose norm is 2 or 3 and an order $\mathbf{Z}[\omega_i]$ containing β_i. Explicitly, they are: $\beta_1 := 1 + \sqrt{-1}$, $\beta_2 := \sqrt{-2}$, $\beta_3 := 1 + \sqrt{-2}$, $\beta_4 := \beta_5 := \sqrt{-3}$, $\beta_6 := \frac{-1+\sqrt{-7}}{2}$, $\beta_7 := \frac{-1+\sqrt{-11}}{2}$, and $\omega_1 := \sqrt{-1}$, $\omega_2 := \omega_3 := \sqrt{-2}$,

$\omega_4 := \frac{-1+\sqrt{-3}}{2}$, $\omega_5 := \sqrt{-3}$, $\omega_6 := \beta_6$, and $\omega_7 := \beta_7$. Then, $\{1, \omega_i\}$ forms an integral base of an order of $\mathbf{Q}(\beta_i)$. Using the continued fraction method described in Stark[8], we can find an elliptic curve $E_i/\mathbf{Q}$ whose endomorphism ring is $\mathbf{Z}[\omega_i]$ and an explicit formula for the X-coordinate component for multiplication by β_i. Let $Y^2 = X^3 + A_i X + B_i$ be the Weierstrass equation of E_i. For $c \in \mathbf{Z}[\omega_i]$, we denote by $[c]$ the endomorphism of E_i satisfying $\tau \circ [c] = c\tau + O(\tau^2)$. Let $\Psi_i(X)$, $\Theta_i(X) \in \mathbf{Q}(\omega_i)[X]$ be polynomials such that the X-coordinate of $[\beta_i](x, y)$ is $\Theta_i(x)/\Psi_i(x)$ for $(x, y) \in E_i$, or in the language of rational functions,

$$\xi \circ [\beta_i] = \frac{\Theta_i(\xi)}{\Psi_i(\xi)}.$$

We normalize them so that Θ_i is monic. For completeness, we list them:

β_i	A_i	B_i	$\Theta_i(x)/\Psi_i(x)$
$1+\sqrt{-1}$	5	0	$\dfrac{x^2+5}{2\omega x}$
$\sqrt{-2}$	-30	56	$\dfrac{x^2-4x+18}{-2x+8}$
$1+\sqrt{-2}$	-30	56	$\dfrac{x^3+(-4+2\omega)x^2-(46+28\omega)x+112+80\omega}{(\beta x-4-\omega)^2}$
$\sqrt{-3}$	0	7	$\dfrac{x^3+28}{(\beta x)^2}$
$\sqrt{-3}$	-15	22	$\dfrac{x^3-6x^3+33x-56}{\beta^2(x-3)^2}$
$\dfrac{-1+\sqrt{-7}}{2}$	-35	98	$\dfrac{x^2-(4+\omega)x+(7+21\omega)}{-(2+\omega)x+6+5\omega}$
$\dfrac{-1+\sqrt{-11}}{2}$	-264	1694	$\dfrac{x^3-(24+4\omega)x^2+(396+308\omega)x-2200-2464\omega}{(\beta x+6-10\omega)^2}$

(To ease notation, the subscripts for ω_i and β_i are omitted in the last column.) Put $\nu_i := N_{\mathbf{Q}(\omega_i)/\mathbf{Q}}(\beta_i)$. One can observe that $\deg\Theta_i = \nu_i$ and $\deg\Psi_i = \nu_i - 1$. Moreover, $\Theta_i(X)$ and $\Psi_i(X)$ are relatively prime in $\mathbf{Q}(\omega_i)[X]$ and all the coefficients belong to $\mathbf{Z}[\omega_i]$. In particular, they are algebraic integers. These facts can be proved in general (see e.g. Stark[8]) but for our purpose, the above table is enough.

Remark 1. In the above table, we also observe Ψ_i is square for $\nu_i = 3$. This is not a coincidence. In fact Ψ_i is, up to a constant, $\psi_{\beta_i}^2$ in the notation of [6] when ν_i is odd.

In what follows, we omit the subscript i for simplicity. Let $p \geq 5$ be a prime which splits in $\mathrm{End}(E)$ and choose (and fix) a prime ideal $\mathfrak{p}$ dividing p. We add a bar for objects obtained by the reduction modulo $\mathfrak{p}$. Assume moreover E has a good reduction at $\mathfrak{p}$. Note $\overline{E}$ is an elliptic curve defined over $\mathbf{F}_p$. Let $Q \in \overline{E}(\mathbf{F}_q)$ be a point of order l. Assume $\overline{[\beta]}$ preserves $\langle Q \rangle$. For cryptographically interesting cases (e.g. l is a prime greater than $\sqrt{q}+1$), this condition is automatically satisfied. Let λ be an integer satisfying $\overline{[\beta]}Q = \lambda Q$.

Theorem 1. *Let* $S \subset \{1, \ldots, [l/2]\}$. *Suppose there exists* $f \in \mathbf{F}_q[X]$ *such that*

$$f(\xi(nQ)) = n \quad and \quad f(\xi(\lambda nQ)) = \lambda n \tag{2.1}$$

for all $n \in S$. *Then* $\deg f \geq \frac{1}{\nu}\#S$.

Proof. The equation (2.1) implies

$$f(\xi(\overline{[\beta]}nQ)) = \lambda f(\xi(nQ)).$$

For $n \in S$, the X-coordinate of nQ is a solution of

$$\overline{\Psi}(T)^{\deg f} f\left(\frac{\overline{\Theta}(T)}{\overline{\Psi}(T)}\right) = \lambda \overline{\Psi}(T)^{\deg f} f(T). \tag{2.2}$$

Both the left side and the right side are polynomials of degree $\nu \deg(f)$. We need to show that (2.2) is not an identity. Note

$$\mathrm{ord}_P(\xi - \xi(P)) = \begin{cases} 1 & (P \notin \overline{E}[2]) \\ 2 & (P \in \overline{E}[2], \ P \neq \mathcal{O}) \end{cases}$$

for any $P \in \overline{E} - \{\mathcal{O}\}$. Thus, we see $\mathrm{ord}_P \overline{\Psi} \circ \xi = 2$ for $P \in \mathrm{Ker}\overline{[\beta]}$ regardless of the parity of ν. Lemma 1 implies $\mathrm{ord}_P \overline{\Theta} \circ \xi = 0$, i.e., $\overline{\Theta}(\xi(P)) \neq 0$. Hence, the left side of (2.2) is non-zero at P whereas the right side of (2.2) is zero at P. Thus (2.2) is not an identity and it has $\#S$ roots. Therefore we have $\deg f \geq \frac{1}{\nu}\#S$. □

Remark 2. Contrary to the case of multiplication by two, when n runs over S, the values $\xi(nQ)$ and $\xi(n\overline{[\beta]}Q)$ may not be distinct. Hence, a polynomial f satisfying (2.1) does not always exist. A numerical example: take $Y^2 = X^3 + 5X$ as E, and put $p := 17$, $\mathfrak{p} := (4 - \sqrt{-1})$ and $Q := (2, 1)$. Then $l = 13$. On the other hand, $\overline{[\sqrt{-1}]}(x, y) = (-x, 4y)$ for $(x, y) \in \overline{E}$ and $\overline{[1 + \sqrt{-1}]}Q = (8, 12) = 6Q$. We simply use $\lambda := 6$. Consider a polynomial interpolation for $S := \{1, 2\}$. Then the condition (2.1) implies $f(\xi(Q)) = 1$, $f(\xi(6Q)) = 6$, $f(\xi(2Q)) = 2$ and $f(\xi(12Q)) = 12$. Since the order of Q is 13, the first condition and the last one are clearly incompatible.

One sufficient condition for its existence is as follows: $l \geq 5$ is a prime and for each $n \in S$, either $l \nmid (\lambda n \pm m)$ for all $m \in S$ or $\lambda n \in S$. This restricts the size of S rather small.

We can apply the above technique of using a small degree endomorphism to the polynomial interpolation of Diffie-Hellman mapping due to Kiltz and Winterhof[2, Th. 9]. The (computational) Diffie-Hellman problem on $\langle Q \rangle$ is to find a feasible algorithm which receives mQ and nQ and returns mnQ. For a given $S \subset (\mathbf{Z}/l\mathbf{Z})^{\times 2}$, we evaluate the total degree of polynomial F satisfying

$$F(\xi(mQ), \xi(nQ)) = \xi(mnQ)$$

for all $(m, n) \in S$. In order to obtain a lower bound of degree of F, we need the following simple lemma on the degree of a multivariate polynomial, as in [2]. We use the same notation as above except for that now S is two dimensional.

Lemma 2. *Let R be an integral domain and let A be a finite subset of R. Let $U(X_1, \ldots, X_n) \in R[X_1, \ldots, X_n]$ be a non-zero polynomial of total degree d. Put the number of zeros of U in A^n to m. Then, $m \le d(\#A)^{n-1}$.*

Proof. See von zur Gathen and Gerhard[10, Lemma 6.44]. □

Theorem 2. *Let $Q \in \overline{E}(\mathbf{F}_q)$ be a point of order l. We assume that l is prime to ν and that $l \ge \nu + 3$. Assume $\overline{[\beta]}$ preserves $\langle Q \rangle$ and define $\lambda \in \mathbf{Z}/l\mathbf{Z}$ by $\overline{[\beta]}Q = \lambda Q$. Let $S \subset (\mathbf{Z}/l\mathbf{Z})^{\times 2}$. Suppose $F(X, Y) \in \mathbf{F}_q[X, Y]$ satisfies*

$$F(\xi(mQ), \xi(nQ)) = \xi(mnQ)$$

for all $(m, n) \in S$. Put $S' := \{(m, n) \in (\mathbf{Z}/l\mathbf{Z})^{\times 2} : (m, \lambda n) \in S\}$. Then,

$$\deg F \ge \frac{\#(S \cap S')}{4\lfloor l/2 \rfloor (2\nu - 1)}. \tag{2.3}$$

Proof. The proof is similar to that of Kiltz and Winterhof[2, Th. 9]. Since l is prime to ν, there exist integers v and w satisfying $vl + w\nu = 1$. This gives rise to the equality $vl + w(\mathrm{Tr}(\overline{[\beta]}) - \overline{[\beta]})\overline{[\beta]} = 1$ in $\mathrm{End}(\overline{E})$. Evaluating this at Q, we obtain $w(\mathrm{Tr}(\overline{[\beta]}) - \lambda)\lambda Q = Q$. Therefore, $\lambda \in (\mathbf{Z}/l\mathbf{Z})^{\times}$ and $\overline{[\beta]}$ is an automorphism on $\langle Q \rangle$.

For any m and n, we have

$$F(\xi(mQ), \xi(\lambda nQ)) = F(\xi(mQ), \xi(\overline{[\beta]}nQ)) = F\left(\xi(mQ), \frac{\Theta(\xi(nQ))}{\Psi(\xi(nQ))}\right).$$

On the other hand,

$$\begin{aligned}
F(\xi(mQ), \xi(\lambda nQ)) &= \xi(\lambda mnQ) = \xi(\overline{[\beta]}mnQ) = \frac{\Theta(\xi(mnQ))}{\Psi(\xi(mnQ))} \\
&= \frac{\Theta(F(\xi(mQ), \xi(nQ)))}{\Psi(F(\xi(mQ), \xi(nQ)))}
\end{aligned}$$

for all $(m, n) \in S \cap S'$. Put $d := \deg_Y F$ and

$$U(X, Y) := \Psi(F(X, Y))\Psi(Y)^d \left(F\left(X, \frac{\Theta(Y)}{\Psi(Y)}\right) - \frac{\Theta(F(X, Y))}{\Psi(F(X, Y))}\right).$$

Then U is a bivariate polynomial and $\deg U \le (2\nu - 1)\deg F$ and $U(\xi(mQ), \xi(nQ)) = 0$ for all $(m, n) \in S \cap S'$. Now we prove that U is a non-zero polynomial. First of all we show $\lambda \ne \pm 1$. Otherwise, $(\overline{[\beta]} \pm 1)Q = \mathcal{O}$ and thus $N_{\mathbf{Q}(\beta)/\mathbf{Q}}(\beta \pm 1) \equiv 0 \bmod l$. But the left hand side is $\nu \pm \mathrm{Tr}_{\mathbf{Q}(\beta)/\mathbf{Q}}(\beta) + 1$ which is positive and not greater than $\nu + 3$ (see the table for the possible values of β). Hence $\lambda \ne \pm 1$. This implies that F cannot be a constant unless $S \cap S' = \emptyset$. In the case $S \cap S' = \emptyset$, the inequality (2.3) holds trivially. So we assume $\deg F \ge 1$ in what follows. Then, (2.3) holds in case of $\#S \le 2(l - 1)$. Therefore we have only to prove (2.3) under the condition $\#S \ge 2l - 1$. This in particular implies that there exist $(m_1, n_0), (m_2, n_0) \in S$ satisfying

$m_1 \neq \pm m_2$. Then, $F(\xi(m_1Q), \xi(n_0Q)) \neq F(\xi(m_2Q), \xi(n_0Q))$, which ensures $\deg_X F(X, \xi(n_0Q)) \geq 1$. Hence there exists α in the algebraic closure of $\mathbf{F}_q$ satisfying $F(\alpha, \xi(n_0Q)) = \gamma$ where γ is a root of $\Psi(x) = 0$. Note $\Psi(\xi(n_0Q)) \neq 0$ since otherwise $n_0Q \in \mathrm{Ker}\overline{[\beta]}$, a contradiction. Recall that $\gcd(\Psi(x), \Theta(x)) = 1$ over $\mathbf{F}_q[x]$. Therefore, $\Theta(\gamma) \neq 0$ and by

$$U(\alpha, \xi(n_0Q)) = -\Psi(\xi(n_0Q))^d \Theta(\gamma) \neq 0$$

we conclude that U is a non-zero polynomial. Put $A := \{\xi(mQ) : m \in (\mathbf{Z}/l\mathbf{Z})^\times\}$. The correspondence $(\mathbf{Z}/l\mathbf{Z})^2 \ni (m, n) \to (\xi(mQ), \xi(nQ)) \in A^2$ is at most four to one. By Lemma 2, we obtain $\frac{\#(S \cap S')}{4} \leq (2\nu - 1)[l/2]\deg(F)$. $\square$

Remark 3. In the above proof, in case of $\#S \leq 2(l-1)$, the polynomial U may be zero. Indeed, for $S = \{(\pm1, n) : n \in (\mathbf{Z}/l\mathbf{Z})^\times\}$, we see $F(X, Y) = Y$ and $U(X, Y) = 0$. Note $S = S'$ regardless of λ in this case.

3   Pairing Inversion

Let p be a prime and q a power of p. Throughout this section, we denote by E an elliptic curve

$$Y^2 + a_1 XY + a_3 Y = X^3 + a_2 X^2 + a_4 X + a_6.$$

defined over $\mathbf{F}_q$ by the Weierstrass equation. Let $B \in E(\mathbf{F}_q)$ be a point of odd order l where l is prime to p. Assume $\mathbf{F}_q^\times$ contains the primitive l-th root ζ of the unity. (Otherwise, replace q by its suitable power). Then (replacing ζ if necessary) we have a group isomorphism $\langle B \rangle \to \langle \zeta \rangle$ which sends B to ζ by using the Weil pairing or the Tate pairing. Here we consider polynomial interpolations concerning the inverse isomorphism $t : \langle \zeta \rangle \to \langle B \rangle$. Our aim is to give a lower bound of degree of a polynomial f satisfying $f(z) = \xi \circ t(z)$ for some elements z in $\langle \zeta \rangle$.

Theorem 3. *Let S be a subset of $\mathbf{Z}/l\mathbf{Z} - \{0\}$ whose cardinality is greater than 2. Assume $f(T) \in \mathbf{F}_q[T]$ satisfies*

$$f(\zeta^n) = \xi(nB) \quad and \quad f(\zeta^{2n}) = \xi(2nB) \tag{3.1}$$

for all $n \in S$. Then

$$\deg f \geq \begin{cases} \dfrac{1}{5}\#S & (p \geq 3), \\ \dfrac{1}{2}\#S & (p = 2). \end{cases}$$

Proof. First, we note $2nB \notin E[2]$ for all $n \in S$ since l is odd. For $P := (x, y) \in E - E[2]$, we have

$$\xi(2P) = \frac{U(x)}{V(x)}$$

where

$$U(X) = X^4 - b_4 X^2 - 2b_6 X - b_8, \qquad V(X) = 4X^3 + b_2 X^2 + 2b_4 X + b_6$$

and $b_2, \ldots, b_8$ are constants depending on $a_1, \ldots, a_6$. Then (3.1) implies

$$f(\zeta^{2n}) = \frac{U(f(\zeta^n))}{V(f(\zeta^n))}$$

for all $n \in S$, or equivalently,

$$V(f(T))f(T^2) - U(f(T)) = 0 \tag{3.2}$$

has solutions $T = \zeta^n$ for $n \in S$.

In case that p is odd: We have $\deg V(X) = 3$ and $\deg U(X) = 4$. So, the degree of the left side of (3.2) is exactly $5 \deg f$. Hence $5 \deg f \geq \#S$.

In case of $p = 2$: In this case, $\deg V \leq 2$ and the left side of (3.2) may vanish. We need to prove that this does not happen. Explicitly, $b_2 = a_1^2$, $b_4 = a_1 a_3$, $b_6 = a_3^2$, $b_8 = a_1^2 a_6 + a_1 a_3 a_4 + a_2 a_3^2 - a_4^2$. (Actually, we don't need the explicit form of b_8. What we need on b_8 is the relation between b_8 and the discriminant of E.) Thus $V(X) = (a_1 X + a_3)^2$ and $U(X) = X^4 + a_1 a_3 X^2 + b_8$. We see

$$f(T^2)(a_1 f(T) + a_3)^2 = (f(T)^2 + \sigma^{-1}(a_1 a_3)f(T) + \sigma^{-1}(b_8))^2$$

has solutions $T = \zeta^n$ for $n \in S$. (Recall that σ stands for the Frobenius automorphism.) Then

$$(a_1 \sigma^{-1}(f)(T) + f(T))f(T) + a_3 \sigma^{-1}(f)(T) = \sigma^{-1}(a_1 a_3)f(T) + \sigma^{-1}(b_8) \tag{3.3}$$

has solutions $T = \zeta^n$ for $n \in S$. It's sufficient to show that (3.3) is not an identity. Assume $a_1 \sigma^{-1}(f)(T) + f(T)$ is a constant, say, c. Then $f(T) = a_1 \sigma^{-1}(f)(T) + c$ and we have

$$(ca_1 + a_3 + \sigma^{-1}(a_1 a_3)a_1)\sigma^{-1}(f)(T) = c^2 + \sigma^{-1}(a_1 a_3)c + \sigma^{-1}(b_8).$$

Since $\#S > 2$, the condition (3.1) implies that f is not a constant. Hence

$$ca_1 + a_3 + a_1 \sigma^{-1}(a_1 a_3) = 0, \tag{3.4}$$

$$c^4 + a_1 a_3 c^2 + b_8 = 0. \tag{3.5}$$

Then

$$b_2^2 b_8 \overset{(3.5)}{=} a_1^4 b_8 = (a_1 c)^4 + a_1^3 a_3 (a_1 c)^2$$

$$\overset{(3.4)}{=} a_3^4 + a_1^6 a_3^2 + a_1^3 a_3^3 + a_1^3 a_3 a_1^3 a_3 = a_3^4 + a_1^3 a_3^3.$$

On the other hand, the discriminant Δ of E is

$$\Delta = b_2^2 b_8 + b_6^2 + b_2 b_4 b_6$$

$$= a_3^4 + a_1^3 a_3^3 + a_3^4 + a_1^3 a_3^3$$

$$= 0.$$

This contradicts that E is an elliptic curve. Thus $\deg(a_1\sigma^{-1}(f)(T) + f(T)) \geq 1$ and the degree of the left side of (3.3) is at least $\deg f + 1$. Therefore (3.3) is not an identity and it has at most $2 \deg f$ solutions. Hence $\#S \leq 2 \deg f$.   □

Remark 4. We cannot drop the condition $\#S > 2$. Indeed, it is trivial to see that the case $l = 3$ and $S = \{1, 2\}$ yields a counter example.

We can generalize the statement for $p = 2$ as follows. The proof is essentially the same.

Theorem 4. *Let $p \geq 3$ be a prime. Let B, l and S be as in Theorem 3. Assume $f(X) \in \mathbf{F}_q[X]$ satisfies*

$$f(\zeta^n) = \xi(nB) \quad and \quad f(\zeta^{pn}) = \xi(pnB) \tag{3.6}$$

for all $n \in S$. Then $\deg f \geq \frac{1}{p}\#S$.

Proof. For $m \in \mathbf{N}$, let ψ_m be the m-th division polynomial and put $\theta_m(X) = X\psi_m^2(X) - (\psi_{m-1}\psi_{m+1})(X)$. Then,

$$\xi \circ [p] = \theta_p(\xi)/\psi_p(\xi)^2.$$

As before, $T = \zeta^n$ is a solution of

$$f(T^p)\psi_p(f(T))^2 = \theta_p(f(T)) \tag{3.7}$$

for $n \in S$. Recall that $\theta_p(x)$ and $\psi_p(x)$ are relatively prime and both of them are inseparable by Cassels[1]. Thus there exist α_p, $\beta_p \in \mathbf{F}_q[X]$ satisfying $\theta_p(X) = \alpha_p(X)^p$ and $\psi_p(X) = \beta_p(X)^p$. Thus,

$$(\sigma^{-1}(f))(T)\beta_p(f(T))^2 = \alpha_p(f(T)) \tag{3.8}$$

has solutions $T = \zeta^n$ for $n \in S$. Since $\theta_p(X)$ and $\psi_p(X)$ are relatively prime, so are $\alpha_p(X)$ and $\beta_p(X)$. Moreover f is not a constant by the assumption $\#S > 2$. Thus (3.8) is not an identity. Note $\deg \alpha_p \leq p$ and $\deg \beta_p \leq \frac{p-1}{2}$ since p is odd. Therefore we have $\#S \leq p \deg f$.   □

Acknowledgements. The author would like to thank Tanja Lange and Arne Winterhof for their comments on earlier versions of this paper.

References

[1] Cassels, J. W. S.: A note on the division values of $\wp(u)$. *Proc. Cambridge Philos. Soc.* **45** (1949) 167-172.
[2] Kiltz, E., Winterhof, A.: On the interpolation of bivariate polynomials related to the Diffie-Hellman mapping. *Bull. Austral. Math. Soc.* **69** (2004) 305-315.
[3] Lange, T., Winterhof, A.: Polynomial interpolation of the elliptic curve and XTR discrete logarithm, Proc. COCOON 2002, Lect. Notes Comput Sci., **2387**, 137-143, Berlin Heidelberg: Springer, 2002.

[4] Lange, T., Winterhof, A.: Interpolation of the discrete logarithm in $\mathbf{F}_q$ by boolean functions and by polynomial in several variables modulo a divisor of $q-1$. *Discrete Appl. Math.* **128** (2003) 193-206.

[5] Lange, T., Winterhof, A.: Interpolation of the elliptic curve Diffie-Hellman mapping, Proc. AAECC 2003, Lect. Notes Comput. Sci., **2643**, 51-60, Berlin Heidelberg: Springer, 2003.

[6] Satoh, T.: Generalized division polynomials. *Math. Scand.* **94** (2004) 161-184.

[7] Silverman, J. H.: "The arithmetic of elliptic curves". GTM, 106. Berlin-Heidelberg-New York: Springer 1985.

[8] Stark, H.M.: Class-numbers of complex quadratic fields, Modular functions of one variable, I, (Proc. Internat. Summer School, Univ. Antwerp, Antwerp, 1972), Lect. Notes in Math., **320**, 153-174, Berlin: Springer, 1973.

[9] Verheul, E.R.: Evidence that XTR is more secure than supersingular elliptic curve cryptosystem, Advances in cryptology - EUROCRYPT 2001, **2045**, 195-210, Berlin Heidelberg: Springer-Verlag, 2001.

[10] von zur Gathen, J., Gerhard, J.: "Modern computer algebra (2nd ed.)". Cambridge: Cambridge UP 2003.

Finding Good Differential Patterns for Attacks on SHA-1

Krystian Matusiewicz and Josef Pieprzyk

Centre for Advanced Computing - Algorithms and Cryptography,
Department of Computing, Macquarie University,
Sydney, NSW 2109, Australia
{kmatus, josef}@ics.mq.edu.au

Abstract. In this paper we analyse properties of the message expansion algorithm of SHA-1 and describe a method of finding differential patterns that may be used to attack reduced versions of SHA-1. We show that the problem of finding optimal differential patterns for SHA-1 is equivalent to the problem of finding minimal weight codeword in a large linear code. Finally, we present a number of patterns of different lengths suitable for finding collisions and near-collisions and discuss some bounds on minimal weights of them.

1 Introduction

Most of the modern hash functions used in practice are dedicated ones designed using principles of MD4 [18, 19]. The first attack on MD4 appeared only a year after the publication of the algorithm [7]. Both MD4 and its improved version, called MD5 [20], were broken by Dobbertin [10, 8]. Another hash function from the MD family, called RIPEMD [3] was also shown by Dobbertin [9] to be insecure. The shortest variant of HAVAL [29] has been broken by Van Rompay et al. [21]. Recent results obtained by Wang et al. [24, 25] show that is is possible to find collisions for MD4, MD5, HAVAL-128 and RIPEMD within hours on a generic PC. It looks like the message expansion algorithm based on permuting message words and applying them in a different order in each round is a weak point of all these algorithms as it does not provide enough diffusion of differences.

Another group of hash functions are hash functions from the SHA family. The idea of an extended Feistel permutation that was used in the design of the MD family, is also driving the design of the SHA family but with more complex message expansion algorithms. The first member of that family was SHA-0 [11]. It was promptly replaced by an improved version, SHA-1 [15]. Security concerns that led to the re-design of SHA-0 appeared to be true, as in 1998 Chabaud and Joux presented a theoretical attack on SHA-0 [6], which was later implemented and improved allowing to find collisions [12, 13]. Now, one of the most interesting questions in the field of hash function analysis is how secure is the present standard SHA-1, which is different from SHA-0 by only one rotation in the message expansion process.

The same technique used to attack SHA-0 could be applied to launch an attack on SHA-1 provided that there exists a good enough differential pattern. Biham

Ø. Ytrehus (Ed.): WCC 2005, LNCS 3969, pp. 164–177, 2006.

and Chen [2] were able to find patterns that allowed to find collisions for SHA-1 variants reduced to first 34 and 36 steps. Their attack can be extended provided that one can find good differential patterns for longer variants of SHA-1.

In this paper we investigate the problem of finding good differential patterns for SHA-1. First we start with a different presentation of the message expansion algorithm. Next, we show that the problem of finding differential patterns suitable for attacking SHA-1 is equivalent to a problem of finding low-weight codewords in a linear code.

We present the results of our search for the best patterns which can be used in the differential attack and we estimate some bounds on the minimal weight of such patterns.

2   The Differential Attack on SHA

In this section we briefly recall the structure of SHA-1 and describe the basic framework of the differential attack applicable to SHA-0/1.

2.1   Description of the SHA-1 Compression Function

The SHA-1 compression function [15] hashes 512 bit input messages to 160 bit digests. Firstly, 512 bits of the message are divided into 16 32-bit words $W_0, W_1, \ldots, W_{15}$. The rest of 80 words is generated out of the first 16 words according to the following recurrence formula

$$W_i = ROL^1(W_{i-3} \oplus W_{i-8} \oplus W_{i-14} \oplus W_{i-16}) \text{ for } 16 \leq i \leq 79 , \quad (1)$$

where ROL^k denotes rotation of a word by k positions left. If this is the first application of the compression function, five 32-bit registers A, B, C, D, E are initialized to values $A_0 = \text{0x67452301}$, $B_0 = \text{0xefcdab89}$, $C_0 = \text{0x98badcfe}$, $D_0 = \text{0x10325476}$, $E_0 = \text{0xc3d2e1f0}$ accordingly.

Next, the algorithm applies 80 steps ($i = 0, \ldots, 79$). Each step is of the following form:

$$A_{i+1} = ROL^5(A_i) \boxplus f_i(B_i, C_i, D_i) \boxplus E_i \boxplus W_i \boxplus K_i , \quad (2)$$
$$B_{i+1} = A_i ,$$
$$C_{i+1} = ROL^{30}(B_i) ,$$
$$D_{i+1} = C_i ,$$
$$E_{i+1} = D_i ,$$

where $\boxplus$ denotes addition modulo 2^{32} and A_i, B_i, C_i, D_i and E_i denote the values of the registers after i-th iteration. Functions f_i and constants K_i used in each iteration are given in Table 1. The output of the compression function is the concatenation of bits of $A_0 \boxplus A_{80}$, $B_0 \boxplus B_{80}$, $C_0 \boxplus C_{80}$, $D_0 \boxplus D_{80}$ and $E_0 \boxplus E_{80}$.

Table 1. Functions and constants used in SHA-1

step number i	$f_i(B, C, D)$	K_i
$0 - 19$	$(B \wedge C) \oplus (\neg B \wedge D)$	0x5a827999
$20 - 39$	$B \oplus C \oplus D$	0x6ed9eba1
$40 - 59$	$(B \wedge C) \vee (B \wedge D) \vee (C \wedge D)$	0x8f1bbcdc
$60 - 79$	$B \oplus C \oplus D$	0xca62c1d6

2.2   The Differential Attack of Chabaud and Joux

Chabaud and Joux presented in [6] a differential attack on SHA-0. The fundamental observation they made is that a change in the j–th bit of the word W_i can be corrected by complementary changes in the following bits:

○ bit $(j + 6) \bmod 32$ of W_{i+1},
○ bit j of word W_{i+2},
○ bit $(j + 30) \bmod 32$ of W_{i+3},
○ bit $(j + 30) \bmod 32$ of W_{i+4},
○ bit $(j + 30) \bmod 32$ of W_{i+5},

provided that functions $f_{i+1}, \ldots, f_{i+4}$ and additions $\boxplus$ behave like linear functions. That is, a single change of the input to f results in a change of the output of f, a change in two inputs of f leaves the result unchanged and differences propagate through additions without carries. They showed that a one bit disturbance can be corrected by such a pattern with probability between 2^{-2} and 2^{-5} depending on functions $f_i, \ldots, f_{i+4}$, if the disturbance is introduced in the second bit ($j = 1$).

If a disturbance is introduced in the position $j \neq 1$, then there is an additional factor of 2^{-3} caused by 50% chance of inducing a carry in additions in steps $i+3$, $i + 4$, $i + 5$.

The attack is possible due to the property of the message expansion function which does not mix bits in different positions. Thanks to that it was possible to consider the message expansion algorithm as a bit-wise one. Enumeration of all 2^{16} possible bit patterns in the position 1 allowed for choosing a disturbance pattern in the first bit position that led to a global differential pattern δ producing a collision with probability 2^{-61}.

2.3   Improvements

It is possible to improve the attack of Joux and Chabaud by reducing probabilistic behaviour of some initial corrections using a better strategy of selecting messages rather than picking random ones. Biham and Chen proposed in [1] the method of so-called neutral bits. They showed that having a message that behaves correctly for at least 16 first steps after adding a difference δ, it is possible to construct a big set of pairs $(M, M \oplus \delta)$ that have much better probability of a successful correction than the pairs produced from random messages.

3   Analysis of the Message Expansion Algorithm of SHA-1

An additional rotation in the message expansion formula (1) makes finding corrective patterns used in [6] impossible, because now differences propagate to other positions. For SHA-1, a one-bit difference in one of the 16 initial blocks propagates itself to at least 107 bits of the expanded message W. This is illustrated in Fig. 1. However, we were able to find a difference pattern with only 44 bit changes in the expanded message. This suggests that it is interesting to investigate the message expansion algorithm of SHA-1 in a greater detail and check to what extent the differential attack can be applied also to SHA-1.

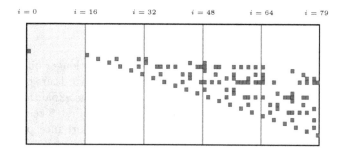

Fig. 1. Propagation of one bit difference in SHA-1 message expansion

The important property of the message expansion process given by the formula (1) is that it is a bijective function producing 16 new words out of 16 old ones. This implies that it is possible to reconstruct the whole expanded message given any 16 consecutive words of it, in particular the first 16. Moreover, if we consider it on a bit level as a function $A : \mathbb{F}^{512} \rightarrow \mathbb{F}^{512}$, it is easy to see that A is $\mathbb{F}_2$-linear as the only operations used are word rotations (which are permutations of bits) and bitwise XOR operations. Then the expansion of the initial message[1] $m \in \mathbb{F}^{512}$ can be expressed as a long vector

$$E_1(m) = \begin{bmatrix} m \\ \hline A(m) \\ \hline A^2(m) \\ \hline A^3(m) \\ \hline A^4(m) \end{bmatrix} \in \mathbb{F}^{2560} \quad . \tag{3}$$

The set of correction masks is built from a disturbance pattern by rotations and delaying the pattern by $1, 2, \ldots, 5$ words in the same way as described in [6]. In order to find disturbance patterns which can give rise to correction patterns one has to look for bit patterns $b \in \mathbb{F}^{2560}$ that satisfy the following conditions:

---

[1] We consider m to be a column vector.

C1. the pattern b has to be of the form (3), i.e. b is the result of the expansion operation,

C2. the pattern b ends with $5 \cdot 32 = 160$ zero bits (the last five words are zero), because each disturbance is corrected in the next 5 steps, so no disturbance may occur after the word 74,

C3. after delaying a pattern by up to 5 words (that is, shifting bits of b down (right) by $5 \cdot 32 = 160$ positions) the shifted pattern must also be the result of the expansion of its first 512 bits, that is

$$[\underbrace{0 \ldots 0}_{160 \text{ bits}} b_0 \, b_1 \, \ldots \, b_{2399}]^T = E_1([0 \ldots 0 \, b_0 \ldots \, b_{351}]^T) \ .$$

C4. b has both the minimal Hamming weight and the maximal number of non-zero bits in position 1.

3.1   Basic Construction

Conditions C1 – C3 imply that in fact we are looking for longer bit sequences of 85 words such that the first 5 words are zero, the next 11 words are chosen in such a way that while the rest of the words are the result of the expansion of the first 16, the last 5 words are zero again. After denoting the first 5 zero words with indices $-5, \ldots, -1$, in positions $0, \ldots, 79$ we get a disturbance pattern which allows for a construction of the corrective pattern.

Using the matrix notation, we are looking for a vector $m \in \mathbb{F}^{512}$ such that $A^4 m$ has 160 trailing zero bits and also $A^{-1}m$ has 160 trailing zeros. As the transformation A is a bijection, this is equivalent to finding a vector

$$v = [v_0, v_1, \ldots, v_{351}, 0, \ldots, 0]^T \in \mathbb{F}^{512} \ ,$$

such that the last 160 bits of $A^{-5}(v)$ contain only zeros, what can be written as

$$
\begin{bmatrix} x_0 \\ x_1 \\ \vdots \\ x_{351} \\ \hline 0 \\ \vdots \\ 0 \end{bmatrix}
=
\begin{bmatrix} a_{0,0} & \cdots & & \cdots & a_{0,511} \\ & \vdots & & & \vdots \\ & & & & \\ a_{352,0} & \cdots & a_{352,351} & & \\ & \vdots & \ddots & \vdots & \vdots \\ a_{511,0} & \cdots & a_{511,351} & \cdots & a_{511,511} \end{bmatrix}
\cdot
\begin{bmatrix} v_0 \\ v_1 \\ \vdots \\ v_{351} \\ 0 \\ \vdots \\ 0 \end{bmatrix}
, \tag{4}
$$

where $A^{-5} = (a_{i,j})_{0 \le i,j \le 511}$.

This condition means that truncated vectors $\bar{v} = [v_0, v_1, \ldots, v_{351}]^T \in \mathbb{F}^{352}$ have to belong to the null-space of the matrix Ω of the form

$$
\Omega = \begin{bmatrix} a_{352,0} & \cdots & a_{352,351} \\ \vdots & \ddots & \vdots \\ a_{511,0} & \cdots & a_{511,351} \end{bmatrix} , \tag{5}
$$

created as a copy of the lower left part of the matrix A^{-5}. It means that the set of all vectors satisfying properties 1– 2 is a linear subspace of $\mathbb{F}^{2560}$ with elements of the form

$$c = [\, A^{-4}(v)^T \,\|\, A^{-3}(v)^T \,\|\, A^{-2}(v)^T \,\|\, A^{-1}(v)^T \,\|\, v^T \,]\,, \tag{6}$$

where $v = [\, \bar{v}^T \,\|\, 0\ldots0\,]^T \in \mathbb{F}^{512}$ and $\bar{v} \in Ker(\Omega)$.

The set of all such vectors c is in fact a linear code C of length $n = 2560$ and, as we have verified that the rank of the matrix Ω is equal to 192, of dimension $k = 192$.

To maximize the probability of a successful correction by the differential pattern, it is necessary to search for the words of minimal Hamming weight and, if possible, for those words with the maximal number of non-zero bits in the position 1.

This is essentially a problem of finding the minimum distance of a linear code, which is known to be NP-hard [22], so there is no easy way of finding optimal corrective patterns. However, there are a number of probabilistic methods [14, 4] that allow for efficient finding of low-weight codewords in big linear codes.

The second part of the condition C4 can be partially achieved using the fact that the expansion process is invariant with respect to the word rotation. The result of the expansion of 16 input words already rotated by a number of bits is the same as the rotation of the result of the expansion of 16 words performed without rotation. Thanks to that, having a pattern of minimal weight it is easy to transform it to a pattern with the maximal number of ones in the position 1 using the word-wise rotation by an appropriate number of positions. Of course, in general this is the problem of finding codewords with the minimal weighted weight, however, our experiments show that this simplified approach gives very good results.

3.2   Reduced Variants

The generalization of the construction presented above can be applied to find good differential patterns for reduced versions of SHA-1.

Assume that we want to find a differential pattern for SHA-1 reduced to $16 < s \le 80$ steps (2). Condition C1 implies that the vector $A^{-1}(m)$ has to have 160 trailing zero bits. If we denote the last 160 rows of the matrix A^{-1} as $A^{-1}[352 :: 511]$ then this condition can be written as

$$0 = A^{-1}[352 :: 511] \cdot m\ . \tag{7}$$

To formulate a simple description of constraints inferred from condition C2, it is convenient to note that the whole message expansion process can be seen as a linear transform $E_1 : \mathbb{F}^{512} \to \mathbb{F}^{2560}$ represented by a matrix of the form

$$E_1 = \begin{bmatrix} I_{512} \\ \hline A \\ \hline A^2 \\ \hline A^3 \\ \hline A^4 \end{bmatrix}\ ,$$

where I_{512} is the identity matrix and A is the linear transform described in Section 3. Now, if we want to find a differential pattern for s steps, 5 words of the expanded message in positions $s - 4$, $s - 3$, ..., s have to be zero. In the matrix notation, 160 entries in the vector $E_1 \cdot m$ have to be zero, precisely these in positions $(s - 4) \cdot 32$, ..., $s \cdot 32 + 31$. If we denote the matrix created by selecting rows of the matrix E_1 with indices $32(s - 4)$, ..., $32s + 31$ by $E_1[32(s - 4) :: 32s + 31]$, then condition C2 can be written as:

$$0 = E_1[32(s - 4) :: 32s + 31] \cdot m \ . \tag{8}$$

Putting together Equations (7) and (8) we obtain the final result. A message $m \in \mathbb{F}^{512}$ gives rise to the corrective pattern if and only if $m \in Ker(\Psi_s)$, where

$$\Psi_s = \left[\frac{A^{-1}[352 :: 511]}{E_1[32(s - 4) :: 32s + 31]} \right] \tag{9}$$

is a matrix of dimensions 320×512 built by placing rows of $E_1[32(s-4) :: 32s+31]$ below rows of $A^{-1}[352 :: 511]$.

4   Search for the Best Patterns

We have shown that the problem of finding disturbance patterns with minimal weights can be seen as a problem of finding minimal weight codewords in a linear code. To find them, we use a simplified version of the algorithm by Leon [14] presented in [5]. We use the parameter $p = 3$ to search for all combinations of up to three rows and for each code we apply at least 100 repetitions of the procedure. The results are presented in Table 2. For each variant of SHA-1 (of length 32 - 85) the second column contains the minimal weight of the pattern found. The results marked with (*) are better than those obtained by Biham and Chen [2]. The patterns we investigate are suitable for attacking only last steps of SHA-1. As the first 20 steps of SHA-1 employ the IF Boolean function, the first 16 words of a disturbance pattern cannot have ones in the same bit position in the two consecutive words. Thus for variants longer than 64, we give only lower bounds on the weight of patterns satisfying the IF condition.

We decided to compute a lower bound because the algorithm we used ensures that there is no codeword of a lower weight with a very high probability. This result is unlikely to be extended in a straightforward way to the case of search for restricted patterns satisfying the IF condition. A way out is finding the lower bound on weights of restricted patterns using unrestricted ones.

According to Biham and Chen [2], it is possible to eliminate the probabilistic behaviour of up to 20 first rounds. Thus the third column (denoted by wt_{20+}) contains minimal weights of patterns where weights of the first 20 steps are not counted.

We are also interested in patterns that do not allow for finding the full collisions but still are suitable for finding near-collisions as this may possibly lead to an easier way of finding multi-block collisions. To obtain them we relax the

Table 2. Hamming weights of the best patterns found. Column wt contains total Hamming weights of patterns, wt_{20+} – weights of patterns with ignored 20 first steps, column wt_n shows total weights of incomplete patterns for near-collisions (patterns ending with only 4 zero blocks).

steps	wt	wt_{20+}	wt_n	steps	wt	wt_{20+}	wt_n	steps	wt	wt_{20+}	wt_n
32	9	2	9	50	35	14	35	68	> 122	> 78	> 90
33	9	2	9	51	35	15	35	69	> 127	> 81	> 127
34	9	2	9	52	35	16	35	70	> 142	> 80	> 124
35	28	4	24	53	35	16	35	71	> 157	> 94	> 142
36	24	5	24	54	78	36	75	72	> 172	> 93	> 139
37	25	5	25	55	80	39*	73	73	> 139	> 111	> 139
38	30	8	30	56	79	41	72	74	> 139	> 98	> 139
39	39	8*	35	57	72	42	72	75	> 142	> 90	> 142
40	41	11	38	58	73	42	55	76	> 187	> 111	> 187
41	41	12	41	59	91	51	66	77	> 184	> 108	> 184
42	41	13	34	60	66	44	66	78	> 198	> 115	> 177
43	41	17	41	61	66	44	66	79	> 220	> 115	> 220
44	50	15	42	62	66	45	66	80	> 172	> 106	> 172
45	45	15	45	63	107	64	87	81	> 255	> 117	
46	56	23	42	64	> 101	> 60	> 96	82	> 242	> 142	
47	56	24*	35	65	> 113	> 66	> 98	83	> 215	> 163	
48	35	14	35	66	> 98	> 58	> 98	84	> 161	> 101	
49	35	14	35	67	> 127	> 69	> 122	85	> 340	> 177	

condition that requires that the last five words must contain zeros only and we allow for non-zero entries in one more block. Weights of the best patterns found this way are listed in the column wt_n.

It is interesting to see that the minimal weights we are able to find are growing in quite an irregular fashion. In fact, after a rapid jump after reaching 35 steps and a steady growth up till the step 47, there is an unexpected downfall to the weight 35 in the step 48. The same pattern, presented in Fig. 6, is suitable for attacks up to 53 steps. After 53 steps, weights get much higher and as we consider patterns without restrictions imposed by the IF function in the first 20 steps of SHA-1, the best pattern for the full SHA-1 will most likely have weight considerably higher than 172.

However, when we relax all the conditions and look only for patterns that result from the expansion process, we are able to find differences with the weight only 44 for the full length message expansion. Such a difference is presented in Table 4.

5   Bounds on Minimal Weights of Short Patterns

Let us discuss some bounds on minimal weights of corrective patterns. Consider the inverse of the transformation (1). It can be written as

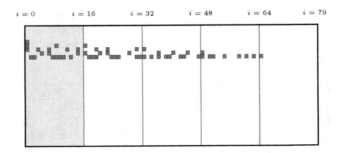

Fig. 2. Inverse propagation of one bit difference applied in the last segment of SHA-1

$$W_i = W_{i+2} \oplus W_{i+8} \oplus W_{i+13} \oplus ROR^1(W_{i+16}), \quad 0 \le i < 64 , \qquad (10)$$

where the last 16 words $W_{64},\ldots,W_{79}$ are set arbitrarily.

Although this formula describes essentially the same transformation, if we consider the fact that the rotation is now applied to only one variable distant by 16 steps, the difference propagation of the expansion process described by Equation (10) is much worse than the original function. In fact, the difference of one bit in one of the last 16 words generates up to 4 changes positioned 55 to 82 bits, what is illustrated in Fig. 2. It is interesting to note that this peculiar behaviour does not depend on the number of positions by which a word is rotated in the algorithm but is rather inherent to the structure of recurrence relations similar to (1).

To estimate the minimal number of ones in the expansion process we divide the set of ones in two groups: these in the same position as the initial bit and those in different positions. The size of the first group can be easily found experimentally, as there are only 2^{16} of all bit sequences generated by the following relation

$$w_i = \begin{cases} m_i & \text{for } 0 \le i < 16, \\ w_{i+2} \oplus w_{i+8} \oplus w_{i+13}, & \text{for } i \ge 16 \end{cases}$$

and much less of them with the first five and the last five elements equal to zero. Minimal weights of such sequences of different lengths are presented in Table 3. Note that to estimate the number of ones for a differential pattern of length s,

Table 3. Minimal weights of sequences of length $s + 5$ with 5 leading and 5 trailing zeros generated by the formula $w_i = w_{i+2} \oplus w_{i+8} \oplus w_{i+13}$

s	32–34	35–38	39,40	41	42,43	44–47	48,49	50	51
min. weight	8	9	11	13	11	14	16	17	16
s	52,53	54–56	57–64	65–67	68–71	72	73–75	76,77	78–85
min. weight	17	18	19	23	22	26	24	29	30

Table 4. A full length difference of weight 44 for unrestricted message expansion of SHA-1

0x00000002	0x00000001	0x00000000	0x00000000	0x00000008
0x00000002	0x00000000	0x00000000	0x00000000	0x00000020
0x00000000	0x00000002	0x00000002	0x00000000	0x00000000
0x00000000	0x00000002	0x00000000	0x00000000	0x00000000
0x00000001	0x00000001	0x00000002	0x00000000	0x00000040
0x00000000	0x00000000	0x00000000	0x00000000	0x00000000
0x80000002	0x00000000	0x00000002	0x00000000	0x00000028
0x00000002	0x00000002	0x00000000	0x00000000	0x00000080
0x80000002	0x00000003	0x00000002	0x00000004	0x00000018
0x00000000	0x00000000	0x00000000	0x00000000	0x00000000
0x00000002	0x00000002	0x00000000	0x00000000	0x00000100
0x00000000	0x00000002	0x00000000	0x00000008	0x00000020
0x00000003	0x00000000	0x00000000	0x00000000	0x000000a0
0x00000000	0x00000000	0x00000000	0x00000000	0x00000200
0x00000002	0x00000002	0x00000000	0x00000010	0x00000020
0x00000002	0x00000000	0x00000000	0x00000000	0x00000000

Table 5. The best differential pattern for the first 34 steps of SHA-1

W[0]= 0x00000002	W[16]= 0x00000000	W[32]= 0x00000000
W[1]= 0x00000000	W[17]= 0x00000000	W[33]= 0x00000000
W[2]= 0x00000002	W[18]= 0x00000000	
W[3]= 0x00000000	W[19]= 0x00000000	
W[4]= 0x00000002	W[20]= 0x00000002	
W[5]= 0x00000000	W[21]= 0x00000000	
W[6]= 0x00000003	W[22]= 0x00000002	
W[7]= 0x00000000	W[23]= 0x00000000	
W[8]= 0x00000000	W[24]= 0x00000000	
W[9]= 0x00000002	W[25]= 0x00000000	
W[10]= 0x00000000	W[26]= 0x00000000	
W[11]= 0x00000000	W[27]= 0x00000000	
W[12]= 0x00000000	W[28]= 0x00000000	
W[13]= 0x00000000	W[29]= 0x00000000	
W[14]= 0x00000002	W[30]= 0x00000000	
W[15]= 0x00000000	W[31]= 0x00000000	

the minimal weight of a sequence of length $s + 5$ has to be considered with 5 leading and 5 trailing zero bits.

The size of the other group of bits cannot be easily estimated. We only can say that it contains at least one element for sequences longer than 16. This makes our estimation work only for variants that are not too long.

As an example, we can consider the differential pattern for 34 steps. The first set for sequences of length 34 contains at least 8 non-zero bits. The second set must contain at least one bit. Thus, we have shown that the pattern presented

Table 6. The best differential pattern for the last 53 steps of SHA-1

	W[32]=0x00000002	W[48]=0x80000000	W[64]=0x00000002
	W[33]=0x80000000	W[49]=0x00000002	W[65]=0x00000000
	W[34]=0x40000003	W[50]=0x80000001	W[66]=0x00000001
	W[35]=0x00000000	W[51]=0x00000000	W[67]=0x00000000
	W[36]=0x00000001	W[52]=0x00000002	W[68]=0x00000000
	W[37]=0x80000002	W[53]=0x00000002	W[69]=0x00000002
	W[38]=0x80000000	W[54]=0x00000000	W[70]=0x00000000
	W[39]=0x00000002	W[55]=0x00000000	W[71]=0x00000000
	W[40]=0x00000001	W[56]=0x00000002	W[72]=0x00000002
	W[41]=0x00000000	W[57]=0x00000000	W[73]=0x00000000
	W[42]=0x80000002	W[58]=0x00000003	W[74]=0x00000000
W[27]=0x00000000	W[43]=0x00000002	W[59]=0x00000000	W[75]=0x00000000
W[28]=0x00000000	W[44]=0x80000002	W[60]=0x00000002	W[76]=0x00000000
W[29]=0x00000000	W[45]=0x00000000	W[61]=0x00000002	W[77]=0x00000000
W[30]=0x40000000	W[46]=0x80000001	W[62]=0x00000002	W[78]=0x00000000
W[31]=0x00000000	W[47]=0x00000000	W[63]=0x00000000	W[79]=0x00000000

in Table 5 is the optimal one for that length. This is the same pattern used by Biham and Chen to find collisions for 34 steps of SHA-1 [2].

6   Conclusions

In this paper we have presented a new characterization of the message expansion process of SHA-1 using linear codes over $\mathbb{F}_2$. This immediately has allowed us to prove that the problem of finding the best differential pattern for SHA-1 is equivalent to the problem of finding the minimum weight codeword in a particular linear code.

Although this problem is hard in general and codes describing message expansion are very long, thanks to an unexpected behaviour of the codes in question, we were able to find differential patterns for reduced versions of SHA-1 of lengths between 34 and 85 steps experimentally.

Our study has shown that minimal weights for reduced variants may vary in an unexpected way. Nevertheless, the longest variant, for which the differential attack of Joux and Chabaud with necessary improvements seems to be possible, is the version of the last 53 steps of SHA-1. We have presented the actual differential for this variant and have improved in a few places weights for shorter variants given by Biham and Chen in [2].

In an effort to establish lower bounds on weights of differences, we have derived some bounds on minimal weights of short differential patterns and have proved that the 34-step differential characteristics used by Biham and Chen is the optimal one for this length. It is interesting to note that all these results are quite general and can be applied to any message expansion structure that uses transformations linear over $\mathbb{F}_2$.

The problem of the resistance of SHA-1 against Chabaud-Joux attack was also studied independently by Rijmen and Oswald [17]. They considered different linear approximations of non-linear Boolean functions and presented results of their search for low-weight differences using a dedicated algorithm they designed. Their results agree with ours, they also report 44 as the minimal weight of the message expansion difference and claim that variants of SHA-1 up to 53 steps can be attacked using disturbance-corrections strategy.

7 Addendum

The year 2005 was very exciting for researchers working on cryptographic hash functions. Shortly after WCC'2005 workshop, Wang et al. presented their final results of the analysis of MD4 [23] and MD5 [27]. Soon after that, they used their techniques of modular differentials to control propagation of differences and message modification to increase the probability of a differential in a practical attack on SHA-0 with the complexity of 2^{39} hash evaluations [28] and the first theoretical attack on the full SHA-1 [26].

Their attack on SHA-1 is based on multi-block near-collisions. Using modular differentials they were able to overcome the problem of consecutive disturbances in the first round by using an irregular differential and controlling differences "by hand" in the first 20 steps. Finding near-collisions instead of full collisions essentially made conditions C2 and C3 (conditions 1 and 2 in Table 2 of their paper [26]) unnecessary and enabled them to use a very low weight disturbance pattern which was a shifted version of the pattern presented in our Table 4.

Another result concerning SHA-1, interesting in the context of this paper, was presented by Pramstaller at al. [16]. Roughly speaking, they represented the differences in chaining variables of a linearized variant of SHA-1 as a linear function of a message difference and tried to find low-weight differences of chaining variables treated as a huge linear code. Later, they derived a set of conditions that made the original function behave like the linear approximation for a selected low-weight difference.

We are sure that coming months will bring even more new results on the analysis of cryptographic hash functions.

References

1. E. Biham and R. Chen. Near collisions of SHA-0. In M. Franklin, editor, *Advances in Cryptology – CRYPTO'04*, volume 3152 of *LNCS*. Springer-Verlag, 2004.
2. E. Biham and R. Chen. New results on SHA-0 and SHA-1. Short talk presented at CRYPTO'04 Rump Session, 2004.
3. A. Bosselaers and B. Preneel, editors. *Integrity Primitives for Secure Information Systems. Final Report of RACE Integrity Primitives Evaluation.*, volume 1007 of *LNCS*. Springer-Verlag, 1995.
4. A. Canteaut and F. Chabaud. A new algorithm for finding minimum-weight words in a linear code: application to McEliece's cryptosystem and to narrow-sense BCH codes of length 511. *IEEE T. Inform. Theory*, 44(1):367–378, 1998.

5. F. Chabaud. On the security of some cryptosystems based on error-correcting codes. In A. D. Santis, editor, *Advances in Cryptology – EUROCRYPT'94*, volume 950 of *LNCS*, pages 131–139, Berlin, 1995. Springer-Verlag.

6. F. Chabaud and A. Joux. Differential collisions in SHA-0. In H. Krawczyk, editor, *Advances in Cryptology – CRYPTO'98*, volume 1462 of *LNCS*, pages 56–71. Springer-Verlag, 1998.

7. B. den Boer and A. Bosselaers. An attack on the last two rounds of MD4. In J. Feigenbaum, editor, *Advances in Cryptology – CRYPTO'91*, volume 576 of *LNCS*, pages 194–203. Springer-Verlag, 1991.

8. H. Dobbertin. The status of MD5 after a recent attack. *CryptoBytes*, 2(2):1,3–6, 1996.

9. H. Dobbertin. RIPEMD with two-round compress function is not collison free. *Journal of Cryptology*, 10(1):51–70, 1997.

10. H. Dobbertin. Cryptanalysis of MD4. *Journal of Cryptology*, 11(4):253–271, 1998.

11. FIPS 180. Secure hash standard (SHS). National Institute of Standards and Technology, May 1993. Replaced by [15].

12. A. Joux. Collisions in SHA-0. Short talk presented at CRYPTO'04 Rump Session, 2004.

13. C. Lemuet. Collision in SHA-0. `sci.crypt` newsgroup message, Message-ID: `cfg007$1h1b$1@io.uvsq.fr`, 12 August 2004.

14. J. S. Leon. A probabilistic algorithm for computing minimum weights of large error-correcting codes. *IEEE T. Inform. Theory*, 34(5):1354–1359, 1988.

15. National Institute of Standards and Technology. Secure hash standard (SHS). FIPS 180-2, August 2002.

16. N. Pramstaller, C. Rechberger, and V. Rijmen. Exploiting coding theory for collision attacks on SHA-1. In N. P. Smart, editor, *Cryptography and Coding: 10th IMA International Conference*, volume 3796 of *LNCS*, pages 78–95. Springer-Verlag, Dec. 2005.

17. V. Rijmen and E. Oswald. Update on SHA-1. In A. Menezes, editor, *Topics in Cryptology – CT-RSA 2005*, volume 3376 of *LNCS*, pages 58–71. Springer-Verlag, Feb. 2005.

18. L. R. Rivest. The MD4 message digest algorithm. In A. J. Menezes and S. A. Vanstone, editors, *Advances in Cryptology – CRYPTO'90*, volume 537 of *LNCS*, pages 303–311. Springer-Verlag, 1991.

19. R. L. Rivest. The MD4 message digest algorithm. Request for Comments (RFC) 1320, Internet Engineering Task Force, April 1992.

20. R. L. Rivest. The MD5 message digest algorithm. Request for Comments (RFC) 1321, Internet Engineering Task Force, April 1992.

21. B. Van Rompay, A. Biryukov, B. Preneel, and J. Vandewalle. Cryptanalysis of 3-pass HAVAL. In C. S. Laih, editor, *Advances in Cryptology – ASIACRYPT'03*, volume 2894 of *LNCS*, pages 228–245. Springer-Verlag, 2003.

22. A. Vardy. The intractability of computing the minimum distance of a code. *IEEE T. Inform. Theory*, 43(6):1757–1766, 1997.

23. X. Wang, X. Lai, D. Feng, H. Chen, and X. Yu. Cryptanalysis of the hash functions MD4 and RIPEMD. In R. Cramer, editor, *Advances in Cryptology – EUROCRYPT'05*, volume 3494 of *LNCS*, pages 1–18. Springer-Verlag, 2005.

24. X. Wang, X. Lai, D. Feng, and H. Yu. Collisions for hash functions MD4, MD5, HAVAL-128 and RIPEMD. Short talk presented at CRYPTO'04 Rump Session, 2004.

25. X. Wang, X. Lai, D. Feng, and H. Yu. Collisions for hash functions MD4, MD5, HAVAL-128 and RIPEMD. Cryptology ePrint Archive, Report 2004/199, August 2004. http://eprint.iacr.org/.

26. X. Wang, Y. L. Yin, and H. Yu. Finding collisions in the full SHA-1. In V. Shoup, editor, *Advances in Cryptology – CRYPTO'05*, volume 3621 of *LNCS*, pages 17–36. Springer-Verlag, Nov. 2005.

27. X. Wang and H. Yu. How to break MD5 and other hash functions. In R. Cramer, editor, *Advances in Cryptology – EUROCRYPT'05*, volume 3494 of *LNCS*, pages 19–35. Springer-Verlag, 2005.

28. X. Wang, H. Yu, and Y. L. Yin. Efficient collision search attacks on SHA-0. In V. Shoup, editor, *Advances in Cryptology – CRYPTO'05*, volume 3621 of *LNCS*, pages 1–16. Springer-Verlag, Nov. 2005.

29. Y. Zheng and J. a. Pieprzyk. HAVAL – a one-way hashing algorithm with variable length of output. In J. Seberry and Y. Zheng, editors, *Advances in Cryptology – AUSCRYPT'92*, volume 718 of *LNCS*, pages 83–104. Springer-Verlag, 1993.

Extending Gibson's Attacks on the GPT Cryptosystem

Raphael Overbeck

GK Electronic Commerce, TU-Darmstadt,
Department of Computer Science,
Cryptography and Computer Algebra Group
overbeck@cdc.informatik.tu-darmstadt.de

Abstract. In this paper we look at the Gabidulin version of the McEliece cryptosystem (GPT). In order to avoid Gibson's attacks on GPT, several variants have been proposed. We cryptanalyze the variant with column scrambler and the one using reducible rank codes. Employing Gibson's attacks as a black box, we get an efficient attack for the parameter sets proposed for GPT with column scrambler. As a countermeasure to our attack, we propose a new variant of the GPT cryptosystem.

1 Introduction

The security of cryptosystems based on error correcting codes is connected to the hardness of the general decoding problem. In 1991 Gabidulin, Paramonov and Tretjakov proposed a variant of the McEliece scheme (GPT) [6] using *rank distance* codes instead of Goppa codes. For the Hamming-metric fast (but exponential) general decoding algorithms are known, but despite of recent advances, there does not exist one for the rank-metric [9]. Thus, smaller public-key sizes may be used for the GPT than for the McEliece cryptosystem using Goppa codes.

Gibson developed two structural attacks for the GPT cryptosystem ([7], [8]) and proved the parameter sets proposed in [6] and [4] to be insecure. Even though, the cryptosystem remained unbroken for large public-keys, as both attacks are exponential in runtime.

Several variants of GPT have been proposed in order to avoid structural attacks (see [2], [1] and [5]). In this paper we take advantage of some nice properties of Gabidulin codes to build structural attacks for two of these variants, namely the "GPT with column scrambler" [2] and the variant using "reducible rank codes" [5].

The paper is structured as follows: First we give a short introduction to rank distance codes and the original GPT cryptosystem. Then we present the two variants this paper addresses and show how to attack the first one. Based on our observations we propose a generalized variant of GPT and finally extend our observations to give a guideline to a possible structural attack for GPT using reducible rank codes.

Ø. Ytrehus (Ed.): WCC 2005, LNCS 3969, pp. 178–188, 2006.

2   Rank Distance Codes

Rank distance codes were presented by Gabidulin in [3]. They are linear codes over the finite field $\mathbb{F}_{q^m}$ for q (power of a) prime and $m \in \mathbb{N}$. As their name suggests they use a special concept of distance.

Definition 1. *Let* $x = (x_1, \cdots, x_n) \in \mathbb{F}_{q^m}^n$ *and* $b_1, \cdots, b_m$ *a basis of* $\mathbb{F}_{q^m}$ *over* $\mathbb{F}_q$. *We can write* $x_i = \sum_{j=1}^m x_{ij} b_j$ *for each* $i = 1, \cdots, n$ *with* $x_{ij} \in \mathbb{F}_q$. *The rank norm* $\| \cdot \|_r$ *is defined as follows:*

$$\|x\|_r := \mathrm{rank}\left((x_{ij})_{1 \leq i \leq n,\, 1 \leq j \leq m} \right) .$$

The rank norm of a vector $x \in \mathbb{F}_{q^m}^n$ is uniquely determined (independent of the choice of basis) and induces a metric, called *rank distance*.

Definition 2. *An* (n, k)-*code* $\mathcal{C}$ *over a finite field* $\mathbb{F}$ *is a* k-*dimensional subvectorspace of the vector space* $\mathbb{F}^n$. *We call the code* $\mathcal{C}$ *an* (n, k, d) *rank distance code if* $d = \min_{x,y \in \mathcal{C}} \|x - y\|_r$. *The matrix* $C \in \mathbb{F}^{k \times n}$ *is a generator matrix for the* (n, k) *code* $\mathcal{C}$ *over* $\mathbb{F}$, *if the rows of* C *span* $\mathcal{C}$ *over* $\mathbb{F}$. *The matrix* $H \in \mathbb{F}^{n \times (n-k)}$ *is called check matrix for the code* $\mathcal{C}$ *if it is the right kernel of* C. *The code generated by* $H^\top$ *is called dual code of* $\mathcal{C}$ *and denoted by* $\mathcal{C}^\perp$.

In [9] Ourivski and Johansson presented an algorithm which solves the general decoding problem in $\mathcal{O}\left((m\frac{d-1}{2})^3 q^{(d-3)(k+1)/2} \right)$ operations over $\mathbb{F}_q$ for (n, k, d) rank distance codes over $\mathbb{F}_{q^m}$. A special class of rank distance codes are the *Gabidulin codes* for which an efficient decoding algorithm exists [3]. We will define these codes by their generator matrix.

Definition 3. *Let* $g \in \mathbb{F}_{q^m}^n$ *be a vector s.t. the components* g_i, $i = 1, \cdots, n$ *are linearly independent over* $\mathbb{F}_q$. *This implies that* $n \leq m$. *The* (n, k, d) *Gabidulin code* $\mathcal{G}$ *is the rank distance code with generator matrix*

$$G = \begin{pmatrix} g_1 & g_2 & \cdots & g_n \\ g_1^q & g_2^q & \cdots & g_n^q \\ \vdots & & \ddots & \vdots \\ g_1^{q^{k-1}} & g_2^{q^{k-1}} & \cdots & g_n^{q^{k-1}} \end{pmatrix} \in \mathbb{F}_{q^m}^{k \times n}. \tag{1}$$

An (n, k) Gabidulin code $\mathcal{G}$ corrects $\lfloor \frac{n-k}{2} \rfloor$ errors and has a minimum distance of $d = n - k + 1$. The dual code of an (n, k) Gabidulin code is a $(n, n - k)$ Gabidulin code (see [4]). The vector g is said to be the *generator vector* of the Gabidulin code $\mathcal{G}$. A decoding algorithm based on the "right Euclidean division algorithm" runs in $\mathcal{O}\left(d \log_2^2 d + dn \right)$ operations over $\mathbb{F}_{q^m}$ for (n, k, d) Gabidulin codes [4].

Throughout this paper we will use the following notation. We write $\mathcal{G} = \langle G \rangle$ if the linear (n, k)-code $\mathcal{G}$ over the field $\mathbb{F}$ has the generator matrix G. We will identify $x \in \mathbb{F}^n$ with $(x_1, \cdots, x_n)$, $x_i \in \mathbb{F}$ for $i = 1, \cdots, n$. For any (ordered) subset $\{j_1, \cdots j_m\} = J \subseteq \{1, \cdots n\}$ we denote the vector $(x_{j_1}, \cdots, x_{j_m}) \in \mathbb{F}^m$ with x_J. Similarly, we denote by $M_{.J}$ the submatrix of a $k \times n$ matrix M consisting of the columns corresponding to the indices of J and $M_{J'.} = \left((M^\top)_{.J'} \right)^\top$ for any (ordered) subset J' of $\{1, \cdots, k\}$. Block matrices will be given in brackets.

3   The GPT Cryptosystem

In this section, we briefly introduce the GPT cryptosystem presented in [6]. In order to better understand the impact of Gibson's attacks we introduce a new security parameter s.

- **System Parameters:** $q, n, m, k, t, s \in \mathbb{N}$, where $n \leq m$, $t < \frac{n-k-1}{2}$ and $s \leq \min\{k, t\}$.
- **Key Generation:** First generate the following matrices over $\mathbb{F}_{q^m}$:
  G: $k \times n$ generator matrix of an (n, k, d) Gabidulin code $\mathcal{G}$ over $\mathbb{F}_{q^m}$.
  S: $k \times k$ random non-singular matrix (the *row scrambler*).
  X: $k \times n$ random matrix with rank s over $\mathbb{F}_{q^m}$ and rank t over $\mathbb{F}_q$.

  Then, compute the $k \times n$ matrix $G' = S(G + X)$ and $e = \frac{n-k}{2} - t$. Further let $\mathcal{D}_\mathcal{G}$ be an efficient decoding algorithm for $\mathcal{G}$.

- **Public Key:** (G', e)
- **Private Key:** $(\mathcal{D}_\mathcal{G}, X, S)$ or (G, X, S) where G is of the form in (1).
- **Encryption:** To encode a plaintext $x \in \mathbb{F}_{q^m}^k$ choose a vector $z \in \mathbb{F}_{q^m}^n$ of rank norm e at random and compute the ciphertext c as follows:

$$c = xG' + z .$$

- **Decryption:** To decode a ciphertext c apply the decoding algorithm $\mathcal{D}_\mathcal{G}$ for $\mathcal{G}$ to it. Since c is at distance less than $(n-k)/2$ from $\mathcal{G}$, we obtain the codeword

$$xSG = \mathcal{D}_\mathcal{G}(c) .$$

Now, we can compute the plaintext x.

The matrix X is called *distortion matrix*. As $\|mSX\|_r \leq t$ the decryption works correctly. In all examples and figures we will choose $n = m$ and $q = 2$. Figure 1 shows public key sizes and approximate workfactors (operations over $\mathbb{F}_q$) for en- and decryption.

Parameters		Size Public	WF	WF
m	k	Key (Bytes)	Encryption	Decryption
36	18	$2,916$	2^{20}	2^{20}
48	24	$6,912$	2^{21}	2^{22}
64	32	$16,384$	2^{23}	2^{24}
128	64	$131,027$	2^{27}	2^{27}

Fig. 1. Parameter sets for the Gabidulin PKC

4   Gibson's Attacks

Gibson presented two structural attacks on the GPT cryptosystem ([7], [8]). They recover an (alternative) private-key from the public-key. We are going to use both attacks as a black box for our new attack. Let $\langle G \rangle$ be an (n, k) Gabidulin code over $\mathbb{F}_{q^m}$, $S \in \mathbb{F}_{q^m}^{k \times k}$ non-singular and $Y \in \mathbb{F}_{q^m}^{k \times n}$ of rank s over $\mathbb{F}_{q^m}$ and rank $0 \leq t \leq n$ over $\mathbb{F}_q$. Then Gibson's attacks return on input of $G' = S(G + Y)$ three matrices $\hat{G}, \hat{X} \in \mathbb{F}_{q^m}^{k \times n}$ and $\hat{S} \in \mathbb{F}_{q^m}^{k \times k}$, s.t.

(i) $\hat{G}$ is a generator matrix of an (n, k) Gabidulin code over $\mathbb{F}_{q^m}$,
(ii) $G' = \hat{S}\left(\hat{G} + \hat{X}\right)$ and
(iii)   the rank of $\hat{X}$ over $\mathbb{F}_q$ is not bigger than t.

Gibson's first attack [7] was developed for the case that the GPT parameter s is 1, but may be adapted to the case where $s \neq 1$ (see [4]). It has runtime

$$\mathcal{O}\left(m^3 (n - k)^3 q^{ms}\right). \tag{2}$$

In [8] Gibson presented a different attack, which is more efficient for larger values of s. It requires that $k + t + 2 \leq n$ (this is a very weak condition) and runs in time

$$\mathcal{O}\left(k^3 + (k + t) f \cdot q^{f(k+2)} + (m - k) t \cdot q^f\right), \tag{3}$$

where $f \approx \max(0, t - 2s, t + 1 - k)$. Note, that this attack runs in polynomial time if $f = 0$ and otherwise is still fast. The success of both attacks is based on some assumptions, which are fulfilled with high probability for random G, S and Y. Figure 2 shows the behavior of the attacks for some sample parameter sets.

Parameters					WF Gibson's attack [7]	WF Gibson's attack [8]	WF attack by J&O [9]
m	k	s	t	e			
36	18	1	6	3	2^{74}	2^{97}	2^{55}
48	24	1	7	5	2^{90}	2^{149}	2^{120}
64	32	2	10	6	2^{173}	2^{224}	2^{188}
128	64	5	20	12	2^{693}	2^{684}	2^{744}

Fig. 2. Attacking the GPT cryptosystem with Gibson's attacks

5   GPT Variants

Apart from a solver for the general decoding problem, structural attacks today are the most severe threat to the GPT cryptosystem. In order to avoid Gibson's attacks, it is possible to choose different parameter sets, since both attacks have exponential runtime. However there have been other attempts to make structural attacks harder to apply or to avoid them completely.

5.1   Applying a Column Scrambler

In order to hide the structure of the public key even more, we may multiply a *column scrambler* T (a non-singular matrix over the base field $\mathbb{F}_q$) from the right to it. Removing the influence of T at decryption does not change the rank of the error vector. Unfortunately Gibson's attacks are still applicable, because GT is a matrix of the form given in equation (1), too. In 2001, Ourivski and Gabidulin suggested to add more redundancy to the code [2], i.e. choose G' in the public key as

$$G' = S\left[Y|G+X\right]T \tag{4}$$

with $Y \in \mathbb{F}_{q^m}^{k \times l}$. Gibson's attacks remain applicable if we guess a set N of n columns out of the matrix G' s.t. $\left(\left[0|G\right]T\right)_{.N}$ has rank n over $\mathbb{F}_q$. This can be done with high probability if T was chosen at random. For carefully chosen Y and T there is no way to choose N s.t. $\left(\left[Y|X\right]T\right)_{.N}$ is of rank $\leq t$. The attacker would still have to guess part of the error even after employing one of Gibson's attacks. This seems impossible, if the search space is big enough. According to [2], a secure choice of parameters could be $q = 2$, $m = 32$, $k = 20$, $l = 8$, $t = 3$ and $e = 3$ with a public key size of approximately $3,200$ bytes.

5.2   Using Reducible Rank Codes

The idea to use *reducible rank codes* (RRC) was first presented in [5]. We want to introduce a slightly different definition.

Definition 4. *Let $\mathcal{C}_i = \langle C_i \rangle$, $i = 1, \cdots, w$ be a family of linear error correcting codes over $\mathbb{F}_{q^m}$ where $\mathcal{C}_i$ is an (n_i, k_i, d_i) code. Then the (linear) code $\mathcal{G}$ given by the generator matrix of the form*

$$G = \begin{bmatrix} C_1 & 0 & \cdots & 0 \\ G_{21} & C_2 & \cdots & 0 \\ \vdots & & \ddots & \vdots \\ G_{w1} & G_{w2} & \cdots & C_w \end{bmatrix}$$

for some matrices $G_{ij} \in \mathbb{F}_{q^m}^{k_i \times n_j}$ is called reducible code. Further, $\mathcal{G}$ has length $n = \sum_{i=1}^{w} n_i$, dimension $k = \sum_{i=1}^{w} k_i$ and minimum distance $d = \min_{1 \leq i \leq w}(d_i)$.

Error correction may be done in sections, starting from the right. If all codes C_i, $i = 1, \cdots w$ are rank distance codes, we call $\mathcal{G}$ a reducible rank code. In [5] all the codes permutation equivalent to such codes are called RRC as well.

Using reducible rank codes for the McEliece cryptosystem is quite a natural extension. In [5] the authors propose to use two (n_i, k_i) Gabidulin codes over $\mathbb{F}_{q^m}$, named $\langle G_i \rangle$, $i \in \{1, 2\}$, and a special matrix $Y \in \mathbb{F}_{q^m}^{k_2 \times n_1}$ to build a reducible rank code. The public key of the cryptosystem may thus be described as

$$G' = S\left(\begin{bmatrix} G_1 & 0 \\ Y & G_2 \end{bmatrix} + X\right)T , \tag{5}$$

where $S \in \mathbb{F}_{q^m}^{k \times k}$ and $T \in \mathbb{F}_q^{n \times n}$ are non-singular and the rank over $\mathbb{F}_q$ of the columns of X corresponding to G_i is less than t_i for $i = 1, 2$. According to [5], the error correction capability of the code defined by G' is $e = \min_{i=1,2} \left\{ \frac{n_i - k_i}{2} - t_i \right\}$, and every parameter set with $n_i \geq 24$ and $e \geq 4$ is considered to provide sufficient security, even if $X = 0$.

6   Extending Gibson's Attacks

In this section we will take advantage of some well known facts and reassemble them to enhance Gibson's attacks by using them as black boxes. First, we show how to correct more errors with a GPT public key as previously claimed to be possible, then we apply our observation to the GPT cryptosystem with column scrambler.

Theorem 1. *Let $\langle G \rangle$ be an (n, k) Gabidulin code over $\mathbb{F}_{q^m}$, $X \in \mathbb{F}_{q^m}^{k \times n}$ of rank $t \leq n - k$ over $\mathbb{F}_q$. Then there exist an invertible Matrix $U \in \mathbb{F}_q^{n \times n}$, a generator matrix $\hat{G}$ of an $(n - t, k)$ Gabidulin code over $\mathbb{F}_{q^m}$ and $\hat{X} \in \mathbb{F}_{q^m}^{k \times t}$ s.t.*

$$(G + X)U = \left[\hat{X} | \hat{G}\right].$$

Further a matrix U satisfying the above condition can be found in $\mathcal{O}\left(n^3\right)$ operations over $\mathbb{F}_q$ if X is known.

Before we prove the theorem we want to give an example of its application. According to [2] an acceptable choice of parameters for the cryptosystem of section 5.1 could be $q = 2$, $n = m = 25$, $k = 15$, $l = 5$, $t = 3$ and $e = 2$ with a public key size of approximately $1,400$ bytes. Let (G, X, Y, S, T) be the private key corresponding to the public key (G', e) of an instance of the GPT cryptosystem with column scrambler, where G' has the form described in equation (4). An attacker could try to apply one of Gibson's attacks to a set J of m columns of G'. This will leave him with a set of three matrices of the following form:

(i) $\dot{G}$ a generator matrix of a (m, k) Gabidulin code over $\mathbb{F}_{q^m}$,
(ii) $\dot{S} \in \mathbb{F}_{q^m}^{k \times k}$ an invertible matrix and
(iii) $\dot{Y} \in \mathbb{F}_{q^m}^{k \times m}$ a distortion matrix of rank $t + l$ over $\mathbb{F}_q$.

The attacker knows that $\dot{S} \left(\dot{G} + \dot{Y} \right) = G'_J$. Applying theorem 1 to our example we can correct errors of rank 1, whereas we would have to correct an error of rank 2. The attacker could try to guess part of the error vector, s.t. the remaining error has rank 1. There are only $q^{m+m-l-t} = 2^{42}$ different error vectors of rank 1 and length $m - l - t$ in our example. This number is small enough to allow a random guess, so the choice of parameters above is insecure. Now we prove the theorem.

Proof. (**Theorem 1**) We exploit the fact that it is easy to determine the linear dependency of the columns of X over $\mathbb{F}_q$. Let $Z \in \mathbb{F}_{q^m}^{k \times t}$ be a set of t linearly independent columns of X. Without loss of generality, we may assume that these are the first t columns. We may solve the linear equation $Zu_i = X_{\cdot i}$ with $u_i \in \mathbb{F}_q^t$ for $i = t+1, \cdots, n$ in time $\mathcal{O}\left(n^3\right)$. Let $\bar{U} := [u_{t+1}|u_{t+2}| \cdots |u_n]$ and Id_κ denote the κ-dimensional identity matrix. Then the last $n - t$ columns of

$$(G + X) \cdot \left[\begin{array}{c|c} \mathrm{Id}_t & -\bar{U} \\ \hline 0 & \mathrm{Id}_{n-t} \end{array} \right] = \left[G_{\cdot\{1,\cdots,t\}} + Z \middle| G \cdot \left[\frac{-\bar{U}}{\mathrm{Id}_{n-t}} \right] \right]$$

form a generator matrix of an $(n - t, k)$ Gabidulin code. Defining the $(k \times t)$ Matrix $\hat{X}$ as $\hat{X} := G_{\cdot\{1,\cdots,t\}} + Z$ proves the theorem.

It follows that the code generated by $G + X$ has an efficient decoding algorithm which decodes errors of rank up to $\frac{n-k-t}{2}$ if X is known. Now we apply this result to break the GPT cryptosystem with column scrambler by viewing it as an instance of the original GPT cryptosystem.

Theorem 2. *Any instance of the GPT cryptosystem with column scrambler with parameters q, n, m, k, t, s and l is equivalent to an instance of the GPT cryptosystem in its basic version with parameters $q, \hat{n} = n + l, \hat{m} = \lceil (n+l)/m \rceil \cdot m$, k, $\hat{t} = t + l$ and $\hat{s} \le \min(k, s + l)$. The GPT with column scrambler may be broken in time $\mathcal{O}$ (Gibson's attack for parameters $q, \hat{n}, \hat{m}, k, \hat{t}, \hat{s}$).*

Proof. Let $(G', e = (n-k)/2 - t)$ with $G' = S\left[Y | G + X\right] T$ be the public key of an instance of the GPT cryptosystem with column scrambler with parameters given above and secret key (G, S, T). If we define $a := \lceil (n+l)/m \rceil$, then every element of $\mathbb{F}_{q^m}$ may be viewed as an element of $\mathbb{F}_{q^{am}}$. Let $(g_1, \cdots, g_n)$ be a generator vector of $\langle G \rangle$. Then we can choose $\hat{g}_1, \cdots, \hat{g}_l \in \mathbb{F}_{q^{am}}$ s.t. $\hat{g}_1, \cdots, \hat{g}_l, g_1, \cdots, g_n$ are linearly independent over $\mathbb{F}_q$. Let $\langle \hat{G} \rangle$ be the $(n+l, k)$ Gabidulin code with generator vector $(\hat{g}_1, \cdots, \hat{g}_l, g_1, \cdots, g_n)$ and $\hat{G}$ of the form in equation (1). If we define

$$\hat{X} := \left[Y - \hat{G}_{\cdot\{1,\cdots,l\}} \middle| X \right],$$

then we have

$$S\left(\hat{G}T + \hat{X}T \right) = S\left(\hat{G} + \hat{X} \right) T = G' \in \mathbb{F}_{q^{am}}^{k \times (n+l)}.$$

This proves the first part of the theorem. We know that the rank of $\hat{X}T$ over $\mathbb{F}_q$ is less than $l + t$ and $k + l + t + 2 \le \hat{n}$, thus we may apply both Gibson's attacks to recover (alternative) S, $\hat{X}T$ and $\hat{G}T$. By theorem 1, we are now able to correct all error vectors of rank less than $(m + l - k - l - t)/2 > (m - k)/2 - t$ efficiently.

We will call the attacks described in theorem 2 "*extended Gibson*" attacks. Note that we don't have any control of $\hat{s}$ using GPT with column scrambler. The size of this parameter of the corresponding GPT instance over an extension

Parameters							WF extended	WF attack
m	k	l	s	$\hat{s}$	t	e	Gibson [8]	by J&O [9]
32	20	8	1	9	3	3	2^{26}	2^{59}
36	18	4	1	3	6	3	2^{26}	2^{55}
48	24	5	1	4	7	5	2^{28}	2^{120}
64	32	6	2	5	10	6	2^{30}	2^{289}
128	64	10	5	10	20	12	2^{35}	2^{744}

Fig. 3. Attacking GPT with column scrambler

field of $\mathbb{F}_{q^m}$ is very likely to be near its upper bound and thus making Gibson's second attack very fast. Figure 3 shows the workfactors of the new attack on GPT with column scrambler, where $\hat{s}$ refers to the expected parameter for the corresponding GPT cryptosystem parameter set.

7   A New Variant of the GPT Cryptosystem

As a consequence from theorem 1, the public key parameter e of the GPT cryptosystem may be chosen larger, than it was proposed originally. The result is a generalized variant for the GPT cryptosystem (GGPT), which includes a column scrambler and new bounds for t and e. While the size of the public key remains the same (compare figure 1), the runtime for decryption decreases to $\mathcal{O}(n^2)$ operations over $\mathbb{F}_{q^m}$.

- **System Parameters:** $q, n, m, k, t, s \in \mathbb{N}$, where $n \leq m$, $t < n - k - 1$ and $s \leq \min\{k, t\}$.
- **Key Generation:** First generate the following matrices over $\mathbb{F}_{q^m}$:
  G: $k \times n$ generator matrix of an (n, k, d) Gabidulin code over $\mathbb{F}_{q^m}$,
  S: $k \times k$ random non-singular matrix (the row scrambler),
  X: $k \times t$ random matrix of rank s over $\mathbb{F}_{q^m}$ and rank t over $\mathbb{F}_q$.

  and T an $n \times n$ random, non-singular matrix over $\mathbb{F}_q$ (the column scrambler). Then compute the $k \times n$ matrix

$$G' = S\left[G_{\cdot\{1,\cdots,t\}} + X\big|G_{\cdot\{t+1,\cdots,n\}}\right]T$$

  and $e = \frac{n-k-t}{2}$. Further let $\mathcal{D}_{\mathcal{G}}$ be an efficient decoding algorithm for the Gabidulin code $\mathcal{G}$ generated by the matrix $G_{\cdot\{t+1,\cdots,n\}}$.
- **Public Key:** (G', e)
- **Private Key:** $(\mathcal{D}_{\mathcal{G}}, S, T)$ or (G, S, T) where G is of the form in (1).
- **Encryption:** To encode a plaintext $x \in \mathbb{F}_{q^m}^k$ choose a vector $z \in \mathbb{F}_{q^m}^m$ of rank norm e at random and compute the ciphertext c as follows:

$$c = xG' + z .$$

– **Decryption:** To decode a ciphertext c apply the decoding algorithm $\mathcal{D}_{\mathcal{G}}$ for $\mathcal{G}$ to $c' = \left(cT^{-1}\right)_{\{t+1,\cdots,n\}}$. Since c' has a rank distance less than $\frac{n-k-t}{2}$ to $\mathcal{G}$, we obtain the codeword

$$xSG_{\{t+1,\cdots,n\}} = \mathcal{D}_{\mathcal{G}}\left(c'\right) \ .$$

Now, we can compute the plaintext x.

Note, that all instances of the GPT cryptosystem, as well as the ones of the GPT cryptosystem with column scrambler may be viewed as instances of the new variant. Figure 4 shows the workfactors of the attacks for some parameter sets of GGPT. Parameter sets were chosen taking into account recent results, which came to light after the WCC05 conference [10].

Parameters					WF extended Gibson [7]	WF extended Gibson [8]	WF attack by J&O [9]
m	k	s	t	e			
64	8	1	40	8	2^{111}	2^{403}	2^{87}
80	8	2	56	8	2^{210}	2^{544}	2^{88}
156	8	8	132	8	2^{1306}	2^{1279}	2^{91}

Fig. 4. Parameter sets for GGPT

8   Attacking GTP with Reducible Rank Codes

In this section, we want to give a hint on how to build a structural attack on the GPT cryptosystem with reducible rank codes. We will show, that if the row scrambler S is generated at random (with no more conditions than being non-singular), the problem of recovering a secret key for GPT with reducible rank codes can be reduced to the problem of recovering a secret key for instances of the GPT variant from the previous section, if the following assumption holds:

Assumption 1. *Let (G', e) be the public key of a random instance of GGPT with parameters q, n, m, k, t and s. Further, let (G, S, T) and $\left(\hat{G}, \hat{S}, \hat{T}\right)$ be two valid secret keys corresponding to (G', e), then with high probability*

$$\left(\hat{T}T^{-1}\right)_{N_1 N_2} = 0 \ ,$$

where $N_1 := \{1, \cdots, t\}$ and $N_2 := \{t+1, \cdots, n\}$.

With other words, we assume, that for most instances of the GPT variant from the previous section, all possible secret keys are closely related to each other: $\hat{G}_{N_2} \left(\hat{T}T^{-1}\right)_{N_2 N_2} = G_{N_2}$. This assumption is corroborated by an observation of Gibson ([7], [8]) for small parameter sets. Gibson states, that the secret key seems to be unique (after some normalization) for most instances of the GPT cryptosystem.

Theorem 3. *Let $\mathcal{A}$ be an oracle which recovers a private key from the public key for any instance of GGPT and assume, that assumption 1 holds. Then we may use $\mathcal{A}$ to recover an alternative private key from the public key for an instance of the GPT cryptosystem with RRC with high probability if every entry of S was chosen uniformly from $\mathbb{F}_{q^m}$.*

Proof. We will give an outline of a proof for the case where the secret RRC is built from two Gabidulin codes of the same dimension. It may be easily extended to all other cases. Let G_1 and G_2 be generator matrices of two (n_i, k), $i \in \{1,2\}$ Gabidulin codes over $\mathbb{F}_{q^m}$. Let (G', e) be the public key corresponding to the private key $(G_1, G_2, X, Y, S$ and $T)$ of an instance of the GPT with RRC, where G' is of the form given in equation (5).

First we guess a set $J \subseteq \{1, \cdots, 2k\}$ s.t. the matrix S_{JK_2} with $K_2 := \{k+1, \cdots, 2k\}$ is invertible. If S is a truly random generated matrix, then the probability that the condition is fulfilled for a random J is not too bad (≥ 0.99) in practice. Let $K_1 := \{1, \cdots, k\}$, then we know:

$$
\begin{aligned}
G'_{J\cdot} &= \left(\left[S_{JK_1}G_1 + S_{JK_2}Y \,\middle|\, S_{JK_2}G_2\right] + (S \cdot X)_{J\cdot}\right)T \\
&= S_{JK_2}\left(\left[S_{JK_2}^{-1}S_{JK_1}G_1 + Y \,\middle|\, G_2\right] + S_{JK_2}^{-1}(S \cdot X)_{J\cdot}\right)T \\
&= S_{Jk_2}\left[S_{JK_2}^{-1}\left(S_{JK_1}G_1 + (S \cdot X)_{JK_1}\right) + Y \,\middle|\, G_2 + S_{JK_2}^{-1}(S \cdot X)_{JK_2}\right]T.
\end{aligned}
$$

Thus, $G'_{J\cdot}$ forms an instance of GGPT. As $S_{JK_2}^{-1}(S \cdot X)_{JK_2}$ is of sufficient small column rank over $\mathbb{F}_q$, we may query $\mathcal{A}$ to obtain an alternative secret key $(\hat{G}_2, \hat{S}_2, \hat{T})$ for $G'_{J\cdot}$ and can compute:

$$
G'\hat{T}^{-1} = S\left(\begin{bmatrix} G_1 & 0 \\ Y & G_2 \end{bmatrix} + X\right)T\hat{T}^{-1} = S\left(\begin{bmatrix} G_1 & 0 \\ Y & G_2 \end{bmatrix}T\hat{T}^{-1} + XT\hat{T}^{-1}\right).
$$

By assumption 1, the last $2e + k$ columns of $XT\hat{T}^{-1}$ and $\begin{bmatrix} G_1 & 0 \end{bmatrix}T\hat{T}^{-1}$ are zero with high probability. Thus, we are able to compute an invertible matrix $\hat{S} \in \mathbb{F}_{q^m}^{2k \times 2k}$, s.t. the last $2e$ columns of $(\hat{S}^{-1}G'\hat{T}^{-1})_{J^C}$ are zero. It follows, that the first $N = n_1 + n_2 - 2e - k$ columns of $(\hat{S}^{-1}G'\hat{T}^{-1})_{J^C}$ build another instance of the GPT variant from the previous section. Thus, on the query $(\hat{S}^{-1}G'\hat{T}^{-1})_{J^C\{1,\cdots,N\}}$, the oracle $\mathcal{A}$ returns three matrices $\bar{G}, \bar{S}, \bar{T} \in \mathbb{F}_q^{N \times N}$, such that

$$
\hat{S}^{-1}G'\hat{T}^{-1}\begin{bmatrix} \bar{T}^{-1} & 0 \\ 0 & \mathrm{Id}_{2e+k} \end{bmatrix} = \begin{bmatrix} Z_1 \, \mathbf{G}_1 & 0 \\ Z_2 \, Z_3 \, \mathbf{G}_2 \end{bmatrix}
$$

for some $(2e + k, k)$ Gabidulin codes $\langle \mathbf{G}_i \rangle$, $i = 1,2$ and some matrices $Z_1, Z_2 \in \mathbb{F}_{q^m}^{2k \times (n_1 + n_2 - 4e - 2k)}$ and $Z_3 \in \mathbb{F}_{q^m}^{k \times (2e+k)}$. (Here, the matrices $\mathbf{G}_i$, $i = 1, 2$ are not necessarily in the form of equation (1).) Now, one can see, that we can correct errors of rank e efficiently in the code defined by G', employing the knowledge of $\hat{T}$, $\bar{T}$ and $\hat{S}$.

Note that the attack proposed in this section runs in oracle-polynomial-time. However, extended Gibson attacks can not be combined with the result of theorem 3 to build an efficient attack on GPT with RRC.

References

1. T.P. Berger and P. Loidreau. Security of the Niederreiter form of the GPT public-key cryptosystem. In *IEEE International Symposium on Information Theory, Lausanne, Suisse*. IEEE, July 2002.

2. E. M. Gabidulin and A. V. Ourivski. Column scrambler for the GPT cryptosystem. *Discrete Applied Mathematics*, 128(1):207–221, 2003.

3. E.M. Gabidulin. Theory of codes with maximum rank distance. *Problems of Information Transmission*, 21, No. 1, 1985.

4. E.M. Gabidulin. On public-key cryptosystems based on linear codes. In *Proc. of 4th IMA Conference on Cryptography and Coding 1993*, Codes and Ciphers. IMA Press, 1995.

5. E.M. Gabidulin, A.V. Ourivski, B. Honary, and B. Ammar. Reducible rank codes and their applications to cryptography. *IEEE Transactions on Information Theory*, 49(12):3289–3293, 2003.

6. E.M. Gabidulin, A.V. Paramonov, and O.V. Tretjakov. Ideals over a non-commutative ring and their applications to cryptography. In *Proc. Eurocrypt '91*, volume 547 of *LNCS*. Springer Verlag, 1991.

7. J. K. Gibson. Severely denting the Gabidulin version of the McEliece public key cryptosystem. *J-DESIGNS-CODES-CRYPTOGR*, 6(1):37–45, July 1995.

8. K. Gibson. The security of the Gabidulin public key cryptosystem. In *Proc. of Eurocrypt'96*, volume 1070 of *LNCS*, pages 212–223. Springer Verlag, 1996.

9. T. Johansson and A.V. Ourivski. New technique for decoding codes in the rank metric and its cryptography applications. *Problems of Information Transmission*, 38, No. 3:237–246, 2002.

10. R. Overbeck. A new structural attack for GPT and variants. In *Proc. of Mycrypt 2005*, volume 3715 of *LNCS*, 2005. to appear.

Reduction of Conjugacy Problem in Braid Groups, Using Two Garside Structures

Maffre Samuel

XLIM, University of Limoges, France
samuel.maffre@unilim.fr

Abstract. We study the *Conjugacy Search Problem* used in braid-based cryptography. We develop an algorithm running in Garside groups generalizing braid groups. The method permits, in some case, to reduce drastically the size of the secret in braid groups. We use the fact that braid groups admit two different Garside structures to improve the efficiency of the reduction. This paper emphasizes the importance of the particular way used to produce *Conjugacy Search Problem* instances. The chosen method influences directly the reduction and then also the security.

1 Introduction

Braid groups, introduced by Artin in [2], are non-abelian groups. *Conjugacy Search Problems* are assumed to be hard, at least on some set of instances, and this is the basis for braid-based cryptography since 1999 [1, 11, 6]. Our work reveals the role played by the random generator of braid. A lot of work exists in the literature on the cryptanalysis of braid cryptosystems. However, they question essentially the choice of instances rather than the protocols themselves. Braid-based cryptography must solve a fundamental problem : to find efficient random generators of braid. In this work, we consider two random generators and show their influence on our reduction method.

Our work is devoted to the *Conjugacy Search Problem* formally : recover a, knowing (x, axa^{-1}). In practice, cryptosystems use essentially a variant of this cryptographic primitive.

The size of a braid group is defined by its number of strands. The size of a braid can be measured by the number of generators or the number of canonical factors which permits to write it. Our method produces an attractive factorization of the secret a in the form of a divisor $(d \prec a)$ and a multiple $(a \prec m)$. The goal is to reduce the length of the secret given by the number of generators. In this way, we produce two reduced conjugacy instances : $(x, d^{-1}axa^{-1}d)$ and $(x, m^{-1}axa^{-1}m)$. The efficiency of this reduction is related to the residual length of the new secrets $d^{-1}a$ and $m^{-1}a$.

Obviously, our reduction is a length attack. It is based on the canonical length. This work generalizes another work from the same author that applies only to Artin's presentation [12] (see also [9]).

We develop an algorithm which runs in a generalization of braid groups : Garside groups. The interesting fact is that braid groups admit two different

Ø. Ytrehus (Ed.): WCC 2005, LNCS 3969, pp. 189–201, 2006.

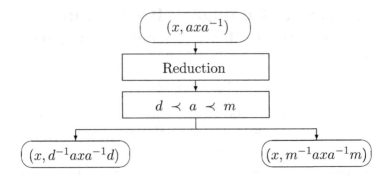

Fig. 1. Reduction of a conjugacy instance

Garside structures : Artin's presentation and the presentation of Birman, Ko and Lee [2, 3]. Then, we apply our algorithm to each Garside structure to improve the efficiency of the reduction. Both presentations can be used in parallel or sequentially.

The algorithm takes as input a conjugacy instance in a Garside group and some information on the secret which can not be considered as security parameters. It is deterministic and polynomial. The complexity is directly related to the complexity of an effective normal form. Though we do not have a theoretical result about the quality of the output divisor and multiple, we can compute them and give experimental results.

For instance, let BKL_{100}, be a braid group with 100 strands on BKL's presentation, and let a classical random generator which produces some braids with a canonical length of 15. With our method, the size of the secret, on average, is reduced from 750 to 8. In some situations, this reduction reaches $1/1000$. It is important to consider that this efficiency depends on the representation of braids. We apply our method on several proposed instances in the literature. Many of them seem broken. However, the protocols themselves are not concerned but our method shows the importance of being careful with the choice of the used random generator of braid. Then, we give criteria to forestall that. This study attempts to establish a more efficient way to use braid-based cryptography.

In Section 2, we introduce the Garside groups that are generalization of braid groups. Then, in Section 3, we present our algorithm that applies to Garside structure. Afterwards, in Section 4, we recall briefly braid-based cryptography. In Section 5, we show how to use our algorithm in braid groups. We analyze in Section 6 the efficiency with simulations.

2   Garside Groups

This introduction to Garside group derives from [5, 7, 8, 14].

Definition 1. *A **monoid** is a couple $(M, \bullet)$, where M is a set, $\bullet$ is a associative law of composition and M has a neutral element for $\bullet$, 1.*

Let M be a monoid, we define a partial order : $a \prec b$ if $\exists c \in M$ such that $ac = b$. One says that a is a *left divisor* of b or that b is a *right multiple* of a. In the same way, a right division is defined, denoted by $\prec^r$.

Definition 2. *Let M be a monoid. The element $x \in M$ is an **atom** if $x \neq 1$ and if $x = yz \Rightarrow (y = 1$ or $z = 1)$. The monoid M is an **atomic monoid** if it is generated by its atoms and, moreover, for every $x \in M$, $\exists N_x \in \mathbb{N}$ such that x can not be written as a product of more than N_x atoms.*

Definition 3. *A monoid M is left-**cancellative** if :*

$$\forall x, y, z \in M, \quad xy = xz \Rightarrow y = z.$$

Definition 4. *A monoid M is a **Gaussian monoid** if it is atomic, cancellative and every pair of elements in M admits a greatest common divisor and a least common multiple.*

In the case of Gaussian monoid, the gcd and lcm are unique. We denote by $\wedge$ the left greatest common divisor and by $\vee$ the right least common multiple.

$$\text{Let } a, b \in M, \quad d = a \wedge b \quad \Leftrightarrow \quad \forall c \in M \quad c \prec a \text{ and } c \prec b \text{ iff } c \prec d$$
$$m = a \vee b \quad \Leftrightarrow \quad \forall c \in M \quad a \prec c \text{ and } b \prec c \text{ iff } m \prec c$$

Definition 5. *A Garside element is an element in the monoid, Δ, the left divisors of which coincide with its right divisors and form a finite subset generating the monoid. A **Garside monoid** is a Gaussian monoid which admits a Garside element.*

Let S be the set of left divisors of Δ, which are called *canonical factors*.

Definition 6. *A group G is a **Garside group** if there exists a Garside monoid of which G is the group of fractions.*

The **Words Problem** appears in Garside groups ; this problem is solved by a normal form which defines a canonical writing for each element of the group.

Definition 7. *Let M be a Garside monoid and G be its group of fraction. For a in G, the **infimum** and the **supremum** of a are respectively*

$$\inf(a) = max\{r \in \mathbb{Z} \; ; \; \Delta^r \prec a\} \text{ and } \sup(a) = min\{r \in \mathbb{Z} \; ; \; a \prec \Delta^r\}$$

*The **canonical length** of a is defined by $cl(a) = \sup(a) - \inf(a)$.*

Definition 8. *Let M be a Garside monoid and G be its group of fraction. For a in G, the **(left)** Δ-**normal form** of a is the unique decomposition $\Delta^p a_1 a_2 \cdots a_{cl(a)}$ with $p = \inf(a)$, $a_1 = \Delta \wedge (\Delta^{-p}a)$ and $a_i = \Delta \wedge ((\Delta^p a_1 a_2 \cdots a_{i-1})^{-1}a)$ for $1 < i \leq cl(a)$.*

We end this section by introducing notations :

Notation 9. *Let a be an element in a Garside group G ; we denote by a^* the* **dual element** *of a, defined by $a^{-1}\Delta^{\sup(a)}$. Moreover, we define both functions :*

$$\tau : G \longrightarrow G \qquad\qquad \partial : S \longrightarrow S$$
$$x \longmapsto \Delta^{-1}x\Delta \qquad\qquad q \longmapsto q^{-1}\Delta$$

The function ∂ is a bijection on S, τ is an automorphism on M. We note that $\tau = \partial^2$ on S.

Let $\Delta^p a_1 a_2 \cdots a_k$ be the *left Δ-normal form* of a, an element of a Garside group G ; thus the *left Δ-normal form* of a^{-1} is given by :

$$a^{-1} = \Delta^{-(p+k)}\tau^{-p-k}\left(\partial(a_k)\right)\tau^{-p-k+1}\left(\partial(a_{k-1})\right)\cdots\tau^{-p-1}\left(\partial(a_1)\right) \qquad (1)$$

3   Scheme of the Reduction

The objective is to reduce the *Conjugacy Search Problem*, giving some information on the secret. Let G be the group of fractions of the Garside monoid M. Let a, x be elements of G. The aim is to determine a, called *secret*, knowing only $(x, x' = axa^{-1})$. Our solution is an algorithm, REDCONJ, that builds a left divisor and a right multiple of the secret :

$$\boxed{\text{find}\quad (d, m) \in G \quad \text{such that}\quad d \prec a \prec m}$$

The algorithm REDCONJ is based on two algorithms : the algorithm RIGHT-MGARSIDE determines a right multiple of the secret ; in a similar way, the algorithm LEFTDGARSIDE determines a left divisor of the secret. The proofs of these algorithms can be found in [12, 13]. Principles of the procedure :

> The algorithm is an iterative process reducing the *canonical length*. It uses the left-right symmetry between a and a^{-1} to reduce on the left and on the right $x' = axa^{-1}$. At each step, we begin by determining a rough right multiple of $a_1 a_2 \cdots a_i$, considering x' on its left ; afterwards, if the canonical length of x' is again large enough, then we simplify x' to the right to obtain a better multiple.

Format of Input Data for RightMGarside, LeftDGarside. We assume that the *canonical length* and the *infimum* of the secret are known ; in braid-based cryptography, we consider that they are not security parameters (see [12]).

$$\underline{\text{INPUT DATA :}}\qquad (X, l_1, l_2, \alpha, \beta) \in M \times \mathbb{N}^2 \times \mathbb{Z}^2$$

$$\text{such that } \exists T, y \in M;\quad \begin{cases} X = \tau^\alpha(T)y\tau^{\alpha+\beta}(T^\star) \\ l_1 = cl(y),\ l_2 = cl(T) \\ \inf(T) = \inf(y) = 0 \end{cases}$$

Algorithm 10. RightMGarside $(X, l_1, l_2, \alpha, \beta)$

Input : Let $X \in M$, $l_1, l_2 \in \mathbb{N}$ and $\alpha, \beta \in \mathbb{Z}$.

Init : $A := 1$, $Y := \tau^{-\alpha}(X)$, $l := 0$, $cond := true$
Compute the left Δ-normal form $Y_1 Y_2 \cdots Y_{\sup(Y)}$ of Y
$B := Y_1 Y_2 \cdots Y_{l_2}$

Loop : While $cond$ and $l < l_2$ do
$\qquad l := l + 1$, $A := AY_1$,
$\qquad Z := Y_1^{-1} Y \tau^{\beta + l_2 - l} \left(\partial(Y_1)^{-1} \right)$
$\qquad Y := Y_1^{-1} Y$
$\qquad$ If $Z \notin M$ then $cond := false$
$\qquad$ Else Compute the left Δ-normal form $Z_1 Z_2 \cdots Z_{\sup(Z)}$ of Z
$\qquad\qquad$ If $\sup(Z) = l_1 + 2(l_2 - l) + 1$ then
$\qquad\qquad\qquad A := A \tau^{-\beta - l_2 + l}(Z_{\sup(Z)})^{-1}$
$\qquad\qquad\qquad Z := \tau^{-\beta - l_2 + l}(Z_{\sup(Z)}) Z Z_{\sup(Z)}^{-1}$
$\qquad\qquad$ EndIf
$\qquad\qquad Y :=$ the left Δ-normal form of Z
$\qquad$ EndIf
$\quad$ EndWhile
$\quad$ If $l < l_2$ then $A := AY_2 Y_3 \cdots Y_{l_2 - l + 1}$ EndIf

Output : $A \wedge B$

Let m be the output of the algorithm RightMGarside satisfying the input format ; therefore $T \prec m$ and $\sup(T) = \sup(m)$.

Algorithm 11. LeftDGarside $(X, l_1, l_2, \alpha, \beta)$

Input : Let $X \in M$, $l_1, l_2 \in \mathbb{N}$ and $\alpha, \beta \in \mathbb{Z}$.

Init : $A := e$, $Y := \tau^{-\alpha}(X)$, $l := 0$, $cond := true$
Compute the right Δ-normal form $Y_1 Y_2 \cdots Y_{\sup(Y)}$ of Y
$B := Y_{\sup(Y) - l_2 + 1} \cdots Y_{\sup(Y)}$

Loop : While $cond$ and $l < l_2$ do
$\qquad l := l + 1$, $A := Y_{\sup(Y)} A$,
$\qquad Z := \tau^{l - \beta - l_2} \left(\partial^{-1}(Y_{\sup(Y)})^{-1} \right) Y Y_{\sup(Y)}^{-1}$
$\qquad Y := Y Y_{\sup(Y)}^{-1}$
$\qquad$ If $Z \notin M$ then $cond := false$
$\qquad$ Else Compute the right Δ-normal form $Z_1 Z_2 \cdots Z_{\sup(Z)}$ of Z
$\qquad\qquad$ If $\sup(Z) = l_1 + 2(l_2 - l) + 1$ then
$\qquad\qquad\qquad A := \tau^{\beta + l_2 - l}(Z_1)^{-1} A$
$\qquad\qquad\qquad Z := Z_1^{-1} Z \tau^{\beta + l_2 - l}(Z_1)$
$\qquad\qquad$ EndIf
$\qquad\qquad Y :=$ the right Δ-normal form of Z
$\qquad$ EndIf
$\quad$ EndWhile
$\quad$ If $l < l_2$ then $A := Y_{\sup(p) - l_2 + l + 1} \cdots Y_{\sup(Y)} A$ EndIf
$\quad C := \tau^{-\beta} (A \wedge B)$

Output : $\Delta^{\sup(C)} C^{-1}$

Property 12. *[12] Let $a, b \in M$ a Garside monoid :*
$$a \prec b \quad \Leftrightarrow \quad b^\star \prec^r a^\star \Delta^{\sup(b)-\sup(a)}$$

The process of LEFTDGARSIDE is similar to the one of RIGHTMGARSIDE but it works on the right, using the *right Δ-normal form*. It produces first a left multiple of T^* ; afterwards, the previous property gives

$$\begin{cases} T^\star \prec^r m \\ \sup(T^\star) = \sup(m) \end{cases} \Rightarrow \Delta^{\sup(m)} m^{-1} \prec T$$

Thus the output of LEFTDGARSIDE is a left divisor of T.

Algorithm 13. RedConj (x, x', r, p)

> *Input : Let* $x, x' \in G, \qquad r, p \in \mathbb{Z}.$
> $\overline{\quad \textit{Init} : X := \Delta^{p-\inf(x)} x' \qquad\qquad\qquad}$
> $\qquad\quad l_1 := cl(x), \quad l_2 := p$
> $\qquad\quad \alpha = \inf(x) - r - p, \quad \beta = -\inf(x)$
>
> $\qquad\quad M := \text{RIGHTMGARSIDE}(X, l_1, l_2, \alpha, \beta)$
> $\qquad\quad D := \text{LEFTDGARSIDE}(X, l_1, l_2, \alpha, \beta)$
> $\overline{\quad \textit{Output} : (\Delta^r D, \Delta^r M) \qquad\qquad\qquad\qquad}$

Let a, x be elements of G. We consider the conjugacy instance $(x, x' = axa^{-1})$ and denote (d, m) the output of the algorithm REDCONJ having as input the 4-tuple $(x, x', \inf(a), cl(a))$. Therefore d is a left divisor of a and m is a right multiple of a with the same *supremum*.

That comes from the following expression :

$$x' = axa^{-1} = \Delta^{\inf(x)-cl(a)} \tau^{\inf(x)-\sup(a)}(T) \tau^{-\sup(a)}(z) \tau^{-\sup(a)}(T^\star)$$

with $T = \Delta^{-\inf(a)} a$, $z = \Delta^{-\inf(x)} x$. We note that $\inf(T) = 0 = \inf(z)$ and $a^{-1} = T^\star \Delta^{-\sup(a)}$.

Measure of the Efficiency of This Reduction

• The complexity of the algorithms RIGHTMGARSIDE, LEFTDGARSIDE and REDCONJ is
$$\mathcal{O}\left(l * \mathcal{O}(\ \Delta\text{-normal form }\,)\right) \tag{2}$$
where l is the canonical length of the secret. Indeed, the most expensive operation is the computation of the normal form and the length of the loop depends clearly on the canonical length of the secret. In braid groups, the computation of the Δ-normal form is efficient, thus this one of the algorithm REDCONJ is too.
• This method takes as input a "conjugacy instance" and returns a left divisor and a right multiple of the secret : $d \prec a \prec m$. The efficiency of the reduction is controlled by two parameters :

– the **absolute knowledge**, $ak(a) = min(l(m) - l(a), l(a) - l(d))$,
– the **relative knowledge**, $rk(a) = l(m) - l(d)$.

The relative positions of writings can be represented by :

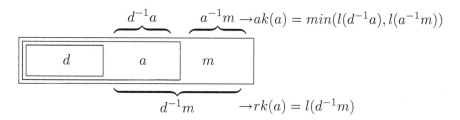

$$d^{-1}a \qquad a^{-1}m \to ak(a) = min(l(d^{-1}a), l(a^{-1}m))$$

$$d^{-1}m \qquad \to rk(a) = l(d^{-1}m)$$

The *absolute knowledge* is not known, because we do not know $l(a)$; it represents the number of generators to complete the secret, by adding the divisor or removing the multiple. In practice, this number permits to measures the cost of completing the search of the secret. The *relative knowledge* is known and gives an overestimation of the *absolute knowledge*, $ak \leq \frac{rk}{2}$.

The cost of a supplementary exhaustive attack determines if a conjugacy instance is broken or not. For a rough method, the cost is :

$$\mathcal{O}\left(\#\mathcal{G}(M)^{ak} * \mathcal{C}(\text{test of identity})\right). \tag{3}$$

4 Outline of Braid-Based Cryptography

The main cryptographic primitive in braid groups is the *Conjugacy Search Problem* and its variants. Here, we consider only the *Conjugacy Search Problem* and its simultaneous variant (for $k = 1$ or 2):

SIMULTANEOUS CONJUGACY PROBLEM : (**SCP**)

> Instance : Let $(x_i, y_i)_{i \in [1,k]} \in G^k$ such that $\exists a \in G$ with $\forall i \in [1, k]$, $y_i = a x_i a^{-1}$.
> Objective : Find $b \in G$ such that $\forall i \in [1, k]$, $y_i = b x_i b^{-1}$.

In practice, the particular problem used for braid-based cryptosystems is the *Generalized Conjugacy Problem of the Diffie-Hellman type* based on the *Generalized conjugacy problem*. These problems are also affected by this reduction [12].

Braids Random Generator

There are essentially two possible representations of a braid : the decomposition of a word as a product of group generators and their inverses ; and the one of normal forms, like the *left Δ-normal form*, as a product of *canonical elements*.

Now, we present two random generators of braid :

PARG POSITIVE ARTIN'S RANDOM GENERATORS : the number of generators, l, is fixed ; we make l random draws on $\mathcal{G}$ (set of group generators) and we add on the left a factor of the type Δ^k, $k \in \mathbb{Z}$.

CRG CANONICAL RANDOM GENERATORS : the *canonical length* is fixed, l ; we make l random draws on S. Afterwards, we reduce the n-braid to its *left Δ-normal form*. While the *canonical length* is smaller than l, we complete with some other random draws. After, we add a factor Δ^k on the left.

5 Double Garside Structures in Braid Groups

The significant fact in braid groups is that they admit two different Garside monoids of which they are the group of fractions. The new idea is to use the two presentations to improve the efficiency of our algorithm. The two Garside presentations of the braid group with n strands are :

- Artin's presentation - 1947 [2]

$$B_n = \left\langle \sigma_1, \sigma_2, \ldots, \sigma_{n-1} \left| \begin{array}{ll} \sigma_i \sigma_j \sigma_i = \sigma_j \sigma_i \sigma_j & \text{if } |i-j| = 1 \\ \sigma_i \sigma_j = \sigma_j \sigma_i & \text{if } |i-j| \geq 2 \end{array} \right. \right\rangle$$

$$\left| \begin{array}{l} \#\mathcal{G}(B_n) = n-1 \quad \text{number of generators} \\ \Delta_{Bn} = (\sigma_1 \sigma_2 \ldots \sigma_{n-1})(\sigma_1 \sigma_2 \ldots \sigma_{n-2}) \ldots (\sigma_1 \sigma_2) \sigma_1 \\ \#S_{Bn} = n! \end{array} \right.$$

- Birman Ko and Lee's presentation - 1998 [3]

$$BKL_n = \left\langle a_{ts} \ (n \geq t > s \geq 1) \left| \begin{array}{ll} a_{ts} a_{rq} = a_{rq} a_{ts} & \text{if } (t-r)(t-q)(s-r)(s-q) > 0 \\ a_{ts} a_{sr} = a_{tr} a_{ts} = a_{sr} a_{tr} & \text{if } n \geq t > s > r \geq 1 \end{array} \right. \right\rangle$$

$$\left| \begin{array}{l} \#\mathcal{G}(BKL_n) = \frac{n(n-1)}{2} \\ \Delta_{BKLn} = a_{n,n-1} a_{n-1,n-2} \ldots a_{2,1} \\ \#S_{BKLn} = \frac{(2n)!}{n!(n+1)!} = C_n \in [3^n, 4^n] (\in o(n!)) \end{array} \right.$$

Remark 14. *We can easily pass from one to the other with polynomial complexity [4]*

$$\sigma_i = a_{i+1,i} \quad \text{and} \quad a_{t,s} = (\sigma_{t-1} \cdots \sigma_{s+1}) \sigma_s \left(\sigma_{s+1}^{-1} \cdots \sigma_{t-1}^{-1} \right)$$

Applying Garside Algorithm

Our algorithm is based on the *canonical length* ; it exploits the density of the writing in the Δ-*normal form*. Then, after one reduction, it is useless to begin again with the same structure on the new instance produced by the output divisor : $(x, d^{-1}axa^{-1}d)$. It is here that the double Garside structure bring a pleasant alternative : we can change the presentation before iteration, or we can use both presentations in parallel.

- **Parallel process:** Consider a conjugacy instance ; we reduce it on both presentations at the same time. Next we transform to BKL's presentation to get some better divisor and multiple, see Figure 2. We can not transform to Artin's presentation because the relation $\prec$ is not respected in this way : a braid admitting a positive word in BKL's presentation does not necessarily admit a positive word in Artin's presentation.
- **Sequential process:** This method exploits information successively on both presentations. We use the output divisor and multiple, obtained in one presentation, to produce two new instances :
$(x, D^{-1}axa^{-1}D)$ and $(x, M^{-1}axa^{-1}M)$. Next, we reduce them in the other presentation. Then, we obtain some better results. See Figure 3.

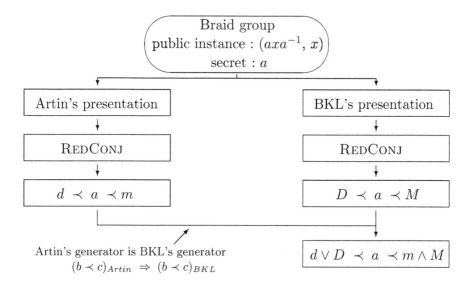

Fig. 2. Parallel process of the reduction

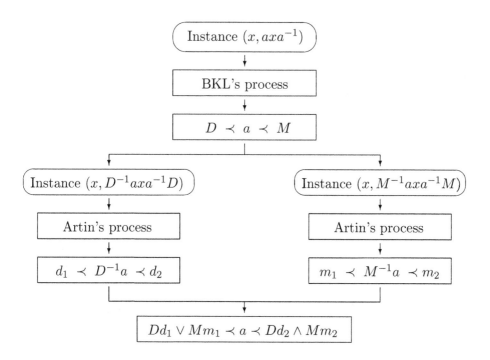

Fig. 3. Sequential process of the reduction

6   Analysis of Efficiency – Simulations

In this Section, we present three tables of results. We give the complexity of the supplementary exhaustive attack. These experimental results average several hundred simulations.

In Table 1 and 2, we propose instances in BKL's presentation to produce the efficiency of the reduction. We consider both random generators : CANONICAL RANDOM GENERATOR and POSITIVE ARTIN'S RANDOM GENERATOR. The complexity is given by $\log_2\left(\frac{n(n-1)}{2}\right) * ak$.

We compare three ways to use our algorithm :

- simply our algorithm
- the parallel process
- the sequential process

We ascertain that both processes (parallel and sequential) complement one another. The representation of the braid has an influence on the efficiency of the reduction. For a braid produced by CRG in BKL's presentation, simply our algorithm is very efficient. The output data are not improved by the parallel process ; only the sequential algorithm let to improve them. In the case of a braid produced by the PARG, its normal structure is not also dense. The both processes bring a distinct improvement and the parallel one seems more efficient.

The efficiency of our algorithm is remarkable. For both random generator, we reduce the size of the secret from several hundred groups generators to only few dozen or even less. The *Simultaneous Conjugacy Problem* is greatly affected. We attain our aim : we give an efficient method to work in braid groups and we show that the choice of the random generator is preponderant in the security of protocol.

In Table 3, we apply the sequential reduction on Artin's presentation to several existing instances from cryptographic literature. Many are broken, but it

Table 1. Reduction on the SCP using the CRG on BKL's presentation

n	$cl(a) = cl(x)$	SCP	type	$l(a)$	$l(a) - l(d)$	$l(m) - l(a)$	$ak(a)$	complexity
60	9	1	simple	265.6	6.7	6.6	5.2	56.1
			parallel		6.7	6.6	5.2	56.1
			sequential		5.2	5.1	3.8	41.0
		2	simple		1.2	1.2	0.6	<10
			parallel		1.2	1.2	0.6	<10
			sequential		0.7	0.6	0.5	<10
100	15	1	simple	743.0	10.0	10.1	8.3	101.8
			parallel		10.0	10.1	8.3	101.8
			sequential		8.3	8.3	6.6	80.9
		2	simple		1.7	1.7	1.0	12.2
			parallel		1.7	1.7	1.0	12.2
			sequential		1.2	1.2	0.8	<10

Table 2. Reduction on the SCP using the PARG on BKL's presentation

n	$l(a) = l(x)$	SCP	type	$l(a) - l(d)$	$l(m) - l(a)$	$ak(a)$
60	1000	1	simple	244.5	1722.1	244.5
			parallel	19.7	1659.0	19.7
			sequential	40.2	200.2	39.9
		2	simple	240.5	1645.6	240.8
			parallel	6.4	1645.6	6.4
			sequential	8.7	96.3	8.7
100	1500	1	simple	299.7	3499.2	299.7
			parallel	34.6	3489.0	34.6
			sequential	73.3	403.0	73.3
		2	simple	274.2	3427.4	274.2
			parallel	11.3	3427.4	11.3
			sequential	16.7	198.1	16.7

Table 3. Sequential reduction using the CRG on Artin's presentation

ref	problem	n	$cl(a)$	$cl(x)$	$l(a)$	$rk(a)$	$ak(a)$	complexity
[4]	GCP	100	15	15	9206.3	31.0	11.5	56.9
		150	20	20	27790.9	44.7	18.0	98.6
		200	30	30	74157.1	57.8	23.7	138.7
[11]	GCP	50	5	3	752.3	18.1	6.1	25.4
		90	12	10	5943.0	28.8	10.7	51.8
[16]	CSP	30	15	15	3410.8	27.9	9.3	44.6
	GCP	60	15	15	3261.6	20.9	7.2	31.5
[10][1]	SCP$_2$	20	4	3	459.5	2.6	0.7	< 10
		24	4	3	654.9	2.5	0.8	< 10
		28	4	3	887.3	2.7	0.9	< 10

([1] we do not use the *Random Super Summit Braid Generator* but only the CANONICAL RANDOM GENERATOR)

is rather the CANONICAL RANDOM GENERATOR which is faulty. Some existing protocols are secure ; however, we must find a procedure to produce some good instances.

The complexity is given by $\log_2(min(n-1, rk)) * ak$ for the *Conjugacy Search Problem* and by $\log_2\left(min(\frac{(n-1)}{2}, rk)\right) * ak$ for the *Generalized Conjugacy Problem*.

7   Conclusion

The interest of this paper is the new proposed method to work in braid groups. We propose an efficient algorithm to reduce the conjugacy problem and its variants : producing a divisor and a multiple of the secret. Moreover, we use the two Garside presentations of braid groups to improve the efficiency of the algorithm.

Even though we present many interesting results, the used random generators are not yet perfect [15]. The representation of braids and their random generators have to be improved to get further results in braid-based cryptography. We began this work in [12], but this development implicates the new proposed canonical random generator. This new work permits to complete an existing list of properties to get an efficient way to produce secure instances. We must meet the following requirements :

- the required properties for the secret a, must be satisfied by its inverse as well.
- $l(a) \approx \frac{l(\Delta)cl(a)}{2}$, where l denotes the number of generators and cl denotes its canonical length.
- the length of the the first canonical factors must be large ($\gg \frac{l(\Delta)}{2}$) whereas the one of the last factors must be short ($\ll \frac{l(\Delta)}{2}$).
- the required properties for the secret must not depend on the presentation.

A further study could be done on the creation of a random generator of braid verifying all existing criteria.

Independently, the new method introduced in this paper on the parallel and sequential work can be improved. A sequential iterative process study could be considered in the future.

Acknowledgments. The author would like to thank François Arnault, Thierry Berger and Philippe Gaborit for their valuable observations.

References

1. I. Anshel, M. Anshel, D. Goldfeld, *An algebraic method for public-key cryptography*, Mathematical Research Letters 6, 1999, 287-291.
2. E. Artin, *Theory of braids*, Annals of Math. 48 (1947), 101-126.
3. J.S. Birman, K.H. Ko, S.J. Lee, *A new approach to the word and conjugacy problems in the braid groups*, Advances in Math. 139-2, 1998, 322-353.
4. J.C. Cha, K.H. Ko, S.J. Lee, J.W. Han, J.H. Cheon, *An efficient implementation of braid groups*, Asiacrypt 2001, LNCS 2248 (2001), 144-156.
5. P. Dehornoy, L. Paris, *Garside groups, a generalization of Artin groups*, Proc. London Math. Soc. 79-3, 1999, 569-604.
6. P. Dehornoy, *Braid-based cryptography*, Contemp. Math., Amer. Math. Soc. 360, 2004.
7. N. Franco, J. Gonzalez-Meneses, *Conjugacy problem for braid groups and Garside groups*, Journal of Algebra 266 (1), 2003, 112-132.
8. F.A. Garside, *The braid group and other groups*, Quart. J. Math. Oxford 20-78, 1969, 235-254.
9. D. Hofheinz, R. Steinwandt, *A practical attack on some braid group cryptographic primitives*, PKC 2003 Proceedings, LNCS 2567, 2003, 187-198.
10. K.H. Ko, D.H. Choi, M.S. Cho, J.W. Lee, *New Signature Scheme Using Conjugacy Problem*, November 2002. (http://eprint.iacr.org/2002/168/)
11. K.H. Ko, S.J. Lee, J.H. Cheon, J.W. Han, J.S. Kang, C. Park, *New public-key cryptosystem using braid groups*, Advances in cryptology - CRYPTO 2000, LNCS 1880, 2000, 166-183.

12. S. Maffre, *A weak key test for braid-based cryptography*, to appear in Designs, Codes and Cryptography.
13. S. Maffre, *Conjugaison et Cyclage dans les groupes de Garside, applications cryptographiques*, Ph.D. Lab. LACO, 2005.
14. M. Picantin, *The conjugacy problem in small Gaussian groups*, Communications in Algebra 29-3, 2001, 1021-1039.
15. H. Sibert, *Algorithmique des groupes de tresses*, Ph.D. Lab. LMNO, 2003.
16. H. Sibert, P. Dehornoy, M. Girault, *Entity authentification schemes using braid word reduction*, WCC 2003, 153-163.
17. The Magma Computational Algebra System for Algebra, Number theory and Geometry. (http://magma.maths.usyd.edu.au/magma)

A New Key Assignment Scheme for
Access Control in a Complete Tree Hierarchy

Alfredo De Santis, Anna Lisa Ferrara, and Barbara Masucci

Dipartimento di Informatica ed Applicazioni,
Università di Salerno 84081 Baronissi (SA), Italy
{ads, ferrara, masucci}@dia.unisa.it

Abstract. A key assignment scheme is a protocol to assign encryption keys and some private information to a set of disjoint user classes in a system organized as a partially ordered hierarchy. The encryption key enables each class to protect its data by means of a symmetric cryptosystem, whereas, the private information allows each class to compute the keys assigned to classes lower down in the hierarchy.

In this paper we consider a particular kind of a hierarchy: the complete rooted tree hierarchy. We propose a key assignment scheme which is not based on unproven specific computational assumptions and that guarantees security against an adversary controlling a coalition of classes of a certain size. Moreover, the proposed scheme is optimal both with respect to the size of the information kept secret by each class and with respect to the randomness needed to set up the scheme.

1   Introduction

The *hierarchical access control problem* deals with the specification of users' access permission and is defined in a scenario where the users of a computer system are organized in a hierarchy formed by a certain number of disjoint classes, called *security classes*. A hierarchy arises from the fact that some users have more access rights than others. The hierarchical access control problem can be solved by using a *key assignment scheme*, that is, a method to assign an encryption key and some private information to each class. The encryption key will be used by each class to protect its data by means of a symmetric cryptosystem. The private information will be used by each class to compute the keys assigned to all classes whose secret data can be accessed by that class. The assignment is carried out by a central authority, the CA, which is active only at the distribution phase.

Akl and Taylor [1] first proposed an elegant solution for the general problem where the hierarchy on security classes is an arbitrary partial order. In their scheme each class is assigned a key that can be used, along with some public parameters generated by the CA, to compute the key assigned to any class lower down in the hierarchy. Subsequently, many researchers have proposed schemes that either have better performances or allow insertion and deletion of classes in the hierarchy (see [2, 8, 9, 11, 12, 13, 14, 15]).

Ø. Ytrehus (Ed.): WCC 2005, LNCS 3969, pp. 202–217, 2006.

The most used approach to key assignment schemes is based on unproven specific assumptions (see $[1, 5, 2, 8, 9, 11, 12, 13, 14, 15]$). For example, Sandhu [15] proposed a key assignment scheme based on the existence of secure symmetric cryptosystems and of one-way functions. Such a scheme has been designed for a particular kind of partially ordered hierarchy, the *rooted tree hierarchy*. The case of a rooted tree hierarchy has also been considered by other researchers (see $[9, 11, 12]$). A different approach, based on information theory and not depending on any unproven specific assumption has been proposed in $[4, 6]$ to design and analyze unconditionally secure key assignment schemes.

In this paper we follow the unconditionally secure approach and propose a key assignment scheme for a particular kind of partially ordered hierarchy, the *complete rooted tree hierarchy*. The paper is organized as follows: in Section 2 we recall the definition of unconditionally secure key assignment schemes given in [4]. In Section 2.1 we prove lower bounds on the size of the private information held by any class and on the amount of random bits needed to set up any key assignment scheme. In Section 3 we describe a key assignment scheme for any rooted tree hierarchy which guarantees security against a single class. Such a scheme has been proposed in [4] and will be used in Section 4 as a starting point to construct a key assignment scheme for any complete tree hierarchy which is secure against a coalition of classes having an arbitrary size. The proposed scheme is optimal both with respect to the size of the private information held by any class and with respect to the amount of random bits needed to set up the scheme.

2   The Model

We consider a scenario where the users of a computer system are divided into a certain number of disjoint classes, called *security classes*. The set of rules that specify the information flow between different user classes in the system defines an *access control policy*. An access control policy can be represented by a directed graph $G = (V, E)$, where the vertex set V corresponds to the set of security classes and there is a directed edge $(u, v) \in E$ if and only if class u can access class v. For each $u \in V$, we define the *accessible set* of u as the set of classes that can be accessed by u, including u itself, i.e., $A_u = \{v \in V : (u, v) \in E\}$. For any subset of classes $X \subseteq V$, we denote by A_X the set $\cup_{v \in X} A_v$. We also define the *forbidden set* of u as the set of classes that cannot access class u, i.e., $F_u = \{v \in V : u \notin A_v\}$.

A *key assignment scheme* for the access control policy $G = (V, E)$ is a method by which a trusted third party, called the central authority (CA), assigns a key and some private information to each class in V. For any class $u \in V$, we denote by p_u the private information sent by the CA to users in class u and by k_u the key assigned to class u, respectively. Moreover, we denote by P_u and K_u the sets of all possible values that p_u and k_u can assume, respectively. Given a set of classes $X = \{u_1, \cdots, u_\ell\}$, where $u_1 < u_2 < \cdots < u_\ell$, we denote by P_X and K_X the sets $P_{u_1} \times \cdots \times P_{u_\ell}$ and $K_{u_1} \times \cdots \times K_{u_\ell}$, respectively.

In this paper, with a boldface capital letter, say $\mathbf{Y}$, we denote a random variable taking values on a set, denoted by the corresponding capital letter Y, according to some probability distribution $\{Pr_\mathbf{Y}(y)\}_{y \in Y}$. The values such a random variable can take are denoted by the corresponding lower case letter. Given a random variable $\mathbf{Y}$, we denote by $H(\mathbf{Y})$ the Shannon entropy of $\{Pr_\mathbf{Y}(y)\}_{y \in Y}$ (we refer the reader to [3] for a complete treatment of Information Theory).

We consider key assignment schemes where the key assigned to each class is unconditionally secure with respect to an adversary controlling a coalition of classes of a limited size. Our schemes are characterized by a security parameter r, the size of the adversary coalition. The maximum value that the security parameter r can assume is equal to the cardinality of the maximum forbidden set, since any adversary coalition for class u can contain at most $|F_u|$ classes. An r-secure key assignment scheme for an access control policy is defined as follows.

Definition 1. ([4]) *Let $G = (V, E)$ be the directed graph that represents an arbitrary access control policy and let $1 \leq r \leq \max_{u \in V} |F_u|$. An r-secure key assignment scheme for G is a method to assign a key to each class in such a way that the following two properties are satisfied:*

1. Any class allowed to access another class can compute the key assigned to that class. *Formally, for any $u \in V$ and any $v \in A_u$, it holds that $H(\mathbf{K}_v|\mathbf{P}_u) = 0$.*
2. Any coalition of at most r classes not allowed to access another class have absolutely no information about the key assigned to that class. *Formally, for any $u \in V$ and any $X \subseteq F_u$ such that $|X| \leq r$, it holds that $H(\mathbf{K}_u|\mathbf{P}_X) = H(\mathbf{K}_u)$.*

In Definition 1 we did not make any assumption on the entropies of the random variables $\mathbf{K}_u$ and $\mathbf{K}_v$, for different classes $u, v \in V$. For example, we could have either $H(\mathbf{K}_u) > H(\mathbf{K}_v)$ or $H(\mathbf{K}_u) \leq H(\mathbf{K}_v)$. Our results apply to the general case of arbitrary entropies of keys, but for clarity we state the next results for the simpler case that all entropies of keys are equal, i.e. $H(\mathbf{K}_u) = H(\mathbf{K}_v)$ for all $u, v \in V$. We denote this common entropy by $H(\mathbf{K})$.

2.1   Lower Bounds

In this section we show lower bounds on the size of the private information held by each class and on the amount of random bits needed to set up any r-secure key assignment scheme. We need the next definition.

Definition 2. ([4]) *Let $G = (V, E)$ be the directed graph that represents an arbitrary access control policy. In any r-secure key assignment scheme for G, a sequence of classes $v_1, \ldots, v_m$ is called r-almost_covered if, either $m = 1$, or $m > 1$ and for any $j = 2, \ldots, m$, there exists a set $X_j \subseteq F_{v_j}$ such that $|X_j| \leq r$ and $\{v_1, \ldots, v_{j-1}\} \subseteq A_{X_j}$.*

The next theorem shows a lower bound on the size of the private information held by each class in any r-secure key assignment scheme.

Theorem 1. *([4]) Let $G = (V, E)$ be the directed graph that represents an arbitrary access control policy. In any r-secure key assignment scheme for G, for any $u \in V$, if there exists an r-almost_covered sequence $v_1, \ldots, v_m$ in A_u, then it holds that $H(\mathbf{P}_u) \geq m \cdot H(\mathbf{K})$.*

As shown by Knuth and Yao [10], the entropy of a random source is related to the average number of independent unbiased random bits necessary to simulate the source. In the following, given a directed graph $G = (V, E)$ representing an arbitrary access control policy, we denote by $H(\mathbf{P}_V)$ the amount of randomness needed by the CA to set up any r-secure key assignment scheme for G.

Theorem 2. *Let $G = (V, E)$ be the directed graph that represents an arbitrary access control policy. In any r-secure key assignment scheme for G, if there exists an r-almost_covered sequence of length m in V, then it holds that $H(\mathbf{P}_V) \geq m \cdot H(\mathbf{K})$.*

Key Assignment Schemes for Rooted Tree Hierarchies. In this section we consider key assignment schemes for an important kind of access control policy: the rooted tree hierarchy. Given a rooted tree $T = (V, E)$, for any class $u \in V$, we denote by h_u the height of the class u, defined by $h_u = 1$ if u is a leaf class and $h_u = 1 + \max_{i=1}^{g_u} h_{u_i}$ otherwise, where g_u is the degree of u and, for $i = 1, \ldots, g_u$, u_i denotes the i-th child of u. We also denote by h the height of the tree, i.e., the height of the root class.

For any two classes $u, v \in V$, class u has access to v's private data if and only if u is an ancestor of v. Therefore, the accessible set A_u of class $u \in V$ consists of the classes in the subtree rooted at u, whereas, the forbidden set F_u consists of the classes that are not ancestors of u. The next lemma shows how to compute the length of an r-almost covered sequence in the accessible set A_u, for any $u \in V$ and any $1 \leq r \leq \max_{u \in V} |F_u|$.

Lemma 1. *([4]) Let $T = (V, E)$ be a rooted tree hierarchy and let $1 \leq r \leq \max_{u \in V} |F_u|$. In any r-secure key assignment scheme for T, for any $u \in V$, there exists an r-almost_covered sequence in A_u, whose length $L(r, h_u)$ is defined by the following recurrence*

$$L(r, h_u) = \begin{cases} 1 & \text{if } u \text{ is a leaf class;} \\ 1 + \sum_{i=1}^{\min\{r, g_u\}} L(r - \min\{r, g_u\} + i, h_{u_i}) & \text{otherwise.} \end{cases} \quad (1)$$

From (1), it is easy to see that $L(r, h_u) \geq L(r', h_u)$, for any $r' = 1, \ldots, r - 1$. The next lemma shows that, given a class $u \in V$, any 1-almost_covered sequence in A_u having length h_u is a 1-almost_covered sequence of maximum length.

Lemma 2. *([4]) Let $T = (V, E)$ be a rooted tree hierarchy. In any 1-secure key assignment scheme for T, for any class $u \in V$, the length of any 1-almost_covered sequence in A_u is less than or equal to h_u.*

The next theorem is an immediate consequence of Theorem 1 and Lemma 1.

Theorem 3. *([4]) Let $T = (V, E)$ be a rooted tree hierarchy and let $1 \leq r \leq \max_{u \in V} |F_u|$. In any r-secure key assignment scheme for T, for any class $u \in V$, it holds that $H(\mathbf{P}_u) \geq L(r, h_u) \cdot H(\mathbf{K})$.*

Remark 1. From Lemma 2 and Theorem 3, the size of the private information held by each class $u \in V$ in any 1-secure key assignment scheme for T is lower bounded by $h_u \cdot H(\mathbf{K})$. In particular, if we consider the root class, it follows that the number of random bits needed by the CA to set up any 1-secure key assignment scheme for T is lower bounded by $h \cdot H(\mathbf{K})$.

Complete Rooted Tree Hierarchies. In the following we consider g-complete rooted tree hierarchies, i.e., such that all leaves of the tree are at the same level and all internal nodes have the same degree g. Given a g-complete rooted tree hierarchy $T = (V, E)$, for each class $u \in V$, the cardinality of the forbidden set F_u is equal to the number of classes in V, that is, $\sum_{i=0}^{h-1} g^i$, minus the number of ancestors of u minus one (the class u), that is, $h - h_u + 1$. Therefore, we have that $|F_u| = \sum_{i=0}^{h-1} g^i - (h - h_u + 1) = \frac{g^h}{g-1} - (h - h_u + 1)$. In particular, for the root class we have $|F_{root}| = \frac{g(g^{h-1}-1)}{g-1}$. Therefore, the maximum value that the security parameter r can assume in a key assignment scheme for a g-complete rooted tree hierarchy is equal to $\frac{g(g^{h-1}-1)}{g-1}$, which is equal to the number of nodes in the tree minus one (the root class).

The next lemma will be a useful tool to show a lower bound on the size of the private information held by each class in any r-secure key assignment scheme for a g-complete rooted tree hierarchy, when $1 < r \leq g$.

Lemma 3. *Let $T = (V, E)$ be a g-complete rooted tree hierarchy and let $1 < r \leq g$. In any r-secure key assignment scheme for T, for any class $u \in V$, it holds that*

$$L(r, h_u) = \sum_{i=1}^{h_u} L(r - 1, i).$$

Proof. The proof is by induction on h_u. Let u be a leaf class. From (1), it follows that $L(r, 1) = 1 = L(r - 1, 1)$.

Assume by inductive hypothesis that $L(r, h'_u) = \sum_{i=1}^{h'_u} L(r - 1, i)$, for any $1 < h'_u < h_u$. Since $g \geq r$, from (1) we get

$$L(r, h_u) = 1 + \sum_{i=1}^{r} L(i, h_u - 1)$$

$$= 1 + \sum_{i=1}^{r-1} L(i, h_u - 1) + L(r, h_u - 1)$$

$$= L(r - 1, h_u) + L(r, h_u - 1) \qquad \text{(from (1))}$$

$$= L(r - 1, h_u) + \sum_{i=1}^{h_u-1} L(r - 1, i) \qquad \text{(from the inductive hypothesis)}$$

$$= \sum_{i=1}^{h_u} L(r-1, i).$$

Hence, the lemma holds.   □

In the next lemma we will use the following equality where a and b are integers, (see equation (5.9) of [7, pag. 159]):

$$\sum_{j=0}^{b} \binom{a+j}{j} = \binom{a+b+1}{b}. \tag{2}$$

Lemma 4. Let $T = (V, E)$ be a g-complete rooted tree hierarchy and let $1 \le r \le g$. In any r-secure key assignment scheme for T, for any class $u \in V$, it holds that

$$L(r, h_u) = \binom{r + h_u - 1}{h_u - 1}.$$

Proof. The proof is by induction on r. Let $r = 1$. From (1) it follows that $L(1, h_u) = 1 + L(1, h_u - 1) = h_u$.
Assume by inductive hypothesis that $L(r', h_u) = \binom{r'+h_u-1}{h_u-1}$, for any $1 < r' < r$. We have that

$$L(r, h_u) = \sum_{i=1}^{h_u} L(r-1, i) \qquad \text{(from Lemma 3)}$$

$$= \sum_{i=1}^{h_u} \binom{r-1+i-1}{i-1} \qquad \text{(from the inductive hypothesis)}$$

$$= \sum_{j=0}^{h_u-1} \binom{r-1+j}{j}$$

$$= \binom{r+h_u-1}{h_u-1} \qquad \text{(from equality (2), setting } a = r-1 \text{ and } b = h_u - 1\text{)}.$$

Hence, the lemma holds.   □

The next theorem is an immediate consequence of Theorem 1 and Lemma 4.

Theorem 4. Let $T = (V, E)$ be a g-complete rooted tree hierarchy and let $1 \le r \le g$. In any r-secure key assignment scheme for T, for any class $u \in V$, it holds that $H(\mathbf{P}_u) \ge \binom{r+h_u-1}{h_u-1} \cdot H(\mathbf{K})$.

In particular, if we consider the root class, it follows that the number of random bits needed by the CA to set up any r-secure key assignment scheme for a g-complete rooted tree hierarchy, where $1 \le r \le g$, is lower bounded by $\binom{r+h-1}{h-1} \cdot H(\mathbf{K})$. The above bounds are both tight. Indeed, in Section 4 we will show an r-secure key assignment scheme for a g-complete tree hierarchy which meets the bounds.

3   A 1-Secure Key Assignment Scheme for Any Rooted Tree Hierarchy

In this section we describe a 1-secure key assignment scheme for any rooted tree hierarchy. Such a scheme has been proposed in [4] and will be used in Section 4 as a starting point to construct r-secure key assignment schemes. Let $T = (V, E)$ be a rooted tree hierarchy with height h and let $q \geq h$ be a prime number. The scheme works as follows: first, the CA randomly chooses a sequence s of h distinct integers in Z_q. These integers will be used to compute the key k_u and the private information p_u associated to each class $u \in V$. Afterwards, for each class $u \in V$, the CA sends the private information p_u to u by means of a secure channel. Such information will be used by each internal class $u \in V$ to compute the key assigned by the CA to any class $v \in A_u$, by iteratively computing the key assigned to any class in the path from u to v. The 1-secure scheme is shown in Figure 1.

Input: A rooted tree $T = (V, E)$.

Let h be the height of T and let $q \geq h$ be a prime number.

Randomly choose a sequence s of h distinct integers in Z_q.

$\{(u, p_u, k_u) : u \in V\} \leftarrow Basic\_Scheme(T, s, q)$

For any $u \in V$, privately send p_u to u.

Fig. 1. A 1-secure key assignment scheme for any rooted tree hierarchy

The scheme used to compute the key k_u and the private information p_u for each class $u \in V$, referred to as the $Basic\_Scheme(T, s, q)$, is described in Figure 2. In the *key generation phase*, starting from the root class, the key for each internal class u is used by the CA to compute the keys for its children $u_1 \ldots, u_{g_u}$, where g_u is the degree of u. In the *private information generation*

$Basic\_Scheme(T, s, q)$

Let h be the height of T and let $s = (y_1, \ldots, y_h)$.

/*Key generation phase*/

Let $root$ be the root of T, then $k_{root} \leftarrow y_h$.

For $j = h$ downto 2 do

    For any $u \in V$ with $h_u = j$ do

        For any $i = 1, \ldots, g_u$, do $k_{u_i} \leftarrow k_u + i \cdot y_{h_{u_i}} \bmod q$.

/*Private information generation phase*/

For any leaf class $u \in V$ do $p_u \leftarrow k_u$.

For any internal class $u \in V$ do $p_u \leftarrow ((y_1, \ldots, y_{h_u-1}) \circ k_u)$.

Return $\{(u, p_u, k_u) : u \in V\}$.

Fig. 2. The basic scheme used by the 1-secure scheme of Figure 1

phase, the CA assigns to each class u the private information p_u, which consists of a sequence of h_u integers in Z_q. The last value of such a sequence is the key k_u. Indeed, if u is a leaf class, then $p_u = k_u$, whereas, if u is an internal class, then $p_u = ((y_1, \ldots, y_{h_u-1}) \circ k_u)$, where the symbol $\circ$ denotes the concatenation of two sequences and $(y_1, \ldots, y_{h_u-1})$ is the sequence of integers needed by u to compute the keys for all classes in its accessible set.

It is easy to see that the scheme is optimal both with respect to the size of the private information held by each class and with respect to the number of random bits needed by the CA to set up the scheme. Indeed, the size of the private information p_u assigned to class u is equal to $h_u \log q$ bits and the amount of random bits is equal to $h \log q$ bits, whereas, the size of the key k_u is equal to $\log q$ bits. Hence, the larger the prime number $q \geq h$, the larger the size of the key, of the private information held by each class and of the number of random bits needed to set up the scheme.

4 An r-Secure Key Assignment Scheme for Any Complete Rooted Tree Hierarchy

In this section we show an r-secure key assignment scheme for any g-complete rooted tree hierarchy. The problem of designing an r-secure scheme for a g-complete rooted tree hierarchy $T = (V, E)$ with height h is reduced to the problem of designing an $(r-1)$-secure scheme for the truncated tree $T_j = (V_j, E_j)$, for $j = 1, \ldots, h$, where T_j is the subtree of T obtained by truncating T at the j-th level, that, is the subtree containing the first j levels of T. The keys and the private information computed by the $(r-1)$-secure schemes on such truncated trees are then combined to produce the keys and the private information computed by the r-secure scheme on T. The recursion bottoms up when $r = 1$: in this case we use the 1-secure scheme presented in Section 3.

The r-secure scheme works as follows: first, the CA randomly chooses a sequence s of $\binom{r+h-1}{h-1}$ distinct integers in Z_q, where $q \geq \binom{r+h-1}{h-1}$ is a prime number. These integers will be used to compute the key k_u and the private information p_u associated to each class $u \in V$. The larger the prime number q, the larger the size of the key and of the private information held by each class. Afterwards, for each class $u \in V$, the CA sends the private information p_u to u by means

---

Input: A g-complete rooted tree $T = (V, E)$ with height h and an integer
$$1 \leq r \leq \frac{g(g^{h-1}-1)}{g-1}.$$
Let $q \geq \binom{r+h-1}{h-1}$ be a prime number.
Randomly choose a sequence s of $\binom{r+h-1}{h-1}$ distinct integers in Z_q.
$\{(u, p_u, k_u) : u \in V\} \leftarrow Scheme(T, r, s, q)$.
For any $u \in V$, privately send p_u to u.

Fig. 3. An r-secure key assignment scheme for any g-complete rooted tree hierarchy

---

$Scheme(T, r, s, q)$

If $r = 1$, then
    $\{(u, p_u, k_u) : u \in V\} \leftarrow Basic\_Scheme(T, s, q)$
    Return $\{(u, p_u, k_u) : u \in V\}$

Let h be the height of T.

Partition the sequence s into h subsequences $s_{[r-1;1]}, \ldots, s_{[r-1;h]}$ such that, for any $j = 1, \ldots, h$, the subsequence $s_{[r-1;j]}$ contains $\binom{r-2+j}{r-1}$ distinct integers.

For any $j = 1, \ldots, h$, let $T_j = (V_j, E_j)$ be the subtree of T obtained by truncating T at the j-th level.

For $j = 2$ to h do
    $\{(u, p_u^{[r-1;j]}, k_u^{[r-1;j]}) : u \in V_j\} \leftarrow Scheme(T_j, r-1, s_{[r-1;j]}, q)$

/*Key generation phase*/
Let $root$ be the root class of T, then $k_{root} \leftarrow s_{[r-1;1]}$.
For $j = h$ downto 2 do
    For any $u \in V$ with $h_u = j$ do
        For any $i = 1, \ldots, g$, do $k_{u_i} \leftarrow k_u + i \cdot k_{u_i}^{[r-1;\ell_u+1]} \bmod q$.

/*Private information generation phase*/
For any leaf class $u \in V$ do $p_u \leftarrow k_u$.
For any internal class $u \in V$ do $p_u \leftarrow p_u^{[r-1;\ell_u+1]} \circ \cdots \circ p_u^{[r-1;h]} \circ k_u$.

Return $\{(u, p_u, k_u) : u \in V\}$.

Fig. 4. The scheme used by the r-secure scheme of Figure 3

of a secure channel. Such information will be used by each class to compute the keys assigned to all classes in its accessible set. The r-secure scheme is shown in Figure 3. The scheme used to compute the key and the private information for each class is referred to as the $Scheme(T, r, s, q)$ and is described in Figure 4. If $r = 1$, the $Scheme(T, r, s, q)$ reduces to the $Basic\_Scheme(T, s, q)$.

If $r > 1$, the scheme proceeds as follows: First, the CA partitions the sequence s into h subsequences $s_{[r-1;1]}, \ldots, s_{[r-1;h]}$, such that, for any $j = 1, \ldots, h$, the sequence $s_{[r-1;j]}$ contains $\binom{r-2+j}{r-1}$ distinct integers. Notice that from equation (2), setting $b = h - 1$ and $a = r - 1$, we have

$$\sum_{j=1}^{h} \binom{r-2+j}{r-1} = \sum_{i=0}^{h-1} \binom{r-1+i}{i} = \binom{r+h-1}{h-1}. \tag{3}$$

Afterwards, for any $j = 1, \ldots, h$, the CA runs the $Scheme$ on inputs $T_j, r - 1, s_{[r-1;j]}$, and q, where $T_j = (V_j, E_j)$ is the subtree of T obtained by truncating T at the j-th level. For any $j = 2, \ldots, h$, let $k_u^{[r-1;j]}$ and $p_u^{[r-1;j]}$ be the key and the private information assigned by such a scheme to a class u in the truncated tree T_j. Starting from the key $k_u^{[r-1;j]}$ and the private information $p_u^{[r-1;j]}$ assigned to the class $u \in T_j$ by the $Scheme$ on inputs $T_j, r - 1, s_{[r-1;j]}$ and q, for any $j = 2, \ldots, h$, the CA computes the key k_u and the private information p_u for any class u in the tree T.

In the *key generation phase* the CA assigns to the root class of T the value $s_{[r-1;1]}$, corresponding to the first subsequence of s, which contains a single element. Starting from the root class, the key k_u for each internal class u at level ℓ_u is used to compute the key for its children $u_1 \ldots, u_g$, which are at level $\ell_u + 1$. In particular, for any $i = 1, \ldots, g$, the computation of the key k_{u_i} also involves the use of the key $k_{u_i}^{[r-1;\ell_u+1]}$ assigned to u_i by the *Scheme* on inputs $T_{\ell_u+1}, r - 1, s_{[r-1;\ell_u+1]}$, and q.

In the *private information generation phase* the CA assigns to each class u the private information p_u, which consists of a sequence of integers in Z_q. The last value of such a sequence is the key k_u. Indeed, if u is a leaf class, then $p_u = k_u$, whereas, if u is an internal class, then $p_u = p_u^{[r-1;\ell_u+1]} \circ \cdots \circ p_u^{[r-1;h]} \circ k_u$, where the symbol $\circ$ denotes the concatenation of two sequences. Each internal class $u \in V$ can use its private information p_u to compute the key assigned by the CA to any class $v \in A_u$, by iteratively computing the key assigned to any class in the path from u to v.

4.1 Analysis of the Scheme

In this section we analyze the scheme proposed in the previous section. We first remark some properties and give some useful definitions in order to prove correctness and security properties. Afterwards, we show that the proposed scheme is optimal both with respect to the size of the private information held by each class and with respect to the randomness needed by the CA to set up the scheme.

We first notice that the scheme of Figure 4 recursively calls itself on different trees and with different security parameters, as shown in the following. Given a tree T having height h, we define the *execution hierarchy* as the tree whose height is equal to r and whose nodes are represented by boxes corresponding to the truncated trees $T_2, \ldots, T_h$ (see the left hand side of Figure 5). The root box T_h corresponds to the entire tree T. Moreover, for any $j = 2, \ldots, h$, any internal box corresponding to the truncated tree T_j has $j - 1$ children, corresponding to the truncated trees $T_2, \ldots, T_j$, respectively.

In the following we define the initialization sequences, the keys and the private information received by the classes during the executions of the schemes on the trees corresponding to all the boxes in the execution hierarchy. For each $i = 2, \ldots, r-1$ and $j_1, j_2, \ldots, j_i \in \{1, \ldots, h\}$, we define $s_{[r-i;j_1,j_2,\ldots,j_i]}$ as the sequence of integers used to set up the scheme on T_{j_i}, where the security parameter is equal to $r - i$ and for each $\alpha = r - 1, \ldots, r - i + 1$, the scheme is executed on T_{j_α} with security

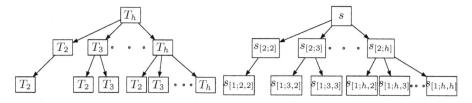

Fig. 5. The execution hierarchy and the corresponding sequences for $r = 3$

parameter equal to α. The right hand side of Figure 5 shows the sequences used to set up the schemes on the trees corresponding to each box of the execution hierarchy drawn on the left hand side of Figure 5. Moreover, for each $i = 2, \ldots, r-1$ and $j_1, j_2, \ldots, j_i \in \{1, \ldots, h\}$, we define the keys and the private information received by a class u during the execution of the scheme on T_{j_i} with security parameter equal to $r-i$ starting by the sequence $s_{[r-i;j_1,j_2,\ldots,j_i]}$, respectively as, $k_u^{[r-i;j_1,j_2,\ldots,j_i]}$ and $p_u^{[r-i;j_1,j_2,\ldots,j_i]}$. We also define $k_u^{[r-i;j_1,j_2,\ldots,j_i]}$ and $p_u^{[r-i;j_1,j_2,\ldots,j_i]}$ as $k_u^{[r-i;j_i]}$ and $p_u^{[r-i;j_i]}$ if $j_1 = j_2 = \ldots = j_i$. For the sake of notational consistency, in the following we refer to k_u and p_u as $k_u^{[r;h]}$ and $p_u^{[r;h]}$, respectively.

Now, we are ready to show that each class, starting from its private information, can compute the keys assigned to all classes in its accessible set.

Theorem 5. *Let $T = (V, E)$ be a g-complete rooted tree hierarchy with height h and let $1 \le r \le \frac{g(g^{h-1}-1)}{g-1}$. In the r-secure scheme each class u can compute the key assigned to each class $v \in A_u$ by using its private information $p_u^{[r;h]}$.*

Proof. The proof is by induction on r. For $r = 1$ the proof follows from the correctness of the 1-secure scheme (see [4]).

Assume by inductive hypothesis that, given a g-complete rooted tree $T = (V, E)$ with height h, in the r-secure scheme, where $2 \le r < r'$, each class u can compute the key assigned to each class $v \in A_u$ by using its private information $p_u^{[r;h]}$.

Let $r = r'$. It is easy to see that, for any $v \in A_u$, the key $k_v^{[r;h]}$ is a function of the key $k_u^{[r;h]}$ and of the keys $k_w^{[r-1;\ell_w]}$, for any class w along the path from u to v. Since $p_u^{[r;h]}$ contains $p_u^{[r-1;\ell_w]}$, from the inductive hypothesis it follows that u can compute the key $k_w^{[r-1;\ell_w]}$ assigned to each class w along the path from u to v. Moreover, $p_u^{[r;h]}$ also contains the key $k_u^{[r;h]}$ assigned to u by the r-secure scheme. Therefore, u can compute the key $k_v^{[r;h]}$ assigned by the r-secure scheme to each class $v \in A_u$, by using its private information $p_u^{[r;h]}$. $\qquad\square$

In order to prove Property 2. of Definition 1, we need the following definition.

Definition 3. *Let $T = (V, E)$ be a g-complete rooted tree hierarchy with height h and let $1 \le r \le \frac{g(g^{h-1}-1)}{g-1}$. A set of classes $\{v_1, v_2, \ldots, v_r\}$ is called an r-strong coalition for a class u if and only if $v_i \in F_u$ and $u \in A_{w_i}$, where w_i is the parent of v_i, for each $i = 1, \ldots, r$.*

The next lemmas will be useful tools to show our results.

Lemma 5 (POLYNOMIAL PROPERTY). *Let $T = (V, E)$ be a g-complete rooted tree hierarchy with height h and let $1 \le r \le \frac{g(g^{h-1}-1)}{g-1}$. Let $u \in V$ be an internal class and let u_i be the i-th child of u, for some $i \in \{1, \ldots, g\}$. In the r-secure scheme, it holds that $k_{u_i} = q_u(i)$, where*

$$q_u(x) = k_u^{[r;h]} + \sum_{j=1}^{r-1} x^{r-j} \cdot k_u^{[j;\ell_u+1]} + x^r \cdot y_{\ell_u+1} \bmod q$$

is a polynomial of degree r and y_{ℓ_u+1} is the last element of the sequence used in the initialization phase of the 1-secure scheme on T_{ℓ_u+1}.

Proof. From the *key generation phase* in the r-secure scheme, it holds that

$$k_{u_i}^{[r;h]} = k_u^{[r;h]} + i \cdot k_{u_i}^{[r-1;\ell_u+1]} \bmod q$$

$$= k_u^{[r;h]} + i \cdot k_u^{[r-1;\ell_u+1]} + i^2 \cdot k_{u_i}^{[r-2;\ell_u+1]} \bmod q$$

$$= \dots$$

$$= k_u^{[r;h]} + \sum_{j=2}^{r-1} i^{r-j} \cdot k_u^{[j;\ell_u+1]} + i^{r-1} \cdot k_{u_i}^{[1;\ell_u+1]} \bmod q$$

$$= k_u^{[r;h]} + \sum_{j=1}^{r-1} i^{r-j} \cdot k_u^{[j;\ell_u+1]} + i^r \cdot y_{\ell_u+1} \bmod q.$$

The last equality follows from the *key generation phase* of the 1-secure scheme, since $k_{u_i}^{[1;\ell_u+1]} = k_u^{[1;\ell_u+1]} + i \cdot y_{\ell_u+1}$. □

Notice that, for each class u, the key k_u is a function of the key held by the root class and of the sequences $s_{[r-1;2]}, \dots, s_{[r-1;\ell_u]}$. Moreover, in the *private information generation phase*, any child of u receives its key and some values in $s_{[r-1;\ell_u+1]}, \dots, s_{[r-1;h]}$. Since the h subsequences $s_{[r-1;1]}, \dots, s_{[r-1;h]}$ are obtained by partitioning the sequence s of distinct integers randomly chosen by the CA, the following property holds:

CHILD_POWER_PROPERTY. *Let $T = (V, E)$ be a g-complete rooted tree hierarchy with height h and let $1 \leq r \leq \frac{g(g^{h-1}-1)}{g-1}$. In the r-secure scheme, given any internal class $u \in V$, each of its children holds a unique information (corresponding to its own key) which can be used to compute the key $k_u^{[r;h]}$.*

Lemma 6. *Let $T = (V, E)$ be a g-complete rooted tree hierarchy with height h, let $1 \leq r \leq \frac{g(g^{h-1}-1)}{g-1}$ and let C be an $(r+1)$-strong coalition of classes in T_{ℓ_u} for a class u. In the r-secure scheme the coalition C is able to compute the key $k_u^{[r;h]}$.*

Proof. The proof is by induction on r. Let $r = 1$ and $C = \{v_1, v_2\}$. We have to distinguish three cases.

1. Let v_1 and v_2 be siblings of u and let z be their parent. From the *key generation phase* in the 1-secure scheme, the keys of the classes in u, v_1 and v_2 are a function of k_z and y_{ℓ_u}. Hence, the coalition is able to compute k_z and y_{ℓ_u}. Afterwards, the coalition is also able to compute the key k_u.
2. Let $\ell_{v_1} > \ell_{v_2}$ and let z be the parent of v_1. Since v_2 holds the value $y_{\ell_{v_1}}$ and the key of z is a function of the key of v_1 and $y_{\ell_{v_1}}$, the coalition is able to compute the key k_z. Notice that z is also an ancestor of u. Since v_2 holds the sequence $(y_{\ell_z+1}, \dots y_{\ell_u})$, from the *key generation phase* in the 1-secure scheme, it is easy to see that the coalition is also able to compute the key k_u.

3. Let $\ell_{v_1} = \ell_{v_2} > \ell_u$ and let z be the parent of v_1 and v_2. Since k_{v_1} and k_{v_2} are a function of the value $y_{\ell_{v_1}}$ and the key of z, the coalition is able to compute both the key k_z and the value $y_{\ell_{v_1}}$. Since v_1 and v_2 also hold the sequence $(y_{\ell_{v_1}+1}, \ldots y_{\ell_u})$, from the *key generation phase* in the 1-secure scheme, it is easy to see that the coalition is also able to compute the key k_u.

Assume by inductive hypothesis that given a g-complete rooted tree hierarchy, in the r-secure scheme, where $2 \le r < r'$, any $(r+1)$-strong coalition of classes in T_{ℓ_u} for a class u is able to compute the key $k_u^{[r;h]}$.

Let $r = r'$. We have to analyze the following two cases:

1. Let the coalition be constituted by $r+1$ siblings of u and let z be their parent. From the POLYNOMIAL_PROPERTY, the keys of the classes in the coalition are the evaluations of the polynomial $q_z(x)$ of degree r, in $r+1$ different points. Hence, the coalition is able to compute the key $k_u^{[r;h]}$.
2. Let the coalition be constituted by i siblings of u and by a set C' of $r+1-i$ classes in T_{ℓ_u-1}, for some $i = 0, \ldots, r$. Let z be the parent of u. From the POLYNOMIAL_PROPERTY, the keys of the siblings of u correspond to i equations in the $r+1$ unknowns $k_z^{[r;h]}$, y_{ℓ_u}, $k_z^{[1;\ell_u]}, \ldots, k_z^{[r-1;\ell_u]}$. From the *private information generation phase*, the coalition holds y_{ℓ_u}. Moreover, for $i \ne r$, from the inductive hypothesis it follows that the $r+1-i$ classes in C' are able to compute the $r-i$ keys $k_z^{[1;\ell_u]}, \ldots, k_z^{[r-i;\ell_u]}$. Since the coalition has i equations in the i unknowns $k_z, k_u^{[r-i+1;\ell_u]}, \ldots, k_u^{[r-1;\ell_u]}$, it is able to compute the key $k_u^{[r;h]}$. □

Since the private information held by each class is contained in the private information held by each of its ancestors, in order to show that any coalition in the forbidden set F_u cannot compute the key of a class u, it is enough to consider only strong coalitions for u.

Theorem 6. *Let $T = (V, E)$ be a g-complete rooted tree hierarchy with height h and let $1 \le r \le \frac{g(g^{h-1}-1)}{g-1}$. In the r-secure scheme, an r-strong coalition for a class u is not able to compute the key $k_u^{[r;h]}$ with probability greater than or equal to $1/q$.*

Proof. The proof is by induction on r. For $r = 1$ the proof follows from the security of the 1-secure scheme (see [4]).

Assume by inductive hypothesis that given a g-complete rooted tree hierarchy, in the r-secure scheme, where $2 \le r < r'$, any r-strong coalition for a class u is not able to compute the key of u with probability greater than or equal to $1/q$.

Let $r = r'$, the proof follows by induction on the level ℓ_u of a class u. Let $\ell_u = 1$, i.e., u is the root class. Any r-strong coalition for u is constituted by r children of u. From the CHILD_POWER_PROPERTY, each class in the coalition holds a unique information (corresponding to its own key) which can be used to compute the key $k_u^{[r;h]}$. From the POLYNOMIAL_PROPERTY, the keys of the classes in the coalition are the evaluations of $q_u(x)$ in r different points. Hence, in order to compute $k_u^{[r;h]}$, the coalition has a system of r equations in $r+1$ unknowns. For any of the q^r possible choices for the r-tuple $(y_{\ell_u+1}, k_u^{[1;\ell_u+1]}, \ldots, k_u^{[r-1;\ell_u+1]})$, there are q^{r-1} corresponding values for the key $k_u^{[r;h]}$. Hence, the probability that the coalition computes the key $k_u^{[r;h]}$ assigned to the class u is equal to $1/q$.

Assume by inductive hypothesis that any r-strong coalition for a class u where $1 < \ell_u < \ell$ is not able to compute the key of u with probability greater than or equal to $1/q$.

Let u be a class at level $\ell_u = \ell$. We have to analyze the following three cases:

Case 1. The coalition is constituted by r children of u. With identical argumentation used to show the basic case where $\ell_u = 1$, we can prove that the probability that the coalition computes the key $k_u^{[r;h]}$ assigned to class u is equal to $1/q$.

Case 2. The coalition is constituted by classes in T_{ℓ_u}. Let z be the parent of u. It is easy to see that the coalition is also an r-strong coalition for z. Assume by contradiction that the coalition is able to compute the key of u with probability greater than or equal to $1/q$. Let C' be the set constituted by the class u and the $0 \leq i \leq r$ classes in the coalition that are also children of z. Let C'' be the set constituted by the classes in the coalition whose levels are less than or equal to ℓ_z. From the POLYNOMIAL PROPERTY, the keys of the classes in C' correspond to $i+1$ equations in the $r+1$ unknowns $k_z^{[r;h]}$, y_{ℓ_u}, $k_z^{[1;\ell_u]}$, $\ldots$, $k_z^{[r-1;\ell_u]}$. Hence, if $i = r$, the coalition is able to compute $k_z^{[r;h]}$. Otherwise, from the *private information generation phase*, the coalition holds y_{ℓ_u} and, from Lemma 6, the $r-i$ classes in C'' are able to compute the $r-i-1$ keys $k_z^{[1;\ell_u]}, \ldots, k_z^{[r-i-1;\ell_u]}$. Since the coalition has $i+1$ equations in the $i+1$ unknowns $k_z, k_z^{[r-i;\ell_u]}, \ldots, k_z^{[r-1;\ell_u]}$, it is able to compute $k_z^{[r;h]}$. This contradicts the inductive hypothesis because the level of z is less than ℓ.

Case 3. The coalition is constituted by $1 \leq i \leq r-1$ children of u and by a set C of $r-i$ classes in T_{ℓ_u}. From the CHILD POWER PROPERTY, each child of u holds a unique information (corresponding to its own key) which can be used to compute the key of u. From the POLYNOMIAL PROPERTY, it follows that those keys correspond to i equations in the $r+1$ unknowns $k_u^{[r;h]}, y_{\ell_u+1}, k_u^{[1;\ell_u+1]}, \ldots, k_u^{[r-1;\ell_u+1]}$. The classes in C hold the value y_{ℓ_u+1} and since they represent an $(r-i)$-strong coalition in T_{ℓ_u} for u, from Lemma 6, they are able to compute the $r-i-1$ keys $k_u^{[1;\ell_u+1]}, \ldots, k_u^{[r-i-1;\ell_u+1]}$ with probability greater than or equal to $1/q$. Hence the coalition has i equations in the $i+1$ unknowns $k_u^{[r;h]}, k_u^{[r-i;\ell_u+1]}, \ldots, k_u^{[r-1;\ell_u+1]}$. In order to compute the key $k_u^{[r;h]}$, the classes in C should be able to compute at least one of those $i+1$ keys, with probability greater than or equal to $1/q$. From Case 2., the classes in C are not able to compute the key $k_u^{[r;h]}$ and from the inductive hypothesis they are not able to compute any information in $(k_u^{[r-i;\ell_u+1]}, \ldots, k_u^{[r-1;\ell_u+1]})$. For any of the q^{i+1} possible choices for the

$(i + 1)$-tuple $(k_u^{[r;h]}, k_u^{[r-i;\ell_u+1]}, \ldots, k_u^{[r-1;\ell_u+1]})$, there are q^i corresponding values for the key $k_u^{[r;h]}$. Hence, the probability that the coalition computes the key $k_u^{[r;h]}$ assigned to class u is equal to $1/q$.   □

It is easy to see that the r-secure scheme is optimal with respect to the randomness needed to set up the scheme when $1 \le r \le g$. The next theorem shows that the scheme is also optimal with respect to the size of the private information held by each class.

Theorem 7. *Let $T = (V, E)$ be a g-complete rooted tree hierarchy with height h and let $1 \le r \le \frac{g(g^{h-1}-1)}{g-1}$. The number of integers contained in the private information distributed to each class u by the r-secure scheme is equal to $\binom{r+h_u-1}{h_u-1}$.*

Proof. The proof is by induction on r. For $r = 1$, it holds that $\binom{r+h_u-1}{h_u-1} = h_u$.

Assume by inductive hypothesis that, for any $r' = 2, \ldots, r - 1$ the private information $p_u^{[r';j]}$ contains $\binom{r'+j-h+h_u-1}{j-h+h_u-1}$ integers. Since $p_u = p_u^{[r-1;\ell_u+1]} \circ \cdots \circ p_u^{[r-1;h]} \circ k_u$, the number of integers contained in p_u is equal to

$$1 + \sum_{j=\ell_u+1}^{h} \binom{r+j-h+h_u-2}{j-h+h_u-1} = 1 + \sum_{j=h-h_u+2}^{h} \binom{r+j-h+h_u-2}{j-h+h_u-1}$$

$$\text{(since } \ell_u = h - h_u + 1\text{)}$$

$$= \sum_{i=0}^{h_u-1} \binom{r-1+i}{i}$$

$$= \binom{r+h_u-1}{h_u-1}.$$

The last equality follows from equation (2), setting $b = h_u - 1$ and $a = r - 1$.   □

References

1. S. G. Akl and P. D. Taylor, *Cryptographic Solution to a Problem of Access Control in a Hierarchy*, ACM Trans. Comput. Syst., 1(3), 239–248, 1983.
2. C. C. Chang, R. J. Hwang, and T. C. Wu, *Cryptographic Key Assignment Scheme for Access Control in a Hierarchy*, Information Systems, 17(3), 243–247, 1992.
3. T. M. Cover, J. A. Thomas, *"Elements of Information Theory"*, John Wiley & Sons, 1991.
4. A. De Santis, A. L. Ferrara, and B. Masucci, *Unconditionally Secure Key Assignment Schemes*, Discrete Applied Mathematics, 154(2), 234–252, February 2006.
5. A. De Santis, A. L. Ferrara, and B. Masucci, *Cryptographic Key Assignment Schemes for Any Access Control Policy*, Inf. Process. Lett., 92(4), 199–205, November 2004.
6. A. L. Ferrara and B. Masucci, *An Information-Theoretic Approach to the Access Control Problem*, in Proc. of The Eighth Italian Conference on Theoretical Computer Science - ICTCS 2003, LNCS, Vol. 2841, 342–354, Springer Verlag, 2003.

7. R. L. Graham, D. E. Knuth, and O. Patashnik, Concrete Mathematics. A Foundation for Computer Science, Addison Wesley, 1988.

8. M. S. Hwang, *A Cryptographic Key Assignment Scheme in a Hierarchy for Access Control*, Math. Comput. Modeling, 26(1), 27–31, 1997.

9. M. S. Hwang, *An Improvement of a Dynamic Cryptographic Key Assignment Scheme in a Tree Hierarchy*, Comput. Math. Appl., 37(3), 19–22, 1999.

10. D. E. Knuth and A. C. Yao, "The Complexity of Nonuniform Random Number Generation", in *Algorithms and Complexity*, Academic Press, 357–428, 1976.

11. H. T. Liaw, S. J. Wang, and C. L. Lei, *A Dynamic Cryptographic Key Assignment Scheme in a Tree Structure*, Comput. Math. Appl., 25(6), 109–114, 1993.

12. H. T. Liaw and C. L. Lei, *An Optimal Algorithm to Assign Cryptographic Keys in a Tree Structure for Access Control*, BIT, 33, 46–56, 1993.

13. C. H. Lin, *Dynamic Key Management Schemes for Access Control in a Hierarchy*, Computer Communications, 20, 1381–1385, 1997.

14. S. J. MacKinnon, P. D. Taylor, H. Meijer, and S. G. Akl, *An Optimal Algorithm for Assigning Cryptographic Keys to Control Access in a Hierarchy*, IEEE Trans. Comput., C-34(9), 797–802, 1985.

15. R. S. Sandhu, *Cryptographic Implementation of a Tree Hierarchy for Access Control*, Inf. Process. Lett., 27, 95–98, 1988.

Multi-Dimensional Hash Chains and Application to Micropayment Schemes

Quan Son Nguyen

Hanoi University of Technology, Hanoi, Vietnam
sonnq@tinhvan.com

Abstract. One-way hash chains have been used in many micropayment schemes due to their simplicity and efficiency. In this paper we introduce the notion of multi-dimensional hash chains, which is a new generalization of traditional one-way hash chains. We show that this construction has storage-computational complexity of $O(\log_2 N)$ per chain element, which is comparable with the best result reported in recent literature. Based on multi-dimensional hash chains, we then propose two cash-like micropayment schemes, which have a number of advantages in terms of efficiency and security. We also point out some possible improvements to PayWord and similar schemes by using multi-dimensional hash chains.

1 Introduction

One-way hash chains are an important cryptographic primitive and have been used as a building block of a variety of cryptographic applications such as access control, one-time signature, electronic payment, on-line auction, etc.

In particular, there are many micropayment schemes based on one-way hash chains, including PayWord [8], NetCard [1], micro-iKP [5] and others.

By definition, micropayments are electronic payments of low value. Other schemes designed for payments of high value normally use a digital signature to authenticate every payment made. Such an approach is not suitable for micropayments because of high computational cost and bank processing cost in comparison with the value of payment.

The use of hash chains in micropayment schemes allows minimizing the use of digital signature, whose computation is far slower than the computation of a hash function (according to [8], hash functions are about 100 times faster than RSA signature verification, and about 10,000 times faster than RSA signature generation). Moreover, because a whole hash chain is authenticated by a single digital signature on the root of chain, successive micropayments can be aggregated into a single larger payment, thus reducing bank processing cost.

There are a variety of improvements to hash chains. For example, in the PayTree payment scheme [7], Jutla and Yung generalized the hash chain to a hash tree. This construction allows the customer to use parts of a tree to pay different vendors. Recently, researchers have proposed a number of improved

Ø. Ytrehus (Ed.): WCC 2005, LNCS 3969, pp. 218–228, 2006.

hash chains, which are more efficient in terms of computational overhead and storage requirement [3, 6, 11, 4].

This paper is organized as follows. In section 2 we introduce the notion of multi-dimensional hash chains (MDHC for short). We also analyze efficiency of this construction and show that RSA modular exponentiations could be used as one-way hash functions of a MDHC. Section 3 describes two cash-like micropayment schemes based on MDHC, which have a number of advantages in terms of efficiency and security. In section 4 we also examine some possible improvements to PayWord and similar schemes. Finally, section 5 concludes the paper.

2   Multi-Dimensional Hash Chain

2.1   Motivation

The notion of MDHC originates from one-way hash chains and one-way accumulators [2]. Here we briefly describe these two constructions.

A hash chain is generated by applying a hash function multiple times. Suppose that we have a one-way hash function $y = h(x)$ and some starting value x_n. A hash chain consists of values $x_0, x_1, x_2, ..., x_n$ where $x_i = h(x_{i+1})$ for $i = 0, 1, ..., n - 1$. The value $x_0 = h^n(x_n)$ is called the root of hash chain. The figure below depicts a hash chain of size n:

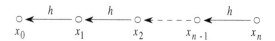

Fig. 1. A one-way hash chain

In contrast, a one-way accumulator is the output of multiple hash functions, each of them applied only once:

$$y = h_1(h_2(...(h_m(x)))) \ . \tag{1}$$

In order to ensure that the output is uniquely determined regardless of the application order, functions $h_1, h_2, ..., h_m$ must be in pairs commutative, i.e. $h_i(h_j(x)) = h_j(h_i(x))$ for any x.

Combining the two constructions described above, we can define a multi-dimensional hash chain as the result of multiple applications of different commutative hash functions, so the root of an m-dimensional hash chain is:

$$X_0 = h_1^{n_1}(h_2^{n_2}(...(h_m^{n_m}(X_N)))) \ . \tag{2}$$

It is necessary to note that MDHC differs from other generalizations of normal hash chain such as hash tree, which is used in PayTree scheme. In particular such trees are generated from multiple leaf nodes, while a MDHC is generated from a single starting value (i.e. the value X_N above).

2.2   Definitions

We begin with necessary definitions.

Definition 1. *Two functions $h_1, h_2 : X \to X$ are called commutative if $h_1(h_2(x)) = h_2(h_1(x))$ for any $x \in X$.*

Definition 2. *A one-way function $h : X \to Y$ is called one-way independent of one-way functions $h_1, h_2, ..., h_m$ of the same domain if for any $x \in X$, computing $h^{-1}(x)$ is intractable even if values $h_1^{-1}(x)$, $h_2^{-1}(x)$, ..., $h_m^{-1}(x)$ are known.*

We now define MDHC as follows.

Definition 3. *Let $h_1, h_2, ..., h_m$ be m one-way hash functions that are in pairs commutative and every of them is one-way independent from all others. An m-dimensional hash chain of size $(n_1, n_2, ..., n_m)$ consists of values $x_{k_1, k_2, ..., k_m}$ where:*

$$x_{k_1, k_2, ..., k_i, ..., k_m} = h_i(x_{k_1, k_2, ..., k_i+1, ..., k_m}) \tag{3}$$

for $i = 1, 2, ..., m$ and $k_i = 0, 1, ..., n_i$.

   The value $X_N = x_{n_1, n_2, ..., n_m}$ is called the starting node, and the value $X_0 = x_{0,0,...0}$ is called the root of the MDHC, which is uniquely determined from X_N due to commutativity of hash functions:

$$X_0 = h_1^{n_1}(h_2^{n_2}(...(h_m^{n_m}(X_N)))) = \prod_{i=1}^{m} h_i^{n_i}(X_N) \ . \tag{4}$$

As an illustration, the figure below depicts a two-dimensional hash chain of size (3,2):

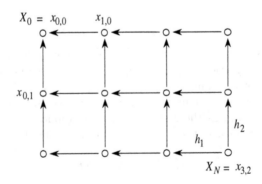

Fig. 2. A two-dimensional hash chain

2.3 Efficiency Analysis

In recent literature, there are a number of improvements to one-way hash chains that aim to be more efficient in terms of computational overhead and storage requirement. A widely used metric for one-way hash chain efficiency is the storage-computational complexity, which is the product of the traversal overhead and the storage required to compute consecutive nodes of the hash chain.

It is easy to see that a linear hash chain size of n has storage-computational complexity of $O(n)$. In fact, if we precompute and store all nodes (storage of $O(n)$), then no computation is needed when a node is requested (traversal of $O(1)$). Alternatively, we can store only the starting value, and compute every node from the beginning each time it is requested. This approach requires storage of $O(1)$ and $O(n)$ computations. Also, if we store each of t nodes, then storage of $O(n/t)$ and $O(t)$ computations are required. So, in any case, the storage-computational complexity of the linear hash chain is $O(n)$.

In [3, 6, 11] the authors have proposed new techniques that make traversal and storage more efficient, which require $O(\log_2 n)$ computations and $O(\log_2 n)$ storage, resulting in storage-computational complexity of $O(\log_2^2 n)$. Recently, Hu et al. [4] have presented a new hierarchical construction for one-way hash chains that requires $O(\log_2 n)$ storage and only $O(1)$ traversal overhead.

In our case of m-dimensional hash chain of size n (for simplicity we assume all dimensions have the same size $n_1 = n_2 = ... = n_m = n$), the number of nodes is $N = (n+1)^m$. If we store only the starting node of the chain (storage of $O(1)$) then maximal number of calculations required to compute any node is $nm = n \log_{n+1} N$, or $\log_2 N$ if we select $n = 1$. In that case the storage-computational complexity of MDHC is $O(\log_2 N)$, which is equivalent to the results in [4].

The advantage of MDHC is its simple implementation that does not rely on the so-called pebbling technique, which is used in the constructions mentioned above. However, the main limitation of this construction is the fact that hash functions have to meet the conditions described in the definition of MDHC. The RSA modular exponentiation is known to meet these conditions, but it is not as fast as the traditional hash functions, e.g. MD5 or SHA.

2.4 RSA Modular Exponentiation

Let consider the function of RSA modular exponentiation:

$$y = x^c \bmod M \tag{5}$$

where c is some constant value and M is an RSA modulus, which is a product of two large primes of equal bit length p and q.

According to [2], the RSA modular exponentiation functions with appropriately selected exponents could meet MDHC requirements.

First, obviously these functions are in pairs commutative:

$$h_i(h_j(x)) = x^{c_i \, c_j} \bmod M = h_j(h_i(x)) \ . \tag{6}$$

Second, one-wayness of these functions is derived from the RSA assumption [9], which states that the problem of finding the modular root $x = y^{1/c} \bmod M$ is intractable.

Finally, regarding one-way independence of functions, Shamir [12] showed that if c is not a divisor of the product $c_1 c_2 \ldots c_m$ then the modular roots $y^{1/c_1} \bmod M$, $y^{1/c_2} \bmod M$, ..., $y^{1/c_m} \bmod M$ are insufficient to compute the value of $y^{1/c} \bmod M$.

Therefore we can use the functions of RSA modular exponentiation as one-way hash functions to construct multi-dimensional hash chains.

In that case we have following recursive expression:

$$x_{k_1,k_2,\ldots,k_i,\ldots,k_m} = (x_{k_1,k_2,\ldots,k_i+1,\ldots,k_m})^{c_i} \bmod M \tag{7}$$

for $i = 1, 2, \ldots, m$, $k_i = 0, 1, \ldots, n_i$ and where $c_1, c_2, \ldots, c_m$ are exponents of RSA functions $h_1, h_2, \ldots, h_m$ respectively.

Note that if one knows the factorization of M (i.e. knows p and q), then one can compute X_0 quickly by using following expression:

$$X_0 = X_N^{\prod_{i=1}^{m} c_i^{n_i} \bmod E} \bmod M \tag{8}$$

where $E = \varphi(M) = (p-1)(q-1)$, and φ denotes the Euler's totient function.

The expression above consists of only one modular exponentiation with modulus M and $\log_2 N$ modular multiplications with modulus E. Since a multiplication is far faster than an exponentiation, this expression allows us to compute X_0 from X_N in a very effective manner.

3   Cash-Like Schemes Based on MDHC

Cash-like payment schemes use the notion of electronic coin, which is an authenticated (by the bank) bit string that is easy to verify, but hard to forge. Examples of such coin are hash collisions (as in MicroMint [8]), or digital signatures (as in Ecash [10]).

Let's recall the definition of MDHC. If we select the size of the hash chain with $n = 1$ then all nodes $X_i = x_{0,0,\ldots,1,\ldots 0}$ (with all $k_{j \neq i} = 0$, except $k_i = 1$) have the same hash value: $h_i(X_i) = X_0$. So we can use a pair (X_i, h_i) as an electronic coin since:

- It is easy to verify by just one hashing.
- It is hard to forge because hash functions h_i are one-way, and their one-way independence assures that coin forgery is impossible even if one knows other coins with the same root X_0.

As a proof of that concept, we suggest two micropayment schemes based on MDHC with the RSA modular exponentiation. We refer to these as S1 and S2 schemes.

3.1 The S1 Scheme

We assume that there are three parties involved in a micropayment scheme, namely a bank (B), a customer (C) and a vendor (V). B is trusted by both C and V.

SETUP:

- B selects an RSA modulus $M = pq$ where p and q are large *safe* primes of equal bit length. A prime p is called safe if $p = 2p' + 1$ where p' is also an odd prime.
- B chooses m constant values $c_1, c_2, ..., c_m$ that satisfy the condition of one-way independence, i.e. each c_i is not a factor of $\prod_{j \neq i} c_j$. These values together with modulus M are public parameters and can be used for multiple coin generations.
- To generate m coins, B picks a random value X_N and computes:

$$C = c_1 c_2 ... c_m \bmod E \quad \text{where } E = (p-1)(q-1) \ , \tag{9}$$

$$X_0 = h_1(h_2(...(h_m(X_N)))) = X_N{}^C \bmod M \ , \tag{10}$$

$$X_i = h_1(h_2(...(h_{i-1}(h_{i+1}(...(h_m(X_N)))))))) = X_N{}^{C\,c_i^{-1} \bmod E} \bmod M \tag{11}$$

for $i = 1, ..., m$.
Now B has m coins (X_i, c_i).
- B keeps X_0 in a public list of coin roots.
- For prevention of double-spending B keeps another list of all unspent coins. In addition, B can also generate vendor-specific as well as customer-specific coins by using some bit portions of constants c_i to form vendor ID and customer ID, similar to the technique used in MicroMint scheme.
- C buys a sufficiently large number of coins from B before making purchases.

PAYMENT:

- C pays a coin (X_i, c_i) to vendor V.
- V verifies the coin by computing $X_0 = X_i^{c_i} \bmod M$, and checks if X_0 is in the list of coin roots. Note that this list is relative small and does not change frequently so C could keep it locally.
- To assure that a coin was not double-spent, V either checks the list of unspent coins on-line with B, or checks (off-line) the list of coins he already received if the coin is vendor-specific.

REDEMPTION:

- V deposits the coins he got from customers to B and receives an amount corresponding to number of coins.

At the end of the coin validity period, C can sell unused coins back to B or exchange them for new coins.

The proposed above scheme has several advantages:

- Coins are hard to forge under the RSA assumption.
- Payment can be made off-line by using vendor-specific coins.
- If customer-specific coins are not used, the scheme is anonymous and un-traceable because coins contain no customer information and there are no links between coins.

However, the disadvantages of this scheme are:

- Generation and verification of coins is not very efficient. Each coin requires one modular exponentiation to generate or verify it, which is much slower than normal hash calculation.
- The list of unspent coins can be very big, though this is a common problem of most coin-based schemes.

To overcome these disadvantages, we propose a modified scheme with larger size hash chains (i.e. with $n > 1$). In this scheme, B generates m chains of coins at once, rather than m single coins. Each coin chain is similar to the hash chain used in the PayWord scheme.

3.2   The S2 Scheme

SETUP:

- B selects public parameters M and $c_1, c_2, ..., c_m$ in the same way as in the S1 scheme. Let n be the size of the hash chains (for simplicity we assume all dimensions have the same size i.e. $n_1 = n_2 = ... = n_m = n$).
- B picks a random value X_N and computes:

$$C = c_1^n c_2^n ... c_m^n \bmod E \quad \text{where } E = (p-1)(q-1) , \tag{12}$$

$$X_0 = X_N{}^C \bmod M , \tag{13}$$

$$X_i = X_N{}^{C c_i^{-n} \bmod E} \bmod M \quad \text{for } i = 1, 2, ..., m . \tag{14}$$

Now B has m coin chains (X_i, c_i). Each of those chains contains exactly n coins $(x_{i,j}, c_i, j)$ for $j = 1, 2, ..., n$ where:

$$x_{i,j} = x_{i,j+1}^{c_i} \bmod M \quad \text{for } i = 1, 2, ..., m \text{ and } j = 0, 1, ..., n-1 , \tag{15}$$

$$x_{i,n} = X_i \quad \text{and} \quad x_{i,0} = X_0 . \tag{16}$$

The coins from one coin chain must be paid to the same vendor.

- For double-spending prevention, now there is no need to keep track of all unspent coins. Instead, B keeps the list of first coins of all unused chains.
- As in the S1 scheme, coin chains can be vendor-specific as well as customer-specific.
- C buys coin chains from B before making purchases.

PAYMENT:

- C pays a vendor V the coins from a coin chain. The first coin of the chain $(x_{i,1}, c_i, 1)$ is verified by computing $X_0 = x_{i,0} = x_{i,1}^{c_i} \bmod M$ and lookup of X_0 in the list of chain roots. It is also checked for double-spending by lookup in the list of unused chains. Any subsequent coin is verified by checking that it hashes to the previous coin in the chain, as in the PayWord scheme:

$$h_i(x_{i,j+1}) = x_{i,j+1}^{c_i} \bmod M \equiv x_{i,j} \ . \tag{17}$$

REDEMPTION:

- V deposits the last coin (i.e. the coin with highest index j) of each coin chain he got from customers to B and receives an amount corresponding to number of coins.

Comparing with the S1 scheme, this modified scheme retains all advantages of S1, but storage requirement is reduced by factor of n. In fact, B keeps track of only the first coins of n-coin chains.

Another advantage of this scheme is more efficient coin generation. Because B knows the factorization of M, he can compute the starting node of a coin chain by just one modular exponentiation. Thus the cost of this computational expensive operation is shared over all coins of the chain. Similarly, B can also verify coin chains that he got from vendors by computing one modular exponentiation per chain.

Generally speaking, the S2 scheme combines the advantages of two different approaches. A first approach uses unrelated coins that are convenient for payments to multiple vendors. Another approach uses chains of coins that are easy to generate and verify. In our scheme different coin chains are unrelated, while coins within a chain are generated and verified only by repeated hashing.

4   Improve PayWord Scheme by Using MDHC

The PayWord scheme has been proposed in [8]. It is based on one-way hash chains described in Sect. 2. In this scheme, before making purchases a customer C generates a hash chain $x_0, x_1, ...x_n$ (that is a chain of paywords) and sends his signature of the root x_0 to the vendor V. The customer then makes a payment to V by revealing the next payword, which can be verified by checking that it hashes to the previous payword.

The PayWord scheme allows a vendor to aggregate successive payments from a customer by sending only last payword he got from the customer to the bank for redemption. However, a vendor cannot aggregate payments of different customers, nor can a customer use the same chain of paywords to make payments to different vendors, because there is no way to merge different hash chains.

By using MDHC, we can improve PayWord scheme in a number of ways. Below we briefly describe two of such possible improvements. Note that some irrelevant details in these descriptions are omitted for convenience.

4.1   Multiple Denominations

In the original PayWord scheme the size of the hash chain must be large enough. For example, if each micropayment is worth 1 cent and total payment is up to $100, then a chain with size of 10,000 must be generated, which requires 10,000 hash calculations.

We can reduce the number of hash calculations by using MDHC instead of linear hash chain. The idea is that every dimension of MDHC will be associated with different weight (or denomination) according to some number system (e.g. decimal or binary).

Suppose we have an m-dimensional hash chain with size of n. If one step in the $(i+1)^{th}$ dimension is equivalent to $(n+1)$ steps in i^{th} dimension, then a node $x_{k_1,k_2,...,k_m}$ corresponds to the value:

$$k_1 + k_2(n+1) + k_3(n+1)^2 + ... + k_m(n+1)^{m-1} . \tag{18}$$

The maximal value that could be represented by this hash chain is $N = (n+1)^m - 1$ and the number of hash calculations required to generate the hash chain is $n\log_{n+1}(N+1)$. In the case of a binary number system (i.e. $n=1$) it is $\log_2(N+1)$.

Returning to the example above, the hash chain now requires just 14 calculations to generate.

Similarly, verification of the payword also requires significantly less calculations than in the case of the original PayWord scheme.

4.2   Multiple Vendors

In the PayWord scheme a hash chain can be used for payments to only one vendor. A customer must generate different hash chains for payment to different vendors.

We can overcome this drawback by using MDHC as well. Let every vendor V_i in the payment system is assigned a different hash function h_i (i.e. a public parameter c_i in the case of RSA modular exponentiation).

Now, in order to make payment to m different vendors, a customer generates an m-dimensional hash chain with their public parameters c_i and signs its root. The customer then makes a payment to V_i by revealing the next payword in the ith dimension, starting from the root of hash chain.

In particular, if the current payword is $x_{k_1,k_2,...,k_i,...,k_m}$, the next payword in i^{th} dimension will be $x_{k_1,k_2,...,k_i+1,...,k_m}$.

At the end of the day, vendors deposit the last paywords they got to the bank for redemption. The bank picks the last payword (which is the one with highest indices) among paywords with certain root (which all come from one customer). Finally, the bank credits vendors V_i by the amount equivalent to k_i, and debits the customer's account accordingly.

There could be other possible improvements to the PayWord scheme by using MDHC. For example we can aggregate payments of different customers into a single MDHC that is generated by the bank, or we can construct a payment scheme with multiple currencies, etc.

5 Conclusion

The proposed multi-dimensional hash chain is a simple and efficient construction for one-way hash chains. Whereas a traditional one-way hash chain has a storage-computational complexity of $O(n)$, our construction achieves a complexity of $O(\log_2 n)$, which is comparable with the best result among other recently proposed constructions.

We show that multi-dimensional hash chains can be very useful in micropayment schemes. In particular, we suggest two cash-like micropayment schemes based on MDHC with RSA modular exponentiation as one-way hash function. The first scheme utilizes coins that are hard to forge under the RSA assumption. This scheme could be also off-line and untraceable. The second scheme has additional advantages including very efficient coin generation/verification and much less storage requirements.

We also point out some possible improvements to PayWord and similar schemes by using MDHC, including payword chains with multiple denominations, and a scheme that allows payment to multiple vendors using the same payword chain.

An open issue for our construction is whether another one-way hash function can be found that meets MDHC requirements, and at the same time is more efficient than RSA modular exponentiation.

References

1. R. Anderson, H. Manifavas, and C. Sutherland. NetCard - a practical electronic cash system. *Proceedings of the 4th Security Protocols International Workshop (Security Protocols)*, pp.49–57, Lecture Notes in Computer Science vol. 1189. Springer-Verlag, Berlin, 1996.
2. J. Benaloh, M. de Mare. One-Way Accumulators: A Decentralized Alternative to Digital Signatures. *Advances in Cryptology – EUROCRYPT '93*. LNCS, vol.765, pp.274–285, Springer-Verlag, 1994.
3. D. Coppersmith and M. Jakobsson. Almost optimal hash sequence traversal. *Proceedings of the Fourth Conference on Financial Cryptography (FC '02)*, Lecture Notes in Computer Science, 2002.
4. Y. Hu, M. Jakobsson and A. Perrig. Efficient Constructions for One-way Hash Chains. *SCS Technical Report Collection*.
   http://reports-archive.adm.cs.cmu.edu/anon/2003/CMU-CS-03-220.ps, 2003.
5. R. Hauser, M. Steiner, and M. Waidner. Micro-payments based on iKP. *Proceedings of the 14th Worldwide Congress on Computer and Communications Security Protection*, pp.67–82, Paris, 1996.
6. M. Jakobsson. Fractal hash sequence representation and traversal. *Proceedings of the 2002 IEEE International Symposium on Information Theory (ISIT '02)*, pp.437–444, 2002.
7. C. Jutla and M. Yung. PayTree: amortized-signature for flexible micropayments. *Proceedings of the 2nd USENIX Workshop on Electronic Commerce*, pp.213–21, Oakland, California, November 1996.

8. R. Rivest and A. Shamir. PayWord and MicroMint: two simple micropayment schemes. *Proceedings of the 4th Security Protocols International Workshop (Security Protocols)*, pp.69–87, Lecture Notes in Computer Science vol. 1189. Springer-Verlag, Berlin, 1996.

9. R. Rivest, A. Shamir, and L.M. Adleman. A method for obtaining digital signatures and public-key cryptosystems. In *Communications of the ACM, 21(2)*. pp.120–126, February 1978.

10. B. Schoenmakers. Security aspects of the Ecash payment system. *State of the Art in Applied Cryptography: Course on Computer Security and Industrial Cryptography - Revised Lectures*, pp.338–52, Lecture Notes in Computer Science vol. 1528. Springer-Verlag, Berlin, 1998.

11. Yaron Sella. On the computation-storage trade-offs of hash chain traversal. *Proceedings of Financial Cryptography 2003 (FC 2003)*, 2003.

12. A. Shamir. On the Generation of Cryptographically Strong Pseudorandom Sequences. *ACM Transactions on Computer Systems (TOCS)*, Volume 1, Issue 1, pp.38–44, 1983.

On the Affine Transformations of HFE-Cryptosystems and Systems with Branches

Patrick Felke*

CITS Research Group
Ruhr-University Bochum
D-44780 Bochum
felke@itsc.rub.de

Abstract. We show how to recover the affine parts of the secret key for a certain class of HFE-Cryptosystems. Further we will show that any system with branches can be decomposed in its single branches in polynomial time on average. The attack on the affine parts generalizes the results from [1, 11] to a bigger class of systems and is achieved by a different approach. Despite the fact that systems with branches are not used anymore (see [11, 6]), our second attack is a still of interest, as it shows that branches belong to the list of algebraic properties, which cannot be hidden by composition with secret affine transformations. We derived both algorithms by considering the cryptosystem as objects from the theory of nonassociative algebras and applying classical techniques from this theory. This general framework might be a useful tool for future investigations of HFE-Cryptosystems, e.g. to detect further invariants, which are not hidden by composition with affine transformations.

Keywords: HFE, finite fields, branches, nonassociative algebra, mixed centralizer, affine transformations.

1 Introduction

At Eurocrypt'88 Imai and Matsumoto (see [7]) proposed a promising cryptosystem called C^* based on multivariate polynomials, especially useful for smartcards. To speed up computation and to enhance security, they introduced the idea of branches. C^* was broken independently by Dobbertin in '93 (unpublished, see [4, 5]) and by Patarin in '95 (see [11]). To repair these systems Dobbertin studied bijective power functions of higher degree, whereas Patarin introduced the HFE-Cryptosystem and also faster variants, which make use of branches (see [11, 12, 13]). The disadvantage of the latter systems is, if an attacker is able to separate the branches, he also benefits from the speed up, because he can attack the single branches separately.

---

* The work described in this paper has been supported in part by the European Commission through the IST Programme under Contract IST-2002-507932 ECRYPT. The information in this document reflects only the author's views, is provided as is and no guarantee or warranty is given that the information is fit for any particular purpose. The user thereof uses the information at its sole risk and liability.

Ø. Ytrehus (Ed.): WCC 2005, LNCS 3969, pp. 229–241, 2006.

In the beginning probabilistic polynomial time attacks to separate the branches were only known for very special systems like C^*. Later more general probabilistic attacks with exponential running time (exponential in the size of the branches, see [6, 11]) were discovered. As a consequence only systems with branches of moderate size could be considered to be secure. Thus the speed up of computation was no longer given and such systems were not used anymore. It remained an open question, if there exists an efficient algorithm to recover big branches for an arbitrary HFE-Cryptosystem. In Section 4 we consider this question from the perspective of nonassociative algebras. This will yield to an algorithm to recover the branches for an arbitrary system in polynomial time on average and thus proving that the answer is no. This gives another item on list of algebraic properties, which cannot be hidden by the HFE-principle.

Section 3 is concerned with the secret affine transformations used to construct the trapdoor. It is an open problem, if the security is affected when linear mappings are chosen instead of affine. At first we briefly describe what we understand by eliminating the affine parts. By applying classical techniques from the theory of nonassociative algebras we show, that the affine parts can be eliminated for certain classes of HFE-systems, including systems like Sflash. This generalizes the results in [1, 11], but we make use of a different approach.

Putting an HFE-system into the perspective of nonassociative algebras requires some technical efforts in the beginning. We will see, that this view finally simplifies finding the invariats that yield to our attacks. We are confident that there might be other invariants, which can be discovered this way.

2   Preliminaries

We assume that the reader is familiar with the theory of finite fields and multivariate polynomials as can be found in [10] for example. In the following we briefly sum up some facts about representations of mappings over finite fields and HFE-Cryptosystems. A detailed description about encryption and signing with HFE-Cryptosystems can be found in [11, 13]. More details about representations of mappings are given in [8].

With $\mathbb{F}_q$, $q = p^m$, we denote the finite field of characteristic p and with $\mathbb{F}_{q^n}$ the extension of degree n. We will often consider $\mathbb{F}_{q^n}$ as an n-dimensional $\mathbb{F}_q$-vector space and via a choice of a basis we will identify it with the vector space $\mathbb{F}_q^n$. Elements $(a_1, \ldots, a_n)$ of $\mathbb{F}_q^n$ will often be denoted by $\underline{a}$. The univariate polynomial ring is denoted by $\mathbb{F}_{q^n}[X]$ and the multivariate polynomial ring by $\mathbb{F}_{q^n}[x_1, \ldots, x_n]$.

Any mapping over $\mathbb{F}_{q^n}$ can be uniquely represented by a polynomial

$$P(X) = \sum_{i=0}^{q^n-1} a_i X^i, a_i \in \mathbb{F}_{q^n}$$

and of course every such polynomial $P(X)$ induces a mapping by $a \mapsto P(a), a \in \mathbb{F}_{q^n}$. Any mapping from $\mathbb{F}_q^n$ into $\mathbb{F}_q^n$ can be uniquely represented by a vector of polynomials

$$(p_1(x_1, \ldots, x_n), \ldots, p_n(x_1, \ldots, x_n)), p_i \in \mathbb{F}_q[x_1, \ldots, x_n]$$

with the property, that if a monomial $\beta x_1^{l_1} \cdots x_n^{l_n}$ occurs in p_k, then $l_i < q$ for $i = 1, \ldots, n$. We will call such a vector reduced. Of course, as above, every such vector induces a mapping on $\mathbb{F}_q^n$.

For any choice of a basis $b_1, \ldots, b_n$ of $\mathbb{F}_{q^n}$, there exists for every mapping F over $\mathbb{F}_{q^n}$ a unique mapping $f = (f_1, \ldots, f_n)$ over $\mathbb{F}_q^n$ with

$$F(a) = F(\sum_{i=1}^{n} \alpha_i b_i) = \sum_{i=1}^{n} f_i(\underline{\alpha}) b_i$$

and vice versa.

Thereby the unique polynomial $P(X)$ of degree $d \leq q^n - 1$ with $F(a) = P(a)$ is called the univariate representation of F. The uniquely determined reduced vector $(p_1(\underline{x}), \ldots, p_n(\underline{x}))$ with $f(\underline{a}) = (p_1(\underline{a}), \ldots, p_n(\underline{a}))$ is called the multivariate representation of F.

We define the degree of a vector of polynomials as $\max\{\deg(p_i) | i = 1, \ldots, n\}$. With this definition the above correspondence is degree preserving in the sense, that if the univariate representation has degree d, then the multivariate representation has degree q-weight of d. We briefly explain, what we understand by the q-weight. Let $d = \sum_{i=0}^{l} c_i q^i, 0 \leq c_i < q$ the q-adic representation of d. Then the q-weight is the sum $\sum_{i=0}^{l} c_i$.

Affine mappings S on $\mathbb{F}_q^n$ will be as usual denoted by $A\underline{x} + \underline{c}$, where A denotes an $n \times n$-matrix, $\underline{x} = (x_1, \ldots, x_n)$ and $\underline{c} \in \mathbb{F}_q^n$. To keep the description in the rest of this paper as simple as possible, we consider the result of a matrix-vector-multiplication as a row vector. Thus $A\underline{x} + \underline{c}$ is already the multivariate representation of the affine mapping S. With this notation the multivariate representation of the composition $S \circ f$, where f is a mapping over $\mathbb{F}_q^n$ with multivariate representation $\underline{p} := (p_1, \ldots, p_n)$ is given by $A\underline{p} + \underline{c}$. Moreover a reduced vector $(p_1, \ldots, p_n)$ will sometimes be identified with the corresponding mapping.

Now we very briefly describe a basic HFE-Cryptosystem with branches. The secret key consists of:

1. $n = n_1 + \cdots + n_l$, a partition of n.
2. Field extensions $\mathbb{F}_{q^{n_k}}$ over a fixed base field $\mathbb{F}_q$ for $k = 1, \ldots, l$. The fields will be represented by the choice of an irreducible polynomial $\mathbb{F}_q[X]$ to construct $\mathbb{F}_{q^{n_k}}$ and an $\mathbb{F}_q$-basis, which determines the isomorphism between $\mathbb{F}_q^{n_k}$ and $\mathbb{F}_{q^{n_k}}$.
3. l HFE-polynomials of degree d_k, that is polynomials of the form $H_k(X) = \sum_{i,j=0}^{n-1} \beta_{ij,k} X^{q^i + q^j} + \sum_i \alpha_{i,k} X^{q^i}$,
   where $\beta_{ij,k}, \alpha_{i,k} \in \mathbb{F}_{q^{n_k}}, k = 1, \ldots, l$.
4. Two affine bijective transformations $S(\underline{x}) = A\underline{x} + \underline{c}, T(\underline{x}) = B\underline{x} + \underline{d}$ of $\mathbb{F}_q^n$.

This constitutes the secret key. The public key is derived by computing the multivariate representation for each of the H_k denoted by

$$
\begin{aligned}
&(h_1(x_1,\ldots,x_{n_1}), &&\ldots, h_{n_1}(x_1,\ldots,x_{n_1}))\\
&(h_{n_1+1}(x_{n_1+1},\ldots,x_{n_1+n_2}), &&\ldots, h_{n_1+n_2}(x_{n_1+1},\ldots,x_{n_1+n_2}))\\
&\qquad\qquad\qquad\vdots\\
&(h_{n-n_l+1}(x_{n-n_l+1},\ldots,x_n), &&\ldots, h_n(x_{n-n_l+1},\ldots,x_n))
\end{aligned}
\tag{1}
$$

Each of these n_i-tuples constitutes a branch. Combining these n_i-tuples gives an n-tuple of polynomials $(h_1,\ldots,h_n)$ in n variables. The public key $(p_1,\ldots,p_n)$ is given by the composition $T \circ (h_1,\ldots,h_n) \circ S$ and consists of n quadratic polynomials in n variables. This implies that the base field $\mathbb{F}_q$ is public. Note, that the polynomials in different branches have different sets of variables and these are mixed up by S,T.

The public key is the multivariate representation of a composition of mappings, where mappings over $\mathbb{F}_{q^{n_i}}, i = 1,\ldots,l$ are involved. This implies, that an encryption can also be considered as chain of compositions like it is given in the following diagram. This different point of view is important for our analysis. Thereby Ψ denotes the canonical isomorphism from $\mathbb{F}_q^n$ into $\mathbb{F}_q^{n_1} \times \cdots \times \mathbb{F}_q^{n_l}$ and ϕ_i the canonical isomorphism from $\mathbb{F}_q^{n_i}$ into $\mathbb{F}_{q^{n_i}}$ given by the chosen basis.

$$
\begin{array}{ccccccc}
\mathbb{F}_q^{n_1} & \xrightarrow{\phi_1} & \mathbb{F}_{q^{n_1}} & \xrightarrow{H_1} & \mathbb{F}_{q^{n_1}} & \xrightarrow{\phi_1^{-1}} & \mathbb{F}_q^{n_1}\\
&&&&&&\\
\vdots &&&&&& \vdots\\
&&&&&&\\
\mathbb{F}_q^{n_l} & \xrightarrow{\phi_l} & \mathbb{F}_{q^{n_l}} & \xrightarrow{H_l} & \mathbb{F}_{q^{n_l}} & \xrightarrow{\phi_l^{-1}} & \mathbb{F}_q^{n_l}
\end{array}
$$

$$
\mathbb{F}_q^n \xrightarrow{S} \mathbb{F}_q^n \quad \overset{\Psi}{\Big\langle} \qquad\qquad\qquad \overset{\Psi^{-1}}{\Big\rangle} \quad \mathbb{F}_q^n \xrightarrow{T} \mathbb{F}_q^n
$$

Now it becomes apparent, that if an attacker is able to recover the branches, he is able to attack every branch.

A basic HFE-Cryptosystem is a system where $l = 1$. The nowadays proposed schemes are variants of this basic system as for example Sflash or Quartz (see [3, 2]).

In some descriptions the univariate polynomials have a constant term. Since these can be captured by T, we skipped it in our description (see also the next section). In the sequel the multivariate and univariate representations are considered with respect to the bases chosen by the designer. For our attacks we do not need to know these bases since we will show, that all necessary information can be computed from the public key.

Now we are going to show how to construct a nonassociative $\mathbb{F}_q$-algebra from an HFE-Cryptosystem. This will be the foundation for the algorithms presented later. By a nonassociative $\mathbb{F}_q$-algebra $\mathcal{U}$ we understand an $\mathbb{F}_q$-vectorspace with a multiplication, which is so that

$$
\lambda(xy) = (\lambda x)y = x(\lambda y) \text{ for all } \lambda \in \mathbb{F}_q, x,y \in \mathcal{U},
$$

and which is also bilinear (i.e. $(x+y)z = xz + yz, z(x+y) = zx + zy$). The associative law is not being assumed. An introduction to this subject can be found in [14].

Given an HFE-Polynomial $H(X) = \sum_{i,j=0}^{n-1} \beta_{ij} X^{q^i+q^j} + \sum_{i=0}^{n-1} \alpha_i X^{q^i}$ we define a multiplication on $\mathbb{F}_{q^n}$ as follows:

$$
M(a,b) := H(a+b) - H(a) - H(b), a,b \in \mathbb{F}_{q^n}.
$$

Since M is given by the sum $\sum_{i,j=0}^{n-1} \beta_{ij}(a^{q^i} b^{q^j} + b^{q^i} a^{q^j})$ this multiplication induces indeed an nonassociative and commutative algebra. Again we can derive n polynomials m_i in $x_1, \ldots, x_n$ and $y_1, \ldots, y_n$, which give the multivariate representation of the mapping M. This is achieved similar to the univariate case, but here we have the defining relation

$$M(\sum_{i=1}^{n} \alpha_i b_i, \sum_{i=1}^{n} \beta_i b_i) = \sum_{i=1}^{n} m_i(\underline{\alpha}, \underline{\beta}) b_i,$$

where $b_1, \ldots, b_n$ is a basis of $\mathbb{F}_{q^n}$. If L_1, L_2 denote the linear part of the secret key of an HFE-Cryptosystem considered over $\mathbb{F}_{q^n}$, then

$$M'(a, b) := L_2(M(L_1(a), L_1(b))) \text{ induces a second algebra.}$$

Note, that the multivariate representation of M' can be calculated from the public key by computing $p_i(\underline{x} + \underline{y}) - p_i(\underline{x}) - p_i(\underline{y})$. We will see, that this still holds when the secret transformations S, T are affine (see Section 4).

3   Eliminating the Affine Parts of S,T

Recall that a polynomial $q(x_1, \ldots, x_n)$ is called homogeneous of degree d, if all monomials that occur have the exact degree d. We start with a lemma, which is crucial for our algorithm. It shows that the affine parts of S, T are not mixed up properly by the application of S, T, if the polynomial $H(X)$ is also homogenous in the sense, that all monomials that occur are of the form $\beta_{ij} X^{q^i + q^j}$.

Lemma 1. *Let $\mathbb{F}_q \neq \mathbb{F}_2$. Further let $S(\underline{x}) = A\underline{x} + \underline{c}, T(\underline{x}) = B\underline{x} + \underline{d}$ be bijective affine mappings over $\mathbb{F}_{q^n}$ with univariate representation $L_1 + c, L_2 + d$ and $H(X) = \sum_{i,j=0}^{n-1} \beta_{ij} X^{q^i + q^j}$.*

If $p_1, \ldots, p_n$ denotes the public key of the resulting cryptosystem, then $p_i(\underline{x}) = q_i(\underline{x}) + l_i(\underline{x}) + a_i$, where a_i is a constant, l_i is linear and q_i is homogeneous of degree 2. Furthermore $(q_1, \ldots, q_n)$ is the multivariate representation of $L_2 \circ H \circ L_1$.

Proof. Let $(h_1, \ldots, h_n)$ denote the multivariate representation of H. Then the public key, which is the multivariate representation of $(L_2 + d) \circ H \circ (L_1 + c)$, equals the reduced vector of the composition $T \circ (h_1, \ldots, h_n) \circ S$. Due to the shape of $H(X)$ the polynomials h_i are homogeneous of degree 2, since the base field is not $\mathbb{F}_2$. Otherwise the reduction involved in the computation of the multivariate representation would imply to substitute x_i^2 by x_i and the h_i could also contain monomials of degree 1.

By comparing the public key $p_1, \ldots, p_n$ with the reduced vector of $T \circ (h_1, \ldots, h_n) \circ S$ one sees, that $p_i = q_i(x) + l_i(x) + a_i$, where $(q_1, \ldots, q_n)$ represents $L_2 \circ H \circ L_1$. From this the result follows as requested.

We restrict to explain our attack for a simple mapping with one branch, i.e. a basic HFE-Cryptosystem with a simple hidden polynomial. The attack for an arbitrary system follows straight forward from this special case, but a general description would be very technical. Therefore we will only give briefly some remarks about the generalization at the end of this section.

Thus, we will restrict to show how to eliminate the affine parts of S, T, if the base field $\mathbb{F}_q$ equals $\mathbb{F}_{2^m}$, where $m > 2$ and $H(X) = \beta X^{q^i + q^j}, i \neq j$. By eliminating we understand that we will compute $\underline{d}$ and $A^{-1}(\underline{c})$, because once $\underline{d}$ and $A^{-1}(\underline{c})$ are known it is easy to transform the system $T \circ (h_1, \ldots, h_n) \circ S$ into $B \circ (h_1, \ldots, h_n) \circ A$. Note, that if $(\underline{y}, \underline{x})$ is a plaintext/ciphertext pair of the first system, then $(\underline{y} - \underline{d}, \underline{x} + A^{-1}(\underline{c}))$ is the corresponding pair of the second one and vice versa. Thus the task to decrypt an intercepted ciphertext of the system with affine transformations can be reduced to the task to decrypt a ciphertext of the according system without the translation vectors.

What we showed above is surprising, because one would expect that the elimination of the translations would imply the knowledge of $\underline{c}$ and not only of $A^{-1}(\underline{c})$. This is not the case for our approach. One can show that an algorithm to compute $\underline{c}$ would yield to an algorithm to compute A. Consequently eliminating the translations and computing $\underline{c}$ are different problems and very likely of a different complexity.

Without loss of generality we assume that $H(X) = X^{q^i + 1}, i \neq 0$. Otherwise consider $((L_2 + d) \circ (\beta X^{q^j})) \circ (X^{q^{n-j}} \circ (X^{q^i + q^j})) \circ (L_1 + c)$, which gives an equivalent system, i.e. a system with different S,T but exactly the same public key and a hidden polynomial of the desired form. In the general description of HFE-systems we mentioned that the constant term in the hidden polynomial can be skipped. With the above notion of an equivalent system one can easily explain why this is true. Similarly as above one shows, that an HFE-system with a hidden polynomial $H(X)$ possessing a constant term is equivalent to a system, where the hidden polynomial has no constant term.

For $H(X) = X^{q^i + 1}, i \neq 0$ we have $M(a, b) = a^{q^i} b + b^{q^i} a$. A natural question in the theory of nonassociative algebras is to determine all annihilating elements, i.e. to determine all $a \in \mathbb{F}_{q^n}$ such that the corresponding mappings $M(a, \cdot)$ or $M(\cdot, a)$ (the so called left or right multiplications) vanish on $\mathbb{F}_{q^n}$. Recall that in our case M and M' are always commutative and therefore we can restrict to consider left multiplications. We begin with a simple lemma.

Lemma 2. Let $i \notin \{0, n\}$, i.e. $X^{q^i} \neq X$.

1. If $M(a, b) = a^{q^i} b + b^{q^i} a = 0$ for all $b \in \mathbb{F}_{q^n}$, then $a = 0$.
2. If $M'(a, b) = 0$ for all $b \in \mathbb{F}_{q^n}$, then $a = 0$.

Proof. For $a \neq 0$ the polynomial $a^{q^i} X + X^{q^i} a$ has less than q^n zeros. This proofs the first assertion. The second part follows immediately from the fact that the linear mappings involved in the definition of M' are always bijective.

Now we will show how to relate the problem of eliminating the translations to the problem of finding annihilating elements.

The public polynomials are the multivariate representation of $P(X) := (L_2 + d) \circ X^{q^i} X \circ (L_1 + c)$. From this we can compute the multivariate representation of

$$P(X+Y) + P(X) = L_2(L_1(X)^{q^i} L_1(Y) + L_1(Y)^{q^i} L_1(X) + L_1(c)^{q^i} L_1(Y) + L_1(Y)^{q^i} L_1(c))$$
$$+ L_2(L_1(Y)^{q^i} L_1(Y)).$$

A Usage of Lemma 1 gives the multivariate representation of the last term $L_2(L_1(Y)^{q^i} L_1(Y))$. So we can eliminate this term by subtracting it. This way we get the multivariate representation of

$$L_2(L_1(Y)^{q^i} (L_1(X) + c) + L_1(X + c)^{q^i} L_1(Y)) = M'(X + L_1^{-1}(c), Y).$$

From Lemma 2 we have that $M'(a + L_1^{-1}(c), Y)$ is the zero mapping iff $a = L_1^{-1}(c)$. This yields to the following algorithm to eliminate the translations.

1. Compute the multivariate representation

$$(q_1(x_1, \ldots, x_n, y_1, \ldots, y_n), \ldots, q_n(x_1, \ldots, x_n, y_1, \ldots, y_n))$$

   of $M'(a + L_1^{-1}(c), Y)$ by calculating $p_i(x_1 + y_1, \ldots, x_n + y_n) + p_i(x_1, \ldots, x_n)$ and eliminating the multivariate part describing $L_2(L_1(Y)^{q^i} L_1(Y))$.
2. Compute $q_i(\underline{x}, e_1)$ for $i = 1, \ldots, n$, where e_1 denotes the first canonical basis vector $(1, 0, \ldots, 0)$. This gives an inhomogeneous system of n linear equations. If it has rank n compute the unique solution, which gives $A^{-1}(\underline{c})$. If the rank is $< n$, add the next n equations $q_i(\underline{x}, e_2)$ and so on, until rank n is reached.
3. Once $c' := A^{-1}(\underline{c})$ is computed, compute $p_i'(\underline{x}) = p_i(\underline{x} + \underline{c}')$ for all i. This gives the multivariate representation of $(L_2 + d) \circ X^{q^i} X \circ L_1$.
4. Compute $p_i'(\underline{0})$ for all i. This gives the vector $\underline{d}$, which can be finally eliminated.

This algorithm is dominated by the running time for the Gaussian elimination for a system with at most n^2 linear equations in n variables. Hence the running time is $O(n^4)$.

If this algorithm is applied to a more general HFE-system as for example a system with hidden polynomial $H(X) := \sum_{i,j=0}^{n-1} \beta_{ij} X^{q^i + q^j}, i \neq j$ then the annihilating elements a have to fulfill

$$M(a, b) = \sum_{ij} \beta_{ij} a^{q^i} b^{q^j} + b^{q^i} a^{q^j} = 0 \text{ for all } b \in \mathbb{F}_{q^n}.$$

For $a \neq 0$ this yields to a polynomial of the form

$$\sum_{ij} \beta_{ij} (a^{q^i} X^{q^j} + X^{q^i} a^{q^j}),$$

which has to be constantly zero. When sorted by X this polynomial has the shape $\sum_{l \in I} r_l(a) X^{q^l}$, where all the r_l are polynomials in a and I a proper index set. Thus a is an annihilating element iff $r_l(a) = 0$ for all $l \in I$. By construction this is the case for $a = 0$ and it is reasonable to assume that in most of the cases this is the only solution. This means it makes sense to assume, that the multiplication M or M' respectively of an HFE-Cryptosystem has the properties from lemma 2. We get the following theorem.

Theorem 1. *Given an arbitrary HFE-Cryptosystem, or a $"-"$-system like Sflash, over a field $\mathbb{F}_q \neq \mathbb{F}_2$ with secret affine transformations $S = A\underline{x} + \underline{c}$ and $T = B\underline{x} + \underline{d}$, then $\underline{c}, \underline{d}$ can be eliminated with $O(n^4)$ field operations on average.*

4   A Fast Algorithm for Separating the Branches

In [11] and [12] a probabilistic polynomial time algorithm to separate the branches is described assuming the underlying HFE-polynomials admit special syzygies. This algorithm is based on the Coppersmith-Patarin attack on Dragon-Schemes (see [12]). We will introduce an algorithm which is also based on the Coppersmith-Patarin attack with a similar running time, but does not require syzygies anymore and is therefore applicable to any HFE-Cryptosystem.

We denote as usual with $S(\underline{x}) = A\underline{x} + \underline{c}, T(\underline{x}) = B\underline{x} + \underline{d}$ the affine transformations used to build up the HFE-Cryptosystem. The crucial step of our attack (which is the same as in [11, 12]) is the computation of a matrix $C = A\Lambda A^{-1}$, where

$$\Lambda = \begin{pmatrix} \Lambda_1 & 0 & 0 & \cdots & 0 \\ 0 & \Lambda_2 & 0 & \cdots & 0 \\ \vdots & 0 & \Lambda_3 & \cdots & 0 \\ \vdots & \vdots & 0 & \ddots & 0 \\ 0 & 0 & \cdots & 0 & \Lambda_l. \end{pmatrix}$$

Thereby l denotes the number of branches and Λ_k the representation matrix of a linear mapping $x \mapsto \lambda_k x, \lambda_k \in \mathbb{F}_{q^{n_k}}$, where $\mathbb{F}_{q^{n_k}}$ is the field belonging to the k-th branch, $1 \leq k \leq l$.

Then from C a matrix G is derived, such that

$$AG = \begin{pmatrix} W_1 & 0 & 0 & \cdots & 0 \\ 0 & W_2 & 0 & \cdots & 0 \\ \vdots & 0 & W_3 & \cdots & 0 \\ \vdots & \vdots & 0 & \ddots & 0 \\ 0 & 0 & \cdots & 0 & W_l \end{pmatrix}, \text{ where } W_i \text{ is a block matrix.}$$

To compute G classical linear algebra related to the theorem of Cayley-Hamilton and Jordan-Normal forms is needed, which is not very surprising due to the structure of C. We skip the details, because this is done in Shamir's attack on the Oil&Vinegar-Schemes [9] and also in [11, 12].

Once G is known the separation is rather straightforward as we will briefly indicate now. As mentioned in Section 2, for the k-th branch there exists a unique set of variables

$$V_k = \{x_{\sum_{j=1}^{k-1} n_j + 1}, \ldots, x_{\sum_{j=1}^{k} n_j}\}.$$

At first one computes $p'_i(\underline{x}) := p_i(G\underline{x})$ for $i = 1, \ldots, n$. The polynomials p'_i have the property, that if $x_s x_t$ is a monomial occuring in p'_i then $x_s, x_t \in V_k$ for a proper k (see also the composition a few lines below). Thus the monomials in p'_i reveal the sets V_k and moreover can be grouped according to the unique set V_k containing their variables. Below we noted down the composition $p_i(G\underline{x})$ as the composition of column vectors. This deviates from our usual notation, but the effect of the composition with G is this way much better visualized.

$$(T \circ \begin{pmatrix} h_1(x_1, \ldots, x_{n_1}) \\ \vdots \\ h_{n_1}(x_1, \ldots, x_{n_1}) \\ \vdots \\ h_{n-n_l+1}(x_{n-n_l+1}, \ldots, x_n) \\ \vdots \\ h_n(x_{n-n_l+1}, \ldots, x_n) \end{pmatrix} \circ S) \circ G =$$

$$T \circ \begin{pmatrix} h_1(x_1, \ldots, x_{n_1}) \\ \vdots \\ h_{n_1}(x_1, \ldots, x_{n_1}) \\ \vdots \\ h_{n-n_l}(x_{n-n_l+1}, \ldots, x_n) \\ \vdots \\ h_n(x_{n-n_l+1}, \ldots, x_n) \end{pmatrix} \circ \begin{pmatrix} l_1(x_1, \ldots, x_{n_1}) + c_1 \\ \vdots \\ l_{n_1}(x_1, \ldots, x_{n_1}) + c_{n_1} \\ \vdots \\ l_{n-n_l+1}(x_{n-n_l+1}, \ldots, x_n) + c_{n-n_{l-1}} \\ \vdots \\ l_n(x_{n-n_l+1}, \ldots, x_n) + c_n \end{pmatrix}$$

The above composition shows that this way the different sets V_k are indeed revealed.

Once this composition is computed one gets by applying Gaussian elimination to $p'_1, \ldots, p'_n$ the desired polynomials, where the first n_1 polynomials have only variables from V_1, the next n_2 polynomials have variables from the set V_2 and so on. This is possible due to the above described property of the monomials in p'_i. This completes the separation.

It might happen that the composition with G does not reveal all branches, but more clusters of branches. In this case the clusters are attacked separately afterwards, and the separation is refined step by step.

To prove that such a G can be computed with high probability, we put the key idea from the Coppersmith-Patarin attack into the perspective of nonassociative algebras. In [12] it was used, that the only linear mappings L over a field $\mathbb{F}_{q^n}$ with

$$L(xy) = xL(y) \quad \text{for all } x, y \in \mathbb{F}_{q^n} \tag{2}$$

are multiplications with a field element in $\mathbb{F}_{q^n}$. In the theory of nonassociative algebras one calls the set of linear mappings L fulfilling equation 2 the *multiplication centralizer*.

Our generalization is as follows. At first we define a proper multiplication. Then we introduce the notion of a *mixed multiplication centralizer* and prove that from this a matrix C can be computed, such that $A^{-1}CA$ is a block matrix. In the following we show how the multiplications M and M' from Section 2 extend to systems with branches.

For every field $\mathbb{F}_{q^{n_k}}$ we have a multiplication $M_k(a,b)$, $k = 1,\ldots, l$. We get the desired multiplication $M(a,b)$ on $\mathbb{F}_{q^n}$ as follows. We consider the multiplication on $\prod_{k=1}^{l} \mathbb{F}_{q^{n_k}}$ defined by

$$((a_1,\ldots,a_l),(b_1,\ldots,b_l)) \mapsto (M_1(a_1,b_1),\ldots, M_b(a_l,b_l)).$$

The multiplication M on $\mathbb{F}_{q^n}$ is given by

$$\Psi^{-1} \circ M_1 \times \cdots \times M_l \circ \Psi,$$

where Ψ is the embedding of $\mathbb{F}_{q^n}$ into the product of fields. With $(m_1,\ldots,m_n)$ we denote the multivariate representation of M. The multiplication M' on $\mathbb{F}_{q^n}$ is given by $M'(a,b) := L_2(M(L_1(a), L_1(b)))$. The polynomials

$$m'_i(\underline{x},\underline{y}) := p_i(\underline{x} + \underline{y}) - p_i(\underline{x}) - p_i(\underline{y})$$

are the multivariate representation of M'. If S, T are affine it is easy to see, that we get the representation by skipping the constant parts after the computation of $p_i(\underline{x}+\underline{y})-p_i(\underline{x})-p_i(\underline{y})$. This becomes apparent, if one computes the linearization $P(X + Y) - P(X) - P(Y)$ for the univariate representation $P(X)$ in the same vein as in section 3.

We define the *mixed multiplication centralizer* to be the set of all linear mappings C, C' fulfilling

$$C'(m'_1(\underline{x},\underline{y}),\ldots,m'_n(\underline{x},\underline{y})) = (m'_1(\underline{x}, C\underline{y}),\ldots, m'_n(\underline{x}, C\underline{y})).$$

This can also be written as

$$C'B(m_1(A\underline{x}, A\underline{y}),\ldots, m_n(A\underline{x}, A\underline{y})) = B(m_1(A\underline{x}, AC\underline{y}),\ldots, m_n(A\underline{x}, AC\underline{y})). \quad (3)$$

From this we see, that if C, C' lie in the mixed centralizer, i.e. they solve the equation (3), then $ACA^{-1}, B^{-1}C'B$ solve

$$Z'((m_1,\ldots,m_n)) = (m_1(\underline{x}, Z\underline{y}),\ldots, m_n(\underline{x}, Z\underline{y})), \quad (4)$$

and if Z, Z' solve (4), then $A^{-1}ZA, BZ'B^{-1}$ solve equation (3). Hence the solutions are conjugated to each other. The centralizer of M' can be computed from the public key with Gaussian elimination, when the elements c_{ij}, c'_{ij} are set as unknowns and plaintext/ciphertext pairs are plugged in to get equations in the unknowns. Now we analyze the mixed centralizer and show that it has the desired property. For special cases it can be completely determined. We give an example for fields of characteristic 2.

Theorem 2. *Given the base field $\mathbb{F}_{2^m}$. Let M be the multiplication as above derived from univariate polynomials $H_k = X^{2^{mi_k}+1}$, where $i_k \notin \{0, n_k\}$ and $\gcd(2^{mi_k} + 1, 2^{mn_k} - 1) = 1$ for $k = 1, \ldots, l$. Then the centralizer consists of all pairs $(A^{-1}ZA, BZB^{-1})$, where Z is the representation matrix of the mapping*

$$a \mapsto \Psi^{-1}((\lambda_1 \cdot \Psi(a)_1, \ldots, \lambda_l \cdot \Psi(a)_l)),$$

$a \in \mathbb{F}_{2^n}$ *and* $\lambda_k \in \mathbb{F}_{2^{m\gcd(i_k, n_k)}}$ *for* $k = 1, \ldots, l$.

Proof. We only give a sketch of the proof.

At first we restrict to one branch. W. log. we consider the first branch. Let L_1, L_2 be linear mappings over $\mathbb{F}_{2^{mn_1}}$ such that $L_2(M_1(a, b)) = M_1(a, L_1(b))$ with $a, b \in \mathbb{F}_{2^{mn_1}}$. By choosing $b = a$ the corresponding equation simplifies to

$$M_1(a, (L_1(a))) = a(L_1(a))^{2^{mi_1}} + a^{2^{mi_1}} L_1(a) = 0.$$

For $a \neq 0$ this equivalent to $\left(\frac{L_1(a)}{a}\right)^{2^{mi_1}} = \frac{L_1(a)}{a}$. Thus, $\frac{L_1(a)}{a} = \theta_a \in \mathbb{F}_{2^{m\gcd(n_1, i_1)}}$ and $L_1(a) = \theta_a a$. Due to our assumption on i_1 this intermediate field does not equal $\mathbb{F}_{2^{mn_1}}$.

Assume we are given two elements e_1, e_2, that are linear independent with respect to the intermediate field $\mathbb{F}_{2^{m\gcd(n_1, i_1)}}$. It follows that $L_1(e_1) = \theta_{e_1} e_1, L_1(e_2) = \theta_{e_2} e_2$ and

$$L_1(e_1 + e_2) = \theta_{e_1 + e_2}(e_1 + e_2) = L_1(e_1) + L_1(e_2) = \theta_{e_1} e_1 + \theta_{e_2} e_2.$$

Since the elements are linear independent every linear combination is unique and thus $\theta_{e_1 + e_2} = \theta_{e_1} = \theta_{e_2}$. By substituting e_1 by αe_1 with $\alpha \in \mathbb{F}_{2^{m\gcd(n_1, i_1)}}$ one gets, that

$$L_1(\alpha e_1) = \alpha \theta_{e_1} e_1.$$

Via this two properties one concludes by considering an $\mathbb{F}_{2^{m\gcd(n_1, i_1)}}$-basis that there exist a $\theta \in \mathbb{F}_{2^{m\gcd(n_1, i_1)}}$, such that

$$L_1(\alpha b) = \alpha \theta b$$

for any element b of this basis and $\alpha \in \mathbb{F}_{2^{m\gcd(n_1, i_1)}}$. This proves that $L_1(x) = \theta x$. By noting that the image of $M_1(a, b)$ contains an $\mathbb{F}_{2^m}$-basis of $\mathbb{F}_{2^{mn_1}}$ one concludes rather straightforward that $L_2 = L_1$.

To show that the centralizer for product of fields is as stated in the theorem is tedious and technical. So we skip this part and end our proof.

Understanding the centralizer of an arbitrary HFE-Cryptosystem with branches is a hard problem. But it is easy to see that all block matrices Z, where every block Λ_k represents a multiplication with an element from the base field $\mathbb{F}_q$, lie in the centralizer of M. It is very likely and confirmed by our experiments that these are the only elements when $H(X)$ is not as simple as above. Thus we have the following reasonable conjecture.

Conjecture 1. The elements C of the centralizer for an arbitrary system with l branches are the matrices $A^{-1}ZA$ with

$$Z = \begin{pmatrix} \Lambda_1 & 0 & 0 & \cdots & 0 \\ 0 & \Lambda_2 & 0 & \cdots & 0 \\ \vdots & 0 & \Lambda_3 & \cdots & 0 \\ \vdots & \vdots & 0 & \ddots & 0 \\ 0 & 0 & \cdots & 0 & \Lambda_l \end{pmatrix}.$$

Thereby Λ_k denotes the representation matrix of a multiplication with $\lambda_k \in \mathbb{F}_q$.

Remark 1. The conjecture does not state anything about C'. As C' is only needed to compute C but not to complete the actual separation, no further knowledge about the structure is necessary.

The separation requires the factorization of the characteristic polynomial of C. Assuming Conjecture 1 the matrices C can be diagonalized with only a few possible Eigenvalues. Consequently the factorization is feasible.

The number of recovered branches depends on the number of different Eigenvalues. If only clusters of branches are recovered, the algorithm can be applied separately to the different clusters. We have the following result.

Theorem 3. *The branches for an arbitrary system can be recovered with $O(n^6)$ field operations on average.*

Remark 2. From the proof of theorem 2 we see, that for this special mappings we can compute the matrix C by computing all matrices with $m'_i(x, Cx) = 0$ for all $x \in \mathbb{F}_{2^{nm}}$. From experiments we have that this seems to remain true for arbitrary HFE-systems over arbitrary characteristic. This variant is a bit faster as the number of required plaintext/ciphertext pairs obviously does not increase and the number of variables in the Gaussian elimination is reduced as C' does not have to be computed anymore. Thus it makes sense to start with this variant and extend to the more complex algorithm, if necessary.

5   Conclusion

We showed that if the base field of a given HFE-Cryptosystem is not $\mathbb{F}_2$ then the security is not affected if the secret transformations are chosen to be linear instead of affine. This was achieved by considering the HFE-system as a nonassociative algebra and showing that from this point of view the affine parts can be eliminated. It is very interesting to investigate if this approach can be modified to reveal the secret key. Furthermore we showed that for an arbitrary system with branches the branches can be separated in polynomial time on average. Again this was achieved by employing the theory of nonassociative algebras.

Both results were based on the assumption that certain conditions are satisfied with a very high probability by the algebra we get from a randomly chosen HFE-Cryptosystem. Therefore it is a very challenging task to find conditions on the hidden polynomial $H(X)$, which yield to algebras fulfilling this assumptions. Clearly finding answers to one of the above problems would yield to a much better understanding about the security of HFE-Crypto systems.

References

1. T. Beth, W. Geiselmann, R. Steinwandt, *Revealing 441 Key Bits of SFLASHv2*, Nessie Workshop Munich, November 2002
2. Nicolas Courtois, Louis Goubin, Jacques Patarin, *Quartz, 128-bit long digital signatures*, Cryptographers' Track Rsa Conference 2001, LNCS 2020, pp.282-297, Springer-Verlag
3. N. Courtois, L. Goubin and J. Patarin, *SFLASH^{v3} a fast symmetric signature scheme*, Cryptology ePrint Archive: Report 2003/211, 2003
4. H. Dobbertin, *internal report 93/94*, German Information Security Agency
5. H. Dobbertin, *Analysis of HFE Schemes Based on Power Functions*, invited talk at YACC '02, 03-07 June, 2002
6. Louis Goubin, J. Patarin, *Improved algorithms for Isomorphisms of Polynomials*, Eurocrypt '98, pp.184-200, Springer-Verlag
7. H. Imai, T. Matsumoto, *Public Quadratic Polynomial-tuples for efficient signature-verification and message-encryption*, Eurocrypt '88, pp.419-453, Springer-Verlag
8. A. Kipnis, A. Shamir, *Cryptanalysis of the HFE Public Key Cryptosystem by Re-linearisation*, Crypto '99, pp.19-30 Springer-Verlag
9. A. Kipnis, A. Shamir, *Crypanalysis of the Oil& Vinegar Signature Scheme*, Crypto '98, pp.206-222, Springer-Verlag
10. R. Lidl, H. Niederreiter, *Finite Fields*, Encyclopedia of Mathematics and its Applications, Vol. 20, 2nd Edition, Cambridge University Press, Cambridge, 1997
11. J. Patarin, *Cryptanalysis of the Matsumoto and Imai Public Key Scheme of Eurocrypt '88*, Crypto '95, pp.248-261, Springer-Verlag
12. J. Patarin, *Asymmetric Cryptography with a Hidden Monomial*, Crypto '96, pp.45-60, Springer Verlag
13. J. Patarin, *Hidden Fields Equations (HFE) and Isomorphisms of Polynomials(IP): Two new families of Asymmetric Algorithms*, Eurocrypt '96, pp.33-48, Springer Verlag
14. R. Schafer, *Introduction to Nonassociative Algebras*, Academic Press, 1966

Dimension of the Linearization Equations of the Matsumoto-Imai Cryptosystems

Adama Diene*, Jintai Ding, Jason E. Gower,
Timothy J. Hodges, and Zhijun Yin

Department of Mathematical Sciences
University of Cincinnati
Cincinnati, OH 45221-0025, USA
adiene@centralstate.edu, ding@math.uc.edu, gowerj@math.uc.edu,
timothy.hodges@uc.edu, yinzhi@math.uc.edu

Abstract. The Matsumoto-Imai (MI) cryptosystem was the first multivariate public key cryptosystem proposed for practical use. Though MI is now considered insecure due to Patarin's linearization attack, the core idea of MI has been used to construct many variants such as Sflash, which has recently been accepted for use in the New European Schemes for Signatures, Integrity, and Encryption project. Linearization attacks take advantage of the algebraic structure of MI to produce a set of equations that can be used to recover the plaintext from a given ciphertext. In our paper, we present a solution to the problem of finding the dimension of the space of linearization equations, a measure of how much work the attack will require.

1 Introduction

In the last two decades, public key cryptography has become an indispensable part of most modern communication systems. However, due to the threat that quantum computers pose to cryptosystems based on "hard" number theory problems, there has recently been great effort put into the search for alternative public key cryptosystems. Multivariate cryptosystems provide a promising alternative since solving a set of multivariate polynomial equations over a finite field appears to be difficult, analogous to integer factorization, though it is unknown precisely how difficult either problem is.

One of the first implementations of a multivariate public key cryptosystem was suggested by Matsumoto and Imai [8]. Fixing a finite field k of characteristic two and cardinality 2^q, they suggested using a bijective map M defined over K, a degree n extension of k. By identifying K with k^n, we see that M induces a multivariate polynomial map $\tilde{M}$. We can "hide" this map by composing on the left by L_1 and on the right by L_2, where the $L_i : k^n \longrightarrow k^n$ are invertible affine linear maps. This gives a map $\bar{M} : k^n \longrightarrow k^n$ defined by

$$\bar{M}(x_1, \ldots, x_n) = L_1 \circ \tilde{M} \circ L_2(x_1, \ldots, x_n) = (y_1, \ldots, y_n).$$

---

* *Current address*: Department of Mathematics and Computer Science, Central State University, Wilberforce, OH 45384.

Ø. Ytrehus (Ed.): WCC 2005, LNCS 3969, pp. 242–251, 2006.

The map originally suggested by Matsumoto and Imai is the map

$$M : X \longmapsto X^{1+2^{q\theta}},$$

where $\gcd(2^{q\theta} + 1, 2^{qn} - 1) = 1$. The resulting system is the Matsumoto-Imai (C^* or MI) cryptosystem. The public key for MI is the system of n quadratic polynomials $y_1, \ldots, y_n$.

Even for a large finite field K, MI is efficient. Unfortunately this scheme was proven insecure under an algebraic attack [9] that produces so-called "linearization equations." These linearization equations can be swiftly generated from the public key and known plaintext/ciphertext pairs, and have the form:

$$\sum a_{ij} x_i y_j + \sum b_i x_i + \sum c_j y_j + d = 0,$$

where $x_1, \ldots, x_n$ are the plaintext variables corresponding to the ciphertext variables $y_1, \ldots, y_n$. Once we have found enough of these equations, and hence the a_{ij}, b_i, c_j and d, we can substitute in the ciphertext to produce a system of linear equations in the plaintext variables. Patarin showed that there are enough linearization equations to produce enough linearly independent linear equations in the plaintext variables, which can then be used to find the plaintext.

After introducing his linearization attack, Patarin posed the general question of how we can find the maximum number of linearly independent linearization equations for MI. The answer to this question is necessary for a complete understanding of both MI and the linearization attack, and may provide valuable insight into related systems derived from MI, such as the Sflash signature schemes [1, 2], PMI and PMI+ [3, 4], HFE [10] and others. In this paper, we use the method developed in [6] to attack the HFE cryptosystem (another generalization of MI), to find the exact dimension of the space of linearization equations.

The complete result, given in Theorems 2 and 3, involves a number of exceptional cases. Let us summarize here the result ignoring the case $n = 2\theta$, which has no cryptographic applications, and some exceptional cases when n is $2, 3$ or 6. Let δ be the dimension of the space of linearization equations. If $q > 1$, then

$$\delta = \begin{cases} \frac{2n}{3}, & \text{if } \theta = n/3 \text{ or } 2n/3; \\ n, & \text{otherwise.} \end{cases}$$

On the other hand, if $q = 1$,

$$\delta = \begin{cases} \frac{2n}{3}, & \text{if } \theta = n/3 \text{ or } 2n/3; \\ 2n, & \text{if } \theta = 1, n - 1 \text{ or } (n \pm 1)/2; \\ \frac{3n}{2}, & \text{if } n = 2\theta \pm 2; \\ n, & \text{otherwise.} \end{cases}$$

Computer simulations for the cases $n \leq 15$ have confirmed these results.

Before getting into the technical details we outline the idea of the proof. Let $X \in K$ and let $Y = X^{2^{q\theta}+1}$. Then $Y^{2^{q\theta}-1} = X^{(2^{q\theta}+1)(2^{q\theta}-1)}$. Multiplying each side by XY, we obtain $Y^{2^{q\theta}}X = YX^{2^{2q\theta}}$. Since K has characteristic two, $Y^{2^{q\theta}}$ and $X^{2^{2q\theta}}$ are linear functions of Y and X respectively. This equation is thus a version of a linearization equation for K. Using the identification of K with k^n and looking at coordinate components yields a set of n (not necessarily independent) linearization equations in the above sense. Moreover, for any integer $m = 0, 1, \ldots, n-1$, we further have

$$(XY^{2^{q\theta}} - YX^{2^{2q\theta}})^{2^{qm}} = 0.$$

Looking at coordinate components yields further linearization equations. When $q > 1$, it turns out that all linearization equations arise in this way. When $q = 1$, there are additional identities that arise for certain exceptional values of θ. For instance, if $\theta = 1$, then $XY = X^4$.

The proof proceeds in the following way. We first define a notion of linearization equation for K and use a simple algebraic trick (exactness of the tensor product) to show that the dimension of the space of linearization equations is the same over both k and K. We then show that the equations above span all possible linearization equations for K and count carefully the dimension of this space.

Note that we only need to do this calculation for M. The composition with the invertible affine linear maps L_i does not affect the dimension of the associated space of linearizations equations.

2 The Linearization Problem

We begin by placing the problem in a general context. Let V be a vector space over k and denote by $\mathrm{Fun}(V, V)$ the set of functions from V to V. If V is the plaintext/ciphertext space of a cryptosystem, then a cipher is an element $M \in \mathrm{Fun}(V, V)$.

More generally, for any pair of sets V and W, denote by $\mathrm{Fun}(V, W)$, the set of all functions from V to W. Define a function

$$\psi_M : \mathrm{Fun}(V \times V, k) \to \mathrm{Fun}(V, k)$$

by

$$\psi_M(f)(v) = f(v, Mv).$$

Recall that for any pair of vector spaces V and W, the set $\mathrm{Fun}(V, W)$ is again a vector space in the usual way:

$$(\lambda f)(v) = \lambda f(V), \qquad (f + g)(v) = f(v) + g(v).$$

Thus both $\mathrm{Fun}(V \times V, k)$ and $\mathrm{Fun}(V, k)$ are vector spaces. It is easily checked that ψ_M is a linear transformation between these spaces. Denote by $\mathcal{A}(V)$ the subspace of $\mathrm{Fun}(V, k)$ consisting of affine linear functions (polynomials of degree less than or equal to one). Note that there is a natural embedding of $\mathcal{A}(V) \otimes \mathcal{A}(V)$ into $\mathrm{Fun}(V \times V, k)$ given by $(f \otimes g)(v, v') = f(v)g(v')$.

Definition 1. *The subspace* $\mathcal{L}_M = \ker \psi_M|_{\mathcal{A}(V) \otimes \mathcal{A}(V)}$ *is defined to be the space of* linearization equations *associated to* M.

Let's see how this definition ties up with the usual definition of linearization equation. Let $\{f_i \mid i = 0, 1, \ldots, n-1\}$ be a basis for the dual space V^*. Then $\mathcal{A}(V)$ has a basis consisting of the f_i and the constant function 1. So the $(n+1)^2$ elements $f_i \otimes f_j$, $f_i \otimes 1$, $1 \otimes f_j$, $1 \otimes 1$ form a basis for $\mathcal{A}(V) \otimes \mathcal{A}(V)$ and an arbitrary element of $\mathcal{A}(V) \otimes \mathcal{A}(V)$ is a bi-affine linear function of the form:

$$\eta = \sum a_{ij}(f_i \otimes f_j) + \sum b_i(f_i \otimes 1) + \sum c_j(1 \otimes f_j) + d(1 \otimes 1).$$

Let $x \in V$ have coordinates $x_i = f_i(x)$ and let $y = M(x)$ have coordinates $y_i = f_i(y)$. Thus $\eta \in \mathcal{L}_M$ if and only if for all $x \in V$,

$$\psi_M(\eta)(x) = \sum a_{ij}x_iy_j + \sum b_ix_i + \sum c_jy_j + d = 0.$$

That is, $\eta \in \mathcal{L}_M$ if and only if $\psi_M(\eta)(x) = 0$ is a linearization equation in the usual sense.

We are now in a position to state the problem that we solve in this article.

Linearization problem. Let q be a positive integer, let k be a finite field of order 2^q, and let K be an extension field of k with $[K : k] = n$. Let θ be an integer such that $1 \le \theta < n$, and let $M \colon K \to K$ be the map $M(X) = X^{1+2^{q\theta}}$. Find $\dim \mathcal{L}_M$, the dimension of the space of linearization equations associated to M.

Note that the condition $\gcd(2^{q\theta} + 1, 2^{qn} - 1) = 1$ is required in the MI cryptosystem, though this assumption is not needed in order to calculate the dimension of $\mathcal{L}_M$.

3   Lifting to K

In order to simplify the calculations, we work inside the larger algebra $\mathrm{Fun}(K \times K, K)$ which we can realize as a homomorphic image of the polynomial ring $K[X, Y]$. Let us recall some general facts about this algebra. Since K is finite of cardinality 2^{qn}, the natural homomorphism $K[X] \to \mathrm{Fun}(K, K)$ is surjective and its kernel is the ideal $(X^{2^{qn}} - X)$. Similarly, the natural homomorphism $K[X, Y] \to \mathrm{Fun}(K \times K, K)$ is also surjective and has kernel $(X^{2^{qn}} - X, Y^{2^{qn}} - Y)$. Let G be the Galois group $\mathrm{Gal}(K, k)$. One of the key observations used in [6], is the following standard result from Galois theory (see for instance [5, Theorem 2]).

Lemma 1. *Denote by* $\mathrm{Fun}_k(K, K) \subset \mathrm{Fun}(K, K)$ *the subspace of* k-*linear endomorphisms of* K. *Then* $\mathrm{Fun}_k(K, K)$ *is naturally isomorphic as a vector space to the group algebra* KG.

Similarly the subset of linear functions $\mathrm{Fun}_k(K \times K, K) \subset \mathrm{Fun}(K \times K, K)$ can be identified with $K(G \times G)$. The group G is cyclic, generated by the polynomial function X^{2^q}. The space of affine linear functions from $K \times K$ to K can be viewed, by extension of coefficients, as $K \otimes \mathcal{A}(K) \otimes \mathcal{A}(K)$. From the above discussion, the elements of $K \otimes \mathcal{A}(K) \otimes \mathcal{A}(K)$, viewed as polynomial functions, have the form

$$\sum_{i,j=0}^{n-1} A_{ij} \otimes X^{2^{qi}} \otimes X^{2^{qj}} + \sum_{i=0}^{n-1} B_i \otimes X^{2^{qi}} \otimes 1 + \sum_{j=0}^{n-1} C_j \otimes 1 \otimes X^{2^{qj}} + D \otimes 1 \otimes 1,$$

for some $A_{ij}, B_i, C_j, D \in K$. As above, for any $M \in \mathrm{Fun}(K, K)$, we may define a map $\hat{\psi}_M : \mathrm{Fun}(K \times K, K) \to \mathrm{Fun}(K, K)$ by $\hat{\psi}_M(f)(x) = f(x, M(x))$. Set $\hat{\mathcal{L}}_M = \ker \hat{\psi}_M|_{K \otimes \mathcal{A}(K) \otimes \mathcal{A}(K)}$. This yields the following exact commutative diagram:

$$
\begin{array}{ccccccc}
0 & \longrightarrow & \mathcal{L}_M & \longrightarrow & \mathcal{A}(K) \otimes \mathcal{A}(K) & \xrightarrow{\psi_M} & \mathrm{Fun}(K, k) \\
& & \downarrow & & \downarrow & & \downarrow \\
0 & \longrightarrow & \hat{\mathcal{L}}_M & \longrightarrow & K \otimes \mathcal{A}(K) \otimes \mathcal{A}(K) & \xrightarrow{\hat{\psi}_M} & \mathrm{Fun}(K, K)
\end{array}
$$

Observe that the bottom line is the image of the top line under the exact functor $K \otimes -$. For $\mathrm{Fun}(K, K)$ is naturally isomorphic to $K \otimes \mathrm{Fun}(K, k)$ and under this identification $\hat{\psi}_M$ identifies with $K \otimes \psi_M$, the image of ψ_M under the functor $K \otimes -$. The exactness of $K \otimes -$ implies that $\ker(K \otimes \psi_M) \cong K \otimes \ker(\psi_M)$ (see, for instance [7]); so $\hat{\mathcal{L}}_M \cong K \otimes \mathcal{L}_M$. Hence $\dim_k \mathcal{L}_M = \dim_K \hat{\mathcal{L}}_M$.

4 Statement of Main Theorems

To find the dimension of the space of linearization equations we must find the dimension of the kernel of the map

$$\hat{\psi}_M : K \otimes \mathcal{A}(K) \otimes \mathcal{A}(K) \to \mathrm{Fun}(K, K).$$

This amounts to finding linearly independent identities of the form

$$\sum_{i,j=0}^{n-1} A_{ij} X^{2^{qi}} Y^{2^{qj}} + \sum_{i=0}^{n-1} B_i X^{2^{qi}} + \sum_{j=0}^{n-1} C_j Y^{2^{qj}} + D = 0,$$

where $Y = X^{2^{q\theta}+1}$, $X^{2^n} = X$ and $Y^{2^n} = Y$ and $A_{ij}, B_i, C_j, D \in K$. As noted above, it is easy to see that $Y = X^{2^{q\theta}+1}$ implies that $Y^{2^{q\theta}} X = Y X^{2^{2q\theta}}$ and hence more generally that

$$(XY^{2^{q\theta}} - YX^{2^{2q\theta}})^{2^{qm}} = 0,$$

for $m = 0, \ldots, n-1$. Generically these will be distinct identities. However, if $n = 3\theta$ or $3\theta/2$, the identities $XY^{2^{q\theta}} - YX^{2^{2q\theta}}$, $X^{2^{q\theta}}Y^{2^{2q\theta}} - Y^{2^{q\theta}}X$, and $X^{2^{2q\theta}}Y - Y^{2^{2q\theta}}X^{2^{q\theta}}$ are evidently dependent, yielding only $2n/3$ independent identities. The case $n = 2\theta$ is an exceptional, highly degenerate case. In this situation, $Y^{2^{q\theta}} = Y$, yielding $(n^2 + n)/2$ identities of the form

$$(Y^{2^{q\theta+l}} - Y^{2^{ql}})X^{2^{qm}} = 0,$$

for $l = 0, \ldots n/2 - 1$ and $m = 0, \ldots n - 1$.

When $q > 1$, these identities are independent and span the set of all identities.

Theorem 2. *If $q > 1$, then the dimension of the space of linearization equations is given by*

$$\dim \hat{\mathcal{L}}_M = \begin{cases} 2n/3, & \text{if } \theta = n/3 \text{ or } 2n/3; \\ (n^2 + n)/2, & \text{if } \theta = n/2; \\ n, & \text{otherwise.} \end{cases}$$

When $q = 1$, further identities occur for special values of θ.

- When $\theta = 1$, $XY = XX^{2+1} = X^4$ yielding n identities of the form $(XY - X^4)^{2^m} = 0$ for $m = 0, \ldots, n - 1$.
- When $\theta = n - 1$, $Y^2 = (X^{2^{n-1}+1})^2 = X^{2^n}X^2 = X^3$, yielding n identities of the form $(XY^2 - X^4)^{2^m} = 0$ for $m = 0, \ldots, n - 1$.
- When $n = 2\theta + 1$, $Y^{2^{\theta+1}} = (X^{2^\theta+1})^{2^{\theta+1}} = X^{2^{2\theta+1}+2^{\theta+1}} = X^{2^{\theta+1}+1} = YX^{2^\theta}$ yielding n identities of the form $(YX^{2^\theta} - Y^{2^{\theta+1}})^{2^m} = 0$ for $m = 0, \ldots, n - 1$.
- When $n = 2\theta - 1$, $Y^{2^\theta} = (X^{2^\theta+1})^{2^\theta} = X^{2^{2\theta}+2^\theta} = X^{2\theta}(X^{2^{2\theta-1}})^2 = X^{2\theta}X^2 = XY$, yielding n identities of the form $(Y^{2^\theta} - XY)^{2^m} = 0$ for for $m = 0, \ldots, n - 1$.
- When $n = 2\theta + 2$, $Y^{2^{\theta+2}}X = X^{2^{2\theta+2}+2^{\theta+2}+1} = X^{2^{\theta+2}+2} = X^{2^{\theta+1}}Y^2$, yielding $n/2$ identities of the form $(Y^{2^{\theta+2}}X - X^{2^{\theta+1}}Y^2)^{2^m} = 0$ for $m = 0, \ldots, n/2 - 1$.
- When $n = 2\theta - 2$, $Y^{2^{\theta-1}}X^{2^{\theta-1}} = (X^{2^\theta+1})^{2^{\theta-1}}X^{2^{\theta-1}} = X^{2^{2\theta-1}+2^\theta} = X^{2+2^\theta} = XY$, yielding $n/2$ identities of the form $(Y^{2^{\theta-1}}X^{2^{\theta-1}} - XY)^{2^m}$ for $m = 0, \ldots, n/2 - 1$.

Again these turn out to be all identities and they are linearly independent.

Theorem 3. *If $q = 1$, the dimension of the space of linearization equations is as follows. When $\theta = n/3$ or $2n/3$,*

$$\dim \hat{\mathcal{L}}_M = \begin{cases} 7, & \text{if } n = 6, \ \theta = 2 \text{ or } 4; \\ 8, & \text{if } n = 3, \ \theta = 1 \text{ or } 2; \\ \frac{2n}{3}, & \text{otherwise.} \end{cases}$$

When $\theta = n/2$,

$$\dim \hat{\mathcal{L}}_M = \begin{cases} 5, & \text{if } n = 2, \ \theta = 1; \\ (n^2 + n)/2, & \text{otherwise.} \end{cases}$$

When $\theta \neq n/3, 2n/3, n/2$,

$$
\dim \hat{\mathcal{L}}_M = \begin{cases} 10, & \text{if } n = 4 \text{ and } \theta = 1 \text{ or } 3; \\ 2n, & \text{if } \theta = 1, n-1 \text{ or } (n \pm 1)/2; \\ \frac{3n}{2}, & \text{if } \theta = n/2 \pm 1; \\ n, & \text{otherwise.} \end{cases}
$$

5 Proofs of Main Theorems

An arbitrary element of $K \otimes \mathcal{A}(K) \otimes \mathcal{A}(K)$ is of the form

$$
\sum_{i,j=0}^{n-1} A_{ij} \otimes X^{2^{qi}} \otimes X^{2^{qj}} + \sum_{i=0}^{n-1} B_i \otimes X^{2^{qi}} \otimes 1 + \sum_{j=0}^{n-1} C_j \otimes 1 \otimes X^{2^{qj}} + D \otimes 1 \otimes 1
$$

and its image under $\hat{\psi}_M$ is

$$
\sum_{i,j=0}^{n-1} A_{ij} X^{2^{qi}} (X^{2^{q\theta}+1})^{2^{qj}} + \sum_{i=0}^{n-1} B_i X^{2^{qi}} + \sum_{j=0}^{n-1} C_j (X^{2^{q\theta}+1})^{2^{qj}} + D,
$$

where because of the relation $(X^{2^{qn}} - X)$ in $\mathrm{Fun}(K, K)$, we may consider the exponents as elements of $\mathbb{Z}_{2^{qn}-1}$. If such a polynomial is in the kernel, its constant term must be zero, so it suffices to look at terms of the form

$$
\sum_{i,j=0}^{n-1} A_{ij} \otimes X^{2^{qi}} \otimes X^{2^{qj}} + \sum_{i=0}^{n-1} B_i \otimes X^{2^{qi}} \otimes 1 + \sum_{j=0}^{n-1} C_j \otimes 1 \otimes X^{2^{qj}}.
$$

Lemma 4. *Let* $\mathcal{M} = \{X^{2^{qi}} \otimes X^{2^{qj}}, X^{2^{qi}} \otimes 1, 1 \otimes X^{2^{qj}} \mid i, j = 0, \ldots n-1\}$. *Then* $\dim \hat{\mathcal{L}}_M = n^2 + 2n - |\hat{\psi}_M(\mathcal{M})|$.

Proof. Let $\mathcal{N} = \{X^{2^{qi}} \mid i = 0, \ldots n-1\}$. Then $\mathcal{N} \cup \{1\}$ forms a basis for $K \otimes \mathcal{A}(K)$ and $\mathcal{M} \cup \{1 \otimes 1\}$ forms a basis for $K \otimes \mathcal{A}(K) \otimes \mathcal{A}(K)$. It is clear from the defintion of $\hat{\psi}_M$ that $\hat{\psi}_M(1 \otimes 1) = 1$ and that $\hat{\psi}_M(\mathcal{M}) \subset \mathcal{N}$. Hence

$$
\mathrm{rank}(\hat{\psi}_M) = |\hat{\psi}_M(\mathcal{M} \cup \{1 \otimes 1\})| = |\hat{\psi}_M(\mathcal{M})| + 1
$$

Hence

$$
\begin{aligned}
\dim \hat{\mathcal{L}}_M &= \dim(K \otimes \mathcal{A}(K) \otimes \mathcal{A}(K)) - \mathrm{rank}(\hat{\psi}_M) \\
&= (n+1)^2 - |\hat{\psi}_M(\mathcal{M})| + 1 \\
&= n^2 + 2n - |\hat{\psi}_M(\mathcal{M})|
\end{aligned}
$$

Thus the problem reduces to the calculation of $|\hat{\psi}_M(\mathcal{M})|$. In the case $q > 1$, this calculation is fairly straightforward, but when $q = 1$, it is a little more intricate.

We can reset the problem in the following way. Define $\mathbb{Z}_n^1$ and $\mathbb{Z}_n^2$ to be two copies of $\mathbb{Z}_n$. Define

$$\phi : (\mathbb{Z}_n \times \mathbb{Z}_n) \cup \mathbb{Z}_n^1 \cup \mathbb{Z}_n^2 \to \mathbb{Z}_{2^{qn}-1}$$

by

$$\phi(i,j) = 2^{qi} + 2^{qj} + 2^{q(\theta+j)}, \text{ for } (i,j) \in (\mathbb{Z}_n \times \mathbb{Z}_n)$$
$$\phi(k) = 2^{qk} \text{ for } k \in \mathbb{Z}_n^1$$
$$\phi(l) = 2^{ql} + 2^{q(\theta+l)} \text{ for } l \in \mathbb{Z}_n^2$$

Clearly $|\hat{\psi}_M(\mathcal{M})| = |\text{Im } \phi|$.

The elements of $\mathbb{Z}_{2^{qn}-1}$ can be represented uniquely in a 2^q-ary expansion of length less than or equal to n. It is convenient to represent this expansion as a circular graph with n vertices representing the place holders and the digits of the expansion as labels on these vertices. For example in the case when $n = 8$, the element 02100301 is represented by the labelled graph in figure 1.

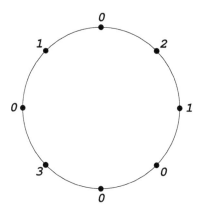

Fig. 1. The representation of the number with 2^q-ary expansion 02100301

Theorem 5. *If $q > 1$, then*

$$|\text{Im } \phi| = \begin{cases} n^2 + 4n/3, & \text{if } n = 3\theta,\ 3\theta/2; \\ (n^2 + 3n)/2, & \text{if } n = 2\theta; \\ n^2 + n, & \text{otherwise.} \end{cases}$$

Proof. The elements of Im ϕ considered as diagrams consist of

1. diagrams with all labels 0 except one label of 1
2. diagrams with all labels 0 except two labels of 1, spaced θ apart.
3. diagrams with all labels 0 except one label of 1 and one label of 2, spaced θ apart.
4. diagrams with all labels 0 except three labels of 1, of which at least one pair is spaced θ apart.

There are n diagrams of type (1). There are n diagrams of type (2) except if $n = 2\theta$ in which case there are only $n/2$ diagrams. There are $2n$ diagrams of type (3), unless $n = 2\theta$, in which case there are only n diagrams.

Consider diagrams of type (4). Consider first diagrams with exactly one pair of labeled vertices spaced θ apart and assume that $n \neq 2\theta$. For each such pair, there are $n - 4$ possible locations for the third labeled vertex if $n \not\equiv 3\theta$ and $n - 3$ locations if $n \equiv 3\theta$. Thus there are $n(n - 4)$ and $n(n - 3)$ total such diagrams respectively. If $n = 2\theta$, there are $n/2$ such pairs and $n - 2$ locations for the third vertex, so there are $n(n - 2)/2$ diagrams.

If n does not divide 3θ and $n \neq 2\theta$, then we can have exactly two pairs of vertices spaced by θ. There are n such diagrams. If $n | 3\theta$, we can have all three vertices spaced by θ and there are $n/3$ of these diagrams.

Thus when $n = 2\theta$, $|\mathrm{Im}\ \phi| = n + n/2 + n + n(n - 2)/2 = (n^2 + 3n)/2$. When $3\theta \equiv 0 \pmod{n}$, $|\mathrm{Im}\ \phi| = n + n + 2n + (n^2 - 3n) + n/3 = n^2 + 4n/3$. Otherwise $|\mathrm{Im}\ \phi| = n + n + 2n + (n^2 - 4n) + n = n^2 + n$.

Theorem 6. *Suppose that $q = 1$. If $3\theta \equiv 0 \pmod{n}$, then*

$$|\mathrm{Im}\ \phi| = \begin{cases} 41, & \text{if } n = 6,\ \theta = 2 \text{ or } 4; \\ 7, & \text{if } n = 3,\ \theta = 1 \text{ or } 2; \\ n^2 + \frac{4n}{3}, & \text{otherwise.} \end{cases}$$

If $2\theta = n$, then

$$|\mathrm{Im}\ \phi| = \begin{cases} 3, & \text{if } n = 2; \\ (n^2 + 3n)/2, & \text{otherwise.} \end{cases}$$

Otherwise,

$$|\mathrm{Im}\ \phi| = \begin{cases} 14, & \text{if } n = 4 \text{ and } \theta = 1 \text{ or } 3; \\ n^2, & \text{if } \theta = 1; \\ n^2, & \text{if } \theta > 1 \text{ and } n = 2\theta \pm 1 \text{ or } \theta + 1; \\ n^2 + \frac{n}{2}, & \text{if } n = 2\theta \pm 2; \\ n^2 + n, & \text{otherwise.} \end{cases}$$

Proof. The elements of $\mathrm{Im}\ \phi$ considered as diagrams consist of essentially the same cases as in the previous proof except that a vertex with a label of 2 transforms into the next vertex moving clockwise around the diagram, labeled with a 1. Thus the possible configurations are now:

1. diagrams with all labels 0 except one label of 1
2. diagrams with all labels 0 except two labels of 1, spaced θ, $\theta - 1$ or $\theta + 1$ apart.
3. diagrams with all labels 0 except three labels of 1, of which at least one pair is spaced θ apart.

The counting of diagrams of type (1) and (3) is the same as in above. Similarly for the diagrams of type (2) spaced θ apart there are again n of these if $n \neq 2\theta$

and $n/2$ if $n = 2\theta$. In the generic case there are an additional n digrams with each of the other two spacing options. However, there are now a number of exceptional cases when a pair of θ, $\theta - 1$ or $\theta + 1$ coincide, or one of them equals $n/2$.

If $\theta = 1$, then the $\theta - 1$ spacing does not occur and when $\theta = n - 1$, the $\theta + 1$ spacing does not occur.

If $n = 2\theta + 2$, then $\theta + 1$ is $n/2$ so there are only $n/2$ diagrams with this spacing. So there are $5n/2$ diagrams of type (2) in total. Similarly, if $n = 2\theta - 2$.

If $n = 2\theta \pm 1$, then the three spacing options become only two and there are $2n$ diagrams of type (2).

The numbers stated in the result are then obtained by adding up the number of diagrams of each type as in the previous proof.

Subtracting these numbers from $n^2 + 2n$ yields the dimension of the space of linearization equations given in the previous section.

References

1. M.-L. Akkar, N. T. Courtois, R. Duteuil and L. Goubin. A Fast and Secure Implementation of Sflash. In *PKC 2003*, LNCS 2567:267–278.
2. N. Courtois, L. Goubin, and J. Patarin. FLASH, a Fast Multivariate Signature Algorithm. In *CT-RSA 2001*, LNCS 2020:298–307.
3. J. Ding. A New Variant of the Matsumoto-Imai Cryptosystem Through Perturbation. In *PKC 2004*, LNCS 2947:305–318.
4. J. Ding and J. E. Gower, Innoculating Multivariate Schemes against Differential Attacks, in Cryptology ePrint archive, report 2005/255 http://eprint.iacr.org/, 2005.
5. N. Jacobson. *Lectures in Abstract Algebra III*, Springer-Verlag, 1964.
6. A. Kipnis and A. Shamir. Cryptanalysis of the HFE Public Key Cryptosystem by Relinearization. In *Crypto 1999*, LNCS 1666:19–30.
7. S. Lang. *Algebra*, Springer-Verlag, 2002.
8. T. Matsumoto and H. Imai. Public Quadratic Polynomial-tuples for Efficient Signature-verification and Message-encryption. In *Eurocrypt 1988*, LNCS 330: 419–453.
9. J. Patarin. Cryptanalysis of the Matsumoto and Imai Public Key Scheme of Eurocrypt'88. In *Crypto 1995*, LNCS 963:248–261.
10. J. Patarin. Hidden Fields Equations (HFE) and Isomorphisms of Polynomials (IP): Two New Families of Asymmetric Algorithms. In *Eurocrypt 1996*, LNCS 1070: 33–48.

RSA-Based Secret Handshakes

Damien Vergnaud

Laboratoire de Mathématiques Nicolas Oresme
Université de Caen, Campus II, B.P. 5186,
14032 Caen Cedex, France
Damien.Vergnaud@math.unicaen.fr

Abstract. A secret handshake mechanism allows two entities, members of a same group, to authenticate each other secretly. This primitive was introduced recently by Balfanz, Durfee, Shankar, Smetters, Staddon and Wong and, so far, all the schemes proposed are based on discrete log systems. This paper proposes three new secret handshake protocols secure against active impersonator and detector adversaries. Inspired by two RSA-based key agreement protocols introduced by Okamoto and Tanaka in 1989 and Girault in 1991, our schemes are, in the random oracle model, provably secure against active adversaries under the assumption that the RSA problem is intractable.

1 Introduction

The concept of *secret handshakes* was introduced in 2003 [1] by Balfanz, Durfee, Shankar, Smetters, Staddon and Wong. The present paper focuses on the proposal of three new practical constructions of secret handshake protocol and the security treatment of them: the schemes are secure against active adversaries assuming the hardness of the RSA problem (and one of its weaker variants for one scheme). These are the first constructs which can be instantiated with the RSA primitive (and thus give a first step towards a problem raised in [5] by Castelluccia, Jarecki and Tsudik)

Background. A secret handshake scheme is a cryptographic primitive recently introduced by Balfanz *et al.* [1] which allows two members of the same group to identify each other secretly, in the sense that each party reveals his/her affiliation to the latter only if the other party is a member of the same group. These protocols model *in silico* the secret handshake in the folklore of the *in vivo* exclusive societies, or guilds.

The protocol proposed in [1] is a simple adaptation of the non-interactive key-agreement scheme of Sakai, Ohgishi and Kasahara [19]. It uses one-time credentials to insure that instances of the handshake protocol performed by the same party cannot be linked. In [22], Xu and Yung proposed secret handshake schemes that achieve *unlinkability* with reusable credentials. However, their schemes offer weaker anonymity to the users who furthermore must be aware of the information of other groups. Recently, Castelluccia *et al.* [5] showed how to build secret

Ø. Ytrehus (Ed.): WCC 2005, LNCS 3969, pp. 252–274, 2006.

handshake protocols using a novel tool they called *CA-oblivious public-key encryption* which is an encryption scheme such that neither the public key, nor the ciphertext reveal any information about the Certification Authority. Their schemes are secure under a standard cryptographic assumption: the hardness of the classical computational Diffie-Hellman problem. However, they also rely on one-time credentials to reach the unlinkability property. In the first announcement of their results, Castelluccia *et al.* stated that they know how to get CA oblivious encryption scheme based on RSA, but the claim was incorrect and, so far, there do not exist protocols based on RSA which offer full anonymity. In [5], the authors explicitly stated as an open problem the construction of a secret handshake protocol based on RSA. The investigation of this issue is the main purpose of the present paper.

Underlying Technique. As mentioned above, in Balfanz *et al.*'s scheme, completing the secret handshake is essentially equivalent to computing a key in Sakai *et al.*'s non-interactive key agreement protocol, that is particular to the two interacting group members. The basic framework of our constructions builds on very similar ideas from *identity-based* and *self-certified* key agreements protocols. In 1984, Shamir [20] introduced the concept of identity-based cryptosystems in which the public keys can be arbitrary bit strings and in particular can be defined as a part of the publicly known identification information. In 1991, Girault [9] refined the concept and introduced self-certified public keys which are computed by both the authority and the user, so that the certificate is "embedded" in the public key itself, and does not take the form of a separate value.

Our first scheme, that we call OT-SH, is based on an identity-based key agreement protocol proposed in 1989 by Okamoto and Tanaka [16] and whose security relies on the RSA problem [11]. The second and the third construction, called Gi+SH and Gi×SH, relies on the non-interactive self-certified key agreement scheme proposed by Girault in his seminal paper [9] whose security is also based on the difficulty of solving RSA [15]. Adding suitable nonces and hash functions in these schemes (along the lines drawn in [8]), we obtained secret handshake protocols whose security against active impersonator and detector adversaries relies, in the ROM, on the difficulty of solving the RSA problem (with the notable exception of the impersonator resistance of Gi+SH which relies on the hardness of the so-called Difference RSA problem [13]). The scheme OT-SH takes four rounds (as the one in [5] based on CA-oblivious encryption), whereas the schemes Gi-SH and Gi×SH take only three rounds (as the original proposal from [1], and the Diffie-Hellman-based scheme from [5]). In fact, these protocols are very similar to Balfanz *et al.*'s original proposal.

Finally, a protocol providing authentication without key exchange is susceptible to an enemy who waits until the authentication is complete and then takes over one end of the communications line. Therefore, we extend the new schemes such that at the end of the handshake, the parties can set up a temporary session key for securing further communication between them.

2   Preliminaries

2.1   Notations

The set of n-bit strings is denoted by $\{0,1\}^n$ and the set of all finite binary strings is denoted by $\{0,1\}^*$. Concatenation of two strings x_1 and x_2 is denoted by $x_1 \| x_2$ and the empty string is denoted ε.

φ denotes the Euler totient function and λ denotes the Carmichael reduced totient function. For any positive integer k, $\mathsf{Prime}(k)$ denotes the set of prime numbers in $[\![2^k, 2^{k+1}]\!]$ and $\mathsf{2Factors}(k)$ the set of integers n such that $n = pq$ with $p < q < 2p$ and $p, q \in \mathsf{Prime}(k)$. An element of $\mathsf{2Factors}(k)$ for any k is called an *RSA-modulus*.

Let $\mathcal{A}$ be a probabilistic Turing machine running in expected polynomial time (a PPTM, for short), and let x be an input for $\mathcal{A}$. The probability space that assigns to a string σ the probability that $\mathcal{A}$, on input x, outputs σ is denoted by $A(x)$. Given a probability space S, a PPTM that samples a random element according to S is denoted by $x \xleftarrow{R} S$. For a finite set X, $x \xleftarrow{R} X$ denotes a PPTM that samples an element uniformly at random from X.

A *two-party protocol* is a pair of interactive probabilistic Turing machines (A, B). An execution of the protocol (A, B) on input x for A and y for B is an alternating sequence of A-rounds and B-rounds, each producing a message m to be delivered to the other party. We denote by "$\rightsquigarrow m$" the transmission of m from one party to the other. The sequence of such message exchanges is called a *transcript* of this execution of the protocol. If, for all x and y, the length of the transcript, as well as the expected running time of A and B on inputs x and y respectively, are polynomial in the length of x and y, then (A, B) is a *polynomial time two-party protocol*.

2.2   Underlying Problems

The security of asymmetric cryptographic tools relies on assumptions about the hardness of certain algorithmic problems. The best known public-key primitive is the RSA function. In order to highlight the fact that our schemes apply to any RSA key generator, we do not pin down any particular generator, but instead parameterize definitions and security results by a choice of generator.

Definition 1. *An* RSA-group generator *is a PPTM that takes a security parameter k as input and outputs a 6-tuple (n, p, q, e, d, g) where $n = pq \in \mathsf{2Factors}(k)$, $e \in [\![3, 2^k]\!]$, $ed \equiv 1 \bmod \varphi(n)$ and $g \in (\mathbb{Z}/n\mathbb{Z})^*$ of order $\lambda(n)$.*

The *RSA assumption* says, roughly speaking, that given a large RSA *modulus* n, an exponent e and α in $(\mathbb{Z}/n\mathbb{Z})^*$, it is hard to find x in $(\mathbb{Z}/n\mathbb{Z})^*$ such that $x^e = \alpha \bmod n$. The *difference RSA assumption* was introduced by Naor in [13]. This non-standard hypothesis deals with the hardness of finding two RSA preimages such that the difference of their images is a given quantity α. The adversary $\mathcal{A}$ has access to a sequence of $m - 1$ pairs $(x_i, y_i) \in [(\mathbb{Z}/n\mathbb{Z})^*]^2$ such that $x_i^e - y_i^e = \alpha \bmod n$ where $\mathcal{A}$ chooses y_i. It should find a new $(x_m, y_m) \in [(\mathbb{Z}/n\mathbb{Z})^*]^2$

such that $x_m^e - y_m^e = \alpha \bmod n$. We denote by $(\alpha + (\cdot)^e)^d$ the oracle that takes input $y \in (\mathbb{Z}/n\mathbb{Z})^*$ and returns $(\alpha+y^e)^d \bmod n$. An adversary solving the difference RSA problem is given oracle access to $(\alpha + (\cdot)^e)^d$.

Definition 2. *Let Gen be an RSA-group generator and let $\mathcal{A}$ be a PPTM. We consider the following random experiments, where $k \in \mathbb{N}$ is a security parameter:*

$Experiment$ $\mathbf{Exp}_{Gen,\mathcal{A}}^{rsa}(k)$	$Experiment$ $\mathbf{Exp}_{Gen,\mathcal{A}}^{diff-rsa}(k)$
$(n,p,q,e,d,g) \xleftarrow{R} Gen(k)$	$(n,p,q,e,d,g) \xleftarrow{R} Gen(k)$
$\alpha \xleftarrow{R} (\mathbb{Z}/n\mathbb{Z})^*$	$\alpha \xleftarrow{R} (\mathbb{Z}/n\mathbb{Z})^*$
$x \xleftarrow{R} \mathcal{A}(n,e,\alpha)$	$(x,y) \xleftarrow{R} \mathcal{A}^{(\alpha+(\cdot)^e)^d}(n,e,y)$
$Return\ 1\ if\ x^e = \alpha \bmod n,$	$Return\ 1\ if\ the\ following\ hold\ and\ 0\ otherwise$
$0\ otherwise$	$(1)\ x^e - y^e = \alpha \bmod n$
	$(2)\ y\ was\ not\ queried\ to\ (\alpha + (\cdot)^e)^d$

We define the success *of $\mathcal{A}$ via $\mathbf{Succ}_{Gen,\mathcal{A}}^{rsa}(k) = \Pr[\mathbf{Exp}_{Gen,\mathcal{A}}^{rsa}(k) = 1]$. (resp. $\mathbf{Succ}_{Gen,\mathcal{A}}^{diff-rsa}(k) = \Pr[\mathbf{Exp}_{Gen,\mathcal{A}}^{diff-rsa}(k) = 1]$).*

Let $q,\tau \in \mathbb{N}^\mathbb{N}$. $\mathcal{A}$ is a a τ-RSA-adversary if for all positive integer k, the experiment $\mathbf{Exp}_{Gen,\mathcal{A}}^{diff-rsa}(k)$ ends in expected time less than $\tau(k)$. $\mathcal{A}$ is a (q,τ)-DIFF-RSA-adversary if for all positive integer k, the experiment $\mathbf{Exp}_{Gen,\mathcal{A}}^{rsa}(k)$ ends in expected time less than $\tau(k)$ and in this experiment $\mathcal{A}$ makes at most $q(k)$ queries to the oracle $(\alpha + (\cdot)^e)^d$.

Let $\varepsilon \in [0,1]^\mathbb{N}$. Gen is said to be (τ,ε)-RSA-secure if for any τ-RSA-adversary $\mathcal{A}$ and any positive integer k, $\mathbf{Succ}_{Gen,\mathcal{A}}^{rsa}(k)$ is smaller than $\varepsilon(k)$. Gen is said to be (q,τ,ε)-DIFF-RSA-secure if for any (q,τ)-DIFF-RSA-adversary $\mathcal{A}$ and any positive integer k, $\mathbf{Succ}_{Gen,\mathcal{A}}^{diff-rsa}(k)$ is smaller than $\varepsilon(k)$.

Finally, the weaker *composite Diffie-Hellman assumption*, related with those described above, is defined as follows:

Definition 3. *Let Gen be an RSA-group generator and let $\mathcal{A}$ be a PPTM. We consider the following random experiments, where $k \in \mathbb{N}$ is a security parameter:*

$Experiment$ $\mathbf{Exp}_{Gen,\mathcal{A}}^{comp-cdh}(k)$
$(n,p,q,e,d,g) \xleftarrow{R} Gen(k)$
$(x,y) \xleftarrow{R} (\llbracket 1, 2^{2k+1} \rrbracket)^2\ ;\ \alpha \leftarrow g^x\ ;\ \beta \leftarrow g^y$
$\gamma \xleftarrow{R} \mathcal{A}(n,g,\alpha,\beta)$
$Return\ 1\ if\ \gamma = g^{xy},\ 0\ otherwise$

Let $\tau \in \mathbb{N}^\mathbb{N}$. $\mathcal{A}$ is a τ-COMP-CDH-adversary if for all positive integer k, the experiment $\mathbf{Exp}_{Gen,\mathcal{A}}^{comp-cdh}(k)$ ends in expected time less than $\tau(k)$. We define the success of $\mathcal{A}$ via $\mathbf{Succ}_{Gen,\mathcal{A}}^{comp-cdh}(k) = \Pr[\mathbf{Exp}_{Gen,\mathcal{A}}^{comp-cdh}(k) = 1]$. Let $\tau \in \mathbb{N}^\mathbb{N}$ and $\varepsilon \in [0,1]^\mathbb{N}$. Gen is said to be (τ,ε)-COMP-CDH-secure if for any τ-COMP-CDH-adversary $\mathcal{A}$ and any positive integer k, $\mathbf{Succ}_{Gen,\mathcal{A}}^{comp-cdh}(k)$ is smaller than $\varepsilon(k)$.

2.3   Ensuring That Handshakes Do Not Reveal the Group

A simple observation that seems to be folklore is that standard RSA does not provide anonymity, even if all *moduli* in the system have the same length. One approach to anonymizing RSA, suggested by Desmedt [7], is to add random multiples of the *modulus* n to the ciphertext. This padding removes any information about the size of n and does not interfere with the reduction of the value modulo n. In the following, we assume that such a technique is adopted and that the adversary gains no information on the RSA *modulus* involved in some protocol (in a statistical sense) from the encoding used in the transcript.

2.4   Proofs of Knowledge of a Discrete Logarithm

In the design of his key agreement protocol, Girault [9] needs a zero-knowledge proof of knowledge of a discrete logarithm in a group of unknown order.

To achieve our security reductions, the executions of the protocol have to be simulated in the ROM. Therefore, in the design of our scheme, we rely on a procedure allowing to prove in a non-transferable way the knowledge of a discrete logarithm without revealing information on their value. We make use of non-interactive designated verifier zero-knowledge proof of knowledge of the discrete logarithm of y in base g (of unknown order). The notation is $\mathrm{DVPK}(a : y = g^a)$. We refer the reader to [9, 10] for further details.

3   Definition of Secret Handshakes

3.1   Syntactic Definition

Roughly speaking, in a secret handshake scheme, there are group authorities having a public key and a matching secret key (CreateGroup). They can provide any user with a pair of keys based on its identity (AddMember) and they manage a certificate revocation list for the group (RemoveMember). The users can then identify themselves in a protocol (Handshake) in which the parties involved begin by knowing only the claimed identities and their own secret key provided by their authority (not necessarily the same). The formal definition of secret handshake schemes proposed by Castelluccia *et al.* and Xu and Yung in [5, 22] is the following:

Definition 4 ([5, 22]). *A secret handshake protocol SH is a 5-tuple (Setup, CreateGroup, AddMember, RemoveMember, Handshake) such that*

- *SH.Setup is a PPTM which takes an integer k as input. The outputs are the public parameters. k is called the security parameter.*
- *SH.CreateGroup is a PPTM which takes the public parameters as input and outputs a pair of group keys $(\mathbf{pk_G}, \mathbf{sk_G})$. It may also output a data structure CRL called a certificate revocation list which is originally empty.*
- *SH.AddMember is a polynomial time two-party protocol (Member, Group) where*

1. *SH.AddMember.Member takes the public parameters, a bit string* ID *and a group public key* $\mathbf{pk_G}$ *as inputs;*
2. *SH.AddMember.Group takes the public parameters,* ID *and the matching group secret key* $\mathbf{sk_G}$ *as inputs.*

SH.AddMember.Member outputs a pair of member keys $(\mathbf{pk}_{ID}, \mathbf{sk}_{ID})$ *associated with* $[\mathbf{pk_G}, ID]$.

- *SH.RemoveMember is a PPTM which takes the public parameters, a bit string* ID, *a group pair of keys* $(\mathbf{pk_G}, \mathbf{sk_G})$ *and the corresponding current CRL as inputs. It outputs an updated CRL which includes the newly revoked certificate* ID.

- *SH.Handshake is a polynomial time two-party protocol (*Init, Match*) where*
    1. *SH.Handshake.Init takes the public parameters, a pair of* (ID_I, ID_M), *a group public key* $\mathbf{pk}_{G_I}$, $(\mathbf{pk}_{ID_I}, \mathbf{sk}_{ID_I})$ *a pair of member keys associated with* $[\mathbf{pk}_{G_I}, ID_I]$ *and a member public key* $\mathbf{pk}_{ID_M}$ *as inputs;*
    2. *SH.Handshake.Match takes the public parameters, the pair* (ID_I, ID_M), *a group public key* $\mathbf{pk}_{G_M}$ *and* $(\mathbf{pk}_{ID_M}, \mathbf{sk}_{ID_M})$ *a pair of member keys associated with* $[\mathbf{pk}_{G_M}, ID_M]$ *and* $\mathbf{pk}_{ID_I}$ *as inputs.*

The algorithms jointly output Accept *if* $\mathbf{pk}_{G_I} = \mathbf{pk}_{G_M} \wedge \{ID_I, ID_M\} \cap CRL = \emptyset$ *and* Reject *otherwise.*

3.2   The Random Oracle Debate

The proof of security for our schemes takes place in the random oracle model, introduced in [2]. In this model, cryptographic protocols are designed and proved secure under the additional assumption that publicly available functions, that are chosen truly at random, exist. These random oracles can only be accessed in a black-box way, by providing an input and obtaining the corresponding output. It has been pointed out that a proof of security in the ROM does not guarantee the existence of an instantiation of the random oracle under which the scheme is secure [4]. However, security proofs in the ROM remain "strong indicators" of the security of an analyzed protocol. Note that, to prove the security of the new schemes, there is no need to assume that all the involved hash functions act as random oracles. Indeed, it is possible to model one of them as a *non-programmable random oracle*. The NPROM is known to be strictly weaker than the ROM [14]. In view of the previous remark, it is a legitimate desire to construct protocols relying as little as possible on random oracles. Therefore, even if our security analysis is carried out in the ROM, we will nevertheless modelize some hash functions as (weaker) non-programmable random oracle.

3.3   Security Requirements

This subsection recalls the security model formally defined in [5, 22]. Roughly speaking, a secret handshake protocol must satisfy the following properties:

1. completeness: if two members engage in the protocol SH.Handshake with valid pair of keys associated with the same group public key, then both parties output Accept at the end of the protocol;

2. impersonator resistance: given a group public key, it is computationally infeasible without the knowledge of some secret key associated with it to successfully execute the protocol SH.Handshake with a member of this group;
3. detector resistance: given a group public key $\mathbf{pk_G}$ and a member public key $\mathbf{pk_{ID}}$, it is computationally infeasible to determine whether $\mathbf{pk_{ID}}$ is associated with $\mathbf{pk_G}$ without the knowledge of a secret key associated with $\mathbf{pk_G}$;
4. indistinguishability to eavesdropper: given two member public keys associated with the same group public key, it is computationally infeasible to distinguish a successful handshake between these members from an unsuccessful one.

In this paper, we consider *active* adversaries against the impersonator resistance (IMP-ACT), the detector resistance (DET-ACT) and the indistinguishability (IND-ACT) of our schemes. The attacker is allowed to run the protocols several times: she can see all exchanged messages, can delete, alter, inject and redirect messages, can initiate communications with another party and can reuse messages from past communications. The adversary is able to trigger several executions of the protocol Handshake. She is also able to interleave these instances, asynchronously and concurrently, with executions with rightful members of the group. The attacker is also allowed to corrupt some users and to ask for additional member keys at any time, and she can do so adaptively.

As usual an adversary is formalized by a PPTM $\mathcal{A}$ that is allowed access to different oracles modeling the capacities mentioned above:

- $\mathcal{O}_{CG}$: it activates a new group authority *via* algorithm SH.CreateGroup. We assume that a group authority is not under $\mathcal{A}$'s control before the new group is established.
- $\mathcal{O}_{AM}$: given as input the identity of a group authority, it executes the protocol SH.AddMember. The algorithm admits an honest user and assigns it with a unique pseudonym ID.
- $\mathcal{O}_{HS}$: given as inputs two pseudonyms ID_I and ID_M, it activates the protocol SH.Handshake between the corresponding members, where none, one or both of them may have been corrupt. A corrupt user will execute what the adversary is pleased of.
- $\mathcal{O}_{RM}$: given as input the identity of a group authority and a bit string ID, it executes SH.RemoveMember to insert ID in the corresponding CRL.
- $\mathcal{O}_{Co}$: given as input the identity of a group authority G and possibly a pseudonym ID associated with G, the oracle returns, to $\mathcal{A}$, the current internal state information (including all secrets) of G or ID's associated member. Once a group or a member is corrupt it will execute what $\mathcal{A}$ is pleased of, until such a corruption is detected. If the corruption of a member is detected, it is excluded from its group via the algorithm SH.RemoveMember. If the corruption of a group is detected, it is excluded from the system.

Resistance to impersonation attacks. This property captures the requirement that an adversary who does not belong to or does not corrupt a member of a group G managed by an uncorrupt group authority, has only a negligible probability of convincing an honest user of G that she is also a member of

G. We consider an adversary $\mathcal{A}$ in the ROM. She takes the public parameters as input, and outputs a triple $(G^\star, \mathsf{ID}^\star, d^\star)$. $\mathcal{A}$ has access to the random oracle(s) $\mathcal{O}_\mathcal{H}$ and to the oracles $\mathcal{O}_{CG}$, $\mathcal{O}_{AM}$, $\mathcal{O}_{HS}$, $\mathcal{O}_{RM}$, and $\mathcal{O}_{Co}$. She succeeds if $\mathsf{ID}^\star$ belongs to $G^\star$; $G^\star$ remains uncorrupt during $\mathcal{A}$'s execution; all corrupt users from $G^\star$ are excluded from $G^\star$ (in particular $\mathsf{ID}^\star$ is uncorrupt) and if the protocol SH.Handshake$(\mathcal{A}, \mathsf{ID}^\star)$ returns Accept when $d^\star = 0$ or if the protocol SH.Handshake$(\mathsf{ID}^\star, \mathcal{A})$ returns Accept when $d^\star = 1$ (i.e. $\mathcal{A}$ initiates the final handshake if and only if $d^\star = 0$).

Definition 5. *Let SH be a secret handshake scheme and let $\mathcal{A}$ be a PPTM. We consider the following random experiment, where $k \in \mathbb{N}$ is a security parameter:*

Experiment $\mathbf{Exp}^{imp\text{-}act}_{SH,\mathcal{A}}(k)$

$\mathcal{P} \xleftarrow{R} SH.\mathsf{Setup}(k)$
$(G^\star, \mathsf{ID}^\star, d^\star) \leftarrow \mathcal{A}^{\mathcal{O}_\mathcal{H}, \mathcal{O}_{CG}, \mathcal{O}_{AM}, \mathcal{O}_{HS}, \mathcal{O}_{RM}, \mathcal{O}_{Co}}(\mathcal{P})$
Return 1 if the following hold and 0 otherwise
   (1) $G^\star$ was obtained by a $\mathcal{O}_{CG}$ query
   (2) $\mathsf{ID}^\star$ was obtained by a $\mathcal{O}_{AM}(G^\star)$ query
   (3) neither $\mathcal{O}_{Co}(G^\star)$ nor $\mathcal{O}_{Co}(G^\star, \mathsf{ID}^\star)$ was queried
   (4) if $\mathcal{O}_{Co}(G^\star, \mathsf{ID})$ for some ID is queried, then $\mathcal{O}_{RM}(G^\star, \mathsf{ID})$ is queried
   (5) if $d^\star = 0$, then SH.Handshake$(\mathcal{A}, \mathsf{ID}^\star)$ returns Accept
      if $d^\star = 1$, then SH.Handshake$(\mathsf{ID}^\star, \mathcal{A})$ returns Accept

---

Let $q_\mathcal{H}, q_{CG}, q_{AM}, q_{HS}, q_{RM}, q_{Co}, \tau \in \mathbb{N}^\mathbb{N}$. $\mathcal{A}$ is a $(q_\mathcal{H}, q_{CG}, q_{AM}, q_{HS}, q_{RM}, q_{Co}, \tau)$-IMP-ACT-adversary if for all positive integer k, the experiment $\mathbf{Exp}^{imp\text{-}act}_{SH,\mathcal{A}}(k)$ ends in expected time less than $\tau(k)$ and in this experiment $\mathcal{A}$ makes at most $q_x(k)$ queries to the oracle $\mathcal{O}_x$ for x in $\{\mathcal{H}, CG, AM, HS, RM, Co\}$. We define the success of the adversary $\mathcal{A}$, via $\mathbf{Succ}^{imp\text{-}act}_{SH,\mathcal{A}}(k) = Pr\left[\mathbf{Exp}^{imp\text{-}act}_{SH,\mathcal{A}}(k) = 1\right]$.

Let $q_\mathcal{H}, q_{CG}, q_{AM}, q_{HS}, q_{RM}, q_{Co}, \tau \in \mathbb{N}^\mathbb{N}$ and let $\varepsilon \in [0,1]^\mathbb{N}$. The scheme SH is said to be $(q_\mathcal{H}, q_{CG}, q_{AM}, q_{HS}, q_{RM}, q_{Co}, \tau, \varepsilon)$-resistant to impersonation against active adversary, if for any $(q_\mathcal{H}, q_{CG}, q_{AM}, q_{HS}, q_{RM}, q_{Co}, \tau)$-IMP-ACT-adversary $\mathcal{A}$ and any positive integer k, the function $\mathbf{Succ}^{imp\text{-}act}_{SH,\mathcal{A}}(k)$ is smaller than $\varepsilon(k)$.

Resistance to detection attacks. This property captures the requirement that an adversary who does not belong to or does not corrupt a member of a group G managed by an uncorrupt group authority, has only a negligible advantage in distinguishing an interplay with an honest member of G from one with a *simulator*. We consider an adversary $\mathcal{A}$ in the random oracle model, which runs in two stages. In the find stage, she takes the public parameters $\mathcal{P}$ as input, and outputs a triple $(G^\star, \mathsf{ID}^\star, d^\star)$, with some state information $\mathcal{I}^\star$. In the guess stage, she gets the information $\mathcal{I}^\star$, executes SH.Handshake with the challenger and outputs a bit $b^\star$.

The adversary $\mathcal{A}$ has access to the random oracle(s) $\mathcal{O}_\mathcal{H}$ and to the oracles $\mathcal{O}_{CG}$, $\mathcal{O}_{AM}$, $\mathcal{O}_{HS}$, $\mathcal{O}_{RM}$, and $\mathcal{O}_{Co}$. In the guess stage, the challenger picks a bit b at random, and the protocol SH.Handshake is executed with $\mathsf{ID}^\star$ if $b = 0$, or executed with a simulator SIM delivering values sampled uniformly at random in the suitable encoding space, if $b = 1$. $\mathcal{A}$ succeeds if $\mathsf{ID}^\star$ belongs to $G^\star$; $G^\star$ remains uncorrupt during $\mathcal{A}$'s execution; all corrupt users from $G^\star$ are excluded from $G^\star$ (in particular $\mathsf{ID}^\star$ is uncorrupt) and if $d^\star = b$.

Definition 6. *Let SH be a secret handshake scheme and let $\mathcal{A}$ be a PPTM. We consider the following random experiment, where $k \in \mathbb{N}$ is a security parameter:*

$$\boxed{\textit{Experiment } \mathbf{Exp}^{\textit{det-act-}b}_{SH,\mathcal{A}}(k)}$$

$\mathcal{P} \xleftarrow{R} SH.Setup(k)$

$(G^\star, \mathsf{ID}^\star, d^\star, \mathcal{I}^\star) \leftarrow \mathcal{A}^{\mathcal{O}_{\mathcal{H}}, \mathcal{O}_{\mathrm{CG}}, \mathcal{O}_{\mathrm{AM}}, \mathcal{O}_{\mathrm{HS}}, \mathcal{O}_{\mathrm{RM}}, \mathcal{O}_{\mathrm{Co}}}(\textit{find}, \mathcal{P})$

$\mathcal{U}_{d^\star} \leftarrow \mathcal{A}^{\mathcal{O}_{\mathcal{H}}, \mathcal{O}_{\mathrm{CG}}, \mathcal{O}_{\mathrm{AM}}, \mathcal{O}_{\mathrm{HS}}, \mathcal{O}_{\mathrm{RM}}, \mathcal{O}_{\mathrm{Co}}}(\textit{find}, \mathcal{I}^\star)$

if $b = 0$ *then* $\mathcal{U}_{1-d^\star} \leftarrow \mathsf{ID}^\star$, *else* $\mathcal{U}_{1-d^\star} \leftarrow \mathsf{SIM}$

Execute $SH.Handshake(\mathcal{U}_0, \mathcal{U}_1)$

$b^\star \leftarrow \mathcal{A}^{\mathcal{O}_{\mathcal{H}}, \mathcal{O}_{\mathrm{CG}}, \mathcal{O}_{\mathrm{AM}}, \mathcal{O}_{\mathrm{HS}}, \mathcal{O}_{\mathrm{RM}}, \mathcal{O}_{\mathrm{Co}}}(\textit{guess}, \mathcal{I}^\star)$

Return 1 if the following hold and 0 otherwise

  (1) $G^\star$ *was obtained by a* $\mathcal{O}_{\mathrm{CG}}$ *query*
  (2) $\mathsf{ID}^\star$ *was obtained by a* $\mathcal{O}_{\mathrm{AM}}(G^\star)$ *query*
  (3) *neither* $\mathcal{O}_{\mathrm{Co}}(G^\star)$ *nor* $\mathcal{O}_{\mathrm{Co}}(G^\star, \mathsf{ID}^\star)$ *was queried*
  (4) *if* $\mathcal{O}_{\mathrm{Co}}(G^\star, \mathsf{ID})$ *for some* ID *is queried, then* $\mathcal{O}_{\mathrm{RM}}(G^\star, \mathsf{ID})$ *is queried*
  (5) $b^\star = b$

Let $q_{\mathcal{H}}, q_{\mathrm{CG}}, q_{\mathrm{AM}}, q_{\mathrm{HS}}, q_{\mathrm{RM}}, q_{\mathrm{Co}}, \tau \in \mathbb{N}^{\mathbb{N}}$. $\mathcal{A}$ *is a* $(q_{\mathcal{H}}, q_{\mathrm{CG}}, q_{\mathrm{AM}}, q_{\mathrm{HS}}, q_{\mathrm{RM}}, q_{\mathrm{Co}}, \tau)$- *DET-ACT-adversary if for all positive integer k, the experiment* $\mathbf{Exp}^{\textit{det-act}}_{SH,\mathcal{A}}(k)$ *ends in expected time less than $\tau(k)$ and in this experiment $\mathcal{A}$ makes at most $q_x(k)$ queries to the oracle $\mathcal{O}_x$ for x in $\{\mathcal{H}, \mathrm{CG}, \mathrm{AM}, \mathrm{HS}, \mathrm{RM}, \mathrm{Co}\}$. We define the advantage of the adversary $\mathcal{A}$, via*

$$\mathbf{Adv}^{\textit{det}-\textit{act}}_{SH,\mathcal{A}}(k) = \left| Pr\left[\mathbf{Exp}^{\textit{det}-\textit{act}-0}_{SH,\mathcal{A}}(k) = 1\right] - Pr\left[\mathbf{Exp}^{\textit{det}-\textit{act}-1}_{SH,\mathcal{A}}(k) = 1\right] \right|.$$

Let $q_{\mathcal{H}}, q_{\mathrm{CG}}, q_{\mathrm{AM}}, q_{\mathrm{HS}}, q_{\mathrm{RM}}, q_{\mathrm{Co}}, \tau \in \mathbb{N}^{\mathbb{N}}$ *and let* $\varepsilon \in [0, 1]^{\mathbb{N}}$. *The scheme SH is said to be* $(q_{\mathcal{H}}, q_{\mathrm{CG}}, q_{\mathrm{AM}}, q_{\mathrm{HS}}, q_{\mathrm{RM}}, q_{\mathrm{Co}}, \tau, \varepsilon)$-*resistant to detection against active adversaries, if for any* $(q_{\mathcal{H}}, q_{\mathrm{CG}}, q_{\mathrm{AM}}, q_{\mathrm{HS}}, q_{\mathrm{RM}}, q_{\mathrm{Co}}, \tau)$-*DET-ACT-adversary $\mathcal{A}$ and any positive integer k, the function* $\mathbf{Adv}^{\textit{det}-\textit{act}}_{SH,\mathcal{A}}(k)$ *is smaller than $\varepsilon(k)$.*

Indistinguishability to eavesdropper. This property captures the requirement that an adversary has only a negligible advantage of distinguishing a successful handshake between uncorrupt members of a group G managed by an uncorrupt group authority from an unsuccessful one. We consider an adversary $\mathcal{A}$ in the ROM, which runs in two stages. In the `find` stage, she takes the public parameters $\mathcal{P}$ as input, and outputs a triple $(G^\star, \mathsf{ID}^\star_I, \mathsf{ID}^\star_M)$ with some state information $\mathcal{I}^\star$. In the `guess` stage, she gets the information $\mathcal{I}^\star$ and $\mathcal{Y}^\star$, generated by the challenger depending on a random bit b. If $b = 0$, then $\mathcal{Y}^\star$ is the transcript of an execution of $SH.Handshake$ between $(\mathsf{ID}^\star_I, \mathsf{ID}^\star_M)$, otherwise $\mathcal{Y}^\star$ is a bit string sampled uniformly at random in the transcript space. Eventually $\mathcal{A}(\mathbf{guess})$ outputs a bit $d^\star$.

The adversary $\mathcal{A}$ has access to the random oracle(s) $\mathcal{O}_\mathcal{H}$ and to the oracles $\mathcal{O}_{CG}$, $\mathcal{O}_{AM}$, $\mathcal{O}_{HS}$, $\mathcal{O}_{RM}$, and $\mathcal{O}_{Co}$. It succeeds if $d^\star = b$, if $\mathsf{ID}_I^\star$ and $\mathsf{ID}_M^\star$ belongs to $G^\star$, and if $G^\star$, $\mathsf{ID}_I^\star$ and $\mathsf{ID}_M^\star$ remain uncorrupt during $\mathcal{A}$'s execution.

Definition 7. *Let SH be a secret handshake scheme and let $\mathcal{A}$ be a PPTM. We consider the following random experiment, where $k \in \mathbb{N}$ is a security parameter:*

$\boxed{Experiment\ \mathbf{Exp}_{SH,\mathcal{A}}^{ind\text{-}act\text{-}b}(k)}$

$\mathcal{P} \xleftarrow{R} SH.\mathsf{Setup}(k)$

$(G^\star, \mathsf{ID}_I^\star, \mathsf{ID}_M^\star, \mathcal{I}^\star) \leftarrow \mathcal{A}^{\mathcal{O}_\mathcal{H}, \mathcal{O}_{CG}, \mathcal{O}_{AM}, \mathcal{O}_{HS}, \mathcal{O}_{RM}, \mathcal{O}_{Co}}(\mathcal{P})$

if $b = 0$ then $\Upsilon^\star \leftarrow SH.\mathsf{Handshake}(\mathsf{ID}_I^\star, \mathsf{ID}_M^\star)$ *else* $\Upsilon^\star \xleftarrow{R} \mathsf{TranscriptSpace}$

Return 1 if the following hold and 0 otherwise
  (1) $G^\star$ *was obtained by a* $\mathcal{O}_{CG}$ *query*
  (2) $\mathsf{ID}_I^\star$ *and* $\mathsf{ID}_M^\star$ *were obtained by a* $\mathcal{O}_{AM}(G^\star)$ *query*
  (3) *neither* $\mathcal{O}_{Co}(G^\star)$ *nor* $\mathcal{O}_{Co}(G^\star, \mathsf{ID}^\star)$ *was queried*
  (4) $d^\star = b$

Let $q_\mathcal{H}, q_{CG}, q_{AM}, q_{HS}, q_{RM}, q_{Co}, \tau \in \mathbb{N}^\mathbb{N}$. $\mathcal{A}$ is a $(q_\mathcal{H}, q_{CG}, q_{AM}, q_{HS}, q_{RM}, q_{Co}, \tau)$-IND-ACT-adversary if for all positive integer k, the experiment $\mathbf{Exp}_{SH,\mathcal{A}}^{ind\text{-}act\text{-}b}(k)$ ($b \in \{0, 1\}$) ends in expected time less than $\tau(k)$ and in this experiment $\mathcal{A}$ makes at most $q_x(k)$ queries to the oracle $\mathcal{O}_x$ for x in $\{\mathcal{H}, CG, AM, HS, RM, Co\}$. We define the advantage of the adversary $\mathcal{A}$, via

$$\mathbf{Adv}_{SH,\mathcal{A}}^{ind-act}(k) = \left| Pr\left[\mathbf{Exp}_{SH,\mathcal{A}}^{ind-act-real}(k) = 1 \right] - Pr\left[\mathbf{Exp}_{SH,\mathcal{A}}^{ind-act-random}(k) = 1 \right] \right|.$$

Let $q_\mathcal{H}, q_{CG}, q_{AM}, q_{HS}, q_{RM}, q_{Co}, \tau \in \mathbb{N}^\mathbb{N}$ and let $\varepsilon \in [0, 1]^\mathbb{N}$. The scheme SH is said to be $(q_\mathcal{H}, q_{CG}, q_{AM}, q_{HS}, q_{RM}, q_{Co}, \tau, \varepsilon)$-indistinguishable against active adversaries, if for any $(q_\mathcal{H}, q_{CG}, q_{AM}, q_{HS}, q_{RM}, q_{Co}, \tau)$-IND-ACT-adversary $\mathcal{A}$ and any positive integer k, the function $\mathbf{Adv}_{SH,\mathcal{A}}^{ind-act}(k)$ is smaller than $\varepsilon(k)$.

4   The New Protocols

Let Gen be an RSA-group generator. For any integer $k \in \mathbb{N}$ (*resp.* $r \in \mathbb{N}$), let $[\{0, 1\}^* \longrightarrow \{0, 1\}^k]$ (*resp.* $[\{0, 1\}^* \longrightarrow \{0, 1\}^r]$) be a hash function family and $\mathcal{H}_k$ (*resp.* $\mathcal{G}_r$) be a random member of this family. Let $f : \mathbb{N} \longrightarrow \mathbb{N}$, we denote in the following $r = f(k)$ (this map is derived from the security proofs; in practice, for a security requirement of 2^{80} operations, we use the values $k = 1024$ and $r = 160$). For all tuple (n, p, q, e, d, g) output by Gen, and all bit strings ID in $\{0, 1\}^*$, with overwhelming probability the integer x, whose binary encoding is $\mathcal{H}_k(\mathsf{ID})$, is invertible modulo n. For the sake of simplification, we suppose that this always happens. Moreover, let us recall that an appropriate mechanism is used in order to ensure that encodings of integers modulo n do not reveal the value of n.

4.1 Heuristic of the Constructions

As described above, the adversaries are afforded enormous power. Therefore, it appears prudent to follow general principles in the design of secret handshake protocols (paraphrased from [8]):

- **asymmetry principle:** *asymmetry in a protocol is desirable.*
  Symmetries in a protocol should be used with caution, due to both the possibility of *reflection attacks,* and attacks in which responses from one party can be re-used within a protocol.
- **chronology principle:** *messages within a particular protocol run should be logically linked or "chained" in some manner.*
  This principle is aimed at precluding *replay attacks* and *interleaving attacks.*
- **"add your own salt" principle:** *a party should be able to incorporate into the data being operated on a reasonable amount of data which he himself randomly selects.*
  In other words, a protocol should not require a party to carry out a cryptographic operation on inputs which may be entirely under the control of an adversary. The objective is to prevent the so called *chosen-ciphertext attacks.*

Moreover, mutual authentication is hardly an end in itself: a party goes through a secret handshake protocol in order to then conduct some transaction that is allowed only to member of his group. In an open setting, the authentication by itself is largely useless because an adversary can "hijack" the ensuing session. Therefore, to have secure transactions, some information from the handshake protocol must be used to authenticate flows in the transaction. We will come back to this issue in section 7, but note *hic et nunc* that this session key exchange should be designed with caution:

- **link principle:** *authentication and key exchange should be linked.*
  If mutual authentication and key exchange are independent, then an attacker could allow two parties to carry out authentication unhindered, and could take over one's party role in key exchange.

4.2 Description of OT-SH

Okamoto and Tanaka's Identity-Based Key Agreement Scheme. The identity-based key exchange protocol proposed by Okamoto and Tanaka in 1989 [16] is described as follows: on input a security parameter k, a trusted authority (the Private Key Generator, PKG) runs the RSA-group generator Gen, publishes (n, e, g), and keeps (p, q, d) secret. Let Alice and Bob be two entities of which identification information is ID_A and ID_B, respectively.

Alice keeps $s_A = \mathcal{H}_k(ID_A)^{-d} \bmod n$ computed by the PKG as her own secret (and so does Bob with $s_B = \mathcal{H}_k(ID_B)^{-d} \bmod n$). The scheme is then basically the Diffie-Hellman scheme implemented over $(\mathbb{Z}/n\mathbb{Z})^*$ (see figure 1). The security of this scheme remained open until 1998 when a solution was given by Mambo and Shizuya [11].

$$\text{Alice} \qquad\qquad\qquad\qquad\qquad\qquad \text{Bob}$$

$$k_A \xleftarrow{R} [\![1, n]\!]$$
$$c_A \leftarrow s_A g^{k_A} \bmod n \xrightarrow{\qquad c_A \qquad}$$

$$\xleftarrow{\qquad c_B \qquad} \quad k_B \xleftarrow{R} [\![1, n]\!]$$
$$\qquad\qquad\qquad\qquad\qquad\qquad c_B \leftarrow s_B g^{k_B} \bmod n$$
$$K = (\mathcal{H}_k(\mathsf{ID}_B) c_B^e)^{k_A} \bmod n \qquad\qquad K = (\mathcal{H}_k(\mathsf{ID}_A) c_A^e)^{k_B} \bmod n$$

Fig. 1. Description of Okamoto-Tanaka identity-based key exchange protocol

Description of OT-SH. Our first new secret handshake scheme OT-SH is a simple variant of the previous protocol. The Setup algorithm establishes as public parameters the RSA-group generator and the hash functions $\mathcal{H}_k$ and $\mathcal{G}_r$. The algorithm CreateGroup is identical to the PKG key generation and the protocol AddMember to a user registration. The 4-round protocol Handshake is designed thanks to appropriate used of nonces and hash values (following the principles exemplified in the previous section) in the key exchange protocol. The scheme is described with all details in figure 2.

Algorithm Setup	Algorithm CreateGroup	Protocol AddMember
Input: $k \in \mathbb{N}$	Input: $\mathcal{P}$	Common Input: $\mathcal{P}, \mathbf{pk_G} = (n, e, g), \mathsf{ID}$
Output: $\mathcal{P}$	Output: $(\mathbf{pk_G}, \mathbf{sk_G})$	Member Input: ε
$\mathcal{P} = (k, \mathsf{Gen}, \mathcal{H}_k)$	$(n, p, q, e, d, g) \xleftarrow{R} \mathsf{Gen}(k)$	Group Input: $\mathbf{sk_G} = d$
	$\mathbf{pk_G} \leftarrow (n, e, g)\ \mathbf{sk_G} \leftarrow d$	Output: $(\mathbf{pk_{ID}}, \mathbf{sk_{ID}})$
		$\mathbf{pk_{ID}} \leftarrow \varepsilon$
		$\mathbf{sk_{ID}} \leftarrow (\mathcal{H}_k(\mathsf{ID}))^{-d} \bmod n$

Protocol Handshake

Common Input: $\mathcal{P}$, $\mathsf{ID_I}$, $\mathsf{ID_M}$, $\mathbf{pk_{ID_I}}$, $\mathbf{pk_{ID_M}}$

Init Input: $\mathbf{pk_{G_I}} = (n_I, e_I, g_I)$, $\mathbf{sk_{ID_I}}$ Match Input: $\mathbf{pk_{G_M}} = (n_M, e_M, g_M)$, $\mathbf{sk_{ID_M}}$

Output: $(b_I, b_M) \in \{\mathtt{Accept}, \mathtt{Reject}\}^2$

① Init's round: $k_I \xleftarrow{R} \{0, 1\}^{n_I}$, $c_I \leftarrow \mathbf{sk_{ID_I}} \cdot g_I^{k_I} \bmod n_I$ $\qquad\qquad \rightsquigarrow c_I$

❶ Match's round: $k_M \xleftarrow{R} \{0, 1\}^{n_M}$, $c_M \leftarrow \mathbf{sk_{ID_M}} \cdot g_M^{k_M} \bmod n_M$ $\qquad \rightsquigarrow c_M$

② Init's round: $V_I \leftarrow \mathcal{G}_r \left[(\mathcal{H}_k(\mathsf{ID_M}) \cdot c_M^{e_I})^{k_I} \bmod n_I \| \mathsf{ID_I} \| \mathsf{ID_M} \| c_I \| c_M \| 0 \right]$ $\qquad \rightsquigarrow V_I$

❷ Match's round

$\qquad$ If $\mathcal{G}_r \left[(\mathcal{H}_k(\mathsf{ID_I}) \cdot c_I^{e_M})^{k_M} \bmod n_M \| \mathsf{ID_I} \| \mathsf{ID_M} \| c_I \| c_M \| 0 \right] = V_I$

$\qquad$ then $b_M = \mathtt{Accept}$; $V_M \leftarrow \mathcal{G}_r((\mathcal{H}_k(\mathsf{ID_I}) \cdot c_I^{e_M})^{k_M} \bmod n_M \| \mathsf{ID_I} \| \mathsf{ID_M} \| c_I \| c_M \| 1)$

$\qquad$ else $b_M = \mathtt{Reject}$; $V_M \xleftarrow{R} \{0, 1\}^r$ $\qquad\qquad\qquad\qquad\qquad\qquad \rightsquigarrow V_M$

• Init's execution ending

$\qquad$ If $\mathcal{G}_r \left[(\mathcal{H}_k(\mathsf{ID_M}) \cdot c_M^{e_I})^{k_I} \bmod n_I \| \mathsf{ID_I} \| \mathsf{ID_M} \| c_I \| c_M \| 1 \right] = V_M$

$\qquad$ then $b_I \leftarrow \mathtt{Accept}$ else $b_I \leftarrow \mathtt{Reject}$

Fig. 2. Description of the scheme OT-SH

4.3   Description of Gi+SH and Gi×SH

Girault's Self-certified Key Agreement Scheme. In 1991, Girault [9] proposed a key agreement which allows to obtain a shared secret key without exchange of messages (but it does require that the public keys of each party are known to the other), based on his new idea of self-certified public key. Rigorous security of this protocol (and variants) has been made clear only recently [15] by Oh, Mambo, Shizuya and Won. With the notations from section 4.2, Alice picks uniformly at random $s_A \in \mathbb{Z}/n\mathbb{Z}$, computes $S = g^{s_A} \bmod n$ and sends S_A together with a designated verifier proof of knowledge of s_A to the authority. If the proof is valid, the authority computes Alice's public key P_A as the RSA signature of the difference (*resp.* quotient) of S_A and $\mathcal{H}_k(\mathsf{ID}_A)$: $P_A = (S_A - \mathcal{H}_k(\mathsf{ID}_A))^d \bmod n$ (*resp.* $P_A = (S_A/\mathcal{H}_k(\mathsf{ID}_A))^d \bmod n$), therefore $g^{s_A} = P_A^e + \mathcal{H}_k(\mathsf{ID}_A) \bmod n$ (*resp.* $g^{s_A} = P_A^e \cdot \mathcal{H}_k(\mathsf{ID}_A) \bmod n$). Similarly, Bob's public key is P_B such that $g^{s_B} = P_B^e + \mathcal{H}_k(\mathsf{ID}_B) \bmod n$ (*resp.* $g^{s_B} = P_B^e \cdot \mathcal{H}_k(\mathsf{ID}_B) \bmod n$) and s_B is known only to Bob. Alice and Bob can thereafter simply exchange an authenticated key by choosing

$$K = (P_A^e + \mathcal{H}_k(\mathsf{ID}_A))^{s_B} = (P_B^e + \mathcal{H}_k(\mathsf{ID}_B))^{s_A} \quad \bmod n$$
$$\left(resp.\ K = (P_A^e \cdot \mathcal{H}_k(\mathsf{ID}_A))^{s_B} = (P_B^e \cdot \mathcal{H}_k(\mathsf{ID}_B))^{s_A} \quad \bmod n\right)$$

This protocol is clearly related to Diffie-Hellman's one, but contrary to it, makes Alice sure that she shares K with Bob and conversely.

Description of Gi+SH and Gi×SH. The new secret handshake schemes Gi-SH and Gi×SH are completely analogous to the one proposed in [1] by Balfanz *et al.*: the non-interactive key-agreement protocol of Sakai *et al.* being replaced by Girault's additive and multiplicative scheme (respectively). They are described in figure 3.

Remark 1. Note that, in the schemes Gi⊕SH (with ⊕ = + or ⊕ = ×), after the run of the protocol Gi⊕SH.AddMember, the group authority does not learn the secret key of the member and thus in contrast to the previous scheme, cannot impersonate that member. However, in [18], Saeednia described a shortcoming of Girault's self-certified model that may be exploited by the authority to compute users' secret keys. This attack applies as well on Gi⊕SH, but it is easy to make this attack ineffective by taking additional precautions. This might be interesting for high-security needs (*e.g.* in traitor tracing).

5   Security Results

The following theorems state that the advantage of any active impersonator, detector or distinguisher adversary against the schemes OT-SH and Gi⊕SH in the ROM can be upper-bounded *via* an explicit function related to the success of solving the RSA problem, the difference RSA problem or the composite Diffie-Hellman problem in Gen. More precisely, the hash function $\mathcal{H}_k$ is modeled by a random oracle, while the hash function $\mathcal{G}_r$ is modeled by a non-programmable random oracle (more precisely, we only suppose that $\mathcal{G}_r$ has the *evaluation point knowledge* (EPK) property: it is not possible to learn the value of what $\mathcal{G}_r(\sigma)$ without knowing the bit string σ).

Algorithm Setup	Algorithm CreateGroup
Input: $k \in \mathbb{N}$	Input: $\mathcal{P}$
Output: $\mathcal{P}$	Output: $(\mathbf{pk_G}, \mathbf{sk_G})$
$\mathcal{P} \leftarrow (k, \mathrm{Gen}, \mathcal{H}_k)$	$(n, p, q, e, d, g) \xleftarrow{R} \mathrm{Gen}(k)$; $\mathbf{pk_G} \leftarrow (n, e, g)$; $\mathbf{sk_G} \leftarrow d$

Protocol AddMember

Common Input: $\mathcal{P}, \mathbf{pk_G}, \mathsf{ID}$ Group Input: $\mathbf{sk_G}$

Output: $(\mathbf{pk_{ID}}, \mathbf{sk_{ID}})$

① Member's round: $\mathbf{sk_{ID}} \xleftarrow{R} [\![2^k, 2^{k+1}]\!]$, $S \leftarrow g^{\mathbf{sk_{ID}}}$, $\mathsf{p} \xleftarrow{R} \mathrm{DVPK}(\gamma | S = g^\gamma) \rightsquigarrow (S, \mathsf{p})$

❶ Group's round: Verify p ; $\mathbf{pk_{ID}} \leftarrow (S \ominus \mathcal{H}_k(\mathsf{ID}))^d \bmod n$ $\rightsquigarrow \mathbf{pk_{ID}}$

Protocol Handshake

Common Input: $\mathcal{P}, \mathsf{ID_I}, \mathsf{ID_M}, \mathbf{pk_{ID_I}}, \mathbf{pk_{ID_M}}$

Init Input: $\mathbf{pk_{G_I}} = (n_I, e_I, g_I)$, $\mathbf{sk_{ID_I}}$

Match Input: $\mathbf{pk_{G_M}} = (n_M, e_M, g_M)$, $\mathbf{sk_{ID_M}}$

Output: $(b_I, b_M) \in \{\mathtt{Accept}, \mathtt{Reject}\}^2$

① Init's round: $r_I \xleftarrow{R} \{0,1\}^r$ $\rightsquigarrow r_I$

❶ Match's round: $r_M \xleftarrow{R} \{0,1\}^r$;

  $V_M \leftarrow \mathcal{G}_r \left[(\mathbf{pk_{ID_I}}^{e_M} \oplus \mathcal{H}_k(\mathsf{ID_I}))^{\mathbf{sk_{ID_M}}} \bmod n_M \| \mathsf{ID_I} \| \mathsf{ID_M} \| r_I \| r_M \| 0 \right]$ $\rightsquigarrow (r_M, V_M)$

② Init's round: $K_I \leftarrow (\mathbf{pk_{ID_M}}^{e_I} \oplus \mathcal{H}_k(\mathsf{ID_M}))^{\mathbf{sk_{ID_I}}} \bmod n_I$

  If $\mathcal{G}_r [K_I \| \mathsf{ID_I} \| \mathsf{ID_M} \| r_I \| r_M \| 0] = V_M$

    then $b_I \leftarrow \mathtt{Accept}$; $V_I \leftarrow \mathcal{G}_r [K_I \| \mathsf{ID_I} \| \mathsf{ID_M} \| r_I \| r_M \| 1]$

    else $b_I \leftarrow \mathtt{Reject}$; $V_I \xleftarrow{R} \{0,1\}^r$ $\rightsquigarrow V_I$

• Match's execution ending

  If $\mathcal{G}_r \left[(\mathbf{pk_{ID_I}}^{e_M} \oplus \mathcal{H}_k(\mathsf{ID_I}))^{\mathbf{sk_{ID_M}}} \bmod n_M \| \mathsf{ID_I} \| \mathsf{ID_M} \| r_I \| r_M \| 1 \right] = V_I$

    then $b_M \leftarrow \mathtt{Accept}$; else $b_M \leftarrow \mathtt{Reject}$

Fig. 3. Description of the schemes Gi+SH and Gi×SH $((\oplus, \ominus) = (+, -)$ and $(\times, /))$

In all cases but two, the proof is more or less routine: the main difficulty of the analysis comes from the introduction of the RSA challenge into the public base. Indeed in the reductionist security proofs, the element g, supposed to be of maximal order in $(\mathbb{Z}/n\mathbb{Z})^*$, is replaced by a random element of $(\mathbb{Z}/n\mathbb{Z})^*$. Lemma 1 asserts that with probability close to $\exp(-1)$, among $(k+1)^2$ such random elements there is an element with order $\lambda(n)$.

Lemma 1. *Let $k \geq 1$ be an integer, and let $n \in 2\mathbf{Factors}(k)$. If an integer g is drawn uniformly at random from $[\![1, n]\!]$, then $\Pr[\mathrm{ord}(g) = \lambda(n)] > 1/(110 \ln k)$.*

Proof. This lemma is a simple consequence of [17, Theorem 15]. □

We illustrate the application of this lemma in the proof of the impersonation resistance of OT-SH (*cf.* Theorem 1). The impersonation resistance (as the detection resistance) of Gi+SH rely on the difference RSA problem. Since this is a less classical assumption, we carry out the arguments in full detail (*cf.* Theorem 2). The remaining four security results are left to the reader and will follow similar proofs for the two given here. In the theorems, $T_{\mathrm{Exp}}(k)$ denotes the time complexity for an exponentiation modulo a $2k + 1$-bit integer.

5.1 Security of the Scheme OT-SH

Theorem 1. *Let Gen be an RSA-group generator, let $f : \mathbb{N} \to \mathbb{N}$ and let OT-SH be the secret handshake protocol instantiated with Gen and f. Let $q_\mathcal{H}, q_{\mathrm{CG}}, q_{\mathrm{AM}}, q_{\mathrm{HS}}, q_{\mathrm{RM}}, q_{\mathrm{Co}}, \tau \in \mathbb{N}^{\mathbb{N}}$.*

1. *Let $\mathcal{A}$ be a $(q_\mathcal{H}, q_{\mathrm{CG}}, q_{\mathrm{AM}}, q_{\mathrm{HS}}, q_{\mathrm{RM}}, q_{\mathrm{Co}}, \tau)$-IMP-ACT-adversary against OT-SH, in the ROM. There exist a τ'-RSA adversary $\mathcal{B}$ such that*

$$\begin{cases} \mathbf{Succ}^{\mathrm{rsa}}_{Gen,\mathcal{B}}(k) \geq \mathbf{Succ}^{\mathrm{imp-act}}_{OT\text{-}SH,\mathcal{A}}(k)/(110\ln(k) \cdot q_{\mathrm{CG}}(k)q_\mathcal{H}(k)) \\ \tau'(k) \leq [\tau(k) + (q_\mathcal{H}(k) + q_{\mathrm{CG}}(k) + q_{\mathrm{AM}}(k) + 3q_{\mathrm{HS}}(k) + O(1))T_{\mathrm{Exp}}(k)] \end{cases}$$

   for all positive integers k.

2. *Let $\mathcal{A}$ be a $(q_\mathcal{H}, q_{\mathrm{CG}}, q_{\mathrm{AM}}, q_{\mathrm{HS}}, q_{\mathrm{RM}}, q_{\mathrm{Co}}, \tau)$-DET-ACT-adversary against OT-SH, in the ROM. There exist a τ'-RSA adversary $\mathcal{B}$ such that*

$$\begin{cases} \mathbf{Succ}^{\mathrm{rsa}}_{Gen,\mathcal{B}}(k) \geq \mathbf{Adv}^{\mathrm{det-act}}_{OT\text{-}SH,\mathcal{A}}(k)/(220\ln(k) \cdot q_{\mathrm{CG}}(k)q_\mathcal{H}(k)) \\ \tau'(k) \leq [\tau(k) + (q_\mathcal{H}(k) + q_{\mathrm{CG}}(k) + q_{\mathrm{AM}}(k) + 3q_{\mathrm{HS}}(k) + O(1))T_{\mathrm{Exp}}(k)] \end{cases}$$

   for all positive integers k.

3. *Let $\mathcal{A}$ be a $(q_\mathcal{H}, q_{\mathrm{CG}}, q_{\mathrm{AM}}, q_{\mathrm{HS}}, q_{\mathrm{RM}}, q_{\mathrm{Co}}, \tau)$-IND-ACT-adversary against OT-SH, in the ROM, making at most $q_\mathcal{G}(k)$ queries to the non-programmable random oracle $\mathcal{O}_\mathcal{G}$ in the experiments $\mathbf{Exp}^{\mathrm{ind\text{-}act\text{-}b}}_{OT\text{-}SH,\mathcal{A}}(k)$ (for $b \in \{0,1\}$ and $k \in \mathbb{N}$). There exist a τ'-COMP-CDH adversary $\mathcal{B}$ such that*

$$\begin{cases} \mathbf{Succ}^{\mathrm{comp\text{-}cdh}}_{Gen,\mathcal{B}}(k) \geq \mathbf{Adv}^{\mathrm{ind\text{-}act}}_{OT\text{-}SH,\mathcal{A}}(k)/[2q_\mathcal{G}(k)] \\ \tau'(k) \leq \tau(k) + (q_\mathcal{H}(k) + 3q_{\mathrm{HS}}(k) + O(1))T_{\mathrm{Exp}}(k) \end{cases}$$

   for all positive integers k.

Proof. We prove only the first part of the theorem. Our method of proof is inspired by Shoup [21]: we define a sequence of games $\mathrm{Game}_0, \dots, \mathrm{Game}_5$ starting from the actual adversary $\mathcal{A}$ and modify it step by step until we reach a final game whose success probability has an upper bound related to solving the RSA problem. All the games operate on the same underlying probability space: the setup, the public and private keys of the groups and the members, the coin tosses of $\mathcal{A}$ and the random oracle.

Let k be a security parameter, let (n, p, q, e, d, g) be a 6-tuple generated by $Gen(k)$ and let $\alpha \in (\mathbb{Z}/n\mathbb{Z})^*$ be a random instance of the RSA problem. We construct a reduction algorithm $\mathcal{B}$ which on input (n, e, α) outputs an element $x \in (\mathbb{Z}/n\mathbb{Z})^*$ aimed to satisfy $x^e = \alpha \bmod n$.

Game_0 We consider a $(q_\mathcal{H}, q_{\mathrm{CG}}, q_{\mathrm{AM}}, q_{\mathrm{HS}}, q_{\mathrm{RM}}, q_{\mathrm{Co}}, \tau)$-IMP-ACT-adversary $\mathcal{A}$, against the scheme OT-SH instantiated with the RSA-group generator Gen, in the random oracle model. $\mathcal{A}$ takes the public parameters as input, and outputs a triple $(G^\star, \mathsf{ID}^\star, d^\star)$. $\mathcal{A}$ has access to the random oracle $\mathcal{O}_\mathcal{H}$ and to the oracles $\mathcal{O}_{CG}$, $\mathcal{O}_{AM}$, $\mathcal{O}_{HS}$, $\mathcal{O}_{RM}$, and $\mathcal{O}_{Co}$. $\mathcal{A}$ succeeds if $\mathsf{ID}^\star$ belongs to $G^\star$, $G^\star$ remains uncorrupt during $\mathcal{A}$'s execution, all corrupt users from $G^\star$ are excluded from $G^\star$ and if in the protocol OT-SH.Handshake$(\mathcal{A}, \mathsf{ID}^\star)$ (*resp.* OT-SH.Handshake$(\mathsf{ID}^\star, \mathcal{A})$), the member $\mathsf{ID}^\star$ returns Accept when $d^\star = 0$ (*resp.* when $d^\star = 1$).

In any game Game_i ($i \in [\![1, 5]\!]$), we denote by Imp_i this event. Without loss of generality we can suppose that any time $\mathcal{A}$ makes a query involving a pseudonym ID to one of the oracles $\mathcal{O}_{CG}$, $\mathcal{O}_{AM}$, $\mathcal{O}_{HS}$, $\mathcal{O}_{RM}$, and $\mathcal{O}_{Co}$, $\mathcal{A}$ has previously queried ID to the random oracle $\mathcal{O}_\mathcal{H}$. In particular, we suppose that $\mathcal{A}$ has queried $\mathsf{ID}^\star$ and $\mathsf{ID}_\mathcal{A}$ (the pseudonym used by $\mathcal{A}$ in the final execution of OT-SH.Handshake) to the random oracle $\mathcal{O}_\mathcal{H}$. By definition, we have $\Pr[\mathsf{Imp}_0] = \mathbf{Succ}^{imp-act}_{OT\text{-}SH, \mathcal{A}}(k)$.

Game_1 The algorithm $\mathcal{B}$ picks uniformly at random an index $\ell_1 \in [\![1, q_{CG}(k)]\!]$ and aborts if the group $G^\star$ was not obtained at the ℓ_1's query to the oracle $\mathcal{O}_{CG}$. We obtain $\Pr[\mathsf{Imp}_1] = \Pr[\mathsf{Imp}_0]/q_{CG}(k)$.

Game_2 $\mathcal{B}$ simulates the $\mathcal{O}_{CG}$ oracle. For the i-th query ($i \in [\![1, q_{CG}(k)]\!] \setminus \{\ell_1\}$), $\mathcal{B}$ executes $\mathsf{Gen}(k)$ and gets $(n_i, p_i, q_i, e_i, d_i, g_i)$. It outputs (n_i, e_i, g_i) as the answer to $\mathcal{A}$'s i-th query for the group denoted by G_i and stores d_i. For the ℓ_1-th query, $\mathcal{B}$ picks uniformly at random $r' \in (\mathbb{Z}/n\mathbb{Z})^*$, computes $r = r^e \bmod n$ and outputs $(n, e, (r\alpha)^e)$. We denote by Λ the event that $(r\alpha)^e$ is of order $\lambda(n)$ in $(\mathbb{Z}/n\mathbb{Z})^*$. If Λ happens, then the distribution of the group authorities keys is unchanged. Therefore, we have $\Pr[\mathsf{Imp}_2 | \Lambda] = \Pr[\mathsf{Imp}_1]$.

Game_3 The algorithm $\mathcal{B}$ picks uniformly at random an index $\ell_2 \in [\![1, q_\mathcal{H}(k)]\!]$. For the i-th bit string ID queried to $\mathcal{O}_\mathcal{H}$, $\mathcal{B}$ picks at random $s_{\mathsf{ID}} \in (\mathbb{Z}/n\mathbb{Z})^*$ and $\mathcal{B}$ sets $h_{\mathsf{ID}} = s_{\mathsf{ID}}^{e^2}$ if $i \neq \ell_2$, otherwise it sets $h_{\mathsf{ID}} = s_{\mathsf{ID}}^e \cdot \alpha$. $\mathcal{B}$ returns h_{ID} as the answer to the oracle call and stores the 4-tuple $(\mathsf{ID}, h_{\mathsf{ID}}, s_{\mathsf{ID}}, i)$ in the H_k-List. $\mathcal{B}$ discards execution such that $\mathsf{ID}_\mathcal{A}$ was not the ℓ_2-th query to the oracle $\mathcal{H}$. We obtain $\Pr[\mathsf{Imp}_3] = \Pr[\mathsf{Imp}_2]/q_\mathcal{H}(k)$.

Game_4 Now, $\mathcal{B}$ simulates the $\mathcal{O}_{AM}$ and the $\mathcal{O}_{Co}$ oracles. The queries $\mathcal{O}_{AM}(G_i)$ for $i \in [\![1, q_{CG}(k)]\!]$ are easily simulated since OT-SH.AddMember outputs only a secret key associated with a pseudonym.

For all $i \in [\![1, q_{CG}(k)]\!] \setminus \{\ell_1\}$, thanks to the knowledge of d_i, $\mathcal{B}$ can perfectly simulate the member key generation protocol for the group G_i, and therefore can reply to $\mathcal{A}$'s queries involving members of G_i. For a member of $G^\star$ with pseudonym ID queried to the $\mathcal{O}_{AM}$ or the $\mathcal{O}_{Co}$ oracle, we know that $\mathsf{ID} \neq \mathsf{ID}_\mathcal{A}$. Therefore $\mathcal{B}$ can retrieve s_{ID} in the H_k-List and computes the d-th power of $\mathcal{O}_\mathcal{H}(\mathsf{ID})^{-1} = h_{\mathsf{ID}}^{-1}$ as s_{ID}^{-e}. Thanks to the knowledge of this value $\mathcal{B}$ can reply to $\mathcal{A}$'s queries involving members of $G^\star$. This perfectly simulates the oracles, therefore we have $\Pr[\mathsf{Imp}_4 | \Lambda] = \Pr[\mathsf{Imp}_3 | \Lambda]$.

Game_5 Finally, $\mathcal{B}$ simulates the handshake protocols for members with pseudonyms say ID_I and ID_M. The only difficulty happens when $\mathsf{ID}_I = \mathsf{ID}_\mathcal{A}$ or $\mathsf{ID}_M = \mathsf{ID}_\mathcal{A}$. In this case, $\mathcal{B}$ picks uniformly at random $u \in [\![1, n]\!]$, sets $c_I = (r\alpha)^{eu} \cdot s_{\mathsf{ID}_I}^{-e}$ or $c_M = (r\alpha)^{eu} \cdot s_{\mathsf{ID}_M}^{-d}$ (respectively). The algorithm $\mathcal{B}$ picks uniformly at random $v \in [\![1, n]\!]$, sets $c_\mathcal{A} = s_{\mathsf{ID}_\mathcal{A}}^{-1} r'(r\alpha)^v$ (so that

$[c_{\mathcal{A}}(s_{\mathsf{ID}_{\mathcal{A}}}\alpha^d)] \in \langle g \rangle)$ and $K = (c_{\mathcal{A}}^e h_{\mathsf{ID}_{\mathcal{A}}})^u$. Querying these suitable values to the non-programmable random oracle $\mathcal{G}_r$, $\mathcal{B}$ can perfectly simulate the $\mathcal{O}_{\mathsf{HS}}$. Therefore, we have $\Pr[\mathsf{Imp}_5|\Lambda] = \Pr[\mathsf{Imp}_4|\Lambda]$.

To summarize, when the Game_5 terminates, $\mathcal{A}$ outputs the triple $(G^\star, \mathsf{ID}^\star, d^\star)$ such that $G^\star$'s public key is $(n, e, (r\alpha)^e)$ and $\mathcal{O}_{\mathcal{H}}(\mathsf{ID}^\star) = s_{\mathsf{ID}^\star}^{e^2}$. Equiped with the pseudonym $\mathsf{ID}_{\mathcal{A}}$, $\mathcal{A}$ interacts with the reduction algorithm $\mathcal{B}$ emulating the member with pseudonym $\mathsf{ID}^\star$ to execute the protocol $\mathsf{OT\text{-}SH.Handshake}(\mathcal{A}, \mathsf{ID}^\star)$ (or $\mathsf{OT\text{-}SH.Handshake}(\mathcal{A}, \mathsf{ID}^\star)$ depending on $d^\star$).

From the simulation, we know that $\mathcal{O}_{\mathcal{H}}(\mathsf{ID}_{\mathcal{A}}) = s_{\mathsf{ID}_{\mathcal{A}}}^e \alpha$. Whatever the bit $d^\star$ is, $\mathcal{B}$ picks uniformly at random $t \in [\![1, n]\!]$ such that t is coprime to e and sets $c_{\mathsf{ID}^\star} = (r\alpha)^t s_{\mathsf{ID}^\star}^e$ in the execution of the handshake protocol.

- If the protocol is successful, thanks to the EPK property of $\mathcal{G}_r$, $\mathcal{B}$ retrieves in its transcript K and $c_{\mathsf{ID}_{\mathcal{A}}}$ such that $K = g^{ek^\star k_{\mathcal{A}}}$, $c_{\mathsf{ID}_{\mathcal{A}}} = g^{k_{\mathcal{A}}} s_{\mathsf{ID}_{\mathcal{A}}}^{-1} \alpha^{-d}$ and
  $$c_{\mathsf{ID}^\star} = g^{k^\star}(s_{\mathsf{ID}^\star}^{-e^2})^d = g^{k^\star} s_{\mathsf{ID}^\star}^{-e} = (r\alpha^e)^{k^\star} s_{\mathsf{ID}^\star}^{-e}.$$
  If Λ happens, then $ek^\star \equiv t \bmod \lambda(n)$, $c_{\mathsf{ID}_{\mathcal{A}}}^t = g^{tk_{\mathcal{A}}} s_{\mathsf{ID}_{\mathcal{A}}}^{-t} \alpha^{-dt} = K s_{\mathsf{ID}_{\mathcal{A}}}^{-t} \alpha^{-dt}$ and thus we have $(c_{\mathsf{ID}_{\mathcal{A}}}^t K^{-1} s_{\mathsf{ID}_{\mathcal{A}}}^t)^e = \alpha^{-t}$. Finally, using the extended Euclidean algorithm $\mathcal{B}$ finds integers u, v such that $ev - tu = 1$, and computes $x = (c_{\mathsf{ID}_{\mathcal{A}}}^t K^{-1} s_{\mathsf{ID}_{\mathcal{A}}}^{-t})^u \alpha^v \bmod n$ such that
  $$\left[(c_{\mathsf{ID}_{\mathcal{A}}}^t K^{-1} s_{\mathsf{ID}_{\mathcal{A}}}^{-t})^u \alpha^v\right]^e = \alpha^{-tu} \alpha^{ev} = \alpha \bmod n.$$

- Otherwise, $\mathcal{B}$ picks uniformly at random $x \in [\![1, n]\!]$.

The output of $\mathcal{B}$ is x and with probability at least $\Pr[\mathsf{Imp}_5|\Lambda] + 2^{-2k-1}$ it is equal to the e-th root of the RSA challenge α. This probability is greater than $\mathbf{Succ}^{\mathsf{imp-act}}_{\mathsf{OT\text{-}SH},\mathcal{A}}(k)/(q_{\mathsf{CG}}(k)q_{\mathcal{H}}(k))$. By lemma 1, we get the claimed bounds for $\mathbf{Succ}^{\mathsf{rsa}}_{\mathsf{Gen},\mathcal{B}}$ and τ'. □

5.2   Security of the Scheme Gi+SH

Theorem 2. *Let* Gen *be an RSA-group generator, let* $f : \mathbb{N} \to \mathbb{N}$ *and let* $\mathsf{Gi{+}SH}$ *be the secret handshake protocol instantiated with* Gen *and* f.
Let $q_{\mathcal{H}}, q_{\mathcal{G}}, q_{\mathsf{CG}}, q_{\mathsf{AM}}, q_{\mathsf{HS}}, q_{\mathsf{RM}}, q_{\mathsf{Co}}, \tau \in \mathbb{N}^{\mathbb{N}}$.

1. *Let* $\mathcal{A}$ *be a* $(q_{\mathcal{H}}, q_{\mathsf{CG}}, q_{\mathsf{AM}}, q_{\mathsf{HS}}, q_{\mathsf{RM}}, q_{\mathsf{Co}}, \tau)$-*IMP-ACT-adversary against* $\mathsf{Gi{+}SH}$, *in the ROM. There exist a* τ'-*DIFF-RSA adversary* $\mathcal{B}$ *such that*
$$\begin{cases} \mathbf{Succ}^{\mathsf{diff-rsa}}_{\mathsf{Gen},\mathcal{B}}(k) \geq \mathbf{Succ}^{\mathsf{imp-act}}_{\mathsf{Gi{+}SH},\mathcal{A}}(k)/[\exp(2)q_{\mathsf{CG}}(k)(q_{\mathsf{AM}}(k) + q_{\mathsf{Co}}(k) + 1)] \\ \tau'(k) \leq \tau(k) + (2q_{\mathcal{H}}(k) + 2q_{\mathsf{CG}}(k) + q_{\mathsf{AM}}(k) + q_{\mathsf{HS}}(k) + O(1))T_{\mathsf{Exp}}(k) \end{cases}$$

   for all positive integers k.

2. *Let* $\mathcal{A}$ *be a* $(q_{\mathcal{H}}, q_{\mathsf{CG}}, q_{\mathsf{AM}}, q_{\mathsf{HS}}, q_{\mathsf{RM}}, q_{\mathsf{Co}}, \tau)$-*DET-ACT-adversary against* $\mathsf{Gi{+}SH}$, *in the ROM. There exist a* $(q_{\mathsf{AM}} + q_{\mathsf{Co}}, \tau')$-*DIFF-RSA adversary* $\mathcal{B}$ *such that*
$$\begin{cases} \mathbf{Succ}^{\mathsf{diff-rsa}}_{\mathsf{Gen},\mathcal{B}}(k) \geq \mathbf{Adv}^{\mathsf{det-act}}_{\mathsf{Gi{+}SH},\mathcal{A}}(k)/[2\exp(2)q_{\mathsf{CG}}(k)(q_{\mathsf{AM}}(k) + q_{\mathsf{Co}}(k) + 1)] \\ \tau'(k) \leq (2q_{\mathcal{H}}(k) + 2q_{\mathsf{CG}}(k) + q_{\mathsf{AM}}(k) + q_{\mathsf{HS}}(k) + O(1))T_{\mathsf{Exp}}(k) \end{cases}$$

   for all positive integers k.

3. *Let $\mathcal{A}$ be a $(q_{\mathcal{H}}, q_{CG}, q_{AM}, q_{HS}, q_{RM}, q_{Co}, \tau)$-IND-ACT-adversary against Gi+SH, in the ROM, making at most $q_{\mathcal{G}}(k)$ queries to the non-programmable random oracle $\mathcal{O}_{\mathcal{G}}$ in the experiments $\mathbf{Exp}^{ind\text{-}act\text{-}b}_{Gi+SH,\mathcal{A}}(k)$ (for $b \in \{0,1\}$ and $k \in \mathbb{N}$). There exist a τ'-COMP-CDH adversary $\mathcal{B}$ such that*

$$\begin{cases} \mathbf{Succ}^{comp\text{-}cdh}_{Gen,\mathcal{B}}(k) \geq \mathbf{Adv}^{ind-act}_{Gi+SH,\mathcal{A}}(k)/[2q_{\mathcal{G}}(k)] \\ \tau'(k) \leq \tau(k) + (q_{\mathcal{H}}(k) + O(1))T_{\mathrm{Exp}}(k) \end{cases}$$

for all positive integers k.

Proof. Again, we prove only the first part of the theorem. Let k be a security parameter, let (n, p, q, e, d, g) be a 6-tuple generated by $\mathsf{Gen}(k)$ and let $\alpha \in (\mathbb{Z}/n\mathbb{Z})^*$ be a random instance of the RSA problem. We construct a reduction algorithm $\mathcal{B}$ which on input (n, e, α) outputs a pair $(x, y) \in [(\mathbb{Z}/n\mathbb{Z})^*]^2$ aimed to satisfy $x^e - y^e = \alpha \bmod n$.

We follow essentially the previous proof with games Game_0 and Game_1 analogous to the corresponding games. With the same notation, we get

$$\Pr[\mathsf{Imp}_1] = \mathbf{Succ}^{imp-act}_{GI+SH,\mathcal{A}}(k)/q_{CG}(k).$$

Game_2 $\mathcal{B}$ simulates the $\mathcal{O}_{CG}$ oracle. For the i-th query ($i \in [\![1, q_{CG}(k)]\!] \setminus \{\ell\}$), $\mathcal{B}$ executes $\mathsf{Gen}(k)$ and gets $(n_i, p_i, q_i, e_i, d_i, g_i)$. It outputs (n_i, e_i, g_i) as the answer to $\mathcal{A}$'s i-th query for the group denoted by G_i and stores d_i. For the ℓ-th query, $\mathcal{B}$ outputs (n, e, g^{e^2}) as $G^\star$'s public key. The distribution of the group authorities keys is unchanged, and therefore, we have $\Pr[\mathsf{Imp}_2] = \Pr[\mathsf{Imp}_1]$.

Following Coron's technique [6], a random coin decides whether $\mathcal{B}$ introduces the challenge α in the hash answer to the oracle $\mathcal{O}_{\mathcal{H}}$, or an element with a known preimage. Let $\delta \in [0,1]$.

Game_3 In this game, for each fresh bit string ID queried to $\mathcal{O}_{\mathcal{H}}$, $\mathcal{B}$ picks at random a bit $t_{\mathsf{ID}} \xleftarrow{R} B_\delta$, where B_δ is the probability distribution over $\{0,1\}$ where 0 is drawn with probability δ and 1 with probability $1 - \delta$. $\mathcal{B}$ stores in a list denoted by H_k-List a 4-tuple $(\mathsf{ID}, \mathcal{O}_{\mathcal{H}}(\mathsf{ID}), \perp, t_{\mathsf{ID}})$. $\mathcal{B}$ discards execution which outputs a triple $(G^\star, \mathsf{ID}^\star, d^\star)$ such that $t_{\mathsf{ID}^\star} = 1$ or $t_{\mathsf{ID}_\mathcal{A}} = 0$.

Since for each bit string ID queried to $\mathcal{O}_{\mathcal{H}}$, the bit t_{ID} is picked independently of the execution of the game Game_4, we have $\Pr[\mathsf{Imp}_3 | \Lambda] = \delta(1-\delta) \Pr[\mathsf{Imp}_2 | \Lambda]$.

Game_4 Now, $\mathcal{B}$ immediately aborts if $\mathcal{A}$ makes a query to $\mathcal{O}_{AM}$ or $\mathcal{O}_{Co}$ involving a pseudonym ID such that $t_{\mathsf{ID}} > 0$. Thus $\Pr[\mathsf{Imp}_4 | \Lambda] = \delta^{q_{AM}(k)+q_{Co}(k)} \Pr[\mathsf{Imp}_3 | \Lambda]$.

Game_5 In this game, $\mathcal{B}$ simulates the random oracle $\mathcal{O}_{\mathcal{H}}$. For each fresh bit string ID queried to $\mathcal{O}_{\mathcal{H}}$, $\mathcal{B}$ picks uniformly at random $m_{\mathsf{ID}} \in [\![1, n]\!]$.

With probability p, it sets $h_{\mathsf{ID}} = m_{\mathsf{ID}}^e \cdot \alpha$ and $t_{\mathsf{ID}} = 0$; otherwise $\mathcal{B}$ picks $t_{\mathsf{ID}} \in [\![1, n]\!]$, coprime to e, and sets $h_{\mathsf{ID}} = g^{e \cdot t_{\mathsf{ID}}} - m_{\mathsf{ID}}^e$. $\mathcal{B}$ returns h_{ID} as the answer to the oracle call and stores the 4-tuple $(\mathsf{ID}, h_{\mathsf{ID}}, m_{\mathsf{ID}}, t_{\mathsf{ID}})$ in the H_k-List. In the random oracle model, this game is clearly identical to the previous one, therefore $\Pr[\mathsf{Imp}_5] = \Pr[\mathsf{Imp}_4]$.

Game$_6$ Now, $\mathcal{B}$ simulates the $\mathcal{O}_{\mathrm{AM}}$ and the $\mathcal{O}_{\mathrm{Co}}$ oracles.

For all $i \in [\![1, q_{\mathrm{CG}}(k)]\!] \setminus \{\ell\}$, thanks to the knowledge of d_i, $\mathcal{B}$ can perfectly simulate the member key generation protocol for the group G_i, and therefore can reply to $\mathcal{A}$'s queries involving members of G_i.

For queries $\mathcal{O}_{\mathrm{AM}}(G^*)$, $\mathcal{B}$ picks uniformly at random a pseudonym ID and queries $\mathcal{O}_{\mathcal{H}}$ on ID, $\mathcal{B}$ act as follows:

- If $t_{\mathsf{ID}} = 0$, then $\mathcal{B}$ picks uniformly at random $s_{\mathsf{ID}} \in [\![1, n]\!]$ and queries $T_{\mathsf{ID}} = (-m_{\mathsf{ID}})^{-e} g^{e^2 s_{\mathsf{ID}}}$ to its oracle $(\alpha + (\cdot)^e)^d$. It gets the value U_{ID}, and sets $p_{\mathsf{ID}} = -U_{\mathsf{ID}} \times m_{\mathsf{ID}} \bmod n$. It outputs p_{ID} as the public key associated with G^* and ID and stores $(p_{\mathsf{ID}}, s_{\mathsf{ID}})$ in a list denoted by KeyList.
- Otherwise, $\mathcal{B}$ outputs m_{ID} as the public key associated with G^* and ID and stores $(m_{\mathsf{ID}}, \perp)$ in a KeyList.

The behaviour of $\mathcal{B}$ is identical for queries $\mathcal{O}_{\mathrm{AM}}(G^*, \mathsf{ID})$ (with always $t_{\mathsf{ID}} = 0$ in this case).

Finally, for members of G^* with pseudonym ID queried to the $\mathcal{O}_{\mathrm{Co}}$ oracle, we know that $t_{\mathsf{ID}} = 0$. $\mathcal{B}$ can retrieve s_{ID} in the KeyList. This perfectly simulates the oracles, therefore we have $\Pr[\mathsf{Imp}_6] = \Pr[\mathsf{Imp}_5]$.

Game$_7$ Finally, $\mathcal{B}$ can easily simulates the handshake protocols for members with pseudonyms say ID_I and ID_M. Again, the only non trivial case occurs when $t_{\mathsf{ID}_I} \neq 0$ and $t_{\mathsf{ID}_M} \neq 0$. In this case, $\mathcal{B}$ sets $K = g^{t_{\mathsf{ID}_I} \cdot t_{\mathsf{ID}_M}}$.

If we denote $k_{\mathsf{ID}_I} = \log_{g^{e^2}}(p_{\mathsf{ID}_I}^e + h_{\mathsf{ID}_I})$ and $k_{\mathsf{ID}_M} = \log_{g^{e^2}}(p_{\mathsf{ID}_M}^e + h_{\mathsf{ID}_M})$, we get $e^2 k_{\mathsf{ID}_I} = e \cdot t_{\mathsf{ID}_I} \bmod \lambda(n)$ and $e^2 k_{\mathsf{ID}_M} = e \cdot t_{\mathsf{ID}_M} \bmod \lambda(n)$, and therefore $K = g^{e^2 k_I k_M}$. Querying this value with suitable nonces to the non-programmable random oracle $\mathcal{G}_r$, $\mathcal{B}$ can perfectly simulate the oracle $\mathcal{O}_{\mathrm{HS}}$, and we have $\Pr[\mathsf{Imp}_7] = \Pr[\mathsf{Imp}_6]$.

Finally, when Game$_7$ terminates, $\mathcal{A}$ outputs a triple $(G^*, \mathsf{ID}^*, d^*)$ such that G^*'s public key is (n, e, g^{e^2}), $\mathcal{O}_{\mathcal{H}}(\mathsf{ID}^*) = g^{et_{\mathsf{ID}}^*} - m_{\mathsf{ID}^*}^e$ and the public key associated with G^* and ID^* is m_{ID^*}. Equipped with the pseudonym $\mathsf{ID}_{\mathcal{A}}$, $\mathcal{A}$ interacts with $\mathcal{B}$ emulating the member with pseudonym ID^* to execute the handshake protocol. From the simulation, we know that $\mathcal{O}_{\mathcal{H}}(\mathsf{ID}_{\mathcal{A}}) = m_{\mathsf{ID}_{\mathcal{A}}}^e \alpha$. Whatever the bit d^* and the nonces are, if the protocol is successful, thanks to the EPK property of $\mathcal{G}_r$, $\mathcal{B}$ retrieves in its transcript $p_{\mathcal{A}}$ the public key of $\mathcal{A}$ and $K = g^{e^2 s^* s_{\mathcal{A}}}$ where $e^2 s^* = e \cdot t_{\mathsf{ID}^*} \bmod \lambda(n)$ and $p_{\mathcal{A}}^e - g^{e^2 s_{\mathcal{A}}} = m_{\mathsf{ID}_{\mathcal{A}}}^e \alpha$. We have $K = g^{et_{\mathsf{ID}^*} s_{\mathcal{A}}} = (g^{e s_{\mathcal{A}}})^{t_{\mathsf{ID}^*}}$, and using the extended Euclidean algorithm $\mathcal{B}$ finds integers u, v such that $t_{\mathsf{ID}^*} u + ev = 1$, and sets $x = p_{\mathcal{A}}/m_{\mathsf{ID}_{\mathcal{A}}}$ and $y = K^u(p_{\mathcal{A}}^e + m_{\mathsf{ID}_{\mathcal{A}}}^e \alpha)^v / m_{\mathsf{ID}_{\mathcal{A}}}$ such that $x^e - y^e = \alpha \bmod n$. It remains to prove the bounds on $\mathbf{Succ}_{\mathrm{Gen}, \mathcal{B}}^{\mathrm{diff-rsa}}$ and τ'. The computation follows readily the one supplied in details in the previous proof. $\qquad\square$

5.3 Security of the Scheme Gi×SH

Theorem 3. *Let Gen be an RSA-group generator, let $f : \mathbb{N} \to \mathbb{N}$ and let Gi×SH be the secret handshake protocol instantiated with Gen and f. Let $q_{\mathcal{H}}, q_{\mathcal{G}}, q_G, q_{CG}, q_{AM}, q_{HS}, q_{RM}, q_{Co}, \tau \in \mathbb{N}^{\mathbb{N}}$.*

1. *Let $\mathcal{A}$ be a $(q_{\mathcal{H}}, q_{CG}, q_{AM}, q_{HS}, q_{RM}, q_{Co}, \tau)$-IMP-ACT-adversary against Gi×SH, in the random oracle model. There exist a τ'-RSA adversary $\mathcal{B}$ such that*

$$\begin{cases} \mathbf{Succ}^{\mathrm{rsa}}_{Gen,\mathcal{B}}(k) \geq \mathbf{Succ}^{\mathrm{imp-act}}_{Gi\times SH,\mathcal{A}}(k)/[\exp(2)q_{CG}(k)(q_{AM}(k) + q_{Co}(k) + 1)] \\ \tau'(k) \leq \tau(k) + (2q_{\mathcal{H}}(k) + 2q_{CG}(k) + q_{AM}(k) + q_{HS}(k) + O(1))T_{\mathrm{Exp}}(k) \end{cases}$$

   for all positive integers k.

2. *Let $\mathcal{A}$ be a $(q_{\mathcal{H}}, q_{CG}, q_{AM}, q_{HS}, q_{RM}, q_{Co}, \tau)$-DET-ACT-adversary against Gi×SH, in the random oracle model. There exist a τ'-RSA adversary $\mathcal{B}$ such that*

$$\begin{cases} \mathbf{Succ}^{\mathrm{rsa}}_{Gen,\mathcal{B}}(k) \geq \mathbf{Adv}^{\mathrm{det-act}}_{Gi\times SH,\mathcal{A}}(k)/[2\exp(2)q_{CG}(k)(q_{AM}(k) + q_{Co}(k) + 1)] \\ \tau'(k) \leq \tau(k) + (2q_{\mathcal{H}}(k) + 2q_{CG}(k) + q_{AM}(k) + q_{HS}(k) + O(1))T_{\mathrm{Exp}}(k) \end{cases}$$

   for all positive integers k.

3. *Let $\mathcal{A}$ be a $(q_{\mathcal{H}}, q_{CG}, q_{AM}, q_{HS}, q_{RM}, q_{Co}, \tau)$-IND-ACT-adversary against Gi×SH, in the random oracle model, making at most $q_G(k)$ queries to the non-programmable random oracle $\mathcal{O}_G$ in the experiments $\mathbf{Exp}^{\mathrm{ind-act-}b}_{Gi\times SH,\mathcal{A}}(k)$ (for $b \in \{0,1\}$ and $k \in \mathbb{N}$). There exist a τ'-COMP-CDH adversary $\mathcal{B}$ such that*

$$\begin{cases} \mathbf{Succ}^{\mathrm{comp-cdh}}_{Gen,\mathcal{B}}(k) \geq \mathbf{Adv}^{\mathrm{ind-act}}_{Gi\times SH,\mathcal{A}}(k)/[2q_G(k)] \\ \tau'(k) \leq \tau(k) + (q_{\mathcal{H}}(k) + O(1))T_{\mathrm{Exp}}(k) \end{cases}$$

   for all positive integers k. □

6 Efficiency Issues

In this section, we compare the performance of all the secret handshake schemes proposed up to now. For concreteness, we assume that our schemes are instantiated with 1024-bits RSA *moduli*, and that the Diffie-Hellman protocols from [5] are instantiated on a 160-bits prime order subgroup of a prime finite field of size 1024 bits. We denote by CJT1 the scheme based on the CA-oblivious encryption and by CJT2 the scheme using the additional proof of knowledge. We assume that the scheme from [1], denoted BDSSSW, is instantiated with the Tate pairing on an elliptic curve of MOV degree 6 on a ground base field of size 171 bits and that computing this bilinear map using Miller's algorithm [12] is 10 times more expensive than computing a discrete exponentiation in a 160 bits subgroup of a prime finite field of size 1024 bits (whose computation time is arbitrarily set to 1). In the table 1, we summarize the schemes' complexity in terms of bits exchanged during the protocol Handshake (in addition to the identification information $\mathsf{ID_I}$ and $\mathsf{ID_M}$) and the computational cost of the protocols AddMember and Handshake. The new schemes compare very favorably in performance with respect to the discrete log systems proposed so far [1, 5] and they can be used over a low bandwidth channel.

Table 1. Efficiency comparison

Scheme	BDSSSW [1]	CJT1 [5]	CJT2 [5]	*OT-SH*	*Gi+SH*	*Gi×SH*
Underlying Problem	CBDH	CDH	CDH	RSA	Diff-RSA	RSA
Number of rounds	3	4	3	4	3	3
Bits exchanged	640	8512	6304	1344	640	640
Computational cost						
AddMember	1.8	1	1	6.4	7.4	7.4
Handshake	22	8	6	4	2	2

7   Authenticated Key Exchange from Secret Handshakes

An authenticated key exchange protocol enables two parties to end up with a shared secret key in a secure and authenticated manner (*i.e.* no adversary can impersonate any party during the protocol or learn any information about the value of the exchanged secret). At the end of the execution of the protocol OT-SH.Handshake (*resp.* Gi⊕SH.Handshake), the parties can set up a temporary session key for securing further communication between them:

$$
\begin{aligned}
\text{session} &= \mathcal{G}_r \left[(\mathcal{H}_k(\mathsf{ID_M}) \cdot c_\mathsf{M}^e)^{k_\mathsf{I}} \bmod n \| \mathsf{ID_I} \| \mathsf{ID_M} \| c_\mathsf{I} \| c_\mathsf{M} \| 2 \right] \\
&= \mathcal{G}_r \left[(\mathcal{H}_k(\mathsf{ID_I}) \cdot c_\mathsf{I}^e)^{k_\mathsf{I}} \bmod n \| \mathsf{ID_I} \| \mathsf{ID_M} \| c_\mathsf{I} \| c_\mathsf{M} \| 2 \right]
\end{aligned}
$$

$$
\left(
\begin{aligned}
\textit{resp. } \text{session} &= \mathcal{G}_r \left[(\mathbf{pk_{ID_I}}^e \oplus \mathcal{H}_k(\mathsf{ID_I}))^{\mathbf{sk_{ID_M}}} \bmod n \| \mathsf{ID_I} \| \mathsf{ID_M} \| n_\mathsf{I} \| n_\mathsf{M} \| 2 \right] \\
&= \mathcal{G}_r \left[(\mathbf{pk_{ID_M}}^e \oplus \mathcal{H}_k(\mathsf{ID_M}))^{\mathbf{sk_{ID_I}}} \bmod n \| \mathsf{ID_I} \| \mathsf{ID_M} \| n_\mathsf{I} \| n_\mathsf{M} \| 2 \right]
\end{aligned}
\right)
$$

As mentioned in [3], a number of desirable attributes of key agreement protocols have been identified:

1. **known session keys:** a protocol still achieves its goal in the face of an adversary who has learned some previous session keys.
2. **(perfect) forward secrecy:** if long-term secrets of one or more entities are compromised, the secrecy of previous session keys is not affected.
3. **unknown key-share:** entity i cannot be coerced into sharing a key with entity j without i's knowledge.
4. **key-compromise impersonation:** the knowledge of i's secret value does not enable an adversary to impersonate other entities to i.
5. **loss of information:** compromise of information that would not ordinarily be available to an adversary does not affect the security of the protocol.
6. **key control:** neither entity should be able to force the session key to a preselected value.

In these *scenarii*, the adversary controls all communication between entities, and can at any time ask an entity to reveal its long-term secret key. Furthermore she may initiate sessions between any two entities, engage in multiple sessions with the same entity at the same time, and in some cases ask an entity to enter a session with itself. In the random oracle model, the *key control* security

Table 2. Security properties of the new AKE-SH

Property	OT-SH	Gi+SH	Gi×SH
known session keys	✔	✔	✔
forward secrecy	✔	✘	✘
unknown key-share	✔	✔	✔
key compromise impersonation	✔	✔	✔
loss of information	✔	✘	✘

requirement of the privacy-preserving authenticated key exchange derived from OT-SH, Gi+SH and Gi×SH is unconditional (in the standard model, forcing a session key to a preselected value is at least as hard as breaking the one-wayness of $\mathcal{G}_r$). The other security requirements against active adversaries are summarized in table 2 (where the symbol ✘ means that the scheme does not reach this security requirement, whereas the symbol ✔ means that in the random oracle model the security requirement reduces to the composite Diffie Hellman problem). The proofs are straightforward adaptations of the proofs of security of the secret handshake protocols and therefore they are left to the reader.

8 Conclusion

A secret handshake protocol is a cryptographic primitive that allow members of a group to authenticate each other secretly. In this paper, we designed three efficient constructions for secret handshake based on the RSA assumption. These constructs are the first fully anonymous protocols relying on the RSA primitive. The new secret handshake protocols can handle roles just as easily as the one in [1, 5]. An interesting open issue is to design simple and efficient fully anonymous secret handshake schemes, relying as little as possible on random oracles (ideally without any).

Acknowledgements. The author expresses his gratitude to Benoît Libert for his suggestions on improving the exposition and for pointing out some errors in an earlier draft of the paper.

References

1. D. Balfanz, G. Durfee, N. Shankar, D. K. Smetters, J. Staddon, and H. C. Wong, *Secret Handshakes from Pairing-Based Key Agreements.*, 2003 IEEE Symposium on Security and Privacy (S&P 2003), IEEE Computer Society, 2003, pp.180–196.
2. M. Bellare and P. Rogaway, *Random Oracles are Practical: A Paradigm for Designing Efficient Protocols.*, Proceedings of the First ACM Conference on Computer and Communications Security, 1993, pp. 62–73.
3. S. Blake-Wilson, D. Johnson, and A. Menezes, *Key Agreement Protocols and their Security Analysis.*, 6th IMA International Conference on Cryptography and Coding (M. Darnell, ed.), Lect. Notes Comput. Sci., vol. 1355, Springer, 1997, pp. 30–45.

4. R. Canetti, O. Goldreich, and S. Halevi, *The Random Oracle Methodology, Revisited.*, J. Assoc. Comput. Mach. **51** (2004), no. 4, 557–594.
5. C. Castelluccia, S. Jarecki, and G. Tsudik, *Secret Handshakes from CA-Oblivious Encryption.*, Advances in Cryptology - ASIACRYPT 2004 (P. J. Lee, ed.), Lect. Notes Comput. Sci., vol. 3329, Springer, 2004, pp. 293–307.
6. J.-S. Coron, *On the Exact Security of Full Domain Hash.*, Advances in Cryptology - CRYPTO 2000 (M. Bellare, ed.), Lect. Notes Comput. Sci., vol. 1880, Springer, 2000, pp. 229–235.
7. Y. Desmedt, *Securing Traceability of Ciphertexts - Towards a Secure Software Key Escrow System.*, Advances in Cryptology - EUROCRYPT'95 (L. C. Guillou and J.-J. Quisquater, eds.), Lect. Notes Comput. Sci., vol. 921, Springer, 1995, pp. 147–157.
8. W. Diffie, P. C. van Oorschot, and M. J. Wiener, *Authentication and Authenticated Key Exchanges.*, Des. Codes Cryptography **2** (1992), no. 2, 107–125.
9. M. Girault, *Self-Certified Public Keys.*, Advances in Cryptology - EUROCRYPT'91 (D. Davies, ed.), Lect. Notes Comput. Sci., vol. 547, Springer, 1991, pp. 490–497.
10. M. Jakobsson, K. Sako, and R. Impagliazzo, *Designated Verifier Proofs and Their Applications.*, Advances in Cryptology - EUROCRYPT'96 (U. M. Maurer, ed.), Lect. Notes Comput. Sci., vol. 1070, Springer, 1996, pp. 143–154.
11. M. Mambo and H. Shizuya, *A Note on the Complexity of Breaking Okamoto-Tanaka ID-Based Key Exchange Scheme.*, First International Workshop on Practice and Theory in Public Key Cryptography, PKC '98 (H. Imai and Y. Zheng, eds.), Lect. Notes Comput. Sci., vol. 1431, Springer, 1998, pp. 258–262.
12. V. S. Miller, *The Weil Pairing, and Its Efficient Calculation.*, J. Cryptology **17** (2004), no. 4, 235–261.
13. M. Naor, *On Cryptographic Assumptions and Challenges.*, Advances in Cryptology - CRYPTO 2003 (D. Boneh, ed.), Lect. Notes Comput. Sci., vol. 2729, Springer, 2003, pp. 96–109.
14. J. B. Nielsen, *Separating Random Oracle Proofs from Complexity Theoretic Proofs: the Non-committing Encryption Case*, Advances in Cryptology - CRYPTO 2002 (M. Yung, ed.), Lect. Notes Comput. Sci., vol. 2442, Springer, 2002, pp. 111–126.
15. S.-H. Oh, M. Mambo, H. Shizuya, and D.-H. Won, *On the Security of Girault Key Agreement Protocols against Active Attacks.*, IEICE Trans. Fundamentals **E86-A** (2003), no. 5, 1181–1189.
16. E. Okamoto and K. Tanaka, *Key Distribution System Based on Identification Information*, IEEE J. Selected Areas in Communications **7** (1989), 481–485.
17. J. Rosser and L. Schoenfeld, *Approximate formulas for some functions of prime numbers*, Ill. J. Math. **6** (1962), 64–94.
18. S. Saeednia, *A Note on Girault's Self-Certified Model.*, Inf. Process. Lett. **86** (2003), no. 3, 323–327.
19. R. Sakai, K. Ohgishi, and M. Kasahara, *Cryptosystems Based on Pairings.*, Proceedings of the Symposium on Cryptography and Information Security (SCIS 2000), 2000.
20. A. Shamir, *Identity-Based Cryptosystems and Signature Schemes.*, Advances in Cryptology - CRYPTO'84 (G. R. Blakley and D. Chaum, eds.), Lect. Notes Comput. Sci., vol. 196, Springer, 1985, pp. 47–53.
21. V. Shoup, *OAEP Reconsidered.*, J. Cryptology **15** (2002), no. 4, 223–249.
22. S. Xu and M. Yung, *k-Anonymous Secret Handshakes with Reusable Credentials.*, Proceedings of the 11th ACM Conference on Computer and Communications Security (V. Atluri, B. Pfitzmann, and P. McDaniel, eds.), ACM, 2004, pp. 158–167.

On a Relation Between Verifiable Secret Sharing Schemes and a Class of Error-Correcting Codes[*]

Ventzislav Nikov[1] and Svetla Nikova[2]

[1] Department of Mathematics and Computing Science,
Eindhoven University of Technology
5P.O. Box 513, 5600 MB, Eindhoven, The Netherlands
vnikov@mail.com
[2] Department Electrical Engineering, ESAT/COSIC,
Katholieke Universiteit Leuven, Kasteelpark Arenberg 10,
B-3001 Heverlee-Leuven, Belgium
svetla.nikova@esat.kuleuven.ac.be

Abstract. In this paper we try to shed a new insight on Verifiable Secret Sharing Schemes (VSS). We first define a new "metric" (with slightly different properties than the standard Hamming metric). Using this metric we define a very particular class of codes that we call *error-set correcting codes*, based on a set of forbidden distances which is a monotone decreasing set. Next we redefine the packing problem for the new settings and generalize the notion of error-correcting capability of the error-set correcting codes accordingly (taking into account the new metric and the new packing). Then we consider burst-error interleaving codes proposing an efficient burst-error correcting technique, which is in fact the well known VSS and Distributed Commitments (DC) pairwise checking protocol and we prove the error-correcting capability of the error-set correcting interleaving codes.

Using the known relationship, due to Van Dijk, between a Monotone Span Program (MSP) and a generator matrix of the code generated by the suitable set of vectors, we prove that the error-set correcting codes in fact has the allowed (opposite to forbidden) distances of the dual access structure of the access structure that the MSP computes. We give an efficient construction for them based on this relation and as a consequence we establish a link between Secret Sharing Schemes (SSS) and the error-set correcting codes.

Further we give a necessary and sufficient condition for the existence of linear SSS (LSSS), to be secure against (Δ, Δ_A)-adversary expressed in terms of an error-set correcting code. Finally, we present necessary and sufficient conditions for the existence of a VSS scheme, based on an error-set correcting code, secure against (Δ, Δ_A)-adversary.

Our approach is general and covers all known linear VSS/DC. It allows us to establish the minimal conditions for security of VSSs. Our main theorem states that the security of a scheme is equivalent to a pure

[*] The work described in this paper has been supported in part by the European Commission through the IST Programme under Contract IST-2002-507932 ECRYPT, and by Concerted Research Action GOA Ambiorix 2005/11 of the Flemish Government.

Ø. Ytrehus (Ed.): WCC 2005, LNCS 3969, pp. 275–290, 2006.
© Springer-Verlag Berlin Heidelberg 2006

geometrical (coding) condition on the linear mappings describing the scheme. Hence the security of all known schemes, e.g. all known bounds for existence of unconditionally secure VSS/DC including the recent result of Fehr and Maurer, can be expressed as certain (geometrical) coding conditions.

1   Preliminaries

The concept of *secret sharing* was introduced by Shamir [20] as a tool to protect a secret from getting exposed or from getting lost. It allows a so-called *dealer* to share a secret among the members of a set $\mathcal{P}$, which are usually called *players* or *participants*, in such a way that only certain specified subsets of players are able to reconstruct the secret (if needed) while smaller subsets have no information about this secret at all (in a strict information theoretic sense).

We call the groups who are allowed to reconstruct the secret *qualified* and the groups who should not be able to obtain any information about the secret *forbidden*. The set of qualified groups is denoted by Γ and the set of forbidden groups by Δ. Denote the participants by P_i, $1 \leq i \leq n$ and the set of all players by $\mathcal{P} = \{P_1, \ldots, P_n\}$. The set Γ is called *monotone increasing* if for each set A in Γ also each set containing A is in Γ. Similarly, Δ is called *monotone decreasing*, if for each set A in Δ also each subset of A is in Δ. A monotone increasing set Γ can be efficiently described by the set Γ^- consisting of the minimal elements in Γ, i.e. the elements in Γ for which no proper subset is also in Γ. Similarly, the set Δ^+ consists of the maximal sets in Δ. The tuple (Γ, Δ) is called an *access structure* if $\Gamma \cap \Delta = \emptyset$. If the union of Γ and Δ is equal to $2^\mathcal{P}$ (so Γ is equal to Δ^c, the complement of Δ), then we say that access structure (Γ, Δ) is *complete* and we denote it just by Γ. In the sequel we shall only consider complete, monotone access structures.

The dual $\Gamma^\perp$ of an access structure Γ, defined on $\mathcal{P}$, is the collection of sets $A \subseteq \mathcal{P}$ such that $\mathcal{P} \setminus A = A^c \notin \Gamma$.

It is common to model cheating by considering an *adversary* $\mathcal{A}$ who may corrupt some of the players. The adversary is characterized by particular subset Δ_A of Δ, called *adversary and privacy structures* [12] respectively, which are monotone decreasing structures. The players which belong to Δ are called also *curious* and the players which belong to Δ_A are called *corrupt* or *bad*.

One can distinguish between *passive* and *active* corruption, see Fehr and Maurer [10] for recent results. Passive corruption means that the adversary obtains the complete information held by the corrupt players, but the players execute the protocol correctly. Active corruption means that the adversary takes full control of the corrupt players. Active corruption is strictly stronger than passive corruption. Both passive and active adversaries may be *static*, meaning that the set of corrupt players is chosen once and for all before the protocol starts, or *adaptive* meaning that the adversary can at any time during the protocol choose to corrupt a new player based on all the information he has at the time, as long as the total set is in Δ_A.

Denote the complement $\Gamma_A = 2^{\mathcal{P}} \setminus \Delta_A = \Delta_A^c$. Its dual access structure $\Gamma_A^{\perp}$ should be called the honest (or good) players structure, since for any set A of corrupt players, i.e. in Δ_A, the complement $A^c = \mathcal{P} \setminus A$ is the set of honest players and vise versa. Note that the set $\{A^c : A \in \Delta_A\}$ is the dual access structure $\Gamma_A^{\perp}$.

Some authors [11] consider also *fail-corrupt* players. To fail-corrupt a player means that the adversary may stop the communication from and to that player at an arbitrary time during the protocol. Once a player is caused to fail, he will not recover the communication. However, the adversary is not allowed to read the internal data of a fail-corrupt player, unless the player is also passively corrupted at the same time. The collection of fail-corrupt players is denoted by $\Delta_F \subseteq \Delta$. Generally we will not consider such kind of corruption, so unless it is exact mentioned we will assume that the adversary cannot fail-corrupt the players.

Definition 1. *[10] An $(\Delta, \Delta_A, \Delta_F)$-adversary is an adversary who can (adaptively) corrupt some players passively and some players actively, as long as the set A of actively corrupt players and the set B of passively corrupt players satisfy both $A \in \Delta_A$ and $(A \cup B) \in \Delta$. Additionally the adversary could fail-corrupt some players in Δ_F. When $\Delta_F = \emptyset$ we will denote it by (Δ, Δ_A), in case $\Delta_A = \Delta$ we will simply say Δ_A-adversary.*

This model is known as *mixed adversary* model. Note that in case of Secret Sharing Schemes we have $\Delta_A = \emptyset$, while for Verifiable Secret Sharing Schemes we have $\Delta_A \neq \emptyset$. In the threshold case we write instead of $(\Delta, \Delta_A, \Delta_F)$-adversary simply (k, k_a, k_f)-adversary. Recently Hirt and Maurer [12] introduced the notion of $\mathcal{Q}^2(\mathcal{Q}^3)$ adversary structure.

Definition 2. *[12] For a given set of players $\mathcal{P}$ and an adversary structure Δ_A, we say that the adversary structure is $\mathcal{Q}^{\ell}$ if no ℓ sets in Δ_A cover the full set $\mathcal{P}$ of players.*

Definition 3. *[17] For any two monotone decreasing sets Δ_1, Δ_2 operation $\uplus$, called element-wise union, is defined as follows: $\Delta_1 \uplus \Delta_2 = \{A = A_1 \cup A_2; A_1 \in \Delta_1, A_2 \in \Delta_2\}$. For any two monotone increasing sets Γ_1, Γ_2 operation $\uplus$ is defined as follows: $\Gamma_1 \uplus \Gamma_2 = \{A = A_1 \cup A_2; A_1 \notin \Gamma_1, A_2 \notin \Gamma_2\}^c$.*

Definition 4. *A secret sharing scheme based on an access structure (Γ, Δ) is a pair (Share and Reconstruct) of protocols (phases) namely, the sharing phase, where the players share a secret $s \in \mathcal{K}$, and the reconstruction phase, where the players try to reconstruct s, such that the following two properties hold:*

- Privacy: *The players of any set $B \in \Delta$ learn nothing about the secret s as a result of the sharing phase.*
- Correctness: *The secret s could be computed by any set of players $A \in \Gamma$.*

Recall that the SSS is called perfect if and only if $\Delta^c = \Gamma$.

2   A Class of Error-Correcting "Codes"

Let $\mathbb{F}$ be a finite field and let the set of secrets for the dealer $\mathcal{D}$ be $\mathcal{K} = \mathbb{F}^{p_0}$. We will only consider the case $p_0 = 1$, even though many of the considerations remain valid in the general case too. Associate with each player P_i ($1 \leq i \leq n$) a positive integer p_i such that the sets of possible shares for player P_i is given by $\mathbb{F}^{p_i}$. Denote by $p = \sum_{i=1}^{n} p_i$ and by $N = p_0 + p$. For the sake of simplicity one could assume that $p_i = 1$ for $0 \leq i \leq n$ in that case $p = n$ and $N = n + 1$ hold.

Now we will recall some definitions from the theory of error-correcting codes. Any non-empty subset $\mathcal{C}$ of $\mathbb{F}^N$ is called a code, the parameter N is called the *length* of the code. Each vector in $\mathcal{C}$ is called *codeword* of $\mathcal{C}$. The *Hamming sphere* (or *ball*) $B_e(\mathbf{x})$ of radius e around a vector $\mathbf{x}$ in $\mathbb{F}^N$ is defined by $B_e(\mathbf{x}) = \{\mathbf{y} \in \mathbb{F}^N : d(\mathbf{x}, \mathbf{y}) \leq e\}$. One of the basic coding theory problems is the so-called *Sphere Packing Problem:* given N and e, what is the maximum number of non-intersecting spheres of radius e that can be placed in $\mathbb{F}^N$, the N-dimensional Hamming space?

Sphere packing is related to *error correction*. The centers of these spheres are at distance at least $2e + 1$ apart from each other and constitute a *code*; these centers are called *codewords* and each corresponds to a possible message that one may want to transmit. Assume now that one of these messages is transmitted and that at most e coordinates are corrupt during the transmission. To decode, i.e., to decide which of the messages was actually sent, compute the Hamming distance between the received vector and all the centers. Since at most e errors occurred, the transmitted word will still be the nearest center, and all errors can be corrected.

Define the *minimum distance* of a code $\mathcal{C} \subseteq \mathbb{F}^N$ as the smallest of all distances between different codewords in $\mathcal{C}$, i.e. $d_{min} = \min_{\mathbf{a}, \mathbf{b} \in \mathcal{C},\ \mathbf{a} \neq \mathbf{b}} d(\mathbf{a}, \mathbf{b})$. It follows from this definition that a code with minimum distance d_{min} can correct $\lfloor (d_{min} - 1)/2 \rfloor$ errors, since spheres with this radius are disjoint (see [16, p.10, Theorem 2]). If d_{min} is even the code can *detect* $d_{min}/2$ errors, meaning that a received word can not have distance $d_{min}/2$ to one codeword and distance less that $d_{min}/2$ to another one. However it may have distance $d_{min}/2$ to more codewords.

Something more actually can be said. Code $\mathcal{C}$ can decode errors and *erasures* simultaneously. An erasure is an ambiguously received coordinate (the value is not 0 or 1 but undecided). Let $\mathcal{C}$ be a code of length N with minimum distance d_{min} and let $e = \lfloor (d_{min} - 1)/2 \rfloor$. Then the code can correct b errors and c erasures as long as $2b + c < d_{min}$ (for more details see [6]). In other words, we should be able to retrieve the transmitted codeword if during the transmission at most c of the symbols in the word are erased and at most b received symbols are incorrect.

If $\mathcal{C}$ is a T-dimensional subspace of $\mathbb{F}^N$, then the code $\mathcal{C}$ is *linear* and is denoted by $[N, T, d_{min}]$. Set $\mathcal{C}^{\perp} = \{\mathbf{y} \mid \langle \mathbf{y}, \mathbf{x} \rangle = 0 \text{ for all } \mathbf{x} \in \mathcal{C}\}$. The set $\mathcal{C}^{\perp}$ is an $(N - T)$-dimensional linear subspace of $\mathbb{F}^N$ and is called the *dual code* of $\mathcal{C}$.

There are two methods to determine a linear code $\mathcal{C}$: a *generator matrix* and a *parity check matrix*. A *generator matrix* of a linear code $\mathcal{C}$ is any $T \times N$ matrix G whose rows form a basis for $\mathcal{C}$. A generator matrix H of $\mathcal{C}^{\perp}$ is called a *parity check matrix* for $\mathcal{C}$. Clearly, the matrix H is of size $(N - T) \times N$. Hence $\mathbf{x} \in \mathcal{C}$ if and only if $H\mathbf{x}^T = \mathbf{0}$, or in other words $HG^T = GH^T = 0$ holds.

When a *sender* wants to send a message (sometimes called *information vector*) say $\mathbf{x}$ to the *receiver* he calculates a codeword of the code by multiplying the information vector with the generator matrix, e.g. $\mathbf{y} = \mathbf{x}G$. The codeword $\mathbf{y}$ is transmitted to the receiver. The receiver decodes the word $\mathbf{z}$ he received, which is the codeword plus errors, i.e. $\mathbf{z} = \mathbf{y} + \mathbf{err}$, if the number of errors is less than a certain number (the error-correcting capabilities of the code). Recall that for each codeword $\mathbf{y}$ the equality $H\mathbf{y}^T = \mathbf{0}$ holds, hence $H\mathbf{z}^T = \mathbf{err}$ (called *syndrome*) holds.

Let for two vectors $\mathbf{x} = (\mathbf{x^0}, \mathbf{x^1}, \ldots, \mathbf{x^n})$ and $\mathbf{y} = (\mathbf{y^0}, \mathbf{y^1}, \ldots, \mathbf{y^n})$ in $\mathbb{F}^N$, where $\mathbf{x^i}, \mathbf{y^i} \in \mathbb{F}^{p_i}$, the set $\delta_p(\mathbf{x}, \mathbf{y})$ is defined by $\delta_p(\mathbf{x}, \mathbf{y}) = \{i : \mathbf{x^i} \neq \mathbf{y^i}\}$. The *p-support* of vector $\mathbf{x}$, denoted by $\sup_p(\mathbf{x})$, is defined by $\sup_p(\mathbf{v}) = \{i : \mathbf{v^i} \neq \mathbf{0}\}$. Hence $\delta_p(\mathbf{x}, \mathbf{y}) = \sup_p(\mathbf{x} - \mathbf{y}) \subseteq \{0, \ldots, n\}$. Considering the properties of the p-support of a vector, we notice some similarities to the properties of the norm. (1) $\sup_p(\mathbf{x}) = \emptyset$ if and only if $\mathbf{x} = \mathbf{0}$, (2) $\sup_p(j\mathbf{x}) = \sup_p(\mathbf{x})$ if $j \neq 0$, and (3) $\sup_p(\mathbf{x} + \mathbf{z}) \subseteq \sup_p(\mathbf{x}) \cup \sup_p(\mathbf{y})$. In their paper [10] Fehr and Maurer pointed out that $\delta_p(\mathbf{x}, \mathbf{y})$ behaves like a metric, as for all vectors $\mathbf{x}, \mathbf{y}, \mathbf{z} \in \mathbb{F}^N$ one has that (1) $\delta_p(\mathbf{x}, \mathbf{x}) = \emptyset$, (2) $\delta_p(\mathbf{x}, \mathbf{y}) = \delta_p(\mathbf{y}, \mathbf{x})$ (symmetry), and (3) $\delta_p(\mathbf{x}, \mathbf{z}) \subseteq \delta_p(\mathbf{x}, \mathbf{y}) \cup \delta_p(\mathbf{y}, \mathbf{z})$, but actually they do not explore this property. Our first step is to use $\delta_p(\mathbf{x}, \mathbf{y})$ instead of the Hamming distance and to explore the properties of the so defined space.

Let Δ be a monotone decreasing collection of subsets of players. Then $B_\Delta(\mathbf{x})$, the Δ-*neighborhood* of pseudo-radii in Δ centered around the vector $\mathbf{x} \in \mathbb{F}^N$, is defined as follows:

$$B_\Delta(\mathbf{x}) = \{\mathbf{y} \in \mathbb{F}^N : \delta_p(\mathbf{x}, \mathbf{y}) \in \Delta\}.$$

In the special case when Γ is an *e-threshold* access structure ($\Delta = \{A : |A| \leq e\}$), the Δ-neighborhood $B_\Delta(\mathbf{x})$ is in fact the Hamming sphere $B_e(\mathbf{x})$. Now we can generalize the classical sphere packing problem:

Generalized Sphere Packing Problem: Given N and Δ, what is the maximum number of non-intersecting Δ-neighborhoods that can be placed in the N-dimensional space?

As usual we will call any non-empty subset $\mathcal{C}$ of $\mathbb{F}^N$ a code. For a code $\mathcal{C}$ the *set of possible (allowed) distances* is defined by

$$\Gamma(\mathcal{C}) = \{A : \text{ there exist } \mathbf{a}, \mathbf{b} \text{ in } \mathcal{C}, \ \mathbf{a} \neq \mathbf{b} \text{ such that } \delta_p(\mathbf{a}, \mathbf{b}) \subseteq A\}$$

and the *set of forbidden distances* is defined by $\Delta(\mathcal{C}) = \Gamma(\mathcal{C})^c$. It is easy to see that $\Delta(\mathcal{C})$ is monotone decreasing and that $\Gamma(\mathcal{C})$ is monotone increasing. Let us call the so-defined codes *error-set correcting codes*. For the classical error-correcting codes all $p_i = 1$ and since the Hamming distance is "symmetric" (because of equivalence of all coordinates) we set $\Delta(\mathcal{C}) = \{A : |A| < d_{min}\}$ keeping the symmetry. Nevertheless for some classical error-correcting codes there are sets A such that $|A| \geq d_{min}$ and there are no codewords $\mathbf{a}$ and $\mathbf{b}$ with property $\delta_P(\mathbf{a}, \mathbf{b}) \subseteq A$. We can define the set of *minimal* codeword support differences as

$$\Gamma(\mathcal{C})^- = \{A : \text{there exist } \mathbf{a}, \mathbf{b} \text{ in } \mathcal{C}, \ \mathbf{a} \neq \mathbf{b} \text{ such that } \delta_p(\mathbf{a}, \mathbf{b}) = A$$
$$\text{but, there is no } \mathbf{c}, \mathbf{d} \in \mathcal{C}, \ \mathbf{c} \neq \mathbf{d}, \ \delta_p(\mathbf{c}, \mathbf{d}) \subsetneqq A\}. \tag{1}$$

We will focus our attention only on linear codes, even though many of the considerations remain valid in non-linear settings too. Using the relation between δ_p and $\sup_p$ we could redefine the notion *minimal codeword* (introduced by Massey [14]) as follows: The codeword $\mathbf{x}$ in $\mathcal{C}$ is *minimal* if $\sup_p(\mathbf{x})$ in $\Gamma(\mathcal{C})^-$.

As noted before, the packing problem is fundamental in error correction. The natural question that arises now is how the new packing problem is related to the theory of error-correcting codes?

In coding theory, any subset of coordinates is equally likely to be in error (and/or erasure). In the model we consider here some subsets of coordinates are assumed to be more likely in error than others. A well-studied model where this situation arises is the so-called *bursty* channel, in which errors occur in clusters. Another related approach are the so-called *D-codes* [9] which have restricted (to some interval) inner distance distribution. Now we will prove that the error-set correcting codes have similar error-correcting capabilities as the classical codes have.

Theorem 1. *An error-set correcting code $\mathcal{C}$ with set of forbidden distances $\Delta(\mathcal{C})$ can correct all errors in Δ if and only if $\Delta \uplus \Delta \subseteq \Delta(\mathcal{C})$.*

Proof. First we will prove that the centers of a new sphere packing constitute a code $\mathcal{C}$ with set of possible distances $\Gamma(\mathcal{C}) \subseteq \Gamma \uplus \Gamma$ (and thus $\Delta \uplus \Delta \subseteq \Delta(\mathcal{C})$). Indeed, let $\mathbf{a}, \mathbf{b}$ be any two distinct centers of $\mathcal{C}$. Any two sets $A, B \in \Delta$ are in the Δ-neighborhoods of say $\mathbf{a}$, resp. $\mathbf{b}$. Since these neighborhoods are non-intersecting we have that $A \cup B \subset \delta_p(\mathbf{a}, \mathbf{b})$. Hence $\delta_p(\mathbf{a}, \mathbf{b}) \notin \Delta \uplus \Delta$. Conversely, suppose that $\delta_p(\mathbf{a}, \mathbf{b}) \in \Delta \uplus \Delta$. Then there exist $A, B \in \Delta$, such that $A \cup B = \delta_p(\mathbf{a}, \mathbf{b})$. By the "triangle inequality" we have that $\delta_p(\mathbf{a}, \mathbf{b}) \subseteq \delta_p(\mathbf{a}, \mathbf{x}) \cup \delta_p(\mathbf{x}, \mathbf{b})$, and equality holds if $\delta_p(\mathbf{a}, \mathbf{x}) \cap \delta_p(\mathbf{b}, \mathbf{x}) = \emptyset$. Now it is easy to see that there exists $\mathbf{x}$ such that $A \cup B = \delta_p(\mathbf{a}, \mathbf{b}) = \delta_p(\mathbf{a}, \mathbf{x}) \cup \delta_p(\mathbf{b}, \mathbf{x})$ and $\delta_p(\mathbf{a}, \mathbf{x}) \subseteq A$, $\delta_p(\mathbf{b}, \mathbf{x}) \subseteq B$. This contradicts the fact that any Δ-neighborhoods of $\mathbf{a}$ and $\mathbf{b}$ are non-intersecting. $\square$

Example 1. Consider the special case with threshold access structure, so $\Delta = \{A : |A| \leq e\}$. Write as above $B_\Delta(\mathbf{x}) = B_e(\mathbf{x})$ (the usual Hamming sphere). Now $\Delta \uplus \Delta = \{A : |A| \leq 2e\} = \Delta(\mathcal{C})$ and so $\Gamma(\mathcal{C}) = \{A : |A| \geq 2e + 1\}$. Hence the minimum distance of $\mathcal{C}$ is $d_{min} = 2e + 1$. In this case, Theorem 1 is equivalent to the classical error-correcting theorem [16, 6].

Remark 1. Assume that a codeword from $\mathcal{C}$ was sent and that some subset of errors $A \in \Delta$ occurred during the transmission. To decode the received vector $\mathbf{z}$, we compute the Δ-neighborhood $B_\Delta(\mathbf{z})$ and check which codeword is in this Δ-neighborhood. In fact, since the error-pattern is a set A in Δ and $\Delta \uplus \Delta \subseteq \Delta(\mathcal{C})$, there will be only one codeword in the Δ-neighborhood of $\mathbf{z}$ and so we can correct the errors.

Something more actually is true: we can decode errors and erasures simultaneously in the generalized setting too. Let $\mathcal{C}$ be a code of length N with set of forbidden distances $\Delta(\mathcal{C})$. Suppose that $\Delta \uplus \Delta \subseteq \Delta(\mathcal{C})$. Then $\mathcal{C}$ can correct all errors in Δ. Moreover, for any $\Delta_c, \Delta_b \subseteq \Delta$ such that $\Delta_c \uplus \Delta_c \uplus \Delta_b \subseteq \Delta(\mathcal{C})$, the code $\mathcal{C}$ can correct all errors in Δ_c and erasures in Δ_b. In fact, the decoding method coincides with the classical method of decoding errors and erasures, see [6] for example.

3   A Burst-Correcting Technique

We will call a *burst* any error pattern consisting of several sub-vectors $\mathbf{x^i}$ of $\mathbf{x} = (\mathbf{x^0}, \mathbf{x^1}, \ldots, \mathbf{x^n})$, which are not necessarily consecutively ordered. First, we will present a standard burst-error correcting technique, which uses error-correcting codes. The idea is to change the order of the coordinates of several consecutive codewords in such a way that a burst is spread out over the various codewords. Let $\mathcal{C}$ be a code of length n and let ℓ be some positive integer. Consider an $\ell \times n$ matrix which has codewords in $\mathcal{C}$ as their rows. Read this matrix column-wise from top to bottom starting with the leftmost column. The resulting codewords have length $n\ell$ and form a so-called *interleaved code* derived from $\mathcal{C}$ at depth ℓ. If $\mathcal{C}$ can correct e-errors then the interleaved code can correct bursts of length $e\ell$.

Let $\mathcal{C}$ be an error-set correcting code of length N, with a set of forbidden distances $\Delta(\mathcal{C})$ and $d \times N$ generator matrix G. The sender wants to send an information matrix $X \in \mathbb{F}^{d \times d}$ (assume for the sake of simplicity that X is symmetric). Note that X could be asymmetric too, in which case X and X^T are encoded. Thus the sender calculates the (array) codeword Y as $Y = XG$, $(Y \in \mathbb{F}^{N \times N})$. Then applying the interleaving approach the sender reads the matrix column-wise. From now on we will consider only interleaved codes at depth d.

Theorem 2. *Let $\mathcal{C}$ be an error-set correcting code of length N, with set of forbidden distances $\Delta(\mathcal{C})$. Then the interleaving error-set correcting code derived from $\mathcal{C}$ of length N can efficiently correct all burst-errors in Δ if and only if $\Delta \uplus \Delta \subseteq \Delta(\mathcal{C})$.*

Proof. Since every row in the array codeword is a codeword of the error-set correcting code $\mathcal{C}$ and the errors are spread we can correct them row by row (see Theorem 1). On the other hand we will show that the known VSS/DC technique called "pair-wise" checking, provides efficient detection of inconsistency in cases with excess of information. Moreover this technique has an additional advantage that all checks can be performed privately (which is of great importance in SSS).

The "pair-wise" technique is applied as follows. The receiver calculates a symmetric consistency $n \times n$ matrix, verifying the equation $G^T Y = G^T X G = Y^T G$. In other words he puts 1 on entry (i, j) if $G_i^T Y_j = Y_i^T G_j$ and 0 otherwise. Using the consistency matrix (as in the VSS/DC protocols, e.g. [5, 17]) and assuming an error pattern in Δ occurs it is easy to find a set in $\Gamma(\mathcal{C})$ which is consistent, therefore uniquely define the codeword.   □

Remark 2. The interleaving error-set correcting code derived from $\mathcal{C}$ of length N can efficiently correct the burst-error patterns in Δ_c and burst-erasure patterns Δ_b if and only if $\Delta_c \uplus \Delta_c \uplus \Delta_b \subseteq \Delta(\mathcal{C})$.

4   SSS as an Example of a Particular Class of "Codes"

First we give a formal definition of a Monotone Span Program.

Definition 5. *[13] A Monotone Span Program (MSP) $\mathcal{M}$ is a quadruple $(\mathbb{F}, M, \varepsilon, \psi)$, where $\mathbb{F}$ is a finite field, M is a matrix (with m rows and $d \leq m$ columns) over $\mathbb{F}$, $\psi : \{1, \ldots, m\} \to \{1, \ldots, n\}$ is a surjective (labelling) function and $\varepsilon = (1, 0, \ldots, 0)^T \in \mathbb{F}^d$ is called target vector.*

As ψ labels each row with a number i from $\{1, \ldots, m\}$ that corresponds to player $P_{\psi(i)}$, we can think of each player as being the "owner" of one or more rows. Also consider a "function" φ from $\{P_1, \ldots, P_n\}$ to $\{1, \ldots, m\}$ which gives for every player P_i the set of rows owned by him (denoted by $\varphi(P_i)$). In some sense φ is "inverse" of ψ. For any set of players $B \subseteq \mathcal{P}$ consider the matrix consisting of rows these players own in M, i.e. $M_{\varphi(B)}$. As is common, we shall shorten the notation $M_{\varphi(B)}$ to just M_B. The reader should stay aware of the difference between M_B for $B \subseteq \mathcal{P}$ and for $B \subseteq \{1, \ldots, m\}$.

An MSP is said *to compute* a (complete) access structure Γ when $\varepsilon \in \text{im}(M_A^T)$ if and only if A is a member of Γ. In other words, the players in A can reconstruct the secret precisely if the rows they own contain in their linear span the target vector of $\mathcal{M}$, and otherwise they get no information about the secret. In other words there exists a so-called *recombination vector* (column) $\boldsymbol{\lambda}$ such that $M_A^T \boldsymbol{\lambda} = \varepsilon$ hence $\langle \boldsymbol{\lambda}, M_A(s, \boldsymbol{\rho})^T \rangle = \langle M_A^T \boldsymbol{\lambda}, (s, \boldsymbol{\rho})^T \rangle = \langle \varepsilon, (s, \boldsymbol{\rho})^T \rangle = s$ for any secret s and any random vector $\boldsymbol{\rho}$. It is easy to check that the vector $\varepsilon \notin \text{im}(M_B^T)$ if and only if there exists a $\boldsymbol{k} \in \mathbb{F}^d$ such that $M_B \boldsymbol{k} = \boldsymbol{0}$ and $\boldsymbol{k}_1 = 1$.

We stress here that

$$A \in \Gamma \iff \exists\ \boldsymbol{\lambda} \in \mathbb{F}^{|\varphi(A)|} \text{ such that } M_A^T \boldsymbol{\lambda} = \varepsilon \tag{2}$$
$$B \notin \Gamma \iff \exists\ \boldsymbol{k} \in \mathbb{F}^d \text{ such that } M_B \boldsymbol{k} = \boldsymbol{0} \text{ and } \boldsymbol{k}_1 = 1.$$

The first property guaranties *correctness* and the second *privacy* of the SSS. Technically the property (2) means that when we consider the restricted matrix M_A for some subset A of $\mathcal{P}$, the first column is linearly dependent to the other columns if and only if $A \notin \Gamma$. Sometimes we will slightly change the first property rewriting it in the following way:

$$A \in \Gamma \iff \exists\ \boldsymbol{\lambda} \in \mathbb{F}^m \text{ such that } M^T \boldsymbol{\lambda} = \varepsilon \text{ and } \sup_p(\boldsymbol{\lambda}) \subseteq A. \tag{3}$$

The latest in fact is the same vector $\boldsymbol{\lambda}$ as in (2), but expanded with zeroes.

Definition 6. *([8, Definition 3.2.2]) Let $\Gamma^- = \{X_1, \ldots, X_r\}$. Then the set of vectors $C = \{\mathbf{c^i} \in \mathbb{F}^m : 1 \leq i \leq r\}$ is said to be* suitable *for the access structure Γ if C satisfies the following properties called $g(\Gamma)$ respectively $d^-(\Delta)$.*

- *$\sup_p(\mathbf{c^i}) = X_i$ for $1 \leq i \leq r$;*
- *For any vector $(\mu_1, \ldots, \mu_r)$ in $\mathbb{F}^r$, such that $\sum_{i=1}^{r} \mu_i \neq 0$, there exists a set $X \in \Gamma = \Delta^c$ satisfying $X \subseteq \sup_p(\sum_{i=1}^{r} \mu_i \mathbf{c^i})$.*

It is easy to verify that the minimal codewords defined by Massey [14] are related to the notion suitable set. In the next theorem due to Van Dijk an important link between a parity check matrix of a code generated as a span of suitable vectors and an MSP matrix is given.

Theorem 3. *([8, Theorem 3.2.5, Theorem 3.2.6]) Let $\Gamma^- = \{X_1, \ldots, X_r\}$. Consider a set of vectors $C = \{\mathbf{c^i} : 1 \leq i \leq r\}$. Let H be a parity check matrix of the code generated by the linear span of the vectors $(1, \mathbf{c^i})$ $1 \leq i \leq r$ and let H be of the form $H = (\varepsilon \mid H')$ (This can be assumed without loss of generality). Then the MSP with the matrix M defined by $M^T = H'$ computes the access structure Γ if and only if the set of vectors C is suitable for Γ.*

There is a tight connection between an access structure and its dual. It turns out that the codes generated by the corresponding sets of suitable vectors are orthogonal.

Theorem 4. *([8, Theorem 3.5.4]) Let $\Gamma^- = \{X_1, \ldots, X_r\}$ be an access structure and $(\Gamma^\perp)^- = \{Z_1, \ldots, Z_t\}$ be its dual. Then there exists a suitable set $C = \{\mathbf{c^i} : 1 \leq i \leq r\}$ for Γ if and only if there exists a suitable set $C^\perp = \{\mathbf{h^j} : 1 \leq j \leq t\}$ for $\Gamma^\perp$.*
Suppose there exists a suitable set C for Γ and a suitable set $C^\perp$ for $\Gamma^\perp$. Let C^* be the code defined by the linear span of vectors $\{(1, \mathbf{c^i}) : 1 \leq i \leq r\}$ and let $C^\perp$ be the code defined by the linear span of vectors of $\{(1, \mathbf{h^j}) : 1 \leq j \leq t\}$. Then the codes C^* and $C^\perp$ are orthogonal to each another.

Lemma 1. *[19] Let $\Gamma^- = \{X_1, \ldots, X_r\}$ be the access structure computed by MSP $\mathcal{M}$. Also let $\boldsymbol{\lambda^i} \in \mathbb{F}^m$ be the recombination vectors that corresponds to X_i see (2) and (3). Then the set of vectors $C = \{\boldsymbol{\lambda^i} : 1 \leq i \leq r\}$ defines a suitable set of vectors for the complete access structure Γ.*

Theorem 5. *[19] Let $\mathcal{M}$ be an MSP program computing Γ, and $\mathcal{M}^\perp$ be an MSP computing the dual access structure $\Gamma^\perp$. Let code $C^\perp$ have the parity check matrix $H^\perp = (\varepsilon \mid (M^\perp)^T)$ and let code C have the parity check matrix $H = (\varepsilon \mid M^T)$. Then for any MSP $\mathcal{M}$ there exists an MSP $\mathcal{M}^\perp$ such that C and $C^\perp$ are dual.*

McEliece and Sarwate [15] reformulated the Shamir's scheme in terms of Reed-Solomon codes instead of in terms of polynomials, adding in this way error-correcting properties. The general relationship between linear codes and secret sharing schemes was established by Massey [14], Blakley and Kabatianskii [2]. In fact, the coding theoretic approach can be reformulated as the vector space construction, which was introduced by Brickel in [3]. This approach was generalized to the so-called generalized vector space construction by Van Dijk [8]. Two approaches of constructing secret sharing schemes based on linear codes could be distinguished.

The first one uses an $[n, k+1, d_{min}]$ linear code $\overline{C}$. Let $\overline{G}$ be a generator matrix of $\overline{C}$, so it is $(k+1) \times n$ matrix. The dealer $\mathcal{D}$ chooses a random information vector $\mathbf{x} \in \mathbb{F}^{k+1}$, subject to $\mathbf{x}_1 = s$ - the secret. Then he calculates the codeword $\mathbf{y}$ corresponding to this information vector as $\mathbf{y} = \mathbf{x}\overline{G}$, $(\mathbf{y} \in \mathbb{F}^n)$. Then $\mathcal{D}$ gives $\mathbf{y}_j$ to player P_j to be his share.

The second approach uses an $[N = n+1, k+1, d_{min}]$ linear code $\widetilde{\mathcal{C}}$. Let $\widetilde{G}$ be a generator matrix of $\widetilde{\mathcal{C}}$, so it is $(k+1) \times (n+1)$. The dealer $\mathcal{D}$ calculates the codeword $\mathbf{y}$ as $\mathbf{y} = \mathbf{x}\widetilde{G}$, $(\mathbf{y} \in \mathbb{F}^N)$, from a random information vector $\mathbf{x} \in \mathbb{F}^{k+1}$, subject to $\mathbf{y}_0 = s$ - the secret. Then $\mathcal{D}$ gives $\mathbf{y}_j$ to player P_j to be his share.

The two kinds of approaches seem different but are related. In the first approach all the shares form a *complete* codeword of the code, while in the second one all the shares form only part of a codeword. But as Van Dijk [8] proved one can simply transform the matrices of the codes, setting $\widetilde{G} = (\varepsilon \mid \overline{G})$. Hence one can consider the code $\overline{\mathcal{C}}$ to be obtained from the code $\widetilde{\mathcal{C}}$ by *puncturing* i.e. by deleting a coordinate [16].

Now we will generalize these approaches to our error-set correcting codes. We will denote the codes and their generator matrices for the first (and the second) approaches by $\overline{\mathcal{C}}$ and $\overline{G}$ ($\widetilde{\mathcal{C}}$ and $\widetilde{G}$, respectively). Let $\overline{\mathcal{C}}$ be a code of length p, with set of forbidden distances $\Delta(\overline{\mathcal{C}})$ and with $d \times p$ generator matrix $\overline{G}$. Analogously let $\widetilde{\mathcal{C}}$ be a code of length N, with set of forbidden distances $\Delta(\widetilde{\mathcal{C}})$ and with $d \times N$ generator matrix $\widetilde{G}$. Recall that $\widetilde{G} = (\varepsilon \mid \overline{G})$ holds.

Lemma 2. *Let* $\mathcal{M} = (\mathbb{F}, M, \varepsilon, \psi)$ *be an MSP computing an access structure* Γ. *Let* $\widetilde{\mathcal{C}}$ *be an error-set correcting code of length* N, *with a set of forbidden distances* $\Delta(\widetilde{\mathcal{C}})$ *and with* $d \times N$ *generator matrix* $\widetilde{G}$ *of the form* $\widetilde{G} = (\varepsilon \mid M^T)$. *Then* $\Delta(\widetilde{\mathcal{C}}) = \Delta^{\perp} \uplus \{\mathcal{D}\}$.

Proof. Let $\mathcal{M}$ be an MSP computing an access structure Γ and $\mathcal{M}^{\perp}$ be its dual MSP. Using $\overline{G} = M^T$ and $\overline{G}^{\perp} = (M^{\perp})^T$ compute the codes $\widetilde{\mathcal{C}}$ and $\widetilde{\mathcal{C}}^{\perp}$. Van Dijk [8] proved that codes $\widetilde{\mathcal{C}}$ and $\widetilde{\mathcal{C}}^{\perp}$ are orthogonal to each other. Moreover Van Dijk showed (see Definition 6 and Theorems 3 and 4) that matrix $\widetilde{G} = (\varepsilon \mid M^T)$ is generated by vectors $(1, \mathbf{h}^j)$ where $\mathbf{h}^j$ are suitable vectors for the dual access structure $\Gamma^{\perp}$. It turns out that the codes $\widetilde{\mathcal{C}}$ and $\widetilde{\mathcal{C}}^{\perp}$ are even dual (see Theorem 5). Thus by Lemma 1 and Definition 6 we have that $sup_p(\mathbf{h}^j) \in (\Gamma^{\perp})^-$. In other words the suitable vectors for $\Gamma^{\perp}$ are the *minimal* codewords for the code $\widetilde{\mathcal{C}}$, see definition (1). Hence we have $\Delta(\widetilde{\mathcal{C}}) = \Delta^{\perp} \uplus \{\mathcal{D}\}$ to be the set of forbidden distances for the code generated by $\widetilde{G}$.   □

Note that the set $\Delta^{\perp} \uplus \{\mathcal{D}\}$ is not monotone decreasing, thus in order to ensure this property we need stronger requirements to hold.

Definition 7. *[19] An MSP is called* Δ-*non-redundant (denoted by* Δ-*rMSP) when* $v \in \ker(M^T) \iff v \neq 0$ *and* $\sup(v) \in \Gamma$ ($\Gamma = \Delta^c$).

Corollary 1. *Let* $\mathcal{M}^{\perp}$ *be a* $\Delta^{\perp}$-*rMSP computing* $\Gamma^{\perp}$ *and let* M *be the matrix of the dual MSP* $\mathcal{M}$ *computing* Γ. *Let* $\widetilde{\mathcal{C}}$ *be an error-set correcting code with a generator matrix* $\widetilde{G}$ *of the form* $\widetilde{G} = (\varepsilon \mid M^T)$. *Then the set of forbidden distances* $\Delta(\widetilde{\mathcal{C}})$ *is equal to* $\Delta^{\perp} \uplus \{\emptyset, \mathcal{D}\}$.

Example 2. In the threshold case $\widetilde{G}$ can be the generator matrix of the extended Reed-Solomon MDS code $[n+1, k+1, n-k+1]$, since $\overline{G}^T$ can be the Vandermonde matrix with rows $(1, \alpha, \alpha^2, \ldots, \alpha^k)$. In other words the extended Reed-Solomon code can be used to generate an (k, n) threshold scheme.

Remark 3. Lemma 2 gives an efficient way to construct error-set correcting codes using MSPs. Note that we do not require any relation between Δ and Δ_A (or for k and k_a).

Now using the results of Theorem 1 and Lemma 2 we obtain the following corollary.

Corollary 2. *An error-set correcting code $\widetilde{\mathcal{C}}$ corrects Δ_A (k_a in the threshold case) errors and one erasure (e.g. $\{\mathcal{D}\}$) if and only if $\Delta_A \uplus \Delta_A \subseteq \Delta^\perp$ (analogously $2k_a < n - k$).*

Remark 4. The main difference between error-set correcting codes and SSS is that the SSS provides *privacy*, meaning that $\Delta \supseteq \Delta_A$ (or $k \geq k_a$).

It was proven in [18] that $\Gamma \uplus \Gamma^\perp = \{\mathcal{P}\}$ holds.

Remark 5. Recall that for a linear $[N, T, d_{min}]$ code $\mathcal{C}$, the Singleton bound $d_{min} \leq N + 1 - T$ holds and that equality is achieved only for MDS codes. It is well known that the dual code $\mathcal{C}^\perp$ is $[N, N - T, d_{min}^\perp]$ and is MDS code if and only if $\mathcal{C}$ is MDS code. Therefore the following inequality holds:

$$d_{min} + d_{min}^\perp \leq N + 2 \tag{4}$$

with equality only for MDS codes. Now we will show that the equality $\Gamma \uplus \Gamma^\perp = \{\mathcal{P}\}$ is a generalization of the classical coding bound (4).

Consider $\widetilde{\mathcal{C}}$ code and its dual $\widetilde{\mathcal{C}}^\perp$. Then by Lemma 2 the relations $\Delta(\widetilde{\mathcal{C}}) = \Delta^\perp \uplus \{\mathcal{D}\}$ and $\Delta(\widetilde{\mathcal{C}}^\perp) = \Delta \uplus \{\mathcal{D}\}$ hold. Thus for the punctured codes $\overline{\mathcal{C}}$ and $\overline{\mathcal{C}}^\perp$ we have $\Delta(\overline{\mathcal{C}}) = \Delta^\perp$ and $\Delta(\overline{\mathcal{C}}^\perp) = \Delta$. Therefore $\Delta(\overline{\mathcal{C}})^c \uplus \Delta(\overline{\mathcal{C}}^\perp)^c = \{\mathcal{P}\}$. Thus, there exist sets A and B such that $A \in \Delta(\overline{\mathcal{C}})^+$, $B \in \Delta(\overline{\mathcal{C}}^\perp)^+$ and $|A \cup B| = n-1$. Hence $(d_{min}(\overline{\mathcal{C}})-1)+(d_{min}(\overline{\mathcal{C}}^\perp)-1) \leq n-1$ holds and thus $d_{min}(\overline{\mathcal{C}})+d_{min}(\overline{\mathcal{C}}^\perp) \leq n+1$. Consider the threshold case. We have that $\widetilde{\mathcal{C}}$, $\widetilde{\mathcal{C}}^\perp$, $\overline{\mathcal{C}}$ and $\overline{\mathcal{C}}^\perp$ are MDS codes. In other words if $\widetilde{\mathcal{C}}$ is an $[n+1, k+1, n+1-k]$ code then $\widetilde{\mathcal{C}}^\perp$ is an $[n+1, n-k, k+2]$ code, $\overline{\mathcal{C}}$ is an $[n, k + 1, n - k]$ code and $\overline{\mathcal{C}}^\perp$ is an $[n, n - k, k + 1]$ code. Now it is easy to check that we have the equality $d_{min}(\overline{\mathcal{C}}) + d_{min}(\overline{\mathcal{C}}^\perp) = n + 1$ in that case.

5   VSS as an Example of a Particular Class of Burst "Codes"

A formal definition of VSS is as follows.

Definition 8. *A* Verifiable Secret Sharing *scheme secure against (Δ, Δ_A)-adversary $\mathcal{A}$ is a pair* (Share-Detect, Reconstruct) *of protocols (phases). At the* *beginning of the* Share-Detect *phase the dealer* $\mathcal{D}$ *inputs to the protocol a secret* $s \in \mathcal{K}$, *at the end of* Share-Detect *phase each player* P_i *is instructed to output either "accept" or "reject". At the end of the* Reconstruct *phase each player* P_i *is instructed to output a value in* $\mathcal{K}$. *The protocol is unconditionally secure if the following properties hold:*

- Termination (Acceptance): *If a honest player* P_i *outputs "reject" at the end of* Share-Detect *then every honest player outputs "reject"; Moreover if the dealer* $\mathcal{D}$ *is not corrupt, then every honest player* P_i *outputs "accept";*
- Correctness (Verifiability): *If a group of honest players* P_i *outputs "accept" at the end of* Share-Detect, *then at this time a value* $s' \in \mathcal{K}$ *has been fixed and at the end of* Reconstruct *all honest players will output the same value* s'. *Moreover if the dealer is not corrupt* $s' = s$.
- Privacy (Unpredictability): *If the secret* s *is chosen randomly from* $\mathcal{K}$, *and the dealer is not corrupt, then any forbidden coalition cannot guess at the end of* Share-Detect *the value* s *with probability better than* $1/|\mathcal{K}|$.

The distributed commitments can be seen as a reduced (weaken) version of VSS, since the VSS schemes provide robustness that the players can reconstruct alone the secret (without dealer's help), while for DC schemes the secret can not be reconstructed without the dealer's help. Note that an SSS with error-correcting capabilities could be considered as an VSS with *honest dealer*, since the *robustness* is guaranteed using the interleaving technique. Therefore we will first revisit the standard approaches described in the literature used to build SSS from codes employing the interleaving technique.

The *first* approach uses an $[n, k+1, d_{min}]$ linear code $\overline{\mathcal{C}}$. Let $\overline{G}$ be a generator matrix of $\overline{\mathcal{C}}$, so its size is $(k+1) \times n$. Now the dealer $\mathcal{D}$ chooses a random information matrix $X \in \mathbb{F}^{(k+1) \times (k+1)}$, except that s (the secret) is in its upper-left corner. Then $\mathcal{D}$ calculates the (array) codeword Y corresponding to this information matrix $Y = X\overline{G}$, ($Y \in \mathbb{F}^{n \times n}$). Note that the rows in Y are the usual codewords of $\overline{\mathcal{C}}$. Using the interleaving approach the dealer $\mathcal{D}$ gives columns $Y_{(j)}$ to the player P_j as his share. Note that the first coordinate in $Y_{(j)}$ corresponds to the first codeword which encodes the secret.

The *second* approach is very similar. Now $\widetilde{\mathcal{C}}$ is an $[N = n+1, k+1, d_{min}]$ linear code. Let $\widetilde{G}$ be a generator matrix of $\widetilde{\mathcal{C}}$, so it is a $(k+1) \times (n+1)$ matrix. The dealer $\mathcal{D}$ calculates the (array) codeword Y as $Y = X\widetilde{G}$, ($Y \in \mathbb{F}^{N \times N}$), from a random information matrix $X \in \mathbb{F}^{(k+1) \times (k+1)}$, except that s (the secret) is in the upper-left corner of Y. Again applying the interleaving approach the dealer $\mathcal{D}$ gives columns $Y_{(j)}$ to player P_j as his share. Note that the first coordinate in $Y_{(j)}$ corresponds to the first codeword which encodes the secret. The zero column $Y_{(0)}$ is the dealer's share.

It is straightforward to generalize these two approaches to error-set correcting codes. In this case $\overline{\mathcal{C}}$ is a code of length p, with a set of forbidden distances $\Delta(\overline{\mathcal{C}})$ and $\overline{G}$ is a $d \times p$ matrix. Analogously $\widetilde{\mathcal{C}}$ is a code of length N, with a set of forbidden distances $\Delta(\widetilde{\mathcal{C}})$ and $\widetilde{G}$ is a $d \times N$ matrix. Recall that $\widetilde{G} = (\varepsilon \mid \overline{G})$ holds. Then $X \in \mathbb{F}^{d \times d}$ and $Y \in \mathbb{F}^{p \times p}$ for the first approach and $Y \in \mathbb{F}^{N \times N}$ for the second. Note that X could be symmetric or asymmetric.

The sharing procedure we have just described coincides with the sharing procedures of the standard VSS/DC protocols [1, 5, 17]. Note that the shares in these protocols are distributed in exactly the same way using the interleaving technique. We will say that the VSS (with honest dealer) is *based on code* $\tilde{\mathcal{C}}$. Now we will translate the results of Lemma 2 and Corollary 2 into the VSS language.

Proposition 1. *Let $\tilde{\mathcal{C}}$ be an error-set correcting code of length N, with a set of forbidden distances $\Delta(\tilde{\mathcal{C}})$. Let consider VSS (with honest dealer) based on this code and $(\Delta, \Delta_A, \Delta_F)$-adversary $((k, k_a, k_f)$-adversary).*

- Correctness:
  Then VSS (with honest dealer) based on this code satisfy the correctness property in Definition 8 if and only if the code $\tilde{\mathcal{C}}$ is able to correct burst-error pattern in Δ_A (k_a in threshold case) and burst-erasure pattern in $\Delta_F \uplus \{\mathcal{D}\}$ ($k_f + 1$), i.e. $\Delta_A \uplus \Delta_A \uplus \{\mathcal{D}\} \uplus \Delta_F \subseteq \Delta(\tilde{\mathcal{C}})$ ($2k_a + k_f + 1 < d_{min}$).
- Privacy:
  Then VSS (with honest dealer) based on this code satisfy the correctness property in Definition 8 if and only if the code $\tilde{\mathcal{C}}$ has $\Delta(\tilde{\mathcal{C}})$ as the set of forbidden distances, i.e. $\Delta(\tilde{\mathcal{C}}) = \Delta^{\perp} \uplus \{\mathcal{D}\}$ ($d_{min} = n - k + 1$).

Proof. The result for a (Δ, Δ_A)-adversary (i.e. without Δ_F) follows directly from Lemma 2 and Corollary 2.

It is straightforward to extend this model to include also *fail-corrupt* players. Recall that to fail-corrupt a player means that the adversary may stop the communication from and to that player at an arbitrary moment during the protocol. From a coding point of view these players are erasures, so the bounds are extended naturally to $\mathcal{P} \notin \Delta_A \uplus \Delta_A \uplus \Delta \uplus \Delta_F$ ($2k_a + k + k_f < n$). □

In coding theory the *Sender* is always assumed to be honest, while in VSS/DC protocol the Dealer could be corrupt. We could simulate the improper behavior of the dealer in the following way.

Let $\tilde{\mathcal{C}}$ be a code of length N, with set of forbidden distances $\Delta(\tilde{\mathcal{C}})$ and $\tilde{G}$ be a $d \times N$ generator matrix for the code. The sender chooses information matrix $X \in \mathbb{F}^{d \times d}$ (using the first approach). Then he computes the array codeword $Y \in \mathbb{F}^{N \times N}$ by $Y = X\tilde{G}$. But instead of distributing the columns of Y to the players as their shares, the dealer introduces a burst-error pattern (not necessarily in Δ_A) obtaining matrix Z from Y in this way. Then he distributes Z as shares. Since after receiving their shares the corrupt players could hand in wrong ones (i.e. introducing another burst-error pattern in Δ_A) in the reconstruction phase we simulate this behavior as *retransmitting* Z to $\tilde{Z}$. Since we are able to correct only the error-patterns in Δ_A, we need to apply twice the decoding algorithm (pairwise checking protocol) in order to correct the errors. But even then we have the problem that the sender could introduce errors not from Δ_A and that the errors he introduced together with the errors that the corrupt players introduced could be not from Δ_A. What the share-detection phase in the VSS/DC protocols (e.g. [5, 17]) achieves more is that the dealer is forced (by the accusation-broadcast mechanism) to defend himself if inconsistent information (not in Δ_A) is found. Thus the honest players have (maybe after being broadcasted by the dealer) consistent shares. This could be simulated by the assumption that Z and Y differ in an error pattern which is a subset of the error pattern between Z and $\widetilde{Z}$. Therefore the difference between Y and $\widetilde{Z}$ is an error pattern from Δ_A. This immediately gives the following requirements for the code in this *retransmitting scenario*.

Theorem 6. *Let $\widetilde{C}$ be an error-set correcting code of length N, with a set of forbidden distances $\Delta(\widetilde{C})$. Let consider VSS based on this code and $(\Delta, \Delta_A, \Delta_F)$-adversary $((k, k_a, k_f)$-adversary).*

- Correctness:
  Then VSS based on this code satisfy the correctness property in Definition 8 if and only if the code $\widetilde{C}$ is able to correct burst-error pattern in Δ_A (k_a in the threshold case) and burst-erasure pattern in $\Delta_F \uplus \{\mathcal{D}\}$ ($k_f + 1$), i.e.
  $$\Delta_A \uplus \Delta_A \uplus \{\mathcal{D}\} \uplus \Delta_F \subseteq \Delta(\widetilde{C}) \ (2k_a + k_f + 1 < d_{min}).$$
- Privacy:
  Then VSS based on this code satisfy the correctness property in Definition 8 if and only if the code $\widetilde{C}$ has $\Delta(\widetilde{C})$ as the set of forbidden distances, i.e.
  $$\Delta(\widetilde{C}) = \Delta^\perp \uplus \{\mathcal{D}\} \ (d_{min} = n - k + 1).$$

The last bounds coincide with the well known bounds in $[1, 11, 5, 10]$, namely, $\Delta_A \uplus \Delta_A \uplus \Delta_F \subseteq \Delta^\perp$ or equivalently $\mathcal{P} \notin \Delta_A \uplus \Delta_A \uplus \Delta_F \uplus \Delta$ (in the threshold case the bound becomes $2k_a + k_f + k < n$).

We recall the following notions [16]: code C is called *weakly self-dual* if and only if $C \subsetneqq C^\perp$, and code C is called *self-dual* if and only if $C = C^\perp$.

Remark 6. It is interesting to look at the following question: How are dual codes and dual access structures linked? On one hand if the codes are weakly self-dual we know the following facts:

- When $\widetilde{C}$ $(\overline{C})$ is weakly self-dual code, i.e. $\widetilde{C} \subsetneqq \widetilde{C}^\perp$, we have $\Gamma(\widetilde{C}) \subsetneqq \Gamma(\widetilde{C}^\perp)$, but from Theorems 3 and 4, it follows that $\Gamma(\widetilde{C}) = \Gamma^\perp$ and $\Gamma(\widetilde{C}^\perp) = \Gamma$. Hence we have $\Gamma^\perp \subsetneqq \Gamma$, i.e. Γ is a $\mathcal{Q}^2$ access structure.
- Let $\widetilde{C}$ $(\overline{C})$ be weakly self-dual code. Taking again into account Theorems 3 and 4, i.e. that $\overline{H} = (M^\perp)^T$, and $\overline{G} = M^T$ we obtain that $D(M^\perp)^T = M^T$, for some non-invertible matrix D. This implies that $\Gamma^\perp \subsetneqq \Gamma$, i.e. Γ is a $\mathcal{Q}^2$ access structure. Note that $\overline{G} \, \overline{H}^T = 0$ implies that $M^T M^\perp = \overline{E}$, where $\overline{E}$ is a zero matrix except for the entry in the upper left corner which is 1.

On the other hand for dual-codes we obtain $\Gamma = \Gamma^\perp$, i.e. the access structure is self-dual, and $D(M^\perp)^T = M^T$ for some invertible matrix D ($M^T M^\perp = \overline{E}$ holds).

Thus weakly self-dual codes correspond to $\mathcal{Q}^2$ access structures, while self-dual codes correspond to self-dual access structures, i.e. to minimal $\mathcal{Q}^2$ access structures.

Several interesting open questions arise.

- Given a $\mathcal{Q}^2$ access structure Γ does there always exist a weakly self-dual error-set correcting code with set of allowed distances Γ?
- Does for any self-dual access structure Γ exist a self-dual error-set correcting code with set of allowed distances Γ?
- One can generalize the notion of *weight* and *distance distribution* of an error-set correcting code (see [16]). It is interesting to check whether the Mac Williams theorem [16] for the weight enumerators of a code and its dual can be generalized to this setting.
- It is well known that for a given access structure Γ (and correspondingly MSP $\mathcal{M}$) the numbers $p_0, p_1, \ldots, p_n$ are the players individual information rate. It would be interesting to see if the invariant theory can be applied to the weight enumerator of self-dual error-set correcting codes (access structures) to find out which numbers are suitable and which are not.

Acknowledgements

The authors would like to thank Henk van Tilborg for the valuable comments and remarks.

References

1. M. Ben-Or, S. Goldwasser, A. Wigderson. Completeness theorems for Non-Cryptographic Fault-Tolerant Distributed Computation, *STOC'88*, pp. 1-10.
2. G. Blakley, G. Kabatianskii. Linear Algebra Aproach to Secret Sharing Schemes, LNCS 829, 1994, pp. 33-40.
3. E. Brickell. Some ideal secret sharing schemes, *J. of Comb. Math. and Comb. Computing* 9, 1989, pp. 105-113.
4. D.Chaum, C.Crepeau, I.Damgard, Multi-Party Unconditionally Secure Protokols, *STOC'88*, pp. 11-19.
5. R. Cramer, I. Damgard, U. Maurer. General Secure Multi-Party Computation from any Linear Secret Sharing Scheme, *EUROCRYPT'00*, LNCS 1807, pp. 316-334.
6. G. Cohen, I. Honkala, S. Litsyn, A. Lobstein. Covering Codes, *Elsevier Science*, Amsterdam, 1997.
7. B. Chor, S. Goldwasser, S. Micali, B. Awerbuch. Verifiable secret sharing and achieving simultaneity in the presence of faults, *FOCS'85*, pp. 383-395.
8. M. van Dijk. Secret Key Sharing and Secret Key Generation, *Ph.D. Thesis*, 1997, TU Eindhoven.
9. P. Delsarte. The Hamming space viewed as an association scheme, *23rd Symp. on Inform. Theory in the Benelux*, 2002, pp. 329-380.
10. S. Fehr, U. Maurer. Linear VSS and Distributed Commitments Based on Secret Sharing and Pairwise Checks, *CRYPTO'02*, LNCS 2442, pp. 565-580.
11. M. Fitzi, M. Hirt, U. Maurer. Trading Correctness for Privacy in Unconditional Multi-Party Computation, *CRYPTO'98*, LNCS 1462, pp. 121-136.
12. M. Hirt, U. Maurer. Player Simulation and General Adversary Structures in Perfect Multiparty Computation, *J. of Cryptology* 13, 2000, pp. 31-60.

13. M. Karchmer, A. Wigderson. On Span Programs, *8th Annual Struct. in Compl, Theory Conf.*, 1993, pp. 102-111.
14. J. Massey. Minimal codewords and secret sharing, *6th Joint Swedish-Russian Int. Workshop on Inform. Theory* 1993, pp. 276-279.
15. R. McEliece, D. Sarwate. On Sharing secrets and Reed-Solomon codes, *Commun. ACM* 24, 1981, pp. 583-584.
16. F. Mac Williams, N. Sloane. The Theory of Error-Correcting Codes, *Elsevier Science*, Amsterdam, 1988.
17. V. Nikov, S. Nikova, B. Preneel, J. Vandewalle. Applying General Access Structure to Proactive Secret Sharing Schemes, *23rd Symp. on Inform. Theory in the Benelux*, 2002, pp. 197-206, *Cryptology ePrint Archive*: Report 2002/141.
18. V. Nikov, S. Nikova, B. Preneel. On Multiplicative Linear Secret Sharing Schemes, *INDOCRYPT'03*, LNCS 2904, pp. 135-147.
19. V. Nikov, S. Nikova, B. Preneel. On the size of Monotone Span Programs, *SCN'04*, LNCS 3352, pp. 252-265.
20. A. Shamir, How to share a secret, *Commun. ACM* 22, 1979, pp. 612-613.

ID-Based Series-Parallel Multisignature Schemes for Multi-Messages from Bilinear Maps

Lihua Wang[1], Eiji Okamoto[1], Ying Miao[1], Takeshi Okamoto[1], and Hiroshi Doi[2]

[1] Graduate School of Systems and Information Engineering,
University of Tsukuba, Tsukuba 305-8573, Japan
`wlh@cipher.risk.tsukuba.ac.jp`, {`okamoto, miao, ken`}`@risk.tsukuba.ac.jp`
[2] Graduate School of Information Security,
Institute of Information Security, Yokohama 221-0835, Japan
`doi@iisec.ac.jp`

Abstract. In this paper series-parallel multisignature schemes for multi-messages are investigated. We propose an ID-based series-parallel multisignature scheme (ID-SP-M4M scheme) based on pairings in which signers in the same subgroup sign the same message, and those in different subgroups sign different messages. Our new scheme is an improvement over the series-parallel multisignature schemes introduced by Doi, Mambo and Okamoto [5] and subsequent results such as the schemes proposed by Burmester et al. [4] and the original protocols proposed by Tada [17, 18], in which only one message is to be signed. Our ID-SP-M4M scheme is secure against forgery signature attack from parallel insiders under the BDH assumption.

1 Introduction

The concept of multisignature schemes was introduced independently by Boyd [3] and Okamoto [15]. A group of n participants generates a multisignature if all n members have to contribute to sign messages. In a multisignature scheme, it is required that the total signature size be smaller than n times of that in the corresponding single signature scheme. Doi et al. [5] considered the case when signers are in different positions and have different responsibilities which often can be reflected by the signing order. In an order-specified multisignature scheme, a multisignature can guarantee not only the set of signers, but also its signing order. Usually, the following two cases are considered: (1) serial signing, in which the signing order can be detected by a verifier from a multisignature; (2) parallel signing, in which the signing order cannot be detected by a verifier from a multisignature.

Multisignature schemes for various group structures composed of serial and parallel signing have been well studied, for example, in [4, 5, 6, 7, 17, 18]. The series-parallel multisignature schemes in [5, 6, 7], [4], and [17, 18] are based on RSA, ElGamal and MOO (Modified Ohta-Okamoto [14]) schemes, respectively. In [4, 5, 6, 7], the public keys corresponding to the signing order have to be registered in advance, while in the schemes proposed in [17, 18], such a task is not necessary. The schemes in [4, 5, 6, 7, 17] are order-specified, whereas the schemes in [18] have the order-flexibility property for a message to be signed. However, unfortunately, almost all of the schemes given above are intended for signing a single message, though series-parallel multisignature schemes for multi-messages are obviously more useful in the real world.

In any company, anyone has the right to launch proposals, but his power is limited according to his rank. Certainly, the higher the rank he holds, the stronger the power he has. If employees want to launch some proposal, then at least two thirds, say, of all the employees should agree to sign it. On the other hand, the employer can sign any message alone. For example, several employees in some project sign a message which requests more members because of the heavy work. The employer examines it, signs another message about the number of new members he would like to recruit, and then submits it to the personnel committee. Upon receipt of this multisignature, the personnel committee verifies its validity, then starts to recruit new members. In this process, the two messages are different.

Sometimes, the employer should sign his message before his employees sign their message. Consider the company in the above example again. Because of the expansion of business, the employer decides to make an investment in equipment, for $10,000, say. He signs this message and then sends it to his employees. On the budget of $10,000, the employees sign a message to order some equipment they need from a firm. On receipt of this multisignature, the person in charge of the firm checks the validity and then takes the order.

As indicated above, in the real world, there is a need to easily change order of signers, add a new signer and exclude a signer. That neither the order of signers nor the signers themselves need to be designated beforehand is an interesting feature which a practical multisignature scheme should have.

In this paper, we propose an ID-based series-parallel multisignature scheme for multi-messages from bilinear maps. This new scheme has the flexibility property for both the number of signers and the order in series, and the public keys corresponding to the signing order have not to be registered in advance. It is a generalization of those schemes mentioned above, for the reason that the signers in different subgroups in the serial chain can sign different messages. This property was called message flexibility in [13].

Mitomi and Miyaji [13] introduced a general model of multisignature schemes (but without any parallel subgroup structure) with message flexibility, order flexibility and order verifiability. By this model, the aim we want to achieve seems to have been achieved. However, in their scheme, the verifier should verify all the signatures signed by the signers one by one, which is clearly inefficient. In our scheme, the verifier needs only to verify the multisignature once.

The case that the messages to be signed in serial signing are different and the multisignature can be verified simultaneously is analogous to the aggregate signature scheme investigated in [2]: given l signatures on l distinct messages from l distinct signers, it is possible to aggregate all these signatures into one single short signature. This single signature (and the l original messages) will convince the verifier that the l users did indeed sign the l original messages, i.e., user i signed message m_i for $i = 1, ..., l$. Clearly, our new scheme is also a generalization of aggregate signature schemes for the reason that no aggregate signature scheme has the parallel subgroup structure.

The rest of this paper is organized as follows. In Section 2, we recap some definitions and difficult mathematical problems. In Section 3, we propose an ID-based series-parallel multisignature scheme for multi-messages from pairings. In Section 4, we give some comparisons among our scheme and previous works. Finally concluding remarks are given in Section 5.

2   Preliminaries

Similarly to Sakai et al.'s [16] and Boneh et al.'s [1] ID-based cryptosystems, our ID-SP-M4M scheme can be built from any bilinear map $\hat{e} : G_1 \times G_1 \longrightarrow G_2$ as long as the bilinear Diffie-Hellman (BDH) problem in $\langle G_1, G_2, \hat{e} \rangle$ is hard. Now we briefly recap some basic concepts and properties related to bilinear maps between groups, and then the BDH assumption.

2.1   Pairings on Elliptic Curves

Let G_1 be an additive cyclic group generated by P, whose order is a large prime q, and G_2 be a multiplicative cyclic group of the same order q. We assume that the discrete logarithm problems in both G_1 and G_2 are hard. Let $\hat{e} : G_1 \times G_1 \longrightarrow G_2$ be a pairing which satisfies the following properties:

(1) *Bilinear.* $\hat{e}(aQ, bR) = \hat{e}(Q, R)^{ab}$ for all $Q, R \in G_1$ and all $a, b \in Z$.
(2) *Nondegenerate.* If $\hat{e}(Q, R) = 1$ for all $Q \in G_1$ then $R = 0$, and also if $\hat{e}(Q, R) = 1$ for all $R \in G_1$ then $Q = 0$.
(3) *Computable.* There is an efficient algorithm to compute $\hat{e}(Q, R)$ for any $Q, R \in G_1$.

We note that the Weil and Tate pairings associated with supersingular elliptic curves or abelian varieties can be modified to create such bilinear maps. According to [1], the modified Weil pairing can be adopted for the convenience of description, although using the Tate pairing would be more efficient [9]. Let $(G_1, +)$ be a subgroup on some elliptic curve $E(F_p)$ (e.g. $E(F_p) : y^2 = x^3 + 1$, for a large prime number $p \equiv 2 \bmod 3$, with q being a large prime factor of $p + 1$, where the point at infinity is denoted as $\mathcal{O}$, see Section 5.1 of [1] or Section 6.8 of [19]) with order q, where q is a large prime number. We note that $(G_1, +)$ is a subgroup of the q-torsion point group $E[q]$. Let $(G_2, \times)$ be the subgroup of $F_{p^2}^*$ of order q. The Weil pairing on the curve $E(F_p)$ is a mapping $e : G_1 \times G_1 \longrightarrow G_2$ (see Appendix A. of [1] or Section 11.2 of [19]). For any $Q, R \in G_1$, the Weil pairing satisfies $e(Q, R) = 1$. In other words, the Weil pairing is degenerate on the group G_1.

To get a nondegenerate map, the modified Weil pairing $\hat{e} : G_1 \times G_1 \longrightarrow G_2$ can be defined as follows: Let $\omega \in F_{p^2}$ be a primitive third root of unity. Define the map $\beta : E(F_{p^2}) \longrightarrow E(F_{p^2}), (x, y) \longmapsto (\omega x, y), \beta(\mathcal{O}) = \mathcal{O}$. Then the modified Weil pairing is defined as

$$\hat{e}(P, Q) = e(P, \beta(Q)),$$

which satisfies the nondegenerate property. Since G_1, G_2 are groups of prime order, this implies that if P is a generator of G_1, then $\hat{e}(P, P)$ is a generator of G_2.

2.2   Security Assumption

Definition 1. *The BDH problem and the computational Diffie-Hellman (CDH) problem are defined as follows:*

*(1) The **BDH problem** in $\langle G_1, G_2, \hat{e} \rangle$: Given $\langle P, aP, bP, cP \rangle$ for some $a, b, c \in Z_q \setminus \{0\}$ where P is a generator of G_1, compute $\hat{e}(P, P)^{abc} \in G_2$.*

*(2) The **CDH problem***
*   *- in the additive cyclic group G_1 of order q: Given $\langle P, aP, bP \rangle$ for some $a, b \in Z_q \setminus \{0\}$ where P is a generator of G_1, compute $abP \in G_1$.*
*   *- in the multiplicative cyclic group G_2 of order q: Given $\langle g, g^a, g^b \rangle$ for some $a, b \in Z_q \setminus \{0\}$ where g is a generator of G_2, compute $g^{ab} \in G_2$.*

Lemma 1 ([1,9]). *The BDH problem in $\langle G_1, G_2, \hat{e} \rangle$ is no harder than the CDH problem in G_1 or G_2, and the CDH problem in G_i is no harder than the discrete logarithm problem in G_i, $i = 1, 2$.*

Moreover, it is known (see [1]) that it is difficult to compute $X \in G_1$ for given $A \in G_1$ and $\hat{e}(X, A) \in G_2$.

Definition 2 (BDH Assumption). *We assume that the BDH problem is hard in $\langle G_1, G_2, \hat{e} \rangle$, which means that there is no efficient algorithm to solve the BDH problem with non-negligible probability.*

Let $Suc^{BDH}(P, aP, bP, cP)$ denote the event that the BDH problem in $\langle G_1, G_2, \hat{e} \rangle$ is solved, that is, if $\langle P, aP, bP, cP \rangle$ for some $a, b, c \in Z_q \setminus \{0\}$ are known where P is a generator of G_1, then $\hat{e}(P, P)^{abc} \in G_2$ can be computed. An algorithm $\mathcal{A}$ has advantage ϵ in solving the BDH problem if

$$Pr[Suc_{\mathcal{A}}^{BDH}(P, aP, bP, cP)] = Pr[\mathcal{A}(P, aP, bP, cP) = \hat{e}(P, P)^{abc}] \geq \epsilon.$$

Then the BDH assumption states that when q is a random k-bit prime, no polynomially bounded algorithm $\mathcal{A}$ has a non-negligible advantage in solving the BDH problem. Here the advantage ϵ is "non-negligible" means that $\epsilon \geq 1/f(k)$ for some polynomial f, where k is the security parameter.

3   An ID-Based Series-Parallel Multisignature Scheme for Multi-Messages from Pairings

In this section, we describe an ID-based series-parallel multisignature scheme for multi-messages (ID-SP-M4M scheme) from pairings on elliptic curves, in which we require that the TA (trusted authority) should be absolutely trusted by the users.

3.1   Series-Parallel Graph

In this paper, we consider the following series-parallel structure (see Figure 1). The signers in the jth subgroup sign the same message m_j by parallel signing, in which the signing order cannot be detected by the verifier from their multisignature. The signers from different subgroups sign different messages, say, $signset^j$ signs m_j, where $m_1, m_2, ..., m_l$ may be different. The signing order of the l subgroups signing these l messages by series signing can be verified by the verifier from their multisignature. This property was called order verifiability in [13]. Since the order needs not to be registered in advance in our scheme, the series signing in our scheme is order-flexible.

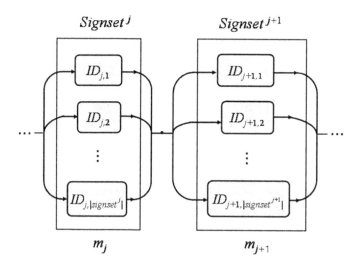

Fig. 1. Series-parallel graph

Notations

* $subgroup^j$: the set of users corresponding to the jth message m_j;
* $valset^j$: a valid subset of members in $subgroup^j$;
* $valset^j_{[I]}$: a $valset^j$ to which user I belongs;
* $signset^j$: the jth signer set, i.e., the set of signing members from $subgroup^j$, which is a $valset^j$;
* $I_{last(j)} \in signset^j$: the last signer in the jth subgroup;
* $T^{(j)}$: the multisignature of $signset^j$ from the jth subgroup.

A subset of $subgroup^j$ is a valid subset if it has:

(1) *the valid number of members*: In a subgroup, the number of signers taking part in the parallel signing is decided according to their ranks they hold. In some case more than 2/3 of the members in the subgroup are required in a parallel multisignature, i.e., $|valset^j| \geq \frac{2}{3}|subgroup^j|$;

(2) *the valid static public key*: $valset^j$'s public key $\sum_{I \in valset^j} Q_I \neq \mathcal{O}$, for $1 \leq j \leq l$.

3.2 Our Protocol

In an ID-SP-M4M scheme, signers are divided into l subgroups. The signers coming from some valid subset of $subgroup^j$ sign the message m_j, where m_i and m_j are usually not identical for $i \neq j$.

For a given security parameter $k \in N$, the TA generates the public system parameters $\wp = (G_1, G_2, q, \hat{e}, P, P_{pub}, H_1, H_2)$ by running a BDH parameter generator (see [1]), choosing two cryptographic hash functions $H_1 : \{0,1\}^* \longrightarrow G_1$ and $H_2 : \{0,1\}^* \longrightarrow Z_q \setminus \{0\}$, secretly selecting $s \in_R Z_q \setminus \{0\}$ and computing $P_{pub} = sP$. Upon request, the TA generates and delivers the private key $S_i = sH_1(ID_i)$ to user i with identity ID_i.

Key Generation and Delivery

(1) **System Key Generation:** The TA generates the system public/private key pair (P_{pub}, s), and publishes the system parameters

$$\wp = (G_1, G_2, q, \hat{e}, P, P_{pub}, H_1, H_2).$$

(2) **Individual Static Key Generation:** The TA generates a private key S_I for user I from his identity ID_I and then distributes $S_I = sH_1(ID_I)$ to this user in a secure way. Then user I has $(H_1(ID_I), S_I)$ as his public/private key pair.

(3) **Individual Dynamic Key Generation:** User I secretly selects $r_I \in_R Z_q \setminus \{0\}$ and computes $V_I = r_I P$. Then $\hat{e}(V_I, \sum_{J \in valset_{[I]}^j} Q_J) \neq 1$ for any $valset_{[I]}^j \subseteq subgroup^j$, where $Q_J = H_1(ID_J)$. In this way, user I obtains his dynamic public/private key pair (V_I, r_I).

We note that signers should update their dynamic keys after they signed any message. However, nothing should be updated if the user hasn't signed any message. The dynamic public key list of $signset^j$ is denoted by $V_{[A \to Z]}^j$, in which the dynamic public keys of users in the jth subgroup are listed alphabetically according to their IDs.

Signing

(1) Set initial phase: $T^{(0)} = T = \mathcal{O}$, and $h_0 = 1$.
(2) The users in some valid subset of $subgroup^j$ form the set $signset^j \in \{valset^j\}$. They compute $Q^{(j)} = \sum_{I \in signset^j} Q_I$ and let $T = T^{(j-1)}$.
(3) User $I \in signset^j$ signs the message m_j in the following way:

$$T_I = h_{j-1} S_I + r_I h_j Q^{(j)},$$

where $h_j = H_2(m_j, V_{[A \to Z]}^j, T^{(j-1)})$.

Note that the coordinates of $V_{[A \to Z]}^j$ and $T^{(j-1)}$ should be changed into strings before hashing. User I computes $T \leftarrow T + T_I$ and appends his identity ID_I in order to avoid any duplication and omission. Then he sends T to the next signer in $signset^j$. Step (3) is repeated until signer $I_{last(j)}$ has signed the message m_j.

(4) Let $T^{(j)} = T$. Signer $I_{last(j)} \in signset^j$ sends the multisignature of the first j subgroups

$$Sign = (m_i, V_{[A \to Z]}^i, T^{(i)} \mid i = 1, ..., j)$$

to $subgroup^{j+1}$ if $j < l$.

(5) Repeat (2), (3) and (4) for $j = 1, 2, ..., l$. Then signer $I_{last^{(l)}}$ submits the series-parallel multisignature

$$Sign = (m_j, V_{[A \to Z]}^j, T^{(j)} \mid j = 1, ..., l)$$

to the verifier.

Verification: On receipt of $Sign$, the verifier computes $V^{(j)} = \sum_{I \in signset^j} V_I$, $Q^{(j)} = \sum_{I \in signset^j} Q_I$, $h_j = H_2(m_j, V_{[A \to Z]}^j, T^{(j-1)})$, for $j = 1, 2, ..., l$, where $T^{(0)} = \mathcal{O}, h_0 = 1$, and then checks

$$\hat{e}(T^{(l)}, P) \overset{?}{=} \prod_{j=1}^{l} \hat{e}(Q^{(j)}, h_{j-1}P_{pub} + h_j V^{(j)}).$$

If the equality holds, then the verifier accepts the multisignature.

We can easily prove that the above ID-SP-M4M scheme can be checked by the verifier efficiently as follows.

Theorem 1. *In the newly proposed ID-SP-M4M scheme, the verifier can efficiently check the validity of the multisignature to the messages $m_1, m_2, ..., m_l$, where $m_1, m_2, ..., m_l$ are signed in order by the signers in $signset^1, signset^2, ...,$ $signset^l$, respectively.*

Proof: From the properties of pairing,

$$\hat{e}(T^{(l)}, P) = \hat{e}(\sum_{j=1}^{l} (\sum_{I \in signset^j} T_I), P)$$

$$= \prod_{j=1}^{l} \hat{e}(\sum_{I \in signset^j} (h_{j-1}S_I + r_I h_j Q^{(j)}), P)$$

$$= \prod_{j=1}^{l} \hat{e}(h_{j-1}s \sum_{I \in signset^j} Q_I + h_j(\sum_{I \in signset^j} r_I)Q^{(j)}, P)$$

$$= \prod_{j=1}^{l} \hat{e}((h_{j-1}s + h_j(\sum_{I \in signset^j} r_I))Q^{(j)}, P)$$

$$= \prod_{j=1}^{l} \hat{e}(Q^{(j)}, (h_{j-1}s + h_j(\sum_{I \in signset^j} r_I))P)$$

$$= \prod_{j=1}^{l} \hat{e}(Q^{(j)}, h_{j-1}P_{pub} + h_j \sum_{I \in signset^j} V_I)$$

$$= \prod_{j=1}^{l} \hat{e}(Q^{(j)}, h_{j-1}P_{pub} + h_j V^{(j)}).$$

In addition, the signature issued by the jth signer set is computed from the signatures of the previous $(j-1)$ signer sets, which guarantees our ID-SP-M4M scheme is order-verifiable.   □

In Appendix, we introduce the attack model, forgery signature attack from parallel insiders (ParaInsiForg Attack Model), and then prove that the above ID-SP-M4M scheme is secure against the forgery signature attack from parallel insiders under the BDH assumption.

4   Discussion

The scheme proposed in this paper is based on pairings. This technique was first used by Sakai et al. [16] to construct an ID-based multisignature scheme. By a slight modification, we can see that their scheme, M-SOK (Modified Sakai-Ohgishi-Kasahara [16]), can become an ID-based series-parallel multisignature scheme for multi-messages (ID-SP-M4M scheme). But similar to the scheme in [2], their hash function used in the signature is from a string to a point on an elliptic curve, which requires much more computation time than that from a string to a number used in our scheme.

As was pointed out in Theorem 1, our ID-SP-M4M scheme is order-verifiable. Meanwhile our scheme also has the order-flexibility property, since the verifier only needs to collect all the static and dynamic public keys of the signer sets for verification, which can be obtained from the multisignature. In fact, not only the order, but also the number of signers are flexible in our scheme.

Comparisons of our new ID-SP-M4M scheme to all of the known series-parallel multisignature schemes and the M-SOK scheme are listed in Table 1.

In the previous related works [4, 5, 6, 7, 17, 18], the scheme proposed by Tada [18] is the only possible series-parallel multisignature scheme which can be modified to sign multi-messages in series by using the technique introduced in [11]. However, similar to all the other previous works [4, 5, 6, 7, 17], it is not an ID-based scheme. Traditional PKI may not provide a good solution in many scenarios. For example, in a tether-less computing architecture, two mobile hosts wanting to communicate might get disconnected from each other and also from the Internet. As exchange of public keys is impossible in this disconnected

Table 1. Comparisons of Series-Parallel Multisignature Schemes

Related Works	Multi-Messages	Order-Flexibility	System	ID-Based	Map to Point ID	Map to Point Message
DMO [6, 7]	No	No	RSA	No	-	-
BDDMOTY [4]	No	No	ElGamal	No	-	-
Tada [18], [17]	Possible,No	Yes	MOO	No	-	-
M-SOK	Possible	Possible	Pairing-Based	Yes	1	1
Our ID-SP-M4M	Yes	Yes	Pairing-Based	Yes	1	0

situation, ID-based cryptosystem fits in with it very well since the public key of the other party can be derived from his identity.

Furthermore, from the efficiency consideration, we analyze our scheme and Tada's scheme [18] in the following two cases:

(1) **Key size.** Tada's original scheme in [18] is order-specified. In order to have order flexibility property, every signer has to prepare a number of private keys which are in different recommended sizes. For example, suppose that there are five signers in a signer set. To keep the security level at the same level for the difficulty of the 1024-IF (integer factoring), apart from a basic private key, each signer has to prepare other four keys with size 1604 bits, 1626 bits, 1641 bits and 1653 bits, respectively (refer to Table 3 in [10]). However, in our scheme, for the same security level, the key size is not required to be bigger than 1024 bits, and each signer needs only one static private key for any possible signing order.

(2) **Verification computation.** The most popular pairing choices are the Weil pairing and the Tate pairing, both of which are computable with Miller's algorithm, but the Tate pairing is usually more efficiently implementable than the Weil pairing. According to some recent results (e.g., [12]), it is estimated that the time for computation of a Weil pairing is approximately 10 times of the exponentiation in G_2. For this reason, the verification computational cost in Tada's scheme is less than that in our scheme if there are few signers. But the situation is changed when the number of signers is large. The verification computational cost of Tada's scheme increases approximately by polynomial curve as the number of signers increases (refer to Table 4. in [10]). However, in our scheme, when the number of signers in parallel increases, the verification computational cost almost does not increase; when the number of signers in series increases, the verification computational cost increases linearly. More precisely, by ignoring the computation of addition between two points in G_1 and hash mapping, we estimate the verification computational cost for a multisignature with l signsets in series as follows. Let *Ver*, *Pair* and *Mult* denote computational costs of verification, pairing and scalar multiplication to a point, respectively, then

$$Ver \approx (l+1)Pair + (2l-1)Mult = (Pair + 2Mult) \cdot l + (Pair - Mult).$$

Therefore, our ID-SP-M4M scheme has a computational advantage in the case that the multisignature is signed by a large number of signers.

5   Conclusion

In this paper, a series-parallel multisignature scheme based on pairings on elliptic curves has been proposed. It is an improvement over the multisignature schemes introduced by Doi, Mambo and Okamoto [5] and subsequently studied by other researchers such as Burmester et al. [4] in which only one message is to be signed. In our protocol, the signers in the same subgroup sign the same message, and those in different subgroups sign different messages. Apart from the above message flexibility, our scheme also has the order flexibility and order verifiability properties. This new scheme is secure against the forgery signature attack from parallel insiders provided that the BDH problem is hard.

References

1. D. Boneh and M. Franklin, *Identity-based encryption from the Weil pairing*, in *Advances in Cryptology - CRYPTO 2001*, LNCS 2139, pp. 213-229, Springer-Verlag, Berlin, 2001.
2. D. Boneh, C. Gentry, B. Lynn and H. Shacham, *Aggregate and verifiably encrypted signatures from bilinear maps*, in *Advances in Cryptology - EUROCRYPT 2003*, LNCS 2656, pp. 416-432, Springer-Verlag, Berlin, 2003.
3. C. Boyd, *Digital multisignatures*, in *Proc. IMA Conf. Crypto. Coding*, Clarendon, Oxford, pp. 241-246, 1989.
4. M. Burmester, Y. Desmedt, H. Doi, M. Mambo, E. Okamoto, M. Tada and Y. Yoshifuji, *A structured ElGamal-type multisignature scheme*, in *Public Key Cryptography 2000*, LNCS 1751, pp. 466-483, Springer-Verlag, Berlin, 2000.
5. H. Doi, M. Mambo and E. Okamoto, *Multisignature scheme with specified order*, in *SCIS94-2A*, Biwako, Japan, 1994.
6. H. Doi, M. Mambo and E. Okamoto, *RSA-based multisignature scheme for various group structure*, Journal of Information Processing Society of Japan, Vol. 41, No. 8, pp. 2080-2091, 2000.
7. H. Doi, M. Mambo and E. Okamoto, *On the security of the RSA-based multisignature scheme for various group structures*, in *ACISP 2000, Information Security and Privacy*, LNCS 1841, pp. 352-367, Springer-Verlag, Berlin, 2000.
8. S. Goldwasser, S. Micali and R. L. Rivest, *A digital signature scheme secure against adaptive chosen-message attacks*, SIAM Journal on Computing, Vol. 17, No. 2, pp. 281-308, 1988.
9. A. Joux, *The Weil and Tate pairings as building blocks for public key cryptosystems*, in *Algorithmic Number Theory*, LNCS 2369, pp. 20-32, Springer-Verlag, Berlin, 2002.
10. K. Kawauchi, Y. Komano, K. Ohta and M. Tada, *Probabilistic multi-signature schemes using a one-way trapdoor permutation*, IEICE Transactions on Fundamentals of Electronics, Communications and Computer Sciences, Vol. E87-A, No. 5, pp. 1141-1153, 2004.
11. K. Kawauchi, H. Minato, A. Miyaji and M. Tada, *A multi-signature scheme with signers' intentions secure against active attacks*, in *ICISC2001*, LNCS 2288, pp. 328-340, Springer-Verlag, Berlin, 2001.
12. P. S. L. M. Barreto, H. Y. Kim, B. Lynn and M. Scott, *Efficient algorithms for pairing-based cryptosystmes*, in *Advances in Cryptology - CRYPTO 2002*, LNCS 2442, pp. 354-368, Springer-Verlag, Berlin, 2002.
13. S. Mitomi and A. Miyaji, *A general model of multisignature schemes with message flexibility, order flexibility, and order verifiability*, IEICE Transactions on Fundamentals of Electronics, Communications and Computer Sciences, Vol. E84-A, No. 10, pp. 2488-2499, 2001.
14. K. Ohta and T. Okamoto, *A digital multisignature scheme based on the Fiat-Shamir scheme*, in *Advances in Cryptology - ASIACRYPT '91*, LNCS 739, pp. 139-148, Springer-Verlag, Berlin, 1993.

15. T. Okamoto, *A digital multisignature scheme using bijective public-key cryptosystems*, ACM Trans. Computer Systems, Vol. 6, pp. 432-441, 1988.
16. R. Sakai, K. Ohgishi and M. Kasahara, *Cryptosystems based on pairing*, in *SCIS2000-C20*, Okinawa, Japan, 2000.
17. M. Tada, *An order-specified multisignature scheme secure against active insider attacks*, in *ACISP 2002, Information Security and Privacy*, LNCS 2384, pp. 328-345, Springer-Verlag, Berlin, 2002.
18. M. Tada, *A secure multisignature scheme with signing order verifiability*, IEICE Transactions on Fundamentals of Electronics, Communications and Computer Sciences, Vol. E86-A, No. 1, pp. 73-88, 2003.
19. L. C. Washington, *Elliptic Curves: Number Theory and Cryptography*, Chapman & Hall/CRC, Boca Raton, 2003.

Appendix: Security Analysis

1 Notions

A generally accepted formal framework for ordinary digital signatures has been given by Goldwasser et al. [8]. Similarly, the definition of a multisignature scheme is described as follows.

Definition 3 (Multisignature Scheme). *A multisignature scheme $(\mathcal{G}, \mathcal{S}, \mathcal{V})$ is a triple of algorithms associated with two finite sets,* $\mathrm{Random}(k), \mathrm{Mspace}(k) \subseteq \{0, 1\}^*$, *for $k \in N$, where:*

- *$\mathcal{G}$, called the key generation algorithm, is a probabilistic algorithm which on input secret parameter 1^k outputs a public/private key pair (pk, sk), where k is the security parameter. For an ID-based multisignature scheme, user's identity ID (or the corresponding hashing value) acts as the public key pk.*
- *$\mathcal{S}$, called the signing algorithm, is an algorithm that, on input of a message $M \in \mathrm{Mspace}(k)$, the signer's private key sk, the secret random number $r \in \mathrm{Random}(k)$ and the previous multisignature, outputs the signature Sign.*
- *$\mathcal{V}$, called the verification algorithm, is an algorithm that, on input of public keys $pk \overset{Def.}{=} pk_1, ..., pk_L$ (or signers' identities IDs in an ID-based scheme), message(s) $M(s)$, and a candidate multisignature Sign, returns 1 if the multisignature is valid, and 0 otherwise.*

Informally, a multisignature scheme is secure against a forgery attack means that there is no polynomial time bounded algorithm $\mathcal{F}$ that can forge a multisignature such that the verification algorithm $\mathcal{V}$ outputs 1 with a non-negligible probability.

The GMR definition [8] identifies four types of forgery against digital signature schemes. They are in order of decreasing strength: total break, universal forgery, selective forgery, and existential forgery. The attacks from adversaries include passive attacks and active attacks for cryptographic signature schemes. An attack is called passive if it is restricted to the access of the verifying key and a number of signed messages (known message attack). An attack is called active if it also accesses the signer to ask for signatures on messages chosen by the attacker (chosen message attack). The attack is successful if the attacker can come up with a signature for a new message, i.e., one for which the signer has not provided a signature before. A stronger active attack is one where each message may be chosen by taking into account the signer's responses to all previously chosen messages (adaptive chosen message attack). Security requirements are then defined as security against a certain type of forgery under a certain type of attack. We define the Forgery Signature Attack from Parallel Insiders (ParaInsiForg attack) model in the next subsection. The security under ParaInsiForg attack corresponds to the security against selective forgery under chosen message attack.

2   Security Against the Forgery Signature Attack from Parallel Insiders

We consider one attack model in which all signers in some valid subset $signset^j$ of $subgroup^j$, except one honest signer I in $signset^j$ and those out of $subgroup^j$, may collude in the attack. The attackers give partial signatures of the message m_j to the honest signer I, where m_j is the successor to $m_1, m_2, ..., m_{j-1}$, and they obtain a valid partial signature T_I of m_j from I, whose dynamic public/private key pair is (V_I, r_I). With this information, the attackers try to obtain the signer I's forgery signature T_I' of m_j' corresponding to the same previous messages $m_1, m_2, ..., m_{j-1}$. We call this attack the forgery signature attack from parallel insiders. These attackers' success is denoted by

$$Suc_{ParaInsiForg}^{(m_1,...,m_{j-1},m_j') \leftarrow (m_1,...,m_{j-1},m_j)} (m_j' \neq m_j),$$

and the attackers are successful if there exists $m_j' \neq m_j$ such that the signer I's forgery signature T_I' can efficiently pass the verification with a non-negligible probability.

Due to the limited space, here we only give a brief discussion about the security of our ID-SP-M4M scheme under the BDH assumption.

Theorem 2. *The newly proposed ID-SP-M4M scheme is secure against the forgery signature attack from parallel insiders (ParaInsiForg attack) provided that the BDH problem is hard in* $\langle G_1, G_2, \hat{e} \rangle$.

Proof (sketch): In the ParaInsiForg attack model, the multisignature corresponding to the previous $(j - 1)$ subgroups is unchanged. Then $T_I = h_{j-1}S_I + r_I h_j Q^{(j)}$, and $T_I' = h_{j-1}S_I + (T_I' - h_{j-1}S_I) = h_{j-1}S_I + r_I h_j' Q^{(j)}$. In order that the attack might be successful, from the additive property of the multisignature generation, we know that the following equality should be satisfied:

$$\hat{e}(T_I', P) = \hat{e}(T_I, P),$$

which implies that

$$\hat{e}(r_I h_j' Q^{(j)}, P) = \hat{e}(r_I h_j Q^{(j)}, P),$$

$$\hat{e}(Q^{(j)}, h_j' V_I) = \hat{e}(Q^{(j)}, h_j V_I),$$

$$\hat{e}(Q^{(j)}, (h_j' - h_j)V_I) = 1.$$

Therefore,

$$Pr[Suc_{ParaInsiForg}^{(m_1,...,m_{j-1},m'_j)\leftarrow(m_1,...,m_{j-1},m_j)}(m'_j \neq m_j)] = Pr[\hat{e}(Q^{(j)}, (h_j - h'_j)V_I) = 1].$$

This probability is negligible since we assumed that the BDH problem is hard in $\langle G_1, G_2, \hat{e}\rangle$.

We go into details. Let $Suc^{BDH}(P, aP, bP, cP)$ denote the event that the BDH problem is solved (see the BDH Assumption in Section 2.2), then

$$Pr[Suc^{BDH}(P, P_{pub}, Q^{(j)}, (h_j - h'_j)V_I)]$$
$$= Pr[\hat{e}(Q^{(j)}, (h_j - h'_j)V_I) = 1]$$
$$\cdot Pr[Suc^{BDH}(P, P_{pub}, Q^{(j)}, (h_j - h'_j)V_I)|\hat{e}(Q^{(j)}, (h_j - h'_j)V_I) = 1]$$
$$+ Pr[\hat{e}(Q^{(j)}, (h_j - h'_j)V_I) \neq 1]$$
$$\cdot Pr[Suc^{BDH}(P, P_{pub}, Q^{(j)}, (h_j - h'_j)V_I)|\hat{e}(Q^{(j)}, (h_j - h'_j)V_I) \neq 1].$$

Note that $P_{pub}, Q^{(j)}, (h_j - h'_j)V_I \in G_1$. So there exist $a, b \in Z_q \setminus \{0\}$ such that $Q^{(j)} = aP$ and $(h_j - h'_j)V_I = bP$, whereas $P_{pub} = sP$. When $\hat{e}(Q^{(j)}, (h_j - h'_j)V_I) = \hat{e}(aP, bP) = \hat{e}(P, P)^{ab} = 1$, the BDH problem $\langle P, P_{pub}, Q^{(j)}, (h_j - h'_j)V_I\rangle$ can be solved since $\hat{e}(P, P)^{sab} = (\hat{e}(P, P)^{ab})^s = 1$. It implies that

$$Pr[Suc^{BDH}(P, P_{pub}, Q^{(j)}, (h_j - h'_j)V_I)|\hat{e}(Q^{(j)}, (h_j - h'_j)V_I) = 1] = 1.$$

Therefore,

$$Pr[Suc^{BDH}(P, P_{pub}, Q^{(j)}, (h_j - h'_j)V_I)]$$
$$= Pr[\hat{e}(Q^{(j)}, (h_j - h'_j)V_I) = 1] + Pr[\hat{e}(Q^{(j)}, (h_j - h'_j)V_I) \neq 1]$$
$$\cdot Pr[Suc^{BDH}(P, P_{pub}, Q^{(j)}, (h_j - h'_j)V_I)|\hat{e}(Q^{(j)}, (h_j - h'_j)V_I) \neq 1]$$
$$\geq Pr[\hat{e}(Q^{(j)}, (h_j - h'_j)V_I) = 1].$$

Clearly, if

$$Pr[Suc_{ParaInsiForg}^{(m_1,...,m_{j-1},m'_j)\leftarrow(m_1,...,m_{j-1},m_j)}(m'_j \neq m_j)]$$

is non-negligible, then

$$Pr[Suc^{BDH}(P, P_{pub}, Q^{(j)}, (h_j - h'_j)V_I)]$$

is also non-negligible. $\square$

A New Public-Key Cryptosystem Based on the Problem of Reconstructing p–Polynomials

Cédric Faure and Pierre Loidreau

Ecole Nationale Supérieure de Techniques Avancées
{Cédric.Faure, Pierre.Loidreau}@ensta.fr

Abstract. In this paper we present a new public key cryptosystem whose security relies on the intractability of the problem of reconstructing p–polynomials. This is a cryptosystem inspired from the Augot–Finiasz cryptosystem published at Eurocrypt 2003. Though this system was broken by Coron, we show However, in our case, we show how these attacks can be avoided, thanks to properties of rank metric and p–polynomials. Therefore, public-keys of relatively small size can be proposed (less than 4000 bits).

1 Introduction

At EUROCRYPT 2003, a cryptosystem based on the so-called problem of *polynomial reconstruction* was presented by Augot and Finiasz [1]. However, this system was broken soon after by Coron, by modifying the Welch–Berlekamp decoding algorithm for Reed–Solomon codes. He managed to argue that in most cases, the system could be broken in Polynomial-time by recovering the valid plaintext from the ciphertext [4]. More recently a result by Kiayias and Yung showed that it was not possible to choose an other way of rescuing the system such as adding more errors and then using Sudan list-decoding algorithm [8]. A different attempt was to use properties of the Trace operator to scramble the structure and design a secure cryptosystem. Once again it was shown that this system could be broken [2, 4].

In this paper we design a cryptosystem based on a new problem called *p–polynomials reconstruction problem*. Whereas the classical polynomial reconstruction problem is closely related to the decoding of Reed–Solomon codes, this problem is closely related to the problem of decoding Gabidulin codes [9]. We show how the attacks investigated in the case of the original Augot–Finiasz cryptosystem can be prevented. Namely, the efficiency of the attacks mainly depends on the metric used in the design of the system. We also construct a public-key cryptosystem with a public-key size of at most 4000 bits, and a security to the state of the art attacks.

The outline of the paper is the following. In section 2, we briefly recall the definitions of rank metric and linear polynomials, mostly in order to fix the notations. In section 3, we present a first, simple adaptation of the Augot-Finiasz system, and we show its vulnerability. In section 4, we introduce the trace operator, and we use it to build the cryptosystem which is the main subject of the paper. Lastly, in section 5, we discuss about the security of this second system.

Ø. Ytrehus (Ed.): WCC 2005, LNCS 3969, pp. 304–315, 2006.

2   Gabidulin Codes and p–Polynomials

Let p be a prime number (practically $p = 2$), $q = p^m$, and $GF(q)$ be the field with q elements. p-polynomials (also called *Linearized polynomials*) over $GF(q)$ were widely investigated by Øre in 1933, 1934 [10, 11]. They are polynomials of the form:

$$P(X) = a_k X^{p^k} + \ldots + a_\ell X^{p^\ell} + \ldots + a_1 X^p + a_0 X,$$

where $a_0, \ldots, a_k$ are elements of $GF(q)$. If $a_k \neq 0$, the integer k is called the p-degree of P. From now on we will denote $[\ell] \stackrel{def}{=} p^\ell$. For any vector $\mathbf{x} \in GF(q)^n$, we denote $P(\mathbf{x}) \stackrel{def}{=} (P(x_1), \ldots, P(x_n))$.

Definition 1. *Let $\mathbf{c} = (c_1, \ldots, c_n)$ be a vector of length n over the field $GF(q) = GF(p^m)$. The rank of $\mathbf{c}$, denoted $Rk(\mathbf{c})$ is the rank of the $m \times n$ p-ary matrix obtained by expanding each coordinate of $\mathbf{c}$ over a basis of $GF(p^m)/GF(p)$.*

Using this definition, we can build the rank distance between two words, by computing the rank (which can move from 0 to n) of their difference. Then, we obtain the rank metric over words of length n in $GF(q)$.

The rank metric is used as a substitute of the Hamming metric, for cryptographic goals mostly. Many properties of the rank metric were widely studied, and it is not the goal of this paper to review them all.

We now assume $k \leq n \leq m$. Let $\mathbf{g} \in GF(q)^n$ a vector of rank n. The Gabidulin code of length n, dimension k and generating vector $\mathbf{g}$ is the set of words obtained by the evaluation of a p-polynomial of degree at most $k - 1$ over $\mathbf{g}$:

$$Gab_k(\mathbf{g}) \stackrel{def}{=} \{(P(g_1), \ldots, P(g_n)) = P(\mathbf{g}) , \ \deg_p(P) \leq k - 1\}.$$

By this construction, it can be shown that the code $Gab_k(\mathbf{g})$ has minimum rank distance $n - k$. Hence it is an optimal code for rank metric [6]. Moreover there exist polynomial-time decoding algorithms that can correct up to the error-correcting capability of the code [7, 9, 13, 14].

3   An Augot–Finiasz Type Cryptosystem

In this section we show how to design a cryptosystem similar to the Augot–Finiasz cryptosystem. We also show that Coron's attacks can equally be adapted, although their complexity is not any more linear in the weight of the error-vector, but exponential.

3.1   Construction of the System

Parameters
Let $n \leq m$, and $k < n$, be integers and let $\mathbf{g} = (g_1, \ldots, g_n) \in GF(q)^n$ be a vector of linearly independent elements over $GF(p)$.

Key Generation

The conceiver of the system picks randomly a p–polynomial P over $GF(q)$ of p–degree $k - 1$. He chooses randomly a vector $\mathbf{E}$ of length n and of rank $W > (n - k)/2$.

- The public key is $\mathbf{K} = P(\mathbf{g}) + \mathbf{E}$.
- The secret key consists of the pair $(P, \mathbf{E})$.

Encryption

The message $\mathbf{m} = (m_0, \ldots, m_{k-2}) \in GF(q)^{k-1}$ can be transformed into the p-polynomial $m(x) = \sum_{i=0}^{k-2} m_i x^{[i]}$. The sender chooses randomly $\alpha \in GF(q)$, and an error-vector $\mathbf{e}$ of rank $\omega \le (n - k - W)/2$. The ciphertext $\mathbf{y}$ is:

$$\mathbf{y} = m(\mathbf{g}) + \alpha\mathbf{K} + \mathbf{e}.$$

Decryption

The receiver projects $\mathbf{y}$ on the subspace of $GF(q)^n$ of dimension $n - W$ that is orthogonal to the vector space generated by the coordinates of $\mathbf{E}$. Since he knows $\mathbf{E}$, he equally knows or can easily compute a non-singular p-ary matrix $\mathbf{R}$ such that $\mathbf{ER}$'s $n - W$ first positions are equal to 0. Hence since $\mathbf{R}$ is p-linear, then $P(\mathbf{g})\mathbf{R} = P(\mathbf{gR})$, and

$$\mathbf{yR} = m(\mathbf{g})\mathbf{R} + \alpha P(\mathbf{g})\mathbf{R} + \alpha\mathbf{ER} + \mathbf{eR} = (m + \alpha P)(\mathbf{gR}) + \alpha\mathbf{ER} + \mathbf{eR}.$$

Let $\widetilde{\mathbf{yR}}$ be the vector of length $n - W$ obtained by removing the last W positions of $\mathbf{yR}$. We obtain that

$$\widetilde{\mathbf{yR}} = (m + \alpha P)(\widetilde{\mathbf{gR}}) + \widetilde{\mathbf{eR}}.$$

But $\mathrm{Rk}(\widetilde{\mathbf{eR}}) \le \mathrm{Rk}(\mathbf{eR}) \le \mathrm{Rk}(\mathbf{e}) \le w \le \frac{n-W-k}{2}$. Therefore, since $(m + \alpha P)$ has p-degree less than k,

By decoding $\widetilde{\mathbf{yR}}$ in the Gabidulin code $Gab_k(\mathbf{g}R)$ one recovers $Q = m + \alpha P$. Since P has p–degree exactly $k - 1$, and since m has degree at most $k - 2$, the field element αP_{k-1} is the leading coefficient of Q. Thus the receiver gets α and finally recovers $m = Q - \alpha P$.

3.2 Investigation on the Security of the System

The security of the system relies on the fact that given the public-key

$$\mathbf{K} = P(\mathbf{g}) + \mathbf{E},$$

it is not computationally feasible to recover $(P, \mathbf{E})$. Since any other possible candidate $(R, \mathbf{F})$ such that $\mathbf{K} = R(\mathbf{g}) + \mathbf{F}$ would lead to wrong decoding, an attacker would have to list all possibilities and try them one by one. It is a kind of list decoding of Gabidulin codes up to the rank W. We can show that such a list-decoding is strictly equivalent to finding all p–polynomial V with $deg_p(V) \leq W$ and f with $deg_p(f) < k$ such that

$$\forall i = 1, \ldots, n \quad V(K_i) = V(f(g_i)). \tag{1}$$

This means that we would have to solve the following problem

Reconstruction$(\mathbf{K}, \mathbf{g} = (g_1, \ldots, g_n), k, W)$
Find the set (V, f) where V is a non-zero p–polynomial of p-degree $\leq W$ and where f is a p–polynomial of p-degree $< k$, such that $V(g_i) = V[f(g_i)]$, for all $i = 1, \ldots, n$.

In rank metric we have no equivalent of the Johnson bound, but our simulation results tend to show that this problem is hard to solve. Actually, we used the MAGMA computational algebra system, which is optimized for fast computations over finite fields. Over small sizes ($p = 2, n = 3$), it is possible to compute the exhaustive list of solutions to the Reconstruction problem. But the size of this list increases dramatically (seemingly exponentionnaly) as soon as W is larger than the error-correcting capability of $Gab_k(\mathbf{g})$. Therefore solving this problem can be considered as being hard, as it implies the manipulation of an oversized list.

The problem on which the security of the Augot-Finiasz cryptosystem relies is also a hard problem. The system was nevertheless broken by Coron who showed that one could generally recover plaintexts by finding roots of a polynomial of degree $\omega + 1$,[4]. This new system does not suffer from the weakness shown by Coron for the original system. Namely, by following the same idea, an attacker would have to find the roots of a polynomial (not a p-polynomial) of degree $(q^{\omega+1} - 1)/(q - 1)$. Therefore this system would provide a much better security than the original one.

Our system suffers from an other weakness, which was shown in [5]. Given a received ciphertext $\mathbf{y}$ an attacker wants to find $m, \alpha, \mathbf{e}$ such that:

$$\mathbf{y} = m(\mathbf{g}) + \alpha \mathbf{K} + \mathbf{e},$$

with $\deg_p(m) \leq k - 2$ and $\mathrm{Rk}(\mathbf{e}) \leq \frac{n-k-W}{2}$.

Using a kind of Welch–Berlekamp technique, as in [4], solving this system is equivalent to finding m, α, V such that:

$$V(\mathbf{y}) = V \circ m(\mathbf{g}) + V(\alpha \mathbf{K}), \tag{2}$$

with $\deg_p(m) \leq k - 2$ and $\deg_p(V) \leq \omega$.

Instead of solving this system, we study a more general system. We search N, V, V' such that:

$$V(\mathbf{y}) = N(\mathbf{g}) + V'(\mathbf{K}),$$

with $\deg_p(N) \leq k + \omega - 2$, $\deg_p(V) \leq \omega$, and $\deg_p(V') \leq \omega$. Considering the coefficients of N, V, V' as unknowns, it is a linear system with n equations and $k + 3\omega + 1$ unknowns. Therefore, it can be solved in polynomial time. Since there is by construction at least one solution to the system, and since we can show that $k + 3w + 1 < n$, this implies that the matrix of the system is degenerate, and the solution space often is of small dimension, typically 1. If it is the case we can find a solution to system (2).

4   System Using the Trace Operator

In this section we were inspired by the attempt to repair the original system by means of the trace operator which was introduced in [2]. In Hamming metric, Coron showed that this could not work. In rank metric however, such attacks cannot be adapted.

Definition 2. *The Trace operator from $GF(q^u)$ to $GF(q)$ is defined by :*

$$\forall x \in GF(q^u), Tr(x) = x + x^q + x^{q^2} + \ldots + x^{q^{u-1}}$$

We can extend this definition to vectors :

$$Tr(\mathbf{x}) \overset{def}{=} (Tr(x_1), \ldots, Tr(x_n))$$

and to linearized polynomials over $GF(q^u)$:

$$Tr(\sum_{i=0}^{k} p_i X^{[i]}) \overset{def}{=} \sum_{i=0}^{k} Tr(p_i) X^{[i]}$$

We will require the following proposition :

Proposition 1. *If $(g_1, \ldots, g_n) \in GF(q)^n$, and P is a p-polynomial over $GF(q^u)$, then $Tr(P(\mathbf{g})) = (Tr(P))(\mathbf{g})$.*

proof
Let $j \in [1, n]$. Since $g_j \in GF(q), \forall x \in GF(q^u)$, we have :
$Tr(xg_j) = g_j Tr(x)$ (by $GF(q)$-linearity).
So $Tr(P(g_j)) = Tr(\sum_{i=0}^{k} p_i g_j^{[i]}) = \sum_{i=0}^{k} g_j^{[i]} Tr(p_i) = Tr(P)(g_j)$.
Hence $Tr(P(\mathbf{g})) = (Tr(P))(\mathbf{g})$.

This leads to the design of a cryptosystem based on the trace operator.

4.1   Design of the System

Parameters
  - We consider $\mathbf{g} = (g_1, \ldots, g_n)$ a vector formed of elements of $GF(q)$ that are linearly independent over $GF(p)$;
  - An extension field $GF(q^u)$ of $GF(q)$;
  - An integer k;
  - An integer $W > \frac{n-k}{2}$. This implies that the linearized reconstruction problem is difficult as it was discussed in section 3.2.

Key generation

We generate randomly a p–polynomial P with coefficients in $GF(q^u)$, of p-degree $k-1$, such that the coefficients $p_{k-1}, \ldots, p_{k-u}$ form a basis of $GF(q^u)$ over $GF(q)$.

We also generate an error vector $\mathbf{E}$, with coefficients in $GF(q^u)$, of rank W.

- The public key is $\mathbf{K} = Tr(P(\mathbf{g})) + \mathbf{E} \in GF(q^u)^n$.
- The secret key is the pair $(P, \mathbf{E})$.

Encryption

Let $m \in GF(q)$ be a plaintext written as a p–polynomial of degree at most $k - u - 1$, that is $m(X) = m_0 X + m_1 X^p + \ldots + m_{k-u-1} X^{p^{k-u-1}}$. We generate randomly $\alpha \in GF(q^u)$ and $\mathbf{e}$ an error vector in $GF(q)$ of rank $\omega \leq (n - W - k)/2$. The ciphertext is:

$$\mathbf{y} = m(\mathbf{g}) + Tr(\alpha \mathbf{K}) + \mathbf{e} \in GF(q)^n$$

The encryption can be done in $O(nk)$ multiplications in $GF(q)$.

Decryption

Without loss of generality we can assume that the receiver knows an invertible p-ary matrix $\mathbf{R}$, such that the first $n - W$ columns of $\mathbf{ER}$ are equal to zero.

Since p-polynomials and the trace operator are $GF(p)$-linear transformations, we have that $Tr(\alpha P(\mathbf{g}))\mathbf{R} = Tr(\alpha P(\mathbf{g}R))$. Since $\mathbf{g}$ has coefficients in $GF(q)$, the vector $\mathbf{g}R$ has also coefficients in $GF(q)$ and by proposition 1, this implies that $Tr(\alpha P(\mathbf{g}R)) = Tr(\alpha P)(\mathbf{g}R)$. Therefore:

$$\mathbf{y}R = m(\mathbf{g})R + Tr(\alpha P(\mathbf{g}))R + Tr(\alpha \mathbf{ER}) + \mathbf{e}R,$$

$$\Leftrightarrow \mathbf{y}R = (m + Tr(\alpha P))(\mathbf{g}R) + Tr(\alpha \mathbf{ER}) + \mathbf{e}R.$$

Let $\widetilde{\mathbf{y}R}$ be the vector obtained by removing the last W positions of $\mathbf{y}R$. We have:

$$\widetilde{\mathbf{y}R} = (m + Tr(\alpha P))(\widetilde{\mathbf{g}R}) + \widetilde{\mathbf{e}R}.$$

But $\text{Rk}(\widetilde{\mathbf{e}R}) \leq \text{Rk}(\mathbf{e}R) \leq \text{Rk}(\mathbf{e}) \leq w \leq \frac{n-W-k}{2}$.

Hence, by decoding $\widetilde{\mathbf{y}R}$ in the code $Gab_k(\widetilde{\mathbf{g}R})$, one recovers the linear polynomial $Q = m + Tr(\alpha P)$. Since the p-degree of m is at most $k - u - 1$, and since $p_{k-1}, \ldots, p_{k-u}$ form a basis of $GF(q^u)$ over $GF(q)$, the receiver recovers α, and then $m = Q - Tr(\alpha P)$.

Using precomputation, the complexity of the decryption phase is then $O(\omega^2(k + \omega) + (k + \omega^2) + u^2 + kn)$ multiplications in $GF(q)$.

5   Security of the System

In the case of Hamming metric, Coron, showed that the cryptosystems could be broken in polynomial-time. It was even shown by Kiayias and Yung that it was illusionary to try to repair the system by adding more errors and then using Sudan algorithm rather than classical decoding algorithms for Reed-Solomon codes [8].

Here, one attack can be translated directly from Coron's approach implying that the parameters have to be carefully chosen, but the second one is too much related with properties of Hamming metric to be adapted to rank metric.

Let $\gamma_1, \ldots, \gamma_u$ be a basis of $GF(q^u)$ over $GF(q)$. If we write $\alpha = \sum_{t=1}^{u} \alpha_t \gamma_t$, we have: $Tr(\alpha \mathbf{K}) = \sum_{t=1}^{u} \alpha_t Tr(\gamma_t \mathbf{K})$. Let $\mathbf{K}_t = Tr(\gamma_t \mathbf{K})$, for $t = 1 \ldots u$.

The vectors $\mathbf{K}_t$ are vectors in $GF(q)$ easily computable from the public key $\mathbf{K}$. Knowing $\mathbf{y}, \mathbf{g}, \mathbf{K}_1, \ldots, \mathbf{K}_u$, recovering the plaintext consists of solving:

$$\exists \mathbf{e}, m, \alpha_1, \ldots, \alpha_u, \begin{cases} \mathbf{y} = m(\mathbf{g}) + \sum_{t=1}^{u} \alpha_t \mathbf{K}_t + \mathbf{e}, \\ \deg_p(m) \leq k - u - 1, Rk(\mathbf{e}) \leq \omega. \end{cases} \tag{3}$$

The rest of this section consists of investigating three ways of solving system (3).

5.1 Decoding Attacks

The decoding attack is a cipher-text only attack. The ciphertext can be seen under the form

$$\mathbf{y} = \mathbf{cG} + \mathbf{e},$$

where $\mathbf{c} = (m_0, \ldots, m_{k-u-1}, \alpha_1, \ldots, \alpha_u)$, and

$$\mathbf{G} = \begin{pmatrix} g_1 & \cdots & g_n \\ \vdots & \ddots & \vdots \\ g_1^{[k-u-1]} & \cdots & g_n^{[k-u-1]} \\ & \mathbf{K}_1 & \\ & \vdots & \\ & \mathbf{K}_u & \end{pmatrix}$$

Hence it can be reduced to a problem of decoding the linear code of dimension k generated by $\mathbf{G}$ up to the rank distance ω. To do this we do not know a better decoder than a general purpose decoder. The most efficient one was designed by Ourivski and Johannson and works in $O((q\omega^3)p^{(k+1)(\omega-1)})$, operations in the base field, [12].

5.2 Attack by Linearization

This attack is made on the ciphertext. The attacker writes the encryption equation, and uses it to obtain a linear system over $GF(q)$.

Proposition 2. *Assuming $k + (u+2)\omega \leq n$, an attacker can recover the plaintext from a given ciphertext in polynomial time with very high probability. If $n > k + (u+2)\omega$, then the same attack can be made in $O(n^3 q^{n-k-(u+2)\omega})$ multiplications with very high probability.*

proof
We consider another kind of Welch–Berlekamp approach. Namely, solving system (3) is equivalent to solving:

$$\exists m, \alpha_1, \ldots, \alpha_u, V, \begin{cases} V(\mathbf{y}) = V \circ m(\mathbf{g}) + \sum_{t=1}^{u} \alpha_t V(\mathbf{K}_t), \\ \deg_p(m) \leq k - u - 1, \deg_p(V) \leq \omega \end{cases}$$

By linearizing the equations, we now obtain the following system:

$$\exists V, R_1, \ldots, R_u, N, \begin{cases} V(\mathbf{y}) = N(\mathbf{g}) + \sum_{t=1}^{u} R_t(\mathbf{K}_t), \\ \deg_p(V) \leq \omega, \deg_p(R_t) \leq \omega, \\ \deg_p(N) \leq k + \omega - u - 1 \end{cases}$$

If one writes the unknown coefficients of polynomials $V, R_1, \ldots, R_u, N$ in a vectorial form, one has to solve :

$$M \begin{pmatrix} V \\ R_1 \\ \vdots \\ R_u \\ N \end{pmatrix} = 0$$

where :

$$M = \begin{pmatrix} y_1 \cdots y_1^{[\omega]} & -K_{11} \cdots -K_{11}^{[\omega]} & \cdots & -K_{u1} \cdots -K_{u1}^{[\omega]} & -g_1 \cdots -g_1^{[k+\omega]-u-1} \\ \vdots \quad \vdots & \vdots \quad \vdots & & \vdots \quad \vdots & \vdots \quad \vdots \\ y_j \cdots y_j^{[\omega]} & -K_{1j} \cdots -K_{1j}^{[\omega]} & \cdots & -K_{uj} \cdots -K_{uj}^{[\omega]} & -g_j \cdots -g_j^{[k+\omega]-u-1} \\ \vdots \quad \vdots & \vdots \quad \vdots & & \vdots \quad \vdots & \vdots \quad \vdots \\ y_n \cdots y_n^{[\omega]} & -K_{1n} \cdots -K_{1n}^{[\omega]} & \cdots & -K_{un} \cdots -K_{un}^{[\omega]} & -g_n \cdots -g_n^{[k+\omega]-u-1} \end{pmatrix}$$

using the notation $[i] = p^i$.

M is $n \times (k + (u+2)\omega + 1)$-matrix in $GF(q)$, we can compute its kernel in polynomial time to find all the solutions for polynomials $V, R_1, \ldots, R_u, N$.

Because there exists a non-trivial solution, the kernel of M is of dimension at least 1. Aside from this condition, practical experiments lead us to think that M is very likely to have maximum rank. Thus, we simply deduce the dimension of $\ker(M)$ from the size of M, and we almost always have : $\dim(\ker(M)) = \max(1, k + (u+2)\omega + 1 - n)$.

From the space of solutions for $V, R_1, \ldots, R_u, N$, we now have to extract a suitable non-trivial solution for $m, \alpha_1, \ldots, \alpha_u, V$, and then this m is the plaintext.

If $k + (u + 2)\omega \le n$, then with high probability $\dim(\ker(M)) = 1$, and thus, any solution of the linearized system allows us to find m by a mere left Euclidian division, the whole attack made in polynomial time.

If this is not the case, one has to check each direction of the solution space, which leads to an attack in roughly $O(n^3 q^{n-k-(u+2)\omega})$ multiplications, [5].

5.3   Algebraic Attacks

This part is devoted to finding the secret element α. Once one recovers the element $\alpha \in GF(q^u)$, it is trivial to recover the plaintext, by computing $\mathbf{y} - Tr(\alpha \mathbf{K}) = m(\mathbf{g}) + \mathbf{e}$ and decoding this vector in the Gabidulin code.

To decrypt we have to solve the following system

$$y_i - Tr(\alpha K_i) = m(g_i) + e_i, \quad \forall i = 1, \ldots, n,$$

where the unknowns are $\mathbf{e} = (e_1, \ldots, e_n)$ of rank ω, $\alpha \in GF(q^u)$, and the p-polynomial m. Solving this system is still equivalent to solving

$$V(y_i - Tr(\alpha K_i)) = V \circ m(g_i), \quad \forall i = 1, \ldots, n, \tag{4}$$

where the unknowns are m, α and the coefficients of a p-polynomial V of degree ω. Now to solve this system we consider the system

$$V(y_i - Tr(\alpha K_i)) = N(g_i), \quad \forall i = 1, \ldots, n \tag{5}$$

where the unknowns are the element α, and the coefficients of two p-polynomials V of degree ω and N of degree $k + \omega - u - 1$. If (α, V, m) is a solution of (4), then $(\alpha, V, N = V \circ m)$ is a solution of (5). There are two manners to solve this system:

Univariate case

Let us define the following $n \times (k + 2\omega - u + 1)$ matrix

$$M(x) = \begin{pmatrix} y_1 - Tr(xK_1) & \cdots & (y_1 - Tr(xK_1))^{[\omega]} & g_1 & \cdots & g_1^{[k+\omega-u-1]} \\ \vdots & \ddots & \vdots & \vdots & \ddots & \vdots \\ y_n - Tr(xK_n) & \cdots & (y_n - Tr(xK_n))^{[\omega]} & g_n & \cdots & g_n^{[k+\omega-u-1]} \end{pmatrix},$$

where $[i] = p^i$.

Provided α is known, $M(\alpha)$ is the matrix of system (5). Since by construction $(k + 2\omega - u + 1) \le n$ and since that we know that there is a non-zero solution, the matrix $M(\alpha)$ is not of full rank. Therefore, the determinant of every square submatrix $\widetilde{M}(x)$ of $M(x)$ satisfies the equation

$$Det(\widetilde{M}(\alpha)) = 0.$$

The Trace operator is a polynomial of degree q^{u-1}, therefore any determinant is a polynomial over $GF(q^u)$ of degree at most $q^{u-1}(p^{\omega+1} - 1)/(p - 1)$. Hence, a way to find the value of α would be to search for common divisors between some of the obtained determinants. However for practical cases, as we will see further, the quantity $q^{u-1}(p^{\omega+1} - 1)/(p - 1) > 2^{80}$.

Since such an approach is not very practical, we skip to another more interesting case.

Multivariate case

If we set $x = \sum_{j=1}^{u} \gamma_j x_j$, where $\gamma_1, \ldots, \gamma_u$ is a basis of $GF(q^u)/GF(q)$, by setting $K_{i,t} \stackrel{def}{=} Tr(\gamma_t K_i)$ for $t = 1, \ldots, u$ and $i = 1, \ldots, n$, we now define

$$M(x_1, \ldots, x_u) \stackrel{def}{=} \begin{pmatrix} y_1 - \sum_{t=1}^{u} K_{1,t} x_t & \cdots & (y_1 - \sum_{t=1}^{u} K_{1,t} x_t)^{[\omega]} & g_1 & \cdots & g_1^{[k+\omega-u-1]} \\ \vdots & \ddots & \vdots & \vdots & \ddots & \vdots \\ y_n - \sum_{t=1}^{u} K_{n,t} x_t & \cdots & (y_n - \sum_{t=1}^{u} K_{n,t} x_t)^{[\omega]} & g_n & \cdots & g_n^{[k+\omega-u-1]} \end{pmatrix}.$$

Once again if $\alpha \stackrel{def}{=} \sum_{t=1}^{u} \gamma_t \alpha_t$, then $M(\alpha_1, \ldots, \alpha_u)$ is the matrix of the linear system (5). Hence, the determinant of every square submatrix $\widetilde{M}(x_1, \ldots, x_u)$ of M satisfies the equation

$$Det(\widetilde{M}(\alpha_1, \ldots, \alpha_u)) = 0.$$

The determinants $\widetilde{M}(x_1, \ldots, x_u)$ are multivariate polynomials of degree at most $(p^{(\omega+1)} - 1)/(p - 1)$. in u variables. We can construct up to $\binom{n}{k+2\omega-u+1}$ different determinants by choosing exactly $k + 2\omega - u + 1$ lines out of n.

We made some simulations in the MAGMA language by using the algorithms finding first Gröbner bases and then solving the equations. Every time we succeeded in computing the Gröbner Basis of the system we obtained only one solution that was exactly the element α of the private key.

Simulation results can be found in Table 1. For these computations we used an OPTERON processor 2,2Ghz with 8Gb of memory.

Table 1. Simulations of attacks made on parameters $n = 36$, $q = 2^{36}$, $k = 10$, $W = 14$

Number of variables	Error-Rank	Degree	Magma 2.11-2/F4
	$\omega = 2$	7	0.01s
	$\omega = 3$	15	0.340s
$u = 2$	$\omega = 4$	31	9s
	$\omega = 5$	63	500s
	$\omega = 6$	127	11 hours
	$\omega = 2$	7	0.06s
$u = 3$	$\omega = 3$	15	54s
	$\omega = 4$	31	15 hours

5.4   Discussion About the Choice of Parameters

Now we propose the following set of parameters:

- *Extension field:* $q = 2^{36}$, and $u = 3$ which implies $q^u = 2^{108}$.
- *Length of the code:* $n = 36$.
- *Public-key:* vector of length 36 over $GF(2^{108})$, that is $36 \times 108 = 3888$ bits.

The different attacks we described in the paper give the following results:

- *Decoding attacks*: The best general purpose decoding algorithm was designed by Ourivski and Johannson [12]. The complexity of recovering a vector of rank ω in a code of dimension $k + u$ is equal to $(n\omega)^3 2^{(k+u+1)(\omega-1)} > 2^{91}$ binary operations.
- *Attacks by linearization*: We have to check $q^{(k+(u+2)\omega-n)} = 2^{144}$ solutions of a linear system to recover the plaintext.
- *Algebraic attacks*: Table 1 shows that these parameters are well beyond what is feasible for now. Namely the system to solve consists of $\approx 2^{32}$ cubic equations of degree 127 over $GF(2^{36})$. In the univariate case, one has to compute *gcd's* of polynomials of degree 2^{100}.

An implementation of the system in the MAGMA language on a 1200 MHz processor gives the following average times (1000 tests). The decoding algorithm used is described in [5, 9].

- *Key generation:* 72 ms.
- *Precomputation:* 16 ms.
- *Cipher:* 23 ms.
- *Decipher:* 7.9 ms.

This corresponds to a the transmision of $(k - u)m = 252$ information bits, encapsulated in a message of $nm = 1296$ bits, the useful transmission rate is so about 11 kb/s on this computing speed. Therefore, we can greatly increase the speed of the algorithms by using an efficient language, for example C language. It is also possible to transmit more information by putting some information on the error codeword. On how to do this, see for example [3].

References

1. D. Augot and M. Finiasz. A public key encryption scheme bases on the polynomial reconstruction problem. In *EUROCRYPT 2003*, pages 222–233, 2003.
2. D. Augot, M. Finiasz, and P. Loidreau. Using the trace operator to repair the polynomial reconstruction based cryptosystem presented at eurocrypt 2003. Cryptology ePrint Archive, Report 2003/209, 2003. http://eprint.iacr.org/.
3. T. P. Berger and P.Loidreau. Designing an efficient and secure public-key cryptosystem based on reducible rank codes. In *Proceedings of INDOCRYPT 2004*, 2004.

4. J.-S. Coron. Cryptanalysis of a public-key encryption scheme based on the polynomial reconstruction problem. In F. Bao, R. Deng, and J. Zhou, editors, *7th International Workshop on Theory and Practice in Public Key Cryptography, PKC 2004*, volume 2947, pages 14–28. Springer, 2004.

5. C. Faure. Etude d'un système de chiffrement à clé publique fondé sur le problème de reconstruction de polynômes linéaires. Master's thesis, Université Paris 7, 2004.

6. E. M. Gabidulin. Theory of codes with maximal rank distance. *Problems of Information Transmission*, 21:1–12, July 1985.

7. E. M. Gabidulin. A fast matrix decoding algorithm for rank-error correcting codes. In G. Cohen, S. Litsyn, A. Lobstein, and G. Zémor, editors, *Algebraic coding*, volume 573 of *LNCS*, pages 126–133. Springer-Verlag, 1991.

8. A. Kiayias and M. Yung. Cryptanalyzing the polynomial-reconstruction based public-key system under optimal parameter choice. Cryptology ePrint Archive, Report 2004/217, 2004. http://eprint.iacr.org/.

9. P. Loidreau. Sur la reconstruction des polynômes linéaires : un nouvel algorithme de décodage des codes de Gabidulin. *Comptes Rendus de l'Académie des Sciences : Série I*, 339(10):745–750, 2004.

10. Ö. Øre. On a special class of polynomials. *Transactions of the American Mathematical Society*, 35:559–584, 1933.

11. Ö. Øre. Contribution to the theory of finite fields. *Transactions of the American Mathematical Society*, 36:243–274, 1934.

12. A. Ourivski and T. Johannson. New technique for decoding codes in the rank metric and its cryptography applications. *Problems of Information Transmission*, 38(3):237–246, September 2002.

13. G. Richter and S. Plass. Error and erasure decoding of rank-codes with a modified Berlekamp-Massey algorithm. In *5th Int. ITG Conference on Source and Channel Coding (SCC 04)*, 2004.

14. R. M. Roth. Maximum-Rank array codes and their application to crisscross error correction. *IEEE Transactions on Information Theory*, 37(2):328–336, March 1991.

On the Wagner–Magyarik Cryptosystem

Françoise Levy-dit-Vehel and Ludovic Perret

ENSTA, 32 Boulevard Victor, 75739 Paris cedex 15
{levy, lperret}@ensta.fr

Abstract. We investigate a monoid variant of the scheme based on the word problem on groups proposed by Wagner and Magyarik at Crypto'84, that has the advantage of being immune to reaction attacks so far. We study the security of this variant. Our main result is a complexity-theoretic one: we show that the problem underlying this cryptosystem, say WM, is NP-hard. We also present an algorithm for solving WM. Its complexity permits to shed light on the size of the parameters to choose to reach a given level of security.

1 Introduction

At Crypto'84, Wagner and Magyarik [10] have outlined a general construction for public key cryptosystems based on the word problem. They also proposed a concrete example of such a system based on finitely presented groups. In 2004, Gonzalez-Vasco and Steinwandt [5] have proven that this particular example is vulnerable to a so-called reaction attack. In this attack, an adversary can, by observing the reaction of a legitimate user, recover the secret key. It has to be mentionned that this attack works under a small additional assumption concerning the public key of the Wagner Magyarik scheme. We refer the reader to [5] for further details.

Wagner and Magyarik also suggest in [10] to replace finitely presented groups by finitely presented monoids, with similar expected performances. The purpose of this paper is to investigate this variant. The main motivation is that, in such a setting, this cryptosystem is so far immune to the reaction attack described in [5]. This is due to the fact that, in the group variant, the recovering of the secret key is done by means of recovering words that are congruent to the empty word. In the monoid setting, such an approach no longer holds, as the only word congruent to the empty word is the empty word itself.

The paper is organized as follows: after having described the monoid-based Wagner Magyarik scheme[1] (sections 3.1 and 3.2), we focus on the underlying problem (section 3.3), that we call WM. We show that a problem proven to be NP-hard in [1] - namely the one of finding a so-called interpretation morphism mapping a finitely presented monoid to a free partially commutative one (TMMI problem) - polynomially reduces to WM. A consequence of this is that WM is NP-hard (there is no such complexity-theoretic result known for the WM problem on groups), but what this reduction also proves is that TMMI is not

---

[1] We will simply call it the "Wagner Magyarik scheme" in the sequel.

Ø. Ytrehus (Ed.): WCC 2005, LNCS 3969, pp. 316–329, 2006.

more difficult than WM, i.e. introducing such a morphism does not add any difficulty. Next, we propose an algorithm for solving WM. In essence, given a finitely presented monoid (Δ, R), this algorithm finds a Thue congruence S such that R refines S and Δ^*/S is a free partially commutative monoid. As a side effect, our algorithm can be seen as one that performs a sort of completion (more precisely, finds a set S of which R is a refinement), in the particular case when S is such that Δ^*/S is a free partially commutative monoid.

2   Preliminaries

We here introduce the necessary material in order to present the (monoid-based) Wagner Magyarik scheme. Let Δ be a finite alphabet, and Δ^* be the free monoid over Δ, with λ representing the empty word. Let R be a subset of $\Delta^* \times \Delta^*$. The *Thue congruence* generated by R on Δ^*, denoted by $\overset{*}{\leftrightarrow}_R$, is the reflexive transitive closure of the relation $\leftrightarrow_R$ defined as follows: $\forall u, v \in \Delta^*$, $u \leftrightarrow_R v \Leftrightarrow \exists x, y \in \Delta^*$, $\exists (\ell, r) \in R$, such that $u = x\ell y$ and $v = xry$ or $u = xry$ and $v = x\ell y$. (Thus we shall always assume that if $(u, v) \in R$, then $(v, u) \notin R$.) The congruence class of $z \in \Delta^*$ w.r.t. R is $[z]_R = \{w \in \Delta^*, w \overset{*}{\leftrightarrow}_R z\}$. The monoid $\Delta^*/\overset{*}{\leftrightarrow}_R$, that we shall denote equivalently by Δ^*/R, or by (Δ, R), is then a so-called finitely presented monoid. The *word problem* for R on Δ^* is then the following:

Given two words $u, v \in \Delta^*$, do we have $u \overset{*}{\leftrightarrow}_R v$?

This problem has been proven undecidable for general instances (Δ, R) [9].

The set R induces a *reduction relation* on Δ^* as follows: for $x, y \in \Delta^*$, and $(\ell, r) \in R$, the single reduction step is defined by: $x\ell y \to_R xry$. The reduction relation induced by R, denoted by $\overset{*}{\to}_R$, is then the (reflexive, antisymmetric and) transitive closure of $\to_R$. We shall often say that we reduce a word w.r.t. R, meaning that we reduce it w.r.t. $\overset{*}{\to}_R$. The link between the congruence relation and the reduction relation is obviously the following: $\forall u, v \in \Delta^*$ $u \overset{*}{\to}_S v \Rightarrow u \overset{*}{\leftrightarrow}_S v$.

Let now $\theta \subseteq \Delta \times \Delta$ be a binary reflexive and symmetric relation on Δ. Whenever $(a, b) \in \theta$, each occurrence of ab (resp. ba) in any word $w \in \Delta^*$ can be replaced by ba (resp. ab). If a word $v \in \Sigma^*$ is derived from a word $w \in \Sigma^*$ by such a sequence of replacements, then we denote it by $v \equiv_\theta w$. It is clear that $\equiv_\theta$ as defined is an equivalence relation. Taking the quotient $\Delta^*/\equiv_\theta$, we obtain the so-called *free partially commutative monoid generated by Δ with respect to the concurrency relation θ*. The *word problem for free partially commutative monoids* is the following:
Given two words $v, w \in \Delta^*$, is $v \equiv_\theta w$? Relying on a result of [4], it has been shown [3] that, for fixed Δ, the word problem in free partially commutative monoids is decidable in linear time in the length of the inputs v and w.

3   A Monoid Variant of Wagner Magyarik System

3.1   The Scheme

Public key: a finitely presented monoid $(\Delta = \{x_1, \ldots, x_n\}, R)$, and two words $w_0, w_1 \in \Delta^*$, with $w_0 \overset{*}{\not\leftrightarrow}_R w_1$.

Secret key: a set of relations $S \subseteq \Delta^* \times \Delta^*$, rendering the word problem in Δ^*/S easy, with the property that

$$\forall u, v \in \Delta^*, \ u \overset{*}{\leftrightarrow}_R v \ \Rightarrow \ u \overset{*}{\leftrightarrow}_S v,$$

and satisfying also $w_0 \overset{*}{\not\leftrightarrow}_S w_1$.

Encryption: to encrypt $b \in \{0, 1\}$, choose a word $w \in \Delta^*$ with $w \overset{*}{\leftrightarrow}_R w_b$.

Decryption: solve the easy word problem in Δ^*/S to find b such that $w \overset{*}{\leftrightarrow}_S w_b$.

Typically, the set S is of the form $S = S_1 \cup S_2 \cup S_3$, where S_1 is a set of rules of the form $(x, \lambda) \in \Delta \times \{\lambda\}$, S_2 is a subset of $\Delta \times \Delta$, and S_3 is a set of commuting rules $(x_i x_j, x_j x_i)$, for $x_i, x_j \in \Delta$. The fact that the word problem in Δ^*/S is easy is due to its *free partially commutative monoid* structure. We shall precise it further in the following section.

Note: in the original Wagner Magyarik scheme on groups, S is defined exactly in the same way ($u \overset{*}{\leftrightarrow}_R v \ \Rightarrow \ u \overset{*}{\leftrightarrow}_S v$, $w_0 \overset{*}{\not\leftrightarrow}_S w_1$, and the word problem in the finitely presented group generated by S is polynomial-time solvable; in particular, S is chosen of the form $S = S_1 \cup S_2 \cup S_3$). The only difference is that, as inverses are well-defined, it allows for a somewhat lighter handling of congruences (by the fact that $u \overset{*}{\leftrightarrow}_R v$ is equivalent to $uv^{-1} \overset{*}{\leftrightarrow}_R \lambda$). Indeed, the main difference between Wagner Magyarik over monoids and over groups does not lie in the description of the schemes, but rather in the fact that - due to the non-existence of inverses - one cannot retrieve any useful information about the private key from a reaction attack on Wagner Magyarik scheme on monoids.

3.2   Some Features of the Set S

Let $S = S_1 \cup S_2 \cup S_3$ be a secret key of a Wagner Magyarik scheme. Denote by $\overset{*}{\leftrightarrow}_S$, the Thue congruence generated by S on Δ^*. The set S is such that the word problem in Δ^*/S is easy in the sense that, after having removed the generators involved in S_1 (i.e. elimination relations of the form (x, λ)) and simplified those involved in S_2 (i.e. relations of the form (x_i, x_j)), we come up with a free partially commutative monoid the concurrency relation of which is S_3. Note that it will indeed be the case in the decryption process: first reduce w.r.t. $S_1 \cup S_2$, then solve the easy word problem induced by the concurrency relation S_3.

Removing generators involved in S_1 simply consists in replacing any x where $(x, \lambda) \in S_1$ by λ, and thus does not make any difficulty. On the other hand, simplifying words by rules of S_2 requires a little attention, so as not to introduce any additional hardness, as the overall process of testing equivalence w.r.t. S must remain easy. Therefore, we shall now show how to reduce in a non ambiguous way w.r.t. the set S_2. We set $\Delta_\lambda = \{x \in \Delta, (x, \lambda) \in S_1\}$, and $\Delta' = \Delta \setminus \Delta_\lambda$. The set S_2 consists of pairs (x_i, x_j), for some letters $x_i, x_j \in \Delta'$, $i \neq j$ (there is no relevance in considering S_2 acting on the whole set Δ). Let

$\overset{*}{\rightarrow}_{S_2}$ be the reduction relation induced by S_2. As the word problem in Δ^*/S is easy, it must be the case that $\overset{*}{\rightarrow}_{S_2}$ contains no cycle, i.e. no sequence of the form $x_i \rightarrow_{S_2} x_j \rightarrow_{S_2} \ldots \rightarrow_{S_2} x_i$ with $j \neq i$. Otherwise, one could be in an infinite loop trying to reduce a word $u \in \Delta'^*$ w.r.t. S_2. As there are no cycle, then every sequence of reductions $u \rightarrow_{S_2} v \rightarrow_{S_2} \ldots$ where $u, v \in \Delta'^*$, must be finite[2]. In particular, every sequence of reductions on letters $x_i \rightarrow_{S_2} x_j \rightarrow_{S_2} \ldots$, where $x_i, x_j \in \Delta'$, must be finite. Thus, in each such sequence, there exits an $x \in \Delta'$ that cannot be further reduced by rules of S_2 (i.e. for which there exists no $y \in \Delta'$, such that $x \rightarrow_{S_2} y$). We shall call such an x *irreducible*. Furthermore, if, starting from some $x_i \in \Delta'$, there were to be two distinct (finite) sequences of reductions $x_i \rightarrow_{S_2} x_j \rightarrow_{S_2} \ldots \rightarrow_{S_2} x$ and $x_i \rightarrow_{S_2} x'_j \rightarrow_{S_2} \ldots \rightarrow_{S_2} x'$, with $x \neq x'$, then one might not be able to decide the word problem in Δ^*/S, because of ambiguity in the reduction w.r.t. S_2. Thus, if such two sequences were to exist, then it had to be the case that they become identical at some point, and thus that $x = x'$. In other words, according to the accepted terminology, (Δ', S_2) is a *confluent semi-Thue system*. This, together with the Noetherian property, implies the uniqueness of an irreducible element to which any letter[3] of Δ' reduces. Thus, denoting by Δ_{irr}, the set of irreducible letters of Δ', we have that, for every letter $x_i \in \Delta'$, there exists a unique letter $x \in \Delta_{irr}$, such that $x_i \overset{*}{\rightarrow}_{S_2} x$. Note that this holds whether x_i is such that there exists a letter $x_j \in \Delta'$, with $(x_i, x_j) \in S_2$, or not. If not, then x_i cannot be reduced w.r.t. S_2, and in this case x is equal to x_i itself.

Now let $u \in \Delta^*$, and let u'' be the word of Δ'^* obtained from u by removing every letter of Δ_λ (this amounts to reducing u w.r.t. the reduction relation $\overset{*}{\rightarrow}_{S_1}$ induced by S_1). Now let u' be the word of Δ^*_{irr}, such that $u'' \overset{*}{\rightarrow}_{S_2} u'$. As S_1 and S_2 act on disjoint alphabets, one can reduce $u \in \Delta^*$ w.r.t to S_1 then S_2, or vice-versa, and will obtain the same result. Thus, we shall denote by $\overset{*}{\rightarrow}_{S_1 \cup S_2}$, the reduction relation obtained by applying $\overset{*}{\rightarrow}_{S_1}$ then $\overset{*}{\rightarrow}_{S_2}$, or the converse. Note that, for a word $u \in \Delta^*$, if u' denotes a word of Δ^*_{irr} such that $u \overset{*}{\rightarrow}_{S_1 \cup S_2} u'$, then u' is not only irreducible w.r.t. to S_2, but also w.r.t. to S_1, for it admits no letter of Δ_λ; thus such a u' is unique.

Next, we define a map that we call $\psi : \Delta^* \rightarrow \Delta^*_{irr}$, by: $\psi(\delta) = \lambda$ if $\delta \in \Delta_\lambda$, and for $\delta \in \Delta'$, $\psi(\delta) = x$, where $x \in \Delta_{irr}$ is such that $\delta \overset{*}{\rightarrow}_{S_2} x$.

In the sequel, we shall make use of the following two results:

Property 1. Let $u \in \Delta^*$ and $u' \in \Delta^*_{irr}$. Then

$$u \overset{*}{\rightarrow}_{S_1 \cup S_2} u' \Leftrightarrow \psi(u) = u'.$$

Proof. Let $u \in \Delta^*$, and let $u'' \in \Delta'^*$, be the unique word such that $u \overset{*}{\rightarrow}_{S_1} u''$ (thus, u'' is fully reduced w.r.t. S_1). We shall show:

$$u'' \overset{*}{\rightarrow}_{S_2} u' \Leftrightarrow \psi(u'') = u'. \tag{1}$$

---

[2] In other words, the reduction-step relation is Noetherian.
[3] And any word of Δ'^*.

The desired result will then follow, as reducing w.r.t. $S_1 \cup S_2$ is the same as reducing w.r.t. S_1 then S_2, and by the fact that we obviously have $\psi(u'') = \psi(u)$. By definition, $\forall \delta \in \Delta'$, $\psi(\delta)$ is defined as the only $\delta_0 \in \Delta_{irr}$, such that $\delta \xrightarrow{*}_{S_2} \delta_0$. Thus, (1) plainly holds on letters of Δ'.

Now let $u' \in \Delta_{irr}^*$, with $u'' \xrightarrow{*}_{S_2} u'$. Write $u'' = u_1'' \cdots u_\ell''$, and, for $1 \leq j \leq \ell$, set $\psi(u_j'') = v_j \in \Delta_{irr}$. As $\psi(u_j'') = v_j \Leftrightarrow u_j'' \xrightarrow{*}_{S_2} v_j, 1 \leq j \leq \ell$, it comes: $u'' = u_1'' \cdots u_\ell'', \xrightarrow{*}_{S_2} v_1 \cdots v_\ell = \psi(u_1'') \cdots \psi(u_\ell'') = \psi(u'')$. Unicity of irreducible elements w.r.t. S_2 yields $u' = \psi(u'')$.

Conversely, let $u' = \psi(u'')$. Write $u' = u_1' \cdots u_\ell', u_j' \in \Delta_{irr}, 1 \leq j \leq \ell$. As $u'' \in \Delta'^*$, no letter of u'' maps to λ by ψ. Thus[4] $|\psi(u'')| = |u''|$, and so $|u''| = \ell$. Let then $u'' = u_1'' \cdots u_\ell''$. As $\psi(u'') = \psi(u_1'') \cdots \psi(u_\ell'') = u' = u_1' \cdots u_\ell'$, it follows that[5] $\psi(u_j'') = u_j', \forall 1 \leq j \leq \ell$, which, as (1) holds on letters, is equivalent to $u_j'' \xrightarrow{*}_{S_2} u_j', \forall 1 \leq j \leq \ell$. Finally, $u_1'' \cdots u_\ell'' \xrightarrow{*}_{S_2} u_1' \cdots u_\ell'$, i.e. $u'' \xrightarrow{*}_{S_2} u'$.

Observe that one has, for all $u, v \in \Delta^*$, $u \xrightarrow{*}_{S_1 \cup S_2} v \Rightarrow u \xleftrightarrow{*}_S v$. This yields:

Property 2. Let $u, v \in \Delta^*$, and $u', v' \in \Delta_{irr}^*$, such that $u \xrightarrow{*}_{S_1 \cup S_2} u'$ and $v \xrightarrow{*}_{S_1 \cup S_2} v'$. Then

$$u \xleftrightarrow{*}_S v \Leftrightarrow u' \xleftrightarrow{*}_{S_3} v'. \tag{2}$$

Proof. For $u, v \in \Delta^*$, let $u', v' \in \Delta_{irr}^*$ such that $u \xrightarrow{*}_{S_1 \cup S_2} u'$ and $v \xrightarrow{*}_{S_1 \cup S_2} v'$. By the above observation,

$$u \xleftrightarrow{*}_S v \Rightarrow u' \xleftrightarrow{*}_S v'.$$

As u' and v' are irreducible w.r.t. $S_1 \cup S_2$, only rules of S_3 can intervene in the rewriting $u' \xleftrightarrow{*}_S v'$. Thus, $u' \xleftrightarrow{*}_S v' \Leftrightarrow u' \xleftrightarrow{*}_{S_3} v'$, and the "*if*" part of (2) holds. The converse is also true, as $u' \xleftrightarrow{*}_{S_3} v'$ implies $u \xrightarrow{*}_{S_1 \cup S_2} u' \xleftrightarrow{*}_{S_3} v' \xleftarrow{*}_{S_1 \cup S_2} v$, and thus $u \xleftrightarrow{*}_S v$.

3.3 The Underlying Hard Problem

We shall here show that the (search) problem underlying the Wagner Magyarik scheme on monoids is NP-hard. Let us call this problem WM_s, and let WM_d denote the associated decision problem. To do so, we shall use a problem introduced in [1], that we call[6] Thue Monoid Morphism Interpretation (TMMI) problem. We shall denote by $TMMI_d$, (resp. $TMMI_s$), the decision (resp. search) version of it. First, let us state WM_d and $TMMI_d$.

WM_d:

Input: an alphabet Δ, a Thue system $R \subseteq \Delta^* \times \Delta^*$, and two words $w_0, w_1 \in \Delta^*$ with $w_0 \not\leftrightarrow_R w_1$.

---

[4] $|u|$ denoting the number of letters of the word u.

[5] $\forall 1 \leq j \leq \ell$, $\psi(u_j'') \neq \lambda$ and $|\psi(u_j'')| = 1$.

[6] This problem was not given any name by the authors of [1].

Question: Does there exist a set $S \subseteq \Delta^* \times \Delta^*$ of the form $S = S_1 \cup S_2 \cup S_3$ with S_i defined as above, satisfying:

$$(i) \; \forall x, y \in \Delta^*, x \overset{*}{\leftrightarrow}_R y \Rightarrow x \overset{*}{\leftrightarrow}_S y.$$
$$(ii) \; w_0 \overset{*}{\not\leftrightarrow}_S w_1.$$

The search problem WM_s is simply the one of effectively constructing such a set S.

Note: as said before, the very structure of S implies that the word problem in Δ^*/S is easy. However, in the WM problem, the important thing is not that S have the particular shape $S = S_1 \cup S_2 \cup S_3$, but rather that the resulting monoid Δ^*/S be a free partially commutative monoid where S is such that R refines S. We shall make use of this observation to design an algorithm to break WM in section 4.

TMMI$_d$:

Input: a Thue system T on an alphabet Δ, and two words y_0 and $y_1 \in \Delta^*$.
Question: Does there exist an alphabet Σ, a nontrivial interpretation morphism $g : \Delta^* \to \Sigma^*$, and a concurrency relation θ on Σ, having the following properties:
1. $g(y_0) \not\equiv_\theta g(y_1)$.
2. for each letter $d \in \Delta$, $g(d)$ is either a letter in Σ or the empty word λ.
3. there exists a letter $d \in \Delta$ such that $g(d)$ is a letter in Σ.
4. for every two words u and v in Δ^* with $u \overset{*}{\longleftrightarrow}_T v$, we have $g(u) \equiv_\theta g(v)$.

The search problem TMMI_s is the one of effectively constructing a solution (Σ, g, θ).

The following result will permit us to prove the hardness of WM_s.

Theorem 1. *TMMI$_d$ is poly-time many-one reducible to WM$_d$.*

Proof. We denote by L_{WM} (resp. L_{TMMI}), the language[7] associated to WM_d (resp. TMMI_d). To prove the theorem, one needs to show that there exists a polynomial-time computable function f from the instances of TMMI_d to the instances of WM_d, satisfying:

$$X \in L_{TMMI} \Leftrightarrow f(X) \in L_{WM}. \tag{3}$$

Let $(\Delta = \{x_1, \ldots, x_n\}, R, y_0, y_1) \in L_{TMMI}$, and let (Σ, g, θ) be a solution of it. Consider the instance $(\Delta, R, w_0 = y_0, w_1 = y_1)$ of WM_d (i.e. f is the identity mapping here). We shall construct a solution of this instance. This amounts to show that $f(\Delta, R, y_0, y_1) \in L_{WM}$. Set $\Delta_\lambda = \{x \in \Delta, g(x) = \lambda\}$; $\{g(x), x \in \Delta\} = \{\sigma_1, \ldots, \sigma_k\} \subseteq \Sigma$ (we need not suppose the equality), where the σ's are pairwise distinct; for each i, $1 \le i \le k$, let $C_{\sigma_i} = \{x \in \Delta, g(x) = \sigma_i\}$. Note that, as $\forall x \in \Delta \setminus \Delta_\lambda, \exists! i, 1 \le i \le k$, s.t. $g(x) = \sigma_i$, we have: $C_{\sigma_i} \cap C_{\sigma_j} = \emptyset$ for $i \ne j$, and $\cup_{1 \le i \le k} C_{\sigma_i} = \Delta \setminus \Delta_\lambda$.

For each i, $1 \le i \le k$, let z_i denote an element of C_{σ_i}. Set $\Delta_{irr} = \{z_1, \ldots, z_k\}$ and construct S as $S_1 \cup S_2 \cup S_3$ with

---

[7] i.e. the set of instances admitting a solution.

$$S_1 = \{(x, \lambda),\ x \in \Delta,\ g(x) = \lambda\} = \Delta_\lambda \times \{\lambda\},$$
$$S_2 = \{(x_i, x_j),\ x_i, x_j \in \Delta \setminus \Delta_\lambda,\ x_i \neq x_j,\ g(x_i) = g(x_j)\},\ \text{and}$$
$$S_3 = \{(x_i x_j, x_j x_i),\ x_i, x_j \in \Delta_{irr},\ x_i \neq x_j,\ (g(x_i), g(x_j)) \in \theta\}.$$

Now we shall show that $S = S_1 \cup S_2 \cup S_3$ is a solution of (Δ, R, w_0, w_1). It suffices to show that (i) and (ii) of the WM problem statement hold. As (Σ, g, θ) is a solution of (Δ, R, y_0, y_1), we have

$$u \overset{*}{\leftrightarrow}_R v \Rightarrow g(u) \equiv_\theta g(v).$$

To show (i), it thus suffices to show that $g(u) \equiv_\theta g(v)$ implies $u \overset{*}{\leftrightarrow}_S v$. We shall first prove:

$$\forall u, v \in \Delta^*,\ g(u) \equiv_\theta g(v) \Leftrightarrow \psi(u) \overset{*}{\leftrightarrow}_{S_3} \psi(v), \tag{4}$$

where $\psi : \Delta^* \to \Delta^*_{irr}$ is defined on letters of Δ by:

$$\psi(\delta) = \lambda \quad \text{if}\ \delta \in \Delta_\lambda,$$
$$\psi(\delta) = z_i \quad \text{if}\ \delta \in C_{\sigma_i}.$$

Observe that we have $\psi = \tau \circ g$, where $\tau : g(\Delta)^* \to \Delta^*_{irr}$ is defined by $\tau(\sigma_i) = z_i,\ \forall 1 \leq i \leq k$, and $\tau(\lambda) = \lambda$. It is clear that τ is a monoid isomorphism. Let θ_g be the restriction of θ to pairs (a, b), with $a, b \in g(\Delta) \subseteq \Sigma$, and set $\tau(\theta_g) = \{(\tau(a)\tau(b), \tau(b)\tau(a)),\ (a, b) \in \theta_g\}$. We then have $\tau(\theta_g) = S_3$. Indeed:

$$(x_i x_j, x_j x_i) \in S_3 \Leftrightarrow (g(x_i), g(x_j)) \in \theta \cap g(\Delta) \times g(\Delta),\ \text{with}\ x_i, x_j \in \Delta_{irr} \Leftrightarrow$$

$$(g(x_i), g(x_j)) \in \theta_g,\ \text{with}\ x_i, x_j \in \Delta_{irr} \Leftrightarrow (\tau \circ g(x_i)\tau \circ g(x_j), \tau \circ g(x_j)\tau \circ g(x_i)) \in \tau(\theta_g)$$

$$\Leftrightarrow (\psi(x_i)\psi(x_j), \psi(x_j)\psi(x_i)) \in \tau(\theta_g).$$

As $\psi(x) = x,\ \forall x \in \Delta_{irr}$, we get:

$$x_i, x_j \in \Delta_{irr},\ (x_i x_j, x_j x_i) \in S_3 \Leftrightarrow (x_i x_j, x_j x_i) \in \tau(\theta_g),\ \text{with}\ x_i, x_j \in \Delta_{irr}.$$

Let $u, v \in \Delta^*$. It comes

$$g(u) \equiv_\theta g(v) \Leftrightarrow g(u) \equiv_{\theta_g} g(v) \Leftrightarrow \tau \circ g(u) \equiv_{\tau(\theta_g)} \tau \circ g(v) \Leftrightarrow \psi(u) \overset{*}{\leftrightarrow}_{S_3} \psi(v),$$

proving equivalence (4).

We shall need to use properties 1 and 2. To do so, we shall show that the reduction relation induced by S_2 has the properties needed in order to apply those results, namely, for all $x \in \Delta'^* \overset{def}{=} (\Delta \setminus \Delta_\lambda)^*$, there exists a unique irreducible (w.r.t. S_2) word $z \in \Delta'^*$, such that $x \overset{*}{\to}_{S_2} z$. First note that, by construction of Δ_{irr} and S_2, no two letters of Δ_{irr} can be congruent w.r.t. S_2.

Let $x \in \Delta'$. Then, there exists a unique $i,\ 1 \leq i \leq k$, such that $x \in C_{\sigma_i}$. Let z_i be the unique element[8] of $\Delta_{irr} \cap C_{\sigma_i}$. Then $g(x) = \sigma_i = g(z_i)$, so that $x \to_{S_2} z_i$. Thus, $\forall x \in \Delta',\ \exists! z \in \Delta_{irr}$, such that $x \to_{S_2} z$, and, by transitivity of

---

[8] Note that z_i could be x itself.

the reduction relation, $\forall x \in \Delta'^*$, $\exists! z \in \Delta^*_{irr}$, such that $x \xrightarrow{*}_{S_2} z$. That is to say that Δ_{irr} can be taken as *the set of irreducible letters w.r.t. S_2*. Besides, for the same reasons as in section 3.2, Δ_{irr} is also the set of irreducible letters w.r.t. $S_1 \cup S_2$. Now, the map ψ is defined exactly as the one of section 3.2, mapping Δ^* onto Δ^*_{irr}. Thus we are in a position to apply properties 1 and 2: for any word $u \in \Delta^*$, $\psi(u)$ is the irreducible word to which u reduces w.r.t. $S_1 \cup S_2$, and $u \xleftrightarrow{*}_S v \Leftrightarrow \psi(u) \xleftrightarrow{*}_{S_3} \psi(v)$. This yields finally $\forall u, v \in \Delta^*$, $u \xleftrightarrow{*}_R v \Rightarrow g(u) \equiv_\theta g(v) \Leftrightarrow \psi(u) \xleftrightarrow{*}_{S_3} \psi(v) \Leftrightarrow u \xleftrightarrow{*}_S v$, ending the proof of ($i$). Property ($ii$) follows from (4), as[9]:

$$g(w_0) \not\equiv_\theta g(w_1) \Leftrightarrow \psi(w_0) \not\overset{*}{\leftrightarrow}_{S_3} \psi(w_1) \Rightarrow \psi(w_0) \not\overset{*}{\leftrightarrow}_S \psi(w_1) \Rightarrow w_0 \not\overset{*}{\leftrightarrow}_S w_1.$$

Let us now show the "only if" part of (3).

Let $f(\Delta, R, y_0, y_1) = (\Delta, R, w_0, w_1)$, with $w_0 = y_0$, $w_1 = y_1$, be an instance[10] of L_{WM}. We want to show that then $(\Delta, R, y_0, y_1) \in L_{TMMI}$. Let $S = S_1 \cup S_2 \cup S_3$ be a solution of (Δ, R, w_0, w_1). Define Δ_λ, Δ' and Δ_{irr} from S_1 and S_2 as in section 3.2, and, for $x \in \Delta_{irr}$, set $C_x = \{x_i \in \Delta', x_i \xrightarrow{*}_{S_2} x\}$.

We shall now construct a solution (Σ, g, θ) of the considered instance of $TMMI_d$. Set $\Sigma = \Delta_{irr}$. Define $g : \Delta^* \to \Sigma^*$ with $g(x) = \lambda$ if $x \in \Delta_\lambda$, and $g(x_i) = x$ if $x_i \in C_x$. Note that g is exactly the map ψ.

Finally, set $\theta = \{(x_i, x_j), (x_i x_j, x_j x_i) \in S_3\} \subseteq \Sigma \times \Sigma$. Observe that θ consists only in commuting rules, i.e. θ is a concurrency relation on Σ.

Assuming $S_1 \neq \Delta \times \{\lambda\}$ (otherwise there can be no two words $w_0, w_1 \in \Delta^*$, with $w_0 \not\overset{*}{\leftrightarrow}_R w_1$), it is plain that conditions 2. and 3. of the $TMMI_d$ problem statement are fullfilled. To show 4., observe that, as S is a solution of (Δ, R, w_0, w_1), we have: $\forall u, v \in \Delta^*$, $u \xleftrightarrow{*}_R v \Rightarrow u \xleftrightarrow{*}_S v$.

As $g = \psi$, we get, using property 2: $\forall u, v \in \Delta^*$, $u \xleftrightarrow{*}_S v \Leftrightarrow g(u) \xleftrightarrow{*}_{S_3} g(v)$. But then condition 4. follows, as, by definition of θ, $g(u) \xleftrightarrow{*}_{S_3} g(v) \Leftrightarrow g(u) \equiv_\theta g(v)$. Condition 1. is an easy consequence of property 2: as $(\Delta, R, w_0, w_1) \in L_{WM}$, we have $w_0 \not\overset{*}{\leftrightarrow}_R w_1$, and, S being a solution of WM_d, we have $w_0 \not\overset{*}{\leftrightarrow}_S w_1$. Thus, $g(y_0) \not\equiv_\theta g(y_1)$.

In [1], it has been shown that $TMMI_d$ poly-time many one reduces to the satisfiability problem SAT_d (the reduction actually being between solutions of corresponding search problems). For the sake of completeness, we quote the following results, proving the NP-completeness of $TMMI_d$ and WM_d, as well as NP-hardness of WM_s:

Proposition 1. *$TMMI_d \in NP$.*

Proof. It suffices to check that a solution of any instance (Δ, T, y_0, y_1) of[11] L_{TMMI} can be checked in time polynomial in the size of (Δ, T, y_0, y_1), namely

---

[9] Remember that $w_0 = y_0$, and $w_1 = y_1$.

[10] Any instance of $TMMI_d$ is such that $y_0 \not\overset{*}{\leftrightarrow}_R y_1$. Thus it makes sense to consider the instance $(\Delta, R, w_0 = y_0, w_1 = y_1)$ of WM_d.

[11] Notation being as in the proof of theorem 1.

in $|\Delta| + ||T|| + |y_0| + |y_1|$, where $||T|| = \sum_{(\ell,r)\in T}(|\ell| + |r|)$. Let us thus prove that each of conditions 1. to 4. of the TMMI$_d$ problem statement can be checked in polynomial time.

First, conditions 2. and 3. can be checked by evaluating $g(\delta)$ for each $\delta \in \Delta$, and see whether we get a letter of Σ or the empty word. This can be done in $O(|\Delta|)$. It is easy to check that θ defines a concurrency relation on Σ: one simply has to verify that θ only contains rules of the form (a, b), for $a, b \in \Sigma$. This can be done in $O(|\Sigma|^2)$. Then, according to proposition 3 given in the appendix, we know that the congruence $g(y_0) \not\equiv_\theta g(y_1)$ can be tested in time $O(|\Sigma|^2(|y_0| + |y_1|))$.

To check condition 4., namely $\forall u, v \in \Delta^*$, $u \overset{*}{\leftrightarrow}_T v \implies g(u) \equiv_\theta g(v)$, observe that it is equivalent to

$$(u, v) \in T \implies g(u) \equiv_\theta g(v). \tag{5}$$

Indeed, (condition 4.) $\Rightarrow$ (5) is straightforward. The converse follows from the transitivity properties of $\overset{*}{\leftrightarrow}_T$ and $\equiv_\theta$: indeed, suppose (5) is true. Let $u, v \in \Delta^*$, with $u \overset{*}{\leftrightarrow}_T v$. Then there exists a finite sequence $u_0, u_1, \dots, u_k$, $k \in \mathbb{N}^*$, of words of Δ^* such that

$$u_0 = u \leftrightarrow_T u_1 \leftrightarrow_T u_2 \leftrightarrow_T \dots \leftrightarrow_T u_{k-1} \leftrightarrow_T u_k = v.$$

For $0 \leq j \leq k-1$, $u_j \leftrightarrow_T u_{j+1}$ means that there exists $(\ell, r) \in T$, and $x, y \in \Delta^*$, with $u_j = x\ell y$ and $u_{j+1} = xry$. This yields $g(u_j) = g(x)g(\ell)g(y)$ and $g(u_{j+1}) = g(x)g(r)g(y)$. As (5) holds, we have $g(\ell) \equiv_\theta g(r)$, so that $g(u_j) \equiv_\theta g(u_{j+1})$. Thus, we have the corresponding sequence

$$g(u) \equiv_\theta g(u_1) \equiv_\theta g(u_2) \equiv_\theta \dots \equiv_\theta g(u_{k-1}) \equiv_\theta g(v),$$

so that condition 4. is true. This proves that condition 4. can be checked in $O(|\Sigma|^2||T||)$.

Corollary 1. *TMMI$_d$ is NP-complete.*

Proof. It is an easy consequence of proposition 1 and the property proved in [1], namely SAT$_d \leq^p_m$ TMMI$_d$.

Proposition 2. *WM$_d \in$ NP.*

Proof. The proof is essentially the same as the one of prop.1: Let (Δ, R, w_0, w_1) be any instance of L_{WM}. The size of (Δ, R, w_0, w_1) is $|\Delta| + ||R|| + |w_0| + |w_1|$, with $||R|| = \sum_{(\ell,r)\in R}(|\ell| + |r|)$. Let S be a solution of this instance. First, to check S has the right form can be done in time $O(|\Delta|^2)$. Once one has verified this, one knows that S induces a free partially commutative monoid structure on Δ^*.

Property (i) in WM$_d$ problem statement can be checked in time $O(|\Delta|^2||R||)$, as it is equivalent to $(u, v) \in R \implies u \overset{*}{\leftrightarrow}_S v$, and by the fact that Δ^*/S is a free partially commutative monoid. For the same reason, (ii) can be checked in time $O(|\Delta|^2(|y_0| + |y_1|))$.

Corollary 2. (i) WM_d is NP-complete.
(ii) WM_s is NP-hard.

Proof. (i) follows from proposition 2, theorem 1 and corollary 1.
(ii) follows from theorem 1 and corollary 1.

4   An Algorithm for WM

In [7], an algorithm has been designed to solve TMMI, that is, to find a solution of TMMI or conclude there is no. We shall here call this algorithm TMMI-Alg and, for the sake of completeness, quote it in the appendix. In the previous section, we have proposed a reduction from the TMMI problem to the WM problem. Furthermore, this reduction is plainly tight, as the mapping between instances of these two problems is indeed the identity. It then seems natural to use TMMI-Alg to design an algorithm for WM.

In essence, the idea is as follows: for each integer i in the range $[1, |\Delta|]$, and every possible morphism g from Δ^* onto $g(\Delta^*)$, with $g(\Delta)$ of size i, we execute algorithm TMMI-Alg until a solution is found or until $i = |\Delta| + 1$.

WM-Alg
Input: (Δ, R, w_0, w_1), an instance of WM.
Output: a solution $S = S_1 \cup S_2 \cup S_3$ of WM - if any - or Error.
$i = 1$
While $i < |\Delta| + 1$ **do**

    For $g \in \mathrm{Hom}(\Delta^*, \Delta^*)$ with $|g(\Delta)| = i$ **do**
        run TMMI-Alg on $(\Delta, g, g(\Delta), R, w_0, w_1)$
        If the output of TMMI-Alg is θ **then** set
            $\{\sigma_1, \ldots, \sigma_k\} = g(\Delta)$
            $C_{\sigma_i} = \{x \in \Delta, g(x) = \sigma_i\}$ and, for each i, choose $z_i \in C_{\sigma_i}$
            $\Delta_{irr} = \{z_1, \ldots, z_k\}$
            $S_1 = \{(x, \lambda),\ x \in \Delta,\ g(x) = \lambda\}$
            $S_2 = \{(x_i, x_j),\ x_i, x_j \in \Delta,\ x_i \neq x_j,\ g(x_i) = g(x_j) \neq \lambda\}$
            $S_3 = \{(x_i x_j, x_j x_i),\ x_i, x_j \in \Delta_{irr},\ x_i \neq x_j,\ (g(x_i), g(x_j)) \in \theta\}$
        Return $S = S_1 \cup S_2 \cup S_3$
      EndIf
    EndFor
    $i = i + 1$

EndWhile
Return Error

Proof of correctness of the algorithm

Theorem 2. *Let R be a Thue system on an alphabet Δ. Then,*

(i) If the algorithm outputs "Error", then there is no solution to the WM problem on the instance (Δ, R, w_0, w_1).
(ii) if the algorithm returns S, then S is a solution of WM on the instance (Δ, R, w_0, w_1).

Proof. (i) If the algorithm outputs "Error", then it is the case that there is no integer i in the range $[1, \Delta]$ for which TMMI-Alg could find a solution on the instance $(\Delta, g, g(\Delta), R, w_0, w_1)$, for any morphism g with $|g(\Delta)| = i$. In other words, no concurrency relation could be found fitting (g, R), for any $g \in \text{Hom}(\Delta^*, \Delta^*)$, $|g(\Delta)| = i$, and any $i \in [1, \Delta]$. This means that $(\Delta, R, w_0, w_1) \notin L_{TMMI}$. According to theorem 1, this is equivalent to $f(\Delta, R, w_0, w_1) \notin L_{WM}$, where f is the identity mapping used in the proof of theorem 1.

(ii) If the algorithm returns S, it means that there exists an integer $i \in [1, \Delta]$, and a morphism $g \in \text{Hom}(\Delta^*, \Delta^*)$, with $|g(\Delta)| = i$, such that TMMI-Alg finds a concurrency relation θ fitting its input $(\Delta, g, g(\Delta), R, w_0, w_1)$. According to the theorem given in the appendix, this means that $(\Delta, R, w_0, w_1) \in L_{TMMI}$, the solution of this instance being $(g, g(\Delta), \theta)$. The definitions of the σ_is', C_{σ_i}s', Δ_{irr}, S set in the algorithm exactly follow the deriving of a solution of WM from a solution of TMMI given in the proof of theorem 1. Thus $f(\Delta, R, w_0, w_1) = (\Delta, R, w_0, w_1) \in L_{WM}$, and the set S is a solution of this instance.

A note on the case $i = |\Delta|$
In this case, g is an isomorphism on Δ (otherwise, it must be that $|g(\Delta)| < |\Delta|$). Thus, the conditions of TMMI-Alg on the projections π_a and $\pi_{\{a,b\}}$ (for a, $b \in g(\Delta)$) of the images $g(u)$ and $g(v)$ (for rules $(u, v) \in R$) are true if, and only if, they are true on u and v. In other words, the concurrency relation that we are looking for can be constructed from rules of R only. Thus there is no need to choose any g in the while loop, for $i = |\Delta|$. We nevertheless put $i < |\Delta| + 1$ as a stopping condition of the while loop, because the case $i = |\Delta|$ - i.e. instances of WM admitting a solution corresponding to a solution of TMMI with g being an isomorphism - obviously has to be handled.

Complexity considerations
The complexity of TMMI-Alg is $O(i^2 |T|)$ operations, where i is the size of $\Sigma_i = g_i(\Delta)$, and where an operation is here an evaluation of π or of g_i. In the worst case (no solution, or a solution for $|g(\Delta)| = |\Delta|$), TMMI-Alg will be iterated $\sum_{i=1}^{|\Delta|-1} |\text{Hom}(\Delta, \Sigma_i)|$ times. With $|\text{Hom}(\Delta, \Sigma_i)| = (i+1)^{|\Delta|}$, and using the rough bound $(i+1)^{|\Delta|} \le |\Delta|^{|\Delta|}$, for all $1 \le i \le |\Delta| - 1$, we get that the number of iterations of the for loop of WM-Alg is $O(|\Delta|^{|\Delta|+1})$. Thus, to solve[12] WM one needs to perform $O(|R||\Delta|^{|\Delta|+3})$ operations.

Note that the cryptographic instances of the WM problem all admit a solution. Thus, in a cryptanalytic setting, it is always the case that there exists an integer

---

[12] To find a solution or conclude that there is no.

k, $1 \leq k \leq |\Delta|$, and a morphism g with $|g(\Delta)| = k$, such that WM-alg returns a set S solution of the considered instance. The complexity of the algorithm in this context is then $O\big(|R|k^3(k+1)^{|\Delta|}\big)$.

Practical complexity : For instances of the TMMI problem admitting a solution - say (g, Σ, θ), with $k = |\Sigma|$ - observe that any isomorphism $\pi \in \mathrm{Isom}(\Sigma^*, \Sigma^*)$, will also yield a solution, namely $(\pi \circ g, \Sigma, \pi(\theta))$, with

$$\pi(\theta) = \{(\pi(a), \pi(b)),\ (a, b) \in \theta\}.$$

Thus the expected number of iterations of the for loop of WM-Alg is $C_k = \sum_{i=1}^{k-1}(i+1)^{|\Delta|} + ((k+1)^{|\Delta|}/k!)$, yielding an overall complexity of $O\big(|R|k^2 C_k\big)$.

Impact on the parameter choice
It seems difficult to devise suitable parameter sizes for Wagner Magyarik system from the sole complexity of the algorithm above. This is due to the fact that many questions still remain open concerning the practicality of the scheme; in particular, the size of Δ and R for which the encryption procedure is both efficient and secure. Thus, we think that an important direction of research (in word problem-based cryptosystems in general) would be the investigation of secure and efficient encryption. Then, the parameter sizes could be tuned w.r.t. to the above complexity in order to twart attacks on the underlying problem.

5 Conclusion

Transposing a group-based scheme due to Wagner and Magyarik to the monoid setting, we have proven a theoretical complexity result about its underlying problem WM. Our proof also showed that a related scheme due to Abisha, Thomas and Subramanian, based on an apparently harder problem (the one of finding an interpretation morphism g), reduces to the problem underlying Wagner Magyarik system. The algorithm to solve WM that we have presented gives a hint on how to choose the parameter sizes in order to reach a given level of security. To our knowledge, it is the first work that does so. More generally, our algorithm can be viewed as a sort[13] of completion procedure [6] that always terminates, in the case when the input is a refinement of a string rewriting system inducing a free partially commutative monoid.

References

1. P.J. Abisha, D. G. Thomas, K. G. Subramanian. *Public Key Cryptosystems Based on Free Partially Commutative Monoids and Groups*. Proceedings of INDOCRYPT 2003, LNCS 2904, Springer, pp.218-227.
2. R.V. Book. *Confluent and Other Types of Thue Systems*. Journal of the ACM 29, 1982, pp.171-182.

---

[13] In the sense it does not produce an equivalent system.

3. R.V. Book, H.N. Liu. *Rewriting Systems and Word Problems in a Free Partially Commutative Monoid.* Inform. Proc. Letters 26, 1987/88, pp.29-32.

4. R. Cori, D. Perrin. *Automates et commutations partielles.* R.A.I.R.O. Informatique théorique 19, 1985, pp.21-32.

5. M.I. González-Vasco, R. Steinwandt. *A Reaction Attack on a Public Key Cryptosystem Based on the Word Problem.* AAECC 14(5), 2004, pp.335-340.

6. D.E. Knuth, P.B. Bendix. *Simple Word Problems in Universal Algebras.* Computational Problems in Abstract Algebra, pp. 263-297, Pergamon Press, N.Y., 1970.

7. F. Levy-dit-Vehel, L. Perret. *Attacks on Public Key Cryptosystems Based on Free Partially Commutative Monoids and Groups.* Proceedings of INDOCRYPT 2004, LNCS 3348, Springer, pp. 275-289.

8. Robert McNaughton.*Contributions of Ronald V. Book to the Theory of String Rewriting Systems.* Rensselaer Polytechnic Institute T.R. n° 96 − 19, 1996.

9. A.M. Turing *The Word Problem in Semi-groups with Cancellation.* Annals of Math.(2) vol.52, 1950, 491-505.

10. N. R. Wagner, M. R. Magyarik. *A Public Key Cryptosystem Based on the Word Problem.* Proceedings of CRYPTO'84, LNCS 96, Springer-Verlag, pp.19-36.

Appendix: An Algorithm to Break the TMMI Problem

Our algorithm relies heavily on the following result of Cori and Perrin, which is indeed the core of the easiness of the word problem in free partially commutative monoids.

In what follows, Σ is as usual a finite alphabet, and $\theta \subseteq \Sigma \times \Sigma$ is a concurrency relation on Σ. For $B \subseteq \Sigma$, let π_B denote the monoid morphism from Σ^* onto B^*, defined by $\pi_B(b) = b$ if $b \in B$, $\pi_B(b) = \lambda$ if $b \in \Sigma \setminus B$.

Proposition 3. *[4] Let $u, v \in \Sigma^*$. We have $u \equiv_\theta v$ if, and only if:*

$$(i) \quad \pi_{\{a\}}(u) = \pi_{\{a\}}(v), \ \forall a \in \Sigma, \ and$$
$$(ii) \quad \pi_{\{a,b\}}(u) = \pi_{\{a,b\}}(v), \ \forall (a,b) \notin \theta.$$

The proof can be found in [4].

The algorithm

We here only quote the algorithm and its correctness result. Underlying ideas, as well as proof of correctness can be found in [7].

For each possible cardinality i of some alphabet Σ_i, and for every morphism g_i from Δ^* onto Σ_i^*, we execute the algorithm below until a solution is found or until $i = |\Delta| + 1$. This algorithm constructs a concurrency relation - say θ_i - fitting T and g_i, if any, or outputs Error otherwise. If the algorithm returns Error, it means that g_i is not a suitable interpretation morphism; thus the algorithm stops and has to be rerun with another g_i. If, for a given i, no suitable g_i has been found, we increment i and rerun the algorithm. The index i is here in the integer range $[1, |\Delta|]$.

TMMI-Alg
Input: Δ, $g_i \in \mathrm{Hom}(\Delta^*, \Sigma_i^*)$, Σ_i, $T \subseteq \Delta^* \times \Delta^*$, y_0 and $y_1 \in \Delta^*$.
Output: A concurrency relation $\theta_i \subseteq \Sigma_i \times \Sigma_i$ or Error.

$\theta_i \leftarrow \emptyset$
For $(u, v) \in T$ **do**
   Compute $g_i(u)$ and $g_i(v)$
   For $a \in \Sigma_i$ **do**
      If $\pi_{\{a\}}(g_i(u)) \neq \pi_{\{a\}}(g_i(v))$ **then** Error
   EndFor
   For $(a, b) \in \Sigma_i \times \Sigma_i$ **do**
      If $\pi_{\{a,b\}}(g_i(u)) \neq \pi_{\{a,b\}}(g_i(v))$
         If (a, b) or (b, a) are not in θ_i **then** $\theta_i \leftarrow \theta_i \cup \{(a, b)\}$
   EndFor
EndFor
If $g_i(y_0) \equiv_{\theta_i} g_i(y_1)$ **then** Error
Return θ_i

Proof of correctness of the algorithm

Theorem 3. *Let T be a Thue system on an alphabet Δ. Then*

i) If, for each i, $1 \leq i \leq |\Delta|$, and for each $g_i \in \mathrm{Hom}(\Delta^*, \Sigma_i^*)$, TMMI-Alg outputs "Error", then there is no solution to the TMMI problem for (Δ, T).

ii) If there exists an i for which, on input $(\Delta, g_i, \Sigma_i, T, y_0, y_1)$, TMMI-Alg returns a concurrency relation θ_i, then $(g_i, \Sigma_i, \theta_i)$ is a solution of the TMMI problem for the instance (Δ, T, y_0, y_1).

Constructions of Complementary Sequences for Power-Controlled OFDM Transmission

Kai-Uwe Schmidt and Adolf Finger

Communications Laboratory
Technische Universität Dresden
01062 Dresden, Germany
schmidtk@ifn.et.tu-dresden.de

Abstract. We present constructions of polyphase sequences suitable for the use as codewords in orthogonal frequency-division multiplexing (OFDM) with strictly bounded peak-to-mean envelope power ratio (PMEPR). Our first construction establishes that each polyphase sequence of length 2^m lies in a complementary set, whose size depends on a special property of its associated generalized Boolean function. Thus we identify a large family of sequences with PMEPR at most 2^{k+1}, where k is a non-negative integer. Our second construction yields sequences that lie in so-called almost complementary pairs and have PMEPR at most 3. A number of coding schemes for OFDM with low PMEPR is then presented. These schemes extend and complement previously proposed coding options.

1 Introduction

Davis and Jedwab [1] established a link between Golay's complementary sequences [2] and certain second-order cosets of a generalized first-order Reed–Muller code. The union of these cosets yields a powerful code, which can be used to perform error correction and ensures a PMEPR not exceeding 2. Its main disadvantage is that the code rate rapidly decreases for larger block lengths. Therefore Davis and Jedwab proposed to include further cosets in order to increase the rate of the codes at the cost of a slightly higher PMEPR [1]. This raised the problem of finding explicit constructions for such cosets. Paterson [3] provided a construction for further second-order cosets comprising sequences lying in so-called complementary sets of size 2^{k+1} ($k > 0$), and thus, the resulting codes have PMEPR at most 2^{k+1}. Parker and Tellambura [4] proposed an elaborate method to construct higher-order cosets comprised of complementary sets. However their construction suffers from the lack of efficient encoding and decoding algorithms. We propose a construction for complementary sets of a given size lying in cosets of a given order. Our construction includes previous constructions in [1] and [3] as special cases. In a straightforward way we then obtain a wide range of coding options for OFDM with low PMEPR. In addition we address the problem of constructing sequences that have low PMEPR and do not necessarily lie in complementary sets. We present a construction for cosets comprised

Ø. Ytrehus (Ed.): WCC 2005, LNCS 3969, pp. 330–345, 2006.

of so-called almost complementary pairs and establish that their PMEPR is at most 3. In this way we prove a conjecture by Davis and Jedwab [1] and Nieswand and Wagner [5] and identify more cosets with PMEPR at most 3. These results further extend possible coding options for OFDM with low PMEPR.

2   Notation and Background

2.1   Problem Statement

We will study an OFDM system with n subcarriers. Let $A = (A_0\, A_1\, \cdots\, A_{n-1})$ be a polyphase codeword of length n that is used to modulate the subcarriers. Its corresponding OFDM signal can be mathematically described by

$$S(A)(\theta) = \sum_{i=0}^{n-1} A_i e^{\sqrt{-1}2\pi(i+\lambda)\theta}, \quad 0 \le \theta < 1,$$

where λ is a positive constant. An important characteristic of such a signal (or of the modulating codeword) is its PMEPR, which is defined to be

$$\mathrm{PMEPR}(A) := \frac{1}{n} \sup_{0 \le \theta < 1} |S(A)(\theta)|^2.$$

Due to engineering reasons there is a high motivation to keep the PMEPR of the transmitted OFDM signals low. A particular elegant solution to solve this power-control issue is to use a special OFDM block code across the subcarriers [6]. By defining the PMEPR for such a code $\mathcal{C}$ to be

$$\mathrm{PMEPR}(\mathcal{C}) := \max_{A \in \mathcal{C}} \mathrm{PMEPR}(A).$$

we can formulate the problem as follows. *Find codes with high code rates and high minimum distances for which the above-defined value is small.*

2.2   Complementary Sequences for OFDM

We begin with recalling and extending some well-known relations between the aperiodic auto-correlation and the PMEPR of a sequence (cf. e.g. [7], [1], [3]). Let $A = (A_0\, A_1\, \cdots\, A_{n-1})$ and $B = (B_0\, B_1\, \cdots\, B_{n-1})$ be two complex-valued sequences. Then the *aperiodic cross-correlation* of A and B at a displacement $\ell \in \mathbb{Z}$ is given by

$$C(A,B)(\ell) := \begin{cases} \displaystyle\sum_{i=0}^{n-\ell-1} A_{i+\ell} B_i^* & 0 \le \ell < n \\ \displaystyle\sum_{i=0}^{n+\ell-1} A_i B_{i-\ell}^* & -n < \ell < 0 \\ 0 & \text{otherwise,} \end{cases}$$

where $()^*$ denotes complex conjugation. The *aperiodic auto-correlation* of A at a displacement $\ell \in \mathbb{Z}$ is then conveniently written as

$$A(A)(\ell) := C(A, A)(\ell).$$

In the sequel the following lemma will be essential for the construction of sequence sets for OFDM with low PMEPR.

Lemma 1. *Suppose a set of N polyphase sequences of length n is given by $\{A^0 A^1 \cdots A^{N-1}\}$. Then the PMEPR of each individual sequence in the set is at most*

$$N + \frac{2}{n} \sum_{\ell=1}^{n-1} \left| \sum_{i=0}^{N-1} A(A^i)(\ell) \right|.$$

In particular, if the set is a complementary set, each sequence has PMEPR at most N [3].

Proof. It is well known (cf. e.g. [7], [1]) and straightforward to show that

$$|S(A)(\theta)|^2 = A(A)(0) + 2 \sum_{\ell=1}^{n-1} \Re\{A(A)(\ell)\, e^{\sqrt{-1}\, 2\pi\ell\theta}\}.$$

Hence

$$\sum_{i=0}^{N-1} |S(A^i)(\theta)|^2 = \sum_{i=0}^{N-1} \left(A(A^i)(0) + 2 \sum_{\ell=1}^{n-1} \Re\{A(A^i)(\ell)\, e^{\sqrt{-1}\, 2\pi\ell\theta}\} \right)$$

$$= Nn + 2 \sum_{\ell=1}^{n-1} \Re\left\{ \sum_{i=0}^{N-1} A(A^i)(\ell)\, e^{\sqrt{-1}\, 2\pi\ell\theta} \right\}$$

$$\leq Nn + 2 \sum_{\ell=1}^{n-1} \left| \sum_{i=0}^{N-1} A(A^i)(\ell) \right|,$$

where we used the fact that $A(A)(0) = n$ for polyphase sequences. The lemma follows then with the definition of the PMEPR. $\square$

The above result motivates the construction of sequences lying in sets of sequences of small size, where the sum of the aperiodic auto-correlation sidelobes of all sequences in the set is small for all nonzero shifts. In this paper we shall particularly study two types of such sequences sets. The first one are the so-called complementary sets, which are defined as follows.

Definition 2. *A set of N sequences is called a* complementary set of size N *if the aperiodic auto-correlations of its members sum up to zero except for the zero displacement. If $N = 2$, the two sequences are commonly termed a* Golay complementary pair *[2].*

By Lemma 1 the PMEPR of each polyphase sequence lying in a complementary set of size N is at most N. A construction for such sequence sets will be established in Section 3.

Definition 3. *A pair of sequences is called an* almost complementary pair *if the sum of the aperiodic auto-correlations of its members is zero except for the zero shift and for at most two more shifts (τ and $-\tau \mid 1 \leq \tau < n$).*

It follows from Lemma 1 that the PMEPR of a polyphase sequence lying in an almost complementary pair is upper-bounded by 6. A tighter upper bound on the PMEPR can be obtained by taking the height of the out-of-phase peak in the auto-correlation sum into account. In Section 4 we shall construct almost complementary pairs comprising sequences with PMEPR at most 3.

2.3 Generalized Boolean Functions and Associated Sequences

A *generalized Boolean function* f is defined as a mapping $f : \mathbb{Z}_2^m \to \mathbb{Z}_q$, where throughout this paper q is assumed to be an even integer. Such a function can be written uniquely in its *algebraic normal form*, i.e., f is the sum of 2^m weighted *monomials*

$$f = f(x_0, x_1, \ldots, x_{m-1}) = \sum_{i=0}^{2^m-1} c_i \prod_{\alpha=0}^{m-1} x_\alpha^{i_\alpha},$$

where the weights $c_0, \ldots c_{2^m-1}$ are in $\mathbb{Z}_q$, and $(i_0 i_1 \ldots i_{m-1})$ is the binary expansion of $0 \leq i < 2^m$, such that $i = \sum_{j=0}^{m-1} i_j 2^j$. The *order of the ith monomial* is defined to be $\sum_{j=0}^{m-1} i_j$, and the *order, or algebraic degree, of a generalized Boolean function* f, denoted by $\deg(f)$, is equal to the highest order of the monomials with a nonzero coefficient in the algebraic normal form of f.

A generalized Boolean function may be equally represented by sequences of length 2^m. We shall define the sequence $(f_0 \, f_1 \cdots f_{2^m-1})$ as the $\mathbb{Z}_q$-*valued sequence associated with* f and the sequence $(\xi^{f_0} \, \xi^{f_1} \ldots \xi^{f_{2^m-1}})$ as the *polyphase sequence associated with* f. Here we denote $f_i = f(i_0, i_1, \cdots, i_{m-1})$, where $(i_0 i_1 \cdots i_{m-1})$ is the binary expansion of the integer $0 \leq i < 2^m$.

In the remainder of this subsection we recall the technique of the restriction of polyphase sequences of length 2^m and its application to the expansion of correlations of sequences. For details see [3] and [8].

Definition 4. *[3] Let $f : \mathbb{Z}_2^m \to \mathbb{Z}_q$ be a generalized Boolean function in the variables $x_0, x_1, \cdots, x_{m-1}$, and let F be its associated polyphase sequence. Suppose*

$$0 \leq j_0 < j_1 < \cdots < j_{k-1} < m$$

is a list of k indices and write $x = (x_{j_0} x_{j_1} \cdots x_{j_{k-1}})$. We shall call the entries of x the restricting variables. *Let $d = (d_0 d_1 \cdots d_{k-1}) \in \mathbb{Z}_2^k$, and let $(i_0 i_1 \cdots i_{m-1})$ be the binary expansion of the integer i. Then the* restricted sequence $F|_{x=d}$ *is a sequence of length 2^m with its elements $(F|_{x=d})_i$ being defined as*

$$(F|_{x=d})_i := \begin{cases} F_i & \text{if} \quad (i_{j_0} i_{j_1} \cdots i_{j_{k-1}}) = (d_0 d_1 \cdots d_{k-1}) \\ 0 & \text{if} \quad (i_{j_0} i_{j_1} \cdots i_{j_{k-1}}) \neq (d_0 d_1 \cdots d_{k-1}) \end{cases},$$

where $i = 0, 1, \cdots, 2^m - 1$. For the case $k = 0$ we fix $F|_{x=d} = F$.

Following [3] it is a consequence of the above definition that

$$F = \sum_{d \in \mathbb{Z}_2^k} F|_{x=d}.$$ (1)

A sequence that is restricted in k variables comprises $2^m - 2^{m-k}$ zero entries and 2^{m-k} nonzero entries. Those nonzero entries are determined by a function, which is denoted as $f|_{x=d}$ and called a *restricted function*. This function is a Boolean function in $m - k$ variables and is obtained by replacing the variables x_{j_α} by d_α for all $0 \le \alpha < k$ in the original function f. The restricted sequence $F|_{x=d}$ is then found by associating a polyphase sequence of length 2^{m-k} with $f|_{x=d}$ and inserting $2^m - 2^{m-k}$ zeros at the corresponding positions. Similarly to a disjunctive normal form of a Boolean function [9, Chapter 13], the original function f can be reconstructed from the functions $f|_{x=d}$ by

$$f = \sum_{d \in \mathbb{Z}_2^k} f|_{x=d} \prod_{\alpha=0}^{k-1} x_{j_\alpha}^{d_\alpha} (1 - x_{j_\alpha})^{(1-d_\alpha)}.$$ (2)

The following lemma will be useful to expand correlations of sequences of length 2^m.

Lemma 5. *[3],[8] Let $f, g : \mathbb{Z}_2^m \to \mathbb{Z}_q$ be two generalized Boolean functions in the variables $x_0, x_1, \cdots, x_{m-1}$, and let F and G be their associated polyphase sequences, respectively. Let $J = \{j_0, j_1, \cdots, j_{k-1}\}$ and $I = \{i_0, i_1, \cdots, i_{k'-1}\}$ be two sets of indices, such that $I \cap J = \varnothing$ and $I \cup J \subseteq \{0, 1, \cdots, m-1\}$. Write $x = (x_{j_0} x_{j_1} \cdots x_{j_{k-1}})$ and $x' = (x_{i_0} x_{i_1} \cdots x_{i_{k'-1}})$. Suppose d, d_1, d_2 are binary words of length k and c, c_1, c_2 are binary words of length k'. Then we have*

$$C(F|_{x=d_1}, G|_{x=d_2})(\ell) = \sum_{c_1, c_2} C(F|_{xx'=d_1 c_1}, G|_{xx'=d_2 c_2})(\ell)$$

and

$$A(F|_{x=d})(\ell) = \sum_{c} A(F|_{xx'=dc})(\ell) + \sum_{c_1 \ne c_2} C(F|_{xx'=dc_1}, F|_{xx'=dc_2})(\ell).$$

2.4   Generalized Reed–Muller Codes

We recall (slightly modified) definitions and some basic properties of the generalized Reed–Muller codes $RM_q(r, m)$ and $ZRM_q(r, m)$ (cf. [1] and [3]).

Definition 6. *(a) For $0 \le r \le m$ the code $RM_q(r, m)$ is defined as the set of sequences associated with a generalized Boolean function $\mathbb{Z}_2^m \to \mathbb{Z}_q$ of order at most r. (b) For $q \ge 4$ and $1 \le r \le m$ the code $ZRM_q(r, m)$ is defined as the set of sequences associated with a generalized Boolean function $\mathbb{Z}_2^m \to \mathbb{Z}_q$ with algebraic normal form containing monomials of order at most $r - 1$ and two times the monomials of order r.*

Clearly for $q \geq 4$ and $1 \leq r \leq m$ we have $\mathrm{ZRM}_q(r, m) \subset \mathrm{RM}_q(r, m)$. Now recall the classical definitions of the minimum Hamming distance d_H and Lee distance d_L of a code $\mathcal{C} \subseteq \mathbb{Z}_q^n$ (see e.g. [9]), and notice that in the binary case (i.e. $q = 2$) the minimum Hamming and Lee distances coincide. We have:

Result 7. *[1], [3] The minimum Lee distances of* $\mathrm{RM}_q(r, m)$ *and* $\mathrm{ZRM}_q(r, m)$ *are equal to* 2^{m-r} *and* 2^{m-r+1}, *respectively.*

2.5   A Known Construction for Complementary Pairs

We recall a construction for complementary pairs from [3]. With each quadratic form f over $\mathbb{Z}_q$ in the variables $x_{i_0}, x_{i_1}, \cdots, x_{i_{m-1}}$, generally given by

$$\sum_{0 \leq j < k < m} q_{jk} x_{i_j} x_{i_k} + L, \qquad q_{jk} \in \mathbb{Z}_q$$

with L being an affine form over $\mathbb{Z}_q$, one can associate a labeled graph $G(f)$. The vertices of this graph are labeled with $i_0, i_1, \cdots, i_{m-1}$, and the edge between vertex i_j and vertex i_k is labeled with q_{jk}. Such a graph is called a *path* (passing through m vertices) if $m = 1$ (then the graph consists of a single vertex) or if $m \geq 2$ and f is of the form

$$\frac{q}{2} \sum_{\alpha=0}^{m-2} x_{i_{\pi(\alpha)}} x_{i_{\pi(\alpha+1)}},$$

where π is a permutation of $\{0, 1, \cdots, m-1\}$. The indices $i_{\pi(0)}$ and $i_{\pi(m-1)}$ are called *end vertices* of the path. We are now in the position to quote:

Result 8. *[3] Let* $0 \leq j_0 < j_1 < \cdots < j_{k-1} < m$ *be a list of* k *indices, write* $x = (x_{j_0} x_{j_1} \cdots x_{j_{k-1}})$, *and let* $d \in \mathbb{Z}_2^k$. *Suppose* $f : \mathbb{Z}_2^m \to \mathbb{Z}_q$ *is a generalized Boolean function in the variables* $x_0, x_1, \cdots, x_{m-1}$, *such that* $f|_{x=d}$ *is quadratic and* $G(f|_{x=d})$ *is a path (in* $m - k$ *vertices). Let* F *and* F' *be the polyphase sequences associated with* f *and* $f + (q/2)x_a + c'$, *respectively. Then* $F|_{x=d}$ *and* $F'|_{x=d}$ *form a complementary pair. Here* a *is an end vertex of the path* $G(f|_{x=d})$ *and* $c' \in \mathbb{Z}_q$.

In particular, if $k = 0$, the above result identifies $(m!/2)q^{m+1}$ polyphase sequences lying in complementary pairs [3, Corollary 11], [1, Theorem 3], where in the latter reference $q = 2^h$.

3   A Construction for Complementary Sets

Theorem 9. *Let* $f : \mathbb{Z}_2^m \to \mathbb{Z}_q$ *be a generalized Boolean function in* m *variables* $x_0, x_1, \cdots, x_{m-1}$. *Define a list of* k *indices by*

$$0 \leq j_0 < j_1 < \cdots < j_{k-1} < m$$

and write $x = (x_{j_0} x_{j_1} \cdots x_{j_{k-1}})$. *Suppose that for each $d \in \mathbb{Z}_2^k$ the restricted function $f|_{x=d}$ is quadratic and the graph $G(f|_{x=d})$ is a path having an end vertex a_d. Then the polyphase sequences associated with the functions*

$$ f + \frac{q}{2} \left(\sum_{\alpha=0}^{k-1} c_\alpha x_{j_\alpha} + c' e \right) \qquad c_0, \cdots, c_{k-1}, c' \in \mathbb{Z}_2 $$

form a complementary set of size 2^{k+1}, where

$$ e = \sum_{d \in \mathbb{Z}_2^k} x_{a_d} \prod_{\alpha=0}^{k-1} x_{j_\alpha}^{d_\alpha} (1 - x_{j_\alpha})^{(1-d_\alpha)}. $$

Proof. Write $c = (c_0 c_1 \cdots c_{k-1})$ and denote the 2^{k+1} sequences in the set by $F_{cc'}$. We have to show that the sum of auto-correlations $\sum_{c, c'} A(F_{cc'})(\ell)$ is zero for $\ell \neq 0$. We employ Lemma 5 and write

$$ \underbrace{\sum_{c, c'} A(F_{cc'})(\ell) = \sum_{c, c'} \sum_d A(F_{cc'}|_{x=d})(\ell)}_{S_1} + \underbrace{\sum_{c, c'} \sum_{d_1 \neq d_2} C(F_{cc'}|_{x=d_1}, F_{cc'}|_{x=d_2})(\ell)}_{S_2}. $$

We first focus on the term S_1, which becomes

$$ S_1 = \sum_c \sum_d \left(A(F_{c0}|_{x=d})(\ell) + A(F_{c1}|_{x=d})(\ell) \right). $$

Recall that $e|_{x=d} = x_{a_d}$ is an end vertex of the graph $G(f|_{x=d})$. Thus the functions corresponding to $F_{c0}|_{x=d}$ and $F_{c1}|_{x=d}$ are

$$ f|_{x=d} + \frac{q}{2} \sum_{\alpha=0}^{k-1} c_\alpha d_\alpha \qquad \text{and} \qquad f|_{x=d} + \frac{q}{2} \sum_{\alpha=0}^{k-1} c_\alpha d_\alpha + \frac{q}{2} x_{a_d}, $$

respectively. Notice that the sum over α is just a constant occuring in both functions. Hence, by hypothesis and by Result 8, $F_{c0}|_{x=d}$ and $F_{c1}|_{x=d}$ form a complementary pair. It follows that the inner term of S_1 is zero for $\ell \neq 0$, and thus, also S_1 itself is zero for $\ell \neq 0$.

It remains to prove that the term S_2 is zero. This part of the proof follows more or less the same reasoning as the second part of the proof of [3, Theorem 12]. $\square$

Theorem 9 generalizes [3, Theorem 12] from complementary sets that contain sequences corresponding to quadratic generalized Boolean functions to complementary sets comprised of sequences associated with arbitrary generalized Boolean functions. Hence Theorem 9 provides a general upper bound on the PMEPR of polyphase sequences of length 2^m. Moreover this bound remains the same for all words in a coset of $\mathrm{RM}_q(1, m)$.

Corollary 10. *Suppose that* $f : \mathbb{Z}_2^m \to \mathbb{Z}_q$ *is a generalized Boolean function in the variables* $x_0, x_1, \cdots, x_{m-1}$. *If there exists a set of* k *restricting variables* $x = (x_{j_0} x_{j_1} \cdots x_{j_{k-1}})$ *with*

$$0 \le j_0 < j_1 < \cdots < j_{k-1} < m,$$

such that for each $d \in \mathbb{Z}_2^k$, *the restricted function* $f|_{x=d}$ *is quadratic and the graph* $G(f|_{x=d})$ *is a path, then the polyphase sequences in the coset* $f + \mathrm{RM}_q(1,m)$ *have PMEPR at most* 2^{k+1}.

It should be noted that for large k the above corollary appears to be rather weak. Using Corollary 10 and (2), it is now straightforward to find an explicit construction for sequences having PMEPR at most 2^{k+1}. Therefore partition the m indices $\{0, 1, \cdots, m-1\}$ into sets $I = \{i_0, i_1, \cdots, i_{m-k-1}\}$ and $J = \{j_0, j_1, \cdots, j_{k-1}\}$. Let $\pi_0, \pi_1, \cdots, \pi_{2^k-1}$ be 2^k permutations of $\{0, 1, \cdots, m-k-1\}$, and let $g_0, \cdots, g_{m-k-1}, g' : \mathbb{Z}_2^k \to \mathbb{Z}_q$ be $m-k+1$ generalized Boolean functions. Then each sequence associated with

$$\frac{q}{2} \sum_{d \in \mathbb{Z}_2^k} \sum_{\alpha=0}^{m-k-2} x_{i_{\pi_d(\alpha)}} x_{i_{\pi_d(\alpha+1)}} \prod_{\beta=0}^{k-1} x_{j_\beta}^{d_\beta} (1 - x_{j_\beta})^{(1-d_\beta)}$$

$$+ \sum_{\alpha=0}^{m-k-1} x_{i_\alpha} g_\alpha(x_{j_0}, \cdots, x_{j_{k-1}}) + g'(x_{j_0}, \cdots, x_{j_{k-1}}) \quad (3)$$

satisfies Corollary 10 for a particular k and, hence, has PMEPR at most 2^{k+1}. It is apparent that each such a sequence lies inside $\mathrm{RM}_q(k+2, m)$ and particularly inside $\mathrm{ZRM}_q(k+2, m)$ if $q \ge 4$.

In what follows we focus on a particular subset of the sequences associated with forms of type (3). In this way we can obtain a wide range of coding options for OFDM with low PMEPR that allow a trade-off between the size of the sequence set and minimum distance.

Corollary 11. *Let* $g_0, \cdots, g_{m-k-1}, g' : \mathbb{Z}_2^k \to \mathbb{Z}_q$ *be* $m-k+1$ *generalized Boolean functions. Suppose* $2 \le r \le k+1$, $\deg(g_\alpha) \le r-1$ *for* $\alpha = 0, 1, \cdots m-k-1$, *and* $\deg(g') \le r$. *Then the polyphase sequences associated with the forms*

$$\frac{q}{2} \sum_{\alpha=0}^{m-k-2} x_{\pi(\alpha)} x_{\pi(\alpha+1)} + \sum_{\alpha=0}^{m-k-1} x_\alpha g_\alpha(x_{m-k}, \cdots, x_{m-1}) + g'(x_{m-k}, \cdots, x_{m-1}),$$

where π *is a permutation of* $\{0, 1, \cdots, m-k-1\}$, *have PMEPR at most* 2^{k+1}. *Moreover the sequences form cosets of* $\mathrm{RM}_q(1,m)$, *which are contained inside* $\mathrm{RM}_q(r, m)$. *In particular these cosets are contained in* $\mathrm{ZRM}_q(r, m)$ *if (i)* $q \ge 4$ *and* $k = 0$ *or if (ii)* $q \ge 4$ *and all coefficients of the monomials in the algebraic normal forms of* g_α *with degree equal to* $r-1$ *and in the algebraic normal form of* g' *with degree equal to* r *are even.*

Now suppose that k is given. Then a simple counting argument shows that the above corollary identifies $2^{K_{\mathrm{RM}}}$ words inside $\mathrm{RM}_q(r, m)$, where

$$K_{\mathrm{RM}} = \log_2 \frac{(m-k)!}{2} + \left[\binom{k}{r} + (m-k+1) \sum_{i=0}^{r-1} \binom{k}{i} \right] \log_2 q$$

and $2^{K_{\mathrm{ZRM}}}$ words inside $\mathrm{ZRM}_q(r, m)$ with

$$K_{\mathrm{ZRM}} = K_{\mathrm{RM}} - \binom{k}{r} - (m-k) \binom{k}{r-1}.$$

4   A Construction for Almost Complementary Pairs

Theorem 12. *Let $k \leq m - 3$. Suppose $J = \{j_0, j_1, \cdots, j_{k-1}\}$ and $I = \{i_0, i_1, \cdots, i_{m-k-1}\}$ are two sets of indices, such that $I \cap J = \varnothing$ and $I \cup J = \{0, 1, \cdots, m-1\}$. Write $x = (x_{j_0} x_{j_1} \cdots x_{j_{k-1}})$ and let $d \in \mathbb{Z}_2^k$. Let $f : \mathbb{Z}_2^m \to \mathbb{Z}_q$ be a generalized Boolean function in the variables $x_0, x_1, \cdots, x_{m-1}$, such that $f|_{x=d}$ is of the form*

$$\frac{q}{2} \sum_{\gamma=0}^{m-k-2} x_{i_{\pi(\gamma)}} x_{i_{\pi(\gamma+1)}} + \alpha x_b x_d + \beta x_c x_d + \sum_{\gamma=0}^{m-k-1} c_\gamma x_{i_{\pi(\gamma)}} + c,$$

$$c_0, \cdots, c_{m-k-1}, c, \alpha, \beta \in \mathbb{Z}_q,$$

where π is a permutation of $\{0, 1, \cdots, m - k - 1\}$ and $(a\,b\,c\,d)$ is either $(i_{\pi(m-k-1)}\, i_{\pi(0)}\, i_{\pi(1)}\, i_{\pi(2)})$ or $(i_{\pi(0)}\, i_{\pi(m-k-1)}\, i_{\pi(m-k-2)}\, i_{\pi(m-k-3)})$. Let F and F' be the polyphase sequences associated with the functions f and $f' = f + (q/2) x_a + c'$, respectively, where $c' \in \mathbb{Z}_q$. Then, provided that $\alpha - \beta = 0$ or $\alpha + \beta = 0$, $F|_{x=d}$ and $F'|_{x=d}$ form an almost complementary pair. In particular we have

$$|A(F|_{x=d})(\ell) + A(F'|_{x=d})(\ell)| = \begin{cases} 2^{m-k} & \ell = 0 \\ 2^{m-k-1} & \ell = \pm \tau, \quad 1 \leq \tau < n \\ 0 & otherwise. \end{cases}$$

Proof. We take $(a\,b\,c\,d) = (i_{\pi(m-k-1)}\, i_{\pi(0)}\, i_{\pi(1)}\, i_{\pi(2)})$. With a similar reasoning we can perform the proof for $(a\,b\,c\,d) = (i_{\pi(0)}\, i_{\pi(m-k-1)}\, i_{\pi(m-k-2)}\, i_{\pi(m-k-3)})$.

Substitute: $G = F|_{x=d}$, $G' = F'|_{x=d}$, $g = f|_{x=d}$, and $g' = f'|_{x=d}$, and keep in mind that G and G' are in general restricted sequences. We are interested in the expression $A(G)(\ell) + A(G')(\ell)$, which is expanded using Lemma 5

$$\begin{aligned} &A(G)(\ell) + A(G')(\ell) \\ &= A(G|_{x_a=0})(\ell) + A(G|_{x_a=1})(\ell) + A(G'|_{x_a=0})(\ell) + A(G'|_{x_a=1})(\ell) \\ &\quad + C(G|_{x_a=0}, G|_{x_a=1})(\ell) + C(G|_{x_a=1}, G|_{x_a=0})(\ell) \\ &\quad + C(G'|_{x_a=0}, G'|_{x_a=1})(\ell) + C(G'|_{x_a=1}, G'|_{x_a=0})(\ell). \end{aligned} \tag{4}$$

Let us first consider the case $k = m - 3$, and take this as the base case for the proof. Then we proceed to prove the theorem for $k < m - 3$ by induction on k. For $k = m - 3$ we have $a = d$, and the list of restricting indices writes $J = \{0, 1, \cdots, m - 1\}\backslash\{a, b, c\}$. The functions corresponding to the sequences $G|_{x_a=0}$, $G|_{x_a=1}$, $G'|_{x_a=0}$, and $G'|_{x_a=1}$ are then

$$g|_{x_a=0} = r \qquad\qquad\qquad g'|_{x_a=0} = g|_{x_a=0} + c'$$

$$g|_{x_a=1} = r + \left(\alpha + \frac{q}{2}\right)x_b + \beta x_c + c_a \qquad g'|_{x_a=1} = g|_{x_a=1} + c' + \frac{q}{2}$$

with

$$r = \frac{q}{2}x_b x_c + c_b x_b + c_c x_c + c.$$

Since we have $g'|_{x_a=0} = g|_{x_a=0} + c'$ and $g'|_{x_a=1} = g|_{x_a=1} + c' + q/2$, it follows that $G'|_{x_a=0} = \xi^{c'}G|_{x_a=0}$ and $G'|_{x_a=1} = -\xi^{c'}G|_{x_a=1}$. This implies that the cross-correlations in (4) are related as follows

$$C(G|_{x_a=0}, G|_{x_a=1})(\ell) = -C(G'|_{x_a=0}, G'|_{x_a=1})(\ell)$$
$$C(G|_{x_a=1}, G|_{x_a=0})(\ell) = -C(G'|_{x_a=1}, G'|_{x_a=0})(\ell). \qquad (5)$$

Hence the cross-correlations in (4) sum up to zero for all ℓ. We also conclude

$$A(G|_{x_a=0})(\ell) + A(G|_{x_a=1})(\ell) = A(G'|_{x_a=0})(\ell) + A(G'|_{x_a=1})(\ell). \qquad (6)$$

Next we consider the left-hand side of (6). Write $u = u_a u_b u_c$, $v = v_a v_b v_c$, $G_u = G|_{x_a x_b x_c = u_a u_b u_c}$, and $g_u = g|_{x_a x_b x_c = u_a u_b u_c}$. Then, by Lemma 5, we have

$$A(G|_{x_a=0})(\ell) + A(G|_{x_a=1})(\ell) = \sum_u A(G_u)(\ell) + \sum_{\substack{u,v \\ u_a=v_a \\ (u_b u_c) \neq (v_b v_c)}} C(G_u, G_v)(\ell). \qquad (7)$$

The functions corresponding to the sequences G_u are

$$g_{000} = c \qquad\qquad\qquad g_{100} = c$$

$$g_{010} = c + c_b \qquad\qquad g_{110} = c + c_b + \alpha + \frac{q}{2}$$

$$g_{001} = c + c_c \qquad\qquad g_{101} = c + c_c + \beta$$

$$g_{011} = c + c_b + c_c + \frac{q}{2} \qquad g_{111} = c + c_b + c_c + \alpha + \beta.$$

Each of the eight sequences G_u contains exactly one nonzero element occuring at position $\sum_{\gamma=0}^{k-1} d_\gamma 2^{j_\gamma} + u_a 2^a + u_b 2^b + u_c 2^c$. Thus we have

$$\sum_u A(G_u)(\ell) = \begin{cases} 8 & \ell = 0 \\ 0 & \ell \neq 0 \end{cases}. \qquad (8)$$

It follows also that the twelve cross-correlations of type $C(G_u, G_v)(\ell)$ in the second sum of (7) have exactly one nonzero element, which is located at

$(u_b - v_b)2^b + (u_c - v_c)2^c$. We next collect the cross-correlations having nonzero contributions at the same shift.

Shift 2^b $(u_b = 1, v_b = 0, u_c = v_c)$:

$$C(G_{000}, G_{010})(2^b) + C(G_{100}, G_{110})(2^b) + C(G_{001}, G_{011})(2^b) + C(G_{101}, G_{111})(2^b)$$
$$= \xi^{-c_b} + \xi^{\frac{q}{2} - c_b - \alpha} + \xi^{\frac{q}{2} - c_b} + \xi^{-c_b - \alpha} = 0$$

Shift 2^c $(u_b = v_b, u_c = 1, v_c = 0)$:

$$C(G_{000}, G_{001})(2^c) + C(G_{100}, G_{101})(2^c) + C(G_{010}, G_{011})(2^c) + C(G_{110}, G_{111})(2^c)$$
$$= \xi^{-c_c} + \xi^{-c_c - \beta} + \xi^{\frac{q}{2} - c_c} + \xi^{\frac{q}{2} - c_c - \beta} = 0$$

Shift $2^b + 2^c$ $(u_b = u_c = 1, v_b = v_c = 0)$:

$$C(G_{000}, G_{011})(2^b + 2^c) + C(G_{100}, G_{111})(2^b + 2^c) = \xi^{\frac{q}{2} - c_c - c_b} + \xi^{-c_c - c_b - (\alpha + \beta)}$$

Shift $2^b - 2^c$ $(u_b = v_c = 1, u_c = v_b = 0)$:

$$C(G_{001}, G_{010})(2^b - 2^c) + C(G_{101}, G_{110})(2^b - 2^c) = \xi^{c_c - c_b} + \xi^{\frac{q}{2} + c_c - c_b - (\alpha - \beta)}$$

Moreover there are contributions at the shifts -2^b, -2^c, $-2^b - 2^c$, and $-2^b + 2^c$, which are just the complex conjugated values of those at shifts 2^b, 2^c, $2^b + 2^c$, and $2^b - 2^c$, respectively. Now the additional condition $\alpha + \beta = 0$ or $\alpha - \beta = 0$ comes into play. Then the auto-correlations of G and G' cancel out at either $2^b + 2^c$ or $2^b - 2^c$. Considering (6), we have to count the auto-correlation in (8) and all the contributions from the cross-correlations twice in order to calculate the left-hand side of (4). By carefully counting the contributions, we observe that the left-hand side of (4) is nonzero for at most two nonzero shifts and has absolute values of at most 4 at those positions.

Now consider the case where x contains $k < m - 3$ restricting variables, and suppose that the theorem is true for x containing $k + 1$ variables. We focus on the expanded auto-correlations in (4). For $k < m - 3$ the functions corresponding to the sequences $G|_{x_a=0}$, $G|_{x_a=1}$, $G'|_{x_a=0}$, and $G'|_{x_a=1}$ are

$$g|_{x_a=0} = p \qquad\qquad g'|_{x_a=0} = g|_{x_a=0} + c'$$

$$g|_{x_a=1} = p + \frac{q}{2} x_{i_{\pi(m-k-2)}} + c_a \qquad g'|_{x_a=1} = g|_{x_a=1} + c' + \frac{q}{2}$$

with

$$p = \frac{q}{2} \sum_{\gamma=0}^{m-k-3} x_{i_{\pi(\gamma)}} x_{i_{\pi(\gamma+1)}} + \alpha x_b x_d + \beta x_c x_d + \sum_{\gamma=0}^{m-k-2} c_\gamma x_{i_{\pi(\gamma)}} + c.$$

By the same reasoning leading to (5), it turns out that the cross-correlations in (4) sum up to zero for all ℓ. Using the above theorem as a hypothesis, we know that the sequences $G|_{x_a=0}$ and $G|_{x_a=1}$ form an almost complementary pair. Hence the sum of their auto-correlations has at most two nonzero values at nonzero shifts. The same accounts for the pair of sequences $G'|_{x_a=0}$ and $G'|_{x_a=1}$. Note that the position and the phase of the nonzero components are independent of the restricting variables, while their magnitudes depend on the number of nonzero entries in the sequences $G|_{x_a=0}$, $G|_{x_a=1}$, $G'|_{x_a=0}$, and $G'|_{x_a=1}$. Hence the auto-correlations in (4) always superimpose at the considered positions and sum up to absolute values of at most 2^{m-k-1}.   □

Setting $k = 0$ in the above theorem and applying Lemma 1, we obtain the following simple corollary.

Corollary 13. *For $m > 2$ let $f : \mathbb{Z}_2^m \to \mathbb{Z}_q$ be given by*

$$\frac{q}{2} \sum_{i=0}^{m-2} x_{\pi(i)} x_{\pi(i+1)} + \alpha x_{\pi(0)} x_{\pi(2)} + \beta x_{\pi(1)} x_{\pi(2)},$$

where $\alpha, \beta \in \mathbb{Z}_q$, π is a permutation of $\{0, 1, \cdots, m-1\}$, and $\alpha + \beta = 0$ or $\alpha - \beta = 0$. Then the PMEPR of the polyphase sequence in the coset $f + \mathrm{RM}_q(1, m)$ is at most 3. These cosets lie inside $\mathrm{RM}_q(2, m)$ and particularly inside $\mathrm{ZRM}_q(2, m)$ if $q \geq 4$ and α and β are even.

A simple counting argument shows that Corollary 13 identifies $(2q - 3)(m!/2)q^{m+1}$ sequences with PMEPR at most 3. Notice that $(m!/2)q^{m+1}$ of them have PMEPR bounded by 2, since they are also identified by setting $k = 0$ in Corollary 11. In particular, Corollary 13 applies to the $2m! \, q^{m+1}$ ($q = 2^h$, $h \geq 3$) sequences for which it was conjectured in [1] and [5] that their PMEPR is bounded by 3. Corollary 13 provides a proof for this conjecture and identifies further sequences with PMEPR at most 3. We remark that for $q = 2$ Corollary 13 merely restates the construction of complementary pairs in a pure sense [1, Theorem 3] (where $q = 2^h$), [3, Corollary 11].

5   OFDM Coding Schemes from Reed–Muller Codes

In what follows we apply Corollary 11 and Corollary 13 to construct a number of coding options for OFDM with low PMEPR. Since these codes are unions of cosets of $\mathrm{RM}_q(1, m)$, well known algorithms are readily applicable to encode and decode the codes (see [1], [10] and references therein). Alternatively in [11] the structure of the codes obtained from Corollary 11 is further exploited in order to simplify encoding and decoding. This is particularly efficient if the codes contain a large number of cosets of $\mathrm{RM}_q(1, m)$. We define the code rate and the information rate of a code $\mathcal{C}$ of length $n = 2^m$ over a q-ary alphabet to be $\lfloor \log_q |\mathcal{C}| \rfloor / n$ and $\lfloor \log_2 |\mathcal{C}| \rfloor / n$, respectively.

We remark that all our coding options directly arise just from two simple corollaries. This should be compared with the constructions in [1] and [3], which rely on a variety of techniques, including a computational search in [1]. Nevertheless there is some overlap between our codes and those in [1] and [3]. In the subsequent tables the superscript [1] means that the corresponding coding option

Table 1. Binary Coding Options

n	Option	max. PMEPR	# info bits	code rate	info rate	d_L
16	B1[1]	2	8	0.50	0.50	4
	B2[2]	4	9	0.56	0.56	4
	B3	8	12	0.75	0.75	2
	B4[2]	8	10	0.63	0.63	4
32	B1[1]	2	11	0.34	0.34	8
	B2	4	13	0.41	0.41	8
	B3	8	17	0.53	0.53	4
	B4	8	14	0.44	0.44	8
64	B1[1]	2	15	0.23	0.23	16
	B2	4	17	0.27	0.27	16
	B3	8	23	0.36	0.36	8
	B4	8	19	0.30	0.30	16

Table 2. Quaternary Coding Options

n	Option	max. PMEPR	# info bits	code rate	info rate	d_L
16	Q1[1]	2	13	0.41	0.81	8
	Q2	3	15	0.47	0.94	4
	Q3[4]	4	17	0.53	1.06	4
	Q4[2]	4	14	0.44	0.88	8
	Q5	8	24	0.75	1.50	2
	Q6	8	22	0.69	1.37	4
	Q7[2]	8	15	0.47	0.94	8
32	Q1[1]	2	17	0.27	0.53	16
	Q2	3	20	0.31	0.63	8
	Q3[4]	4	23	0.36	0.72	8
	Q4	4	19	0.30	0.59	16
	Q5	8	33	0.52	1.03	4
	Q6	8	30	0.47	0.94	8
	Q7	8	20	0.31	0.63	16
64	Q1[1]	2	22	0.17	0.34	32
	Q2	3	24	0.19	0.38	16
	Q3[4]	4	29	0.23	0.45	16
	Q4	4	24	0.19	0.38	32
	Q5	8	43	0.34	0.67	8
	Q6	8	39	0.30	0.61	16
	Q7	8	26	0.20	0.41	32

is identical to the Davis–Jedwab construction [1], which can be restated by setting $k = 0$ in Corollary 11. A [2] indicates that a better code has been reported in [1], while a [3] means that the same code appears in [1]. These latter results are based on a computational search, which becomes infeasible for large n. Therefore this situation occurs only for $n = 16$. Coding options marked by a [4] are weaker than codes constructed in [3]. A [5] indicates that a better code can be obtained using the techniques in [3], though the code itself has not been explicitly mentioned in [3]. All other coding options seem to be new or outperform previously reported results at least for the lengths considered here.

Table 1 shows possible coding options for binary signaling. Option B1 is the Davis–Jedwab construction. Option B2 is obtained from Corollary 11 by setting $k = 1$, and Options B3 and B4 arise from Corollary 11 by setting $k = 2$ and taking those codewords lying inside $RM_2(3, m)$ and $RM_2(2, m)$, respectively. The choice between the Options B1, B2, and B4 allows a trade-off between information rate and maximum PMEPR of the code, while the minimum Lee distance is the same. Options B3 and B4 provide a trade-off between minimum Lee distance and informations rate, while the PMEPR is constant.

Table 2 contains a list of coding options for quaternary signaling. Option Q1 is the Davis–Jedwab construction. Option Q2 is obtained from Corollary 13. Option Q3 uses Corollary 11 with $k = 1$ to construct a code lying in $RM_4(2, m)$, while Option Q4 takes its subcode in $ZRM_4(2, m)$. Option Q5 is a code inside $RM_4(3, m)$ and obtained by setting $k = 2$ in Corollary 11. Option Q6 and Option Q7 are the subcodes lying in $ZRM_4(3, m)$ and $ZRM_4(2, m)$, respectively. Notice that we may also construct a subcode of the code in Option Q5 inside $RM_4(2, m)$, however, this code contains less codewords than that in Option Q6, while the minimum distances and the maximum PMEPRs are the same. Moving to a quaternary constellation widely extends the possible coding options, which results in an increased number of possible trade-offs between PMEPR, information rate, and minimum distance. Moreover the information rate can be increased up to twice that of the binary coding schemes. However this goes generally at the cost of a smaller minimum Euclidean distance of the codes, which results in an increased transmission error probability.

Table 3 shows a number of coding options for octary phase-shift keying. These options are obtained in a similar fashion as the quaternary coding options. Option O1 is the Davis–Jedwab construction, Option O2 uses Corollary 13 to construct a code inside $RM_8(2, m)$, while Option O3 takes its subcode in $ZRM_8(2, m)$. Option O4 uses Corollary 11 with $k = 1$ to construct a code inside $RM_8(2, m)$. Option O5 is obtained by taking its subcode contained in $ZRM_8(2, m)$. Option O6 arises by setting $k = 2$ in Corollary 11 to construct a code inside $RM_8(3, m)$, while Options O7 and O8 use its subcodes in $ZRM_8(3, m)$ and $ZRM_8(2, m)$, respectively. Notice that although, based on numerical evaluation of the PMEPR, Option O3 was already mentioned in [1] and [12] for $n = 16$, Corollary 13 settles the theory behind this coding option and validates it for any $m > 2$. By moving to an octary constellation, the number of possible coding options is further extended. Also the information rate can be increased further, which in turn leads to a smaller minimum Euclidean distance of the codes compared to their binary or quaternary counterparts.

Table 3. Octary Coding Options

n	Option	max. PMEPR	# info bits	code rate	info rate	d_L
16	O1[1]	2	18	0.38	1.13	8
	O2	3	22	0.46	1.38	4
	O3[3]	3	20	0.42	1.25	8
	O4[4]	4	25	0.52	1.56	4
	O5[5]	4	22	0.46	1.38	8
	O6	8	36	0.75	2.25	2
	O7	8	34	0.71	2.13	4
	O8	8	25	0.52	1.56	8
32	O1[1]	2	23	0.24	0.72	16
	O2	3	27	0.28	0.84	8
	O3	3	26	0.27	0.81	16
	O4[4]	4	33	0.34	1.03	8
	O5[5]	4	29	0.30	0.91	16
	O6	8	49	0.51	1.53	4
	O7	8	46	0.48	1.44	8
	O8	8	33	0.34	1.03	16
64	O1[1]	2	29	0.15	0.45	32
	O2	3	33	0.17	0.52	16
	O3	3	31	0.16	0.48	32
	O4[4]	4	41	0.21	0.64	16
	O5[5]	4	36	0.19	0.56	32
	O6	8	63	0.33	0.98	8
	O7	8	59	0.31	0.92	16
	O8	8	42	0.22	0.66	32

6   Conclusion

We have established constructions of sequences lying in complementary sets of
a given size (Theorem 9) and in almost complementary pairs (Theorem 12). An
upper bound for the PMEPR of these sequences follows then immediately from
Lemma 1. These results led to a number of coding options for OFDM with low
PMEPR, which extend and complement existing schemes previously reported in
[1], [3], and [12]. Corollary 13 also provides an answer to an open problem stated
by Davis and Jedwab in [1]. Recent results have shown that this corollary in fact
arises in a more general context, and we refer to [13] for details.

Acknowledgment

The authors would like to thank the anonymous reviewers for their valuable
comments. Especially one referee provided very detailed suggestions, which led
to several improvements of the paper.

References

1. Davis, J.A., Jedwab, J.: Peak-to-mean power control in OFDM, Golay complementary sequences, and Reed–Muller codes. IEEE Trans. Inform. Theory **45** (1999) 2397–2417
2. Golay, M.J.E.: Complementary series. IRE Trans. Inform. Theory **7** (1961) 82–87
3. Paterson, K.G.: Generalized Reed–Muller codes and power control in OFDM modulation. IEEE Trans. Inform. Theory **46** (2000) 104–120
4. Parker, M.G., Tellambura, C.: A construction for binary sequence sets with low peak-to-average power ratio. Report No. 242, Department of Informatics, University of Bergen, Norway, http://www.ii.uib.no/~matthew/ (2003)
5. Nieswand, K.M., Wagner, K.N.: Octary codewords with power envelopes of $3 * 2^m$. http://www.mathcs.richmond.edu/~jad/summer.html (1998)
6. Jones, A.E., Wilkinson, T.A.: Combined coding for error control and increased robustness to system nonlinearities in OFDM. Proc. of IEEE 46th Vehicular Technology Conf. (VTC) (1996)
7. Tellambura, C.: Upper bound on the peak factor of n-multiple carriers. IEE Electron. Lett. **33** (1997) 1608–1609
8. Stinchcombe, T.E.: Aperiodic Autocorrelations of Length 2^m Sequences, Complementarity, and Power Control for OFDM. PhD thesis, University of London (2000)
9. MacWilliams, F.J., Sloane, N.J.A.: The Theory of Error-Correcting Codes. North Holland Mathematical Library (1977)
10. Paterson, K.G., Jones, A.E.: Efficient decoding algorithms for generalized Reed–Muller codes. IEEE Trans. Commun. **48** (2000) 1272–1285
11. Schmidt, K.U.: Complementary sets, generalized Reed–Muller codes, and power control for OFDM. submitted to IEEE Trans. Inform. Theory (2005)
12. Paterson, K.G.: Coding techniques for power-controlled OFDM. Proc. of IEEE Int. Symp. on Personal, Indoor and Mobile Radio Commun. (PIMRC) (1998) 801–805
13. Schmidt, K.U.: On cosets of the generalized first-order Reed–Muller code with low PMEPR. submitted to IEEE Trans. Inform. Theory (2005)

A Novel Method for Constructing Almost Perfect Polyphase Sequences*

Xiangyong Zeng[1,3], Lei Hu[2], and Qingchong Liu[3]

[1] The Faculty of Mathematics and Computer Science, Hubei University,
Xueyuan Road 11, Wuhan 430062, China
xzeng@hubu.edu.cn
[2] The State Key Laboratory of Information Security (Graduate School of Chinese
Academy of Sciences), 19A Yuquan Road, Beijing 100049, China
hu@is.ac.cn
[3] Department of Electrical and System Engineering, Oakland University,
Rochester, MI 48309, USA
qliu@oakland.edu

Abstract. This paper proposes a novel method of constructing almost perfect polyphase sequences based on the shift sequence associated with a primitive polynomial $f(x)$ of degree $2J$ over finite field $GF(p)$ (p odd prime, $J = 1, 2, \cdots$) and a pair of almost perfect sequences completely orthogonal. Almost perfect polyphase sequences of length $2(p^J + 1)$ are constructed with phases as any positive even number. New families of almost perfect polyphase sequences in other lengths are also provided. In particular, several new families of almost perfect quadriphase sequences of lengths $m(p^J + 1)$ are attained, where $m = 4$ or 8, and $p^J - 1 \equiv 0 \pmod{m}$.

Keywords: Periodic autocorrelation, almost perfect sequence, quadriphase sequence, binary sequence.

1 Introduction

The periodic autocorrelation function of a sequence is a measure for how much the given sequence differs from its translates. Perfect sequences are complex periodic sequences such that all out-of-phase autocorrelation values are zero. Unfortunately, perfect binary sequences of length $N > 4$ and perfect quadriphase sequences of length $N > 16$ are unknown. Therefore, in several recent publications the construction of "almost-perfect" binary sequences has been discussed. Almost perfect (AP) sequences are complex periodic sequences such that all out-of-phase autocorrelation values are zero except one, whose periodic autocorrelation function (PACF) is as close to perfect as possible. Brown and Goodwin [1] and later Wolfmann [2] presented AP binary sequences based on computer

---

* Lei Hu's work was supported by the National Science Foundation of China (NSFC) under Grants No. 60373041 and No. 60573053. Qingchong Liu's work was supported in part by ARO under Grant W911NF-04-1-0267.

Ø. Ytrehus (Ed.): WCC 2005, LNCS 3969, pp. 346–353, 2006.
© Springer-Verlag Berlin Heidelberg 2006

search. A general construction of these sequences has been given by Langevin [3] and by Pott and Bradley [4]. Using odd-perfect almost binary sequences, Lüke constructed a class of quadriphase sequences of length $N = p^J + 1 \equiv 2 \pmod{4}$ (p odd prime, $J = 1, 2, \cdots$) [5]. It remains unknown whether AP polyphase sequences in any other length exist.

This paper proposes a novel method for constructing AP polyphase sequences. It is a variant of Gong's interleaved construction [6], [7], which uses two completely orthogonal AP sequences and a shift rule defined by a primitive polynomial $f(x)$ over $GF(p)$ of degree $2J$ (with odd prime p and integer J) [8] to generate a new sequence. By the proposed method, AP polyphase sequences of length $m(p^J+1)$ are constructed, where $m = 2, 4, 6$ or 8, and $p^J - 1 \equiv 0 \pmod{m}$. For $m = 2$, the phases of the resulting sequences can be any positive even numbers. AP quadriphase sequences of length $4(p^J + 1)$ or $8(p^J + 1)$ can also be generated in the case of $p^J + 1 \equiv 2 \pmod 4$ or $p^J + 1 \equiv 2 \pmod 8$, respectively. Since the length of all sequences constructed by our method is divisible by 4, all AP quadriphase sequences in this paper are new. Furthermore, other AP polyphase sequences such as six-phase and eight-phase can be constructed by the proposed method.

The remainder of this paper is organized as follows. Section 2 recalls some definitions and introduces a basic method for constructing sequences. Section 3 constructs AP polyphase sequences using the proposed method. Section 4 concludes the study.

2   Preliminaries and a Basic Construction Method

Throughout the paper all sequences we discuss, except shift sequences, are complex-valued polyphase sequences.

Let $a = (a_0, a_1, \cdots, a_{m-1})$ and $b = (b_0, b_1, \cdots, b_{m-1})$ be sequences of length m. Their *cross-correlation function* $\Theta_{a,b}(\tau)$ is defined as

$$\Theta_{a,b}(\tau) = \sum_{i=0}^{m-1} a_i \cdot b^*_{i+\tau}, \quad \tau = 0, 1, \cdots \tag{1}$$

where the symbol $*$ denotes the complex conjugate and the subscript addition is performed *modulo m*. If $a = b$, then $\Theta_a(\tau)$ is called the *autocorrelation function* of a, denoted by $\Theta_a(\tau)$.

The sequence a is said to be *almost perfect* if $\Theta_a(\tau) = 0$ for all $\tau \not\equiv 0 \pmod m$ with exactly one exception.

A pair of AP sequences a and b are said to be *completely orthogonal* if $\Theta_{a,b}(\tau) = 0$ for all τ.

Let p be a positive odd prime and $n = p^J + 1$ for a positive integer J. Let $GF(p^{2J})$ represent the field with p^{2J} elements, and α denote a primitive element of $GF(p^{2J})$, which is constructed from a given primitive polynomial $f(x)$ of degree $2J$ over $GF(p)$. For $k \in Z_{p^{2J}-1}$ define

$$e_k = \begin{cases} \infty, & \text{if } Tr_J^{2J}(\alpha^k) = 0 \\ r, & \text{if } Tr_J^{2J}(\alpha^k) = \gamma^r \end{cases} \tag{2}$$

where $\gamma = \alpha^n$ and the *trace function* $Tr_J^{2J}: GF(p^{2J}) \longrightarrow GF(p^J)$ is defined by $Tr_J^{2J}(x) = x + x^{p^J}$, and the sequence $e = (e_0, e_1, \cdots, e_{n-1})$ is called the *shift sequence* associated with the polynomial $f(x)$. Formula (2) was introduced by Games [8], and it is easy to see that $e_{n+k} = 1 + e_k$ for $0 \leq k \leq n-1$.

The shift sequence e has the following properties.

Lemma 1: [9] Let $e = (e_0, e_1, \cdots, e_{n-1})$ be the shift sequence associated with the primitive polynomial $f(x)$ of degree $2J$ over $GF(p)$. For fixed $k \in Z_n \setminus \{0\}$, the list of differences $(e_{j+k} - e_j) \pmod{(n-2)}: j \in Z_n$ contains each element of Z_{n-2} exactly once and ∞ exactly twice. And the set $\{e_j | j \in Z_n\}$ contains ∞ exactly once.

The *left shift operator* L on a is defined as $L(a) = (a_1, a_2, \cdots, a_{m-1}, a_0)$. For any integer $i > 0$, we make a convention that $L^0(a) = a$ and iteratively define $L^i(a) = L(L^{i-1}(a))$.

A basic method to construct a sequence is given as follows:

(I) For a pair of completely orthogonal sequences a, b and a shift sequence $e = (e_0, e_1, \cdots, e_{n-1})$ constructed by formula (2), we can define an ordered set

$$A = \{A_0, A_1, \cdots, A_{n-1}\}, \text{ where } A_i = \begin{cases} L^{e_i}(a), & \text{if } e_i \neq \infty \\ b, & \text{otherwise} \end{cases}. \tag{3}$$

A is called the set associated with the sequences a, b and e.

(II) Let $U = (U_{i,j})$ be the $m \times n$ matrix whose j-th column is A_j. Listing all entries of U row by row (from left to right and from top to bottom), we obtain a sequence $u = (u_0, u_1, \cdots, u_{mn-1})$ of length mn. u is called the sequence associated with the ordered set A, and U is the matrix form of u.

The original idea of this method was proposed by Gong in 1995 [6], and later she used short p-ary periodic sequences with two-level autocorrelation function and an interleaved structure to construct a set of long p-ary sequences with the desired properties [7]. Different from Gong's construction, a and b in formula (3) are shift distinct (in the terminology of [6]).

The following formula on autocorrelation of sequence u is a basis on which we prove the results in the paper.

Let $0 \leq \tau < mn$ and write $\tau = rn + s$ with $0 \leq r < m$ and $0 \leq s < n$.

Proposition 1:

$$\Theta_u(\tau) = \sum_{j=0}^{n-1} \Theta_{A_j, A_{s+j-\varphi(s+j)n}}(r + \varphi(s+j)) \tag{4}$$

where $\varphi(s+j)$ is 0 if $s+j < n$, and is 1 otherwise.

Proof: Let $T = (T_0, T_1, \cdots, T_{n-1})$ be the matrix form of $L^\tau(u)$. For $0 \leq j < n$, the j-th entry in the sequence $L^\tau(u)$ is a_k, where

$$k = (e_{(s+j) \bmod n} + r + \varphi(s+j)) \bmod m,$$

for $e_{(s+j)\bmod n} \neq \infty$, or is $b_{(r+\varphi(s+j))} \bmod m$ otherwise. It is exactly the first element of T_j. So,

$$T_j = L^{(r+\varphi(s+j))\bmod m}(A_{s+j-\varphi(s+j)n}).$$

Thus, one has

$$
\begin{aligned}
&\Theta_u(\tau) \\
&= \sum_{j=0}^{n-1} \sum_{l=0}^{m-1} U_{j,l} T_{j,l}^* \\
&= \sum_{j=0}^{n-1} \Theta_{A_j, A_{s+j-\varphi(s+j)n}}(r + \varphi(s+j)).
\end{aligned}
$$ □

3 Almost Perfect Polyphase Sequences

In this section, we will use the construction in Section 2 to construct almost perfect polyphase sequences.

Theorem 1: Assume that a and b are a pair of AP sequences which are completely orthogonal, and $\Theta_a(\frac{m}{2}) = -m$. If $m|(n-2)$, then u is an AP sequence of length mn.

Proof: By Proposition 1, one has

$$
\begin{aligned}
&\Theta_u(\tau) \\
&= \sum_{e_j \neq \infty, e_{j+s} \neq \infty} \Theta_a(e_{j+s} - e_j + r) + \\
&\quad \sum_{e_j \neq \infty, e_{j+s} = \infty} \Theta_{a,b}(r + \varphi(s+j) - e_j) + \\
&\quad \sum_{e_j = \infty, e_{j+s} \neq \infty} \Theta_{b,a}(e_{(s+j)\bmod n} + r + \varphi(s+j)) + \\
&\quad \sum_{e_j = \infty, e_{j+s} = \infty} \Theta_b(r + \varphi(s+j))
\end{aligned}
$$

where $0 \leq s, j \leq n-1$.

When $s = 0$, by Lemma 1, one has

$$\Theta_u(\tau) = (n-1)\Theta_a(r) + \Theta_b(r)$$

and then $\Theta_u(\tau) = \Theta_u(rn) = 0$ if and only if $r \not\equiv 0 \pmod{\frac{m}{2}}$.

When $s \neq 0$,

$$
\begin{aligned}
&\Theta_u(\tau) \\
&= \sum_{e_j \neq \infty, e_{j+s} \neq \infty} \Theta_a(e_{j+s} - e_j + r) + \sum_{e_j \neq \infty, e_{j+s} = \infty} \Theta_{a,b}(r + \varphi(s+j) - e_j) + \\
&\quad \sum_{e_j = \infty, e_{j+s} \neq \infty} \Theta_{b,a}(e_{(s+j)\bmod n} + r + \varphi(s+j)).
\end{aligned}
$$

Since $m|(n-2)$ and $\Theta_a(\frac{m}{2}) = -m$,

$$\sum_{e_j \neq \infty, e_{j+s} \neq \infty} \Theta_a(e_{j+s} - e_j + r)$$
$$= \sum_{l=0}^{n-3} \Theta_a(l \bmod m)$$
$$= \frac{n-2}{m} \sum_{l=0}^{m-1} \Theta_a(l)$$
$$= \frac{n-2}{m}(\Theta_a(0) + \Theta_a(\tfrac{m}{2}))$$
$$= 0.$$

By Lemma 1, one has

$$\sum_{e_j \neq \infty, e_{j+s} = \infty} \Theta_{a,b}(r + \varphi(s+j) - e_j)$$
$$= \sum_{e_j = \infty, e_{j+s} \neq \infty} \Theta_{b,a}(e_{(s+j)\bmod n} + r + \varphi(s+j))$$
$$= 0$$

since a and b are completely orthogonal. Therefore, $\Theta_u(\tau) = 0$ for $s \neq 0$.

Combining the above two cases, it is proved that u is an AP sequence.   □

Thus, if there is a pair of completely orthogonal AP sequences of length m such that $m|(p^J - 1)$ for some odd prime p and positive integer J, then one can construct an AP sequence of length $m(p^J + 1)$.

Corollary 1: If $a = (1, -1)$ and $b = (\exp(\frac{2s\pi\sqrt{-1}}{t}), \exp(\frac{2s\pi\sqrt{-1}}{t}))$ where t is a positive even number and $0 \leq s < t$, then u is an AP t-phase sequence of length $2(p^J + 1)$ and its complete PACF is

$$\Theta_u(\tau) = \begin{cases} 2(p^J + 1), & \text{if } \tau \equiv 0 \bmod 2(p^J + 1) \\ -2(p^J + 1) + 4, & \text{if } \tau \equiv p^J + 1 \bmod 2(p^J + 1) \\ 0, & \text{otherwise} \end{cases} \quad (5)$$

In particular, u is an almost balanced binary sequence if $s = 0$ or $s = 1, t = 2$.

Binary AP sequences of length $2(p^J + 1)$ have been constructed in [2] and [4]. Non-binary sequences constructed in this paper are new.

Example 1: Let $p = 3$, $J = 2$, $a = (1, -1)$, $b = (j, j)$ where

$$j \in \{1, \sqrt{-1}, -1, -\sqrt{-1}\} \text{ and } e = (4, 2, 6, 6, 5, \infty, 2, 4, 5, 2).$$

By Corollary 1,

$$u = (1, 1, 1, 1, -1, j, 1, 1, -1, 1, -1, -1, -1, -1, 1, j, -1, -1, 1, -1)$$

with the almost perfect PACF

$$\Theta_u(\tau) = (20, 0, 0, 0, 0, 0, 0, 0, 0, 0, -16, 0, 0, 0, 0, 0, 0, 0, 0, 0).$$

By means of an exhaustive computer search on the possible pairs of AP binary and quadriphase sequences of length $4 \leq m \leq 32$, which satisfy the requirement in Theorem 1, only in length 4 and 8 quadriphase sequence pairs have been found. By applying Theorem 1 to two of these pairs, the following corollary is obtained.

Corollary 2: (1) Let $a = (1, 1, -1, -1)$ and $b = (1, j, 1, j)$ where $j = \sqrt{-1}$. If $4 | (p^J - 1)$, then u is an AP quadriphase sequence of length $4(p^J + 1)$ and has only two elements different from ± 1. Its complete PACF is

$$
\Theta_u(\tau) = \begin{cases} 4(p^J + 1), & \text{if } \tau \equiv 0 \bmod 4(p^J + 1) \\ -4(p^J + 1) + 8, & \text{if } \tau \equiv 2(p^J + 1) \bmod 4(p^J + 1) \\ 0, & \text{otherwise.} \end{cases} \tag{6}
$$

(2) Let $a = (1, 1, j, -j, -1, -1, -j, j)$ and $b = (1, 1, 1, -1, 1, 1, 1, -1)$. If $8 | (p^J - 1)$, then u is an AP quadriphase sequence of length $8(p^J + 1)$ and its complete PACF is

$$
\Theta_u(\tau) = \begin{cases} 8(p^J + 1), & \text{if } \tau \equiv 0 \bmod 8(p^J + 1) \\ -8(p^J + 1) + 16, & \text{if } \tau \equiv 4(p^J + 1) \bmod 8(p^J + 1) \\ 0, & \text{otherwise.} \end{cases} \tag{7}
$$

Example 2: Let $a = (1, 1, -1, -1)$, $b = (1, j, 1, j)$, $p = 17$, $J = 1$ and

$$e = (14, 8, 13, 10, 11, 1, 3, 6, 15, \infty, 8, 0, 14, 13, 8, 8, 12, 8).$$

By Corollary 2 (1),

$$
\begin{aligned}
u = (&-1, 1, 1, -1, -1, 1, -1, -1, -1, 1, 1, 1, -1, 1, 1, 1, 1, 1, -1, \\
&1, -1, -1, 1, -1, 1, -1, 1, j, 1, 1, -1, -1, 1, 1, 1, 1, 1, 1, -1, \\
&-1, 1, 1, -1, 1, 1, 1, 1, -1, -1, 1, -1, -1, -1, -1, -1, 1, \\
&-1, 1, 1, -1, 1, -1, 1, -1, j, -1, -1, 1, 1, -1, -1, -1, -1)
\end{aligned}
$$

with the almost perfect PACF

$$
\begin{aligned}
\Theta_u(\tau) = (&72, 0, 0, 0, 0, 0, 0, 0, 0, 0, 0, 0, 0, 0, 0, 0, 0, 0, 0, 0, 0, 0, 0, 0, \\
&0, 0, 0, 0, 0, 0, 0, 0, 0, 0, 0, 0, 0, -64, 0, 0, 0, 0, 0, 0, 0, 0, 0, 0, \\
&0, 0, 0, 0, 0, 0, 0, 0, 0, 0, 0, 0, 0, 0, 0, 0, 0, 0, 0, 0, 0, 0, 0, 0).
\end{aligned}
$$

Example 3: Let $a = (1, 1, j, -j, -1, -1, -j, j)$, $b = (1, 1, 1, -1, 1, 1, 1, -1)$, $p = 3$, $J = 2$ and $e = (4, 2, 6, 6, 5, \infty, 2, 4, 5, 2)$. By Corollary 2 (2),

$$
\begin{aligned}
u = (&-1, j, -j, -j, -1, 1, j, -1, -1, j, -1, -j, j, j, -j, 1, -j, -1, -j, \\
&-j, -j, -1, 1, 1, j, 1, -1, -j, j, -1, j, -1, 1, 1, 1, -1, -1, j, 1, \\
&-1, 1, -j, j, j, 1, 1, -j, 1, 1, -j, 1, j, -j, -j, j, 1, j, 1, j, j, j, 1, \\
&-1, -1, -j, 1, 1, j, -j, 1, -j, 1, -1, -1, -1, -1, 1, -j, -1, 1)
\end{aligned}
$$

with the almost perfect PACF

$$
\begin{aligned}
\Theta_u(\tau) = (&80, 0, 0, 0, 0, 0, 0, 0, 0, 0, 0, 0, 0, 0, 0, 0, 0, 0, 0, 0, 0, 0, 0, 0, 0, 0, \\
&0, 0, 0, 0, 0, 0, 0, 0, 0, 0, 0, -64, 0, 0, 0, 0, 0, 0, 0, 0, 0, 0, 0, 0, \\
&0, 0, 0, 0, 0, 0, 0, 0, 0, 0, 0, 0, 0, 0, 0, 0, 0, 0, 0, 0, 0, 0, 0, 0, 0, 0).
\end{aligned}
$$

By Corollary 2, one can construct many new AP quadriphase sequences. For example, the proposed method can generate AP quadriphase sequences with length in $\{72, 80, 104, 152, 200\}$, which are not achievable by previous construction methods.

In addition, other AP polyphase sequences can be similarly constructed by the proposed method.

Corollary 3: (1) Let $a = (1, \omega_6, 1, -1, -\omega_6, -1)$ and $b = (1, 1, \omega_6^2, 1, 1, \omega_6^2)$ where ω_6 is a complex primitive sixth root of unity. If $6|(p^J - 1)$, then u is an AP six-phase sequence of length $6(p^J + 1)$ and its complete PACF is

$$
\Theta_u(\tau) = \begin{cases} 6(p^J + 1), & \text{if } \tau \equiv 0 \bmod 6(p^J + 1) \\ -6(p^J + 1) + 12, & \text{if } \tau \equiv 3(p^J + 1) \bmod 6(p^J + 1) \\ 0, & \text{otherwise.} \end{cases} \tag{8}
$$

(2) Let $a = (1, \omega_8, \omega_8^2, -\omega_8^3, -1, -\omega_8, -\omega_8^2, \omega_8^3)$ and $b = (1, 1, 1, -1, 1, 1, 1, -1)$ where ω_8 is a complex primitive eighth root of unity. If $8|(p^J - 1)$, then u is an AP eight-phase sequence of length $8(p^J + 1)$ and its complete PACF is

$$
\Theta_u(\tau) = \begin{cases} 8(p^J + 1), & \text{if } \tau \equiv 0 \bmod 8(p^J + 1) \\ -8(p^J + 1) + 16, & \text{if } \tau \equiv 4(p^J + 1) \bmod 8(p^J + 1) \\ 0, & \text{otherwise.} \end{cases} \tag{9}
$$

Let a polyphase sequence $s = (s_0, s_1, \cdots, s_{m-1})$ and s_i $(0 \leq i \leq l - 1)$ take values in a symbol alphabet Q, the imbalance $I(s)$ of s over Q is defined as

$$
I(s) = \max_{q_1, q_2 \in Q} |N(q_1) - N(q_2)|
$$

where $N(q_1)$ denotes the number of symbol q_1 occuring in s. The sequence s is called balanced if $I(s) = 0$. According to the proposed construction, since the entries of b occur only once in u, the balance property of a dominantly determines that of u. In the case of a being balanced, one has following result:

Proposition 2: If a is balanced, then one has $I(u) = I(b)$.

Remark: (1) By Proposition 2, if a is balanced, the AP sequences constructed are nearly balanced. Sequences derived from Corollaries 2(2) and 3(2) are such ones, while those from Corollary 3(1) suffer a large imbalance. Sequences constructed in Corollaries 1 and 2(1) are almost binary.

(2) An exhaustive computer search has shown that there does not exist any other polyphase AP sequences of length $4 \leq m \leq 32$, which satisfy the requirement in Theorem 1, such that the resulting polyphase sequences u have a better balance property than those sequences constructed as in Corollaries 2(2) and 3(2).

(3) The AP sequences obtained in this paper depend on completely orthogonal AP pairs, which were found by computer search. A general method to find them is not provided in this paper. As pointed out by a reviewer, the condition that each sequence is almost perfect and completely orthogonal with each other looks quite strict if the length of each sequence is large. Thus, we consider it not

optimistic to find an efficient method for generating completely orthogonal AP pairs.

4   Conclusion

Polyphase sequences with almost perfect autocorrelation are attained by using the shift sequence and the completely orthogonal pair. New families of almost perfect quadriphase sequences in lengths $4(p^J + 1)$ and $8(p^J + 1)$ for appropriate parameter p are found. It is possible to find new polyphase sequences by employing other completely orthogonal pairs.

Acknowledgment. The authors thank anonymous reviewers for their helpful comments.

References

[1] Brown R. F., Goodwin G. C.: New class of pseudorandom binary sequences. IEE Electron. Lett. **3** (1967) 198–199
[2] Wolfmann J. : Almost perfect autocorrelation sequence. IEEE Trans. Inform. Theor. **38** (1992) 1412–1418
[3] Langevin P.: Almost perfect binary sequences. Appl. Algebra in Eng., Commun. Comput. **4** (1993) 95–102
[4] Pott A., Bradley S. P.: Existence and nonexistence of almost-perfect autocorrelation sequences. IEEE Trans. Inform. Theor. **41** (1995) 301–303
[5] Lüke A.: Almost-perfect quadriphase sequences. IEEE Trans. Inform. Theor. **47** (2001) 2607–2608
[6] Gong G.: Theory and application of q-ary interleaved sequences. IEEE Trans. Inform. Theor. **41** (1995) 400–411
[7] Gong G.: New designs for signal sets with low cross correlation, balance property, and large linear span, GF(p) case. IEEE Trans. Inform. Theor. **48** (2002) 2847–2867
[8] Games R. A.: Crosscorrelation of M-sequences and GMW sequences with the same primitive polynomial. Discrete Applied Mathematics. **12** (1985) 149–146
[9] Chan A.H., Games R.A.: On the linear span of binary sequences obtained from q-ary m-sequences, q odd. IEEE Trans. Inform. Theor. **36** (1990) 548–552

Linear Filtering of Nonlinear Shift-Register Sequences

Berndt M. Gammel and Rainer Göttfert

Infineon Technologies AG
Munich, Germany

Abstract. Nonlinear n-stage feedback shift-register sequences over the finite field $\mathbb{F}_q$ of period $q^n - 1$ are investigated under linear operations on sequences. We prove that all members of an easily described class of linear combinations of shifted versions of these sequences possess useful properties for cryptographic applications: large periods, large linear complexities and good distribution properties. They typically also have good maximum order complexity values as has been observed experimentally. A running key generator is introduced based on certain nonlinear feedback shift registers with modifiable linear feedforward output functions.

1 Introduction

This article deals with linearly filtered nonlinear shift-register sequences of span n and period $q^n - 1$. More precisely, the underlying shift register is a nonlinear n-stage feedback shift register over $\mathbb{F}_q$ which for any nonzero initial state vector produces a periodic sequence of period $q^n - 1$. Here $\mathbb{F}_q$ denotes the finite field of order q. A linear feedforward function is applied to the stages of the nonlinear feedback shift register to produce the linearly filtered sequence. We show that under easily controlled conditions on the linear filter function, the filtered sequence will have the same period and linear complexity as the original sequence whose linear complexity typically is close to the period length. We prove that the linearly filtered sequences possess good distribution properties. Furthermore, we report experimental results regarding the maximum order complexity of linearly filtered binary nonlinear shift-register sequences.

One purpose of linear filtering a nonlinear shift-register sequence is to increase the maximum order complexity of the sequence up to a value of about twice the logarithm of the period length, a value typical for random sequences (see Jansen [8]). Another objective is to enter variability into a system deploying primitive (see Definition 1) nonlinear feedback shift registers. As an illustration we discuss in Section 5 a configurable running key generator based on primitive binary shift registers endowed with modifiable (possibly key-dependent) linear feedforward logics.

The here discussed concept of linearly filtering nonlinear feedback shift-register sequences mirrors the complementary concept of nonlinearly filtering linear shift register sequences. The latter concept has been treated in the

Ø. Ytrehus (Ed.): WCC 2005, LNCS 3969, pp. 354–370, 2006.

literature extensively. The intent there is to generate sequences of large linear complexity out of maximal period linear feedback shift-register sequences. This is approached by applying certain nonlinear functions to some shifted versions of the given linear feedback shift-register sequence. We mention the work of Groth [7], Key [9], Siegenthaler, Forré and Kleiner [23], Fúster-Sabater and Caballero-Gil [3], Massey and Serconek [13], Paterson [20], Rueppel [21], and Lam and Gong [11]. Two outstanding contributions to the challenging task of creating nonlinear feedback shift registers of maximum cycle lengths are Mykkeltveit [15] and Mykkeltveit, Siu and Tong [16].

2 Preliminaries

Throughout this paper V denotes the $\mathbb{F}_q$-vector space of all inifinite sequences of elements of the finite field $\mathbb{F}_q$. The sum of two sequences $\sigma = (s_i)_{i=0}^{\infty}$ and $\tau = (t_i)_{i=0}^{\infty}$ in V is defined by $\sigma + \tau = (s_i + t_i)_{i=0}^{\infty}$. The product of a sequence $\sigma \in V$ and a scalar $c \in \mathbb{F}_q$ is defined by $c\sigma = (cs_i)_{i=0}^{\infty}$. A useful linear operator on the vector space V is the shift operator T defined by $T\sigma = (s_{i+1})_{i=0}^{\infty}$ for all $\sigma = (s_i)_{i=0}^{\infty}$ in V. If g is an arbitrary polynomial over $\mathbb{F}_q$, then $g(T)$ is again a linear operator on V.

A sequence $\sigma \in V$ is called *periodic* if there is a positive integer r such that $s_{i+r} = s_i$ for $i = 0, 1, \ldots$. The smallest positive integer r with this property is called the *period* of σ, and we write $\text{per}(\sigma) = r$. Let $\sigma \in V$ be periodic and let g be a monic polynomial over $\mathbb{F}_q$. We call g a *characteristic polynomial* of σ if the linear operator $g(T)$ annihilates σ, i.e. $g(T)\sigma = \mathbf{0}$, where $\mathbf{0}$ denotes the zero sequence of V (the sequence all of whose terms are 0). For instance, $g(x) = x^r - 1$ is a characteristic polynomial of σ, if σ is periodic with period r. For any periodic sequence $\sigma \in V$,

$$J_\sigma = \{g \in \mathbb{F}_q[x] : g(T)\sigma = \mathbf{0}\}$$

is a nonzero ideal (called the *T-annihilator* of σ) in the principal ideal domain $\mathbb{F}_q[x]$. The uniquely determined monic polynomial $m_\sigma \in \mathbb{F}_q[x]$ with $J_\sigma = (m_\sigma) = m_\sigma \mathbb{F}_q[x]$ is called the *minimal polynomial* of σ. Thus the characteristic polynomials of σ are precisely the monic polynomials in $\mathbb{F}_q[x]$ that are multiples of m_σ.

The degree of m_σ is called the *linear complexity* $L(\sigma)$ of σ. The linear complexity of σ is zero if and only if σ is the zero sequence. If $L(\sigma) \geq 1$, then $L(\sigma)$ is the length of the shortest linear feedback shift register over $\mathbb{F}_q$ that can generate σ. The majority of periodic sequences in V of given period r have linear complexities close to r (see Dai and Yang [2] and Meidl and Niederreiter [14]).

Another natural approach to the minimal polynomial of a periodic sequence makes use of generating functions. Following Niederreiter [17], [19], we associate with an arbitrary sequence $\sigma = (s_i)_{i=0}^{\infty}$ of V its generating function $G_\sigma = \sum_{i=0}^{\infty} s_i x^{-i-1}$, regarded as an element of the field $\mathbb{F}_q((x^{-1}))$ of formal Laurent series over $\mathbb{F}_q$ in the indeterminate x^{-1}. The field $\mathbb{F}_q((x^{-1}))$ contains the field $\mathbb{F}_q(x)$ of rational functions as a subfield.

Lemma 1. *Let* $\sigma = (s_i)_{i=0}^{\infty}$ *be a sequence of elements of* $\mathbb{F}_q$, *and let* g *be a monic polynomial over* $\mathbb{F}_q$ *with* $g(0) \neq 0$. *Then* σ *is a periodic sequence in* V *with characteristic polynomial* g *if and only if*

$$\sum_{i=0}^{\infty} s_i x^{-i-1} = \frac{h(x)}{g(x)} \tag{1}$$

with $h \in \mathbb{F}_q[x]$ *and* $\deg(h) < \deg(g)$.

Proof. The assertion follows immediately by considering the coefficients in the Laurent series expansion of $g(x) \sum_{i=0}^{\infty} s_i x^{-i-1}$. See Niederreiter [17], [19], or [18, p. 218]. □

The polynomial g in Lemma 1 is the minimal polynomial of σ precisely if the rational function in (1) is in reduced form, i.e. $\gcd(h, g) = 1$. If this is not the case, then we can divide the numerator h and the denominator g by $\gcd(h, g)$ to produce the reduced form. It follows that $m_\sigma = g/\gcd(h, g)$.

Notice that the polynomial $h(x) = g(x) \sum_{i=0}^{\infty} s_i x^{-i-1}$ depends only on g and the first $\deg(g)$ terms of σ. Thus the above method is a way to compute the minimal polynomial of a periodic sequence from a known characteristic polynomial and a suitable number of initial terms of the sequence. The method was first described by Willett [24], supported by a rather complicated proof. The special case for the characteristic polynomial $g(x) = x^r - 1$ appeared earlier in Laksov [10].

Let g be a monic polynomial over $\mathbb{F}_q$ of positive degree n and with $g(0) \neq 0$. The sequence $\rho = (r_i)_{i=0}^{\infty}$ in V that has g as a characteristic polynomial and whose first n terms are $r_0 = \cdots = r_{n-2} = 0$ and $r_{n-1} = 1$, is called the *impulse response sequence* for g. We have $g(x) \sum_{i=0}^{\infty} r_i x^{-i-1} = (x^n + \cdots + g_1 x + g_0)(x^{-n} + r_n x^{-n-1} + \cdots) = 1$, so that the generating function of ρ satisfies

$$\sum_{i=0}^{\infty} r_i x^{-i-1} = \frac{1}{g(x)}, \tag{2}$$

which in particular shows that g is also the minimal polynomial of ρ.

Lemma 2. *Let* σ *be a periodic sequence in* V *and let the monic polynomial* g *with* $\deg(g) = n \geq 1$ *and* $g(0) \neq 0$ *be a characteristic polynomial of* σ. *Let* $\rho \in V$ *be the impulse response sequence for* g, *and let* h *be a polynomial over* $\mathbb{F}_q$ *with* $\deg(h) < \deg(g)$. *The rational generating function of* σ *is* h/g *if and only if* $\sigma = h(T)\rho$.

Proof. If we multiply the left-hand side of equation (2) by x^j, where $0 \leq j \leq n - 1$, we get the generating function of $T^j \rho$. If we multiply the right-hand side of (2) by x^j, we obtain the rational function $x^j/g(x)$. Hence the rational generating function of the sequence $T^j \rho$ is $x^j/g(x)$ for $0 \leq j \leq n - 1$. Consider the rational generating function of σ:

$$\frac{h(x)}{g(x)} = \frac{h_0 + h_1 x + \cdots + h_{n-1} x^{n-1}}{g(x)} = h_0 \frac{1}{g(x)} + h_1 \frac{x}{g(x)} + \cdots + h_{n-1} \frac{x^{n-1}}{g(x)}.$$

This is equivalent to $\sigma = h_0 \rho + h_1 T\rho + \cdots + h_{n-1} T^{n-1}\rho = h(T)\rho$. □

If $\sigma \in V$ is periodic with $\mathrm{per}(\sigma) = r \geq 2$, then the least positive integer n such that the n-tuples $\mathbf{s}_i = (s_i, s_{i+1}, \ldots, s_{i+n-1})$, $0 \leq i \leq r - 1$, are all distinct is called the *span* or the *maximum order complexity* of σ. If $\sigma \in V$ is a constant sequence, then its maximum order complexity is defined to be zero.

The periodic sequences in V of span $n \geq 1$ are precisely the output sequences of nonsingular n-stage feedback shift registers over $\mathbb{F}_q$. An n-stage feedback shift register (FSR) over $\mathbb{F}_q$ is uniquely determined by its feedback function $R : \mathbb{F}_q^n \to \mathbb{F}_q$. A sequence $\sigma = (s_i)_{i=0}^\infty$ in V whose terms satisfy the recurrence relation

$$s_{i+n} = R(s_i, s_{i+1}, \ldots, s_{i+n-1}) \qquad \text{for } i = 0, 1, \ldots$$

is called an *output sequence of the FSR* defined by R. The n-tuple $\mathbf{s}_0 = (s_0, s_1, \ldots, s_{n-1})$ is referred to as the *initial state vector* of the sequence. The FSR is called *nonsingular* if the mapping

$$M : (x_0, \ldots, x_{n-1}) \in \mathbb{F}_q^n \mapsto (x_1, \ldots, x_{n-1}, R(x_0, \ldots, x_{n-1})) \in \mathbb{F}_q^n$$

is bijective. Any output sequence of a nonsingular n-stage FSR over $\mathbb{F}_q$ is periodic. The maximum possible period is $r = q^n$ as there are q^n different n-tuples of elements of $\mathbb{F}_q$.

The FSR is called a *linear feedback shift register* (LFSR) over $\mathbb{F}_q$ if the feedback function R is linear; otherwise the FSR is called a *nonlinear feedback shift register* (NLFSR) over $\mathbb{F}_q$. If R is linear, i.e.

$$R(x_0, x_1, \ldots, x_{n-1}) = a_0 x_0 + a_1 x_1 + \cdots + a_{n-1} x_{n-1}$$

with $a_j \in \mathbb{F}_q$ for $0 \leq j \leq n - 1$, then the associated polynomial $g \in \mathbb{F}_q[x]$ given by

$$g(x) = x^n - R(1, x, \ldots, x^{n-1}) = x^n - a_{n-1} x^{n-1} - \cdots - a_1 x - a_0$$

is called the *characteristic polynomial* of the LFSR. The name is justified by the fact that g is a characteristic polynomial of any possible output sequence of the LFSR.

Definition 1. An n-stage FSR over $\mathbb{F}_q$ (linear or nonlinear) is called *primitive* if for any nonzero initial state vector of $\mathbb{F}_q^n$ the corresponding output sequence has period $q^n - 1$.

It is immediate that each primitive FSR is nonsingular. Furthermore, if R is the feedback function of a primitive n-stage FSR over $\mathbb{F}_q$ and $\mathbf{0} \in \mathbb{F}_q^n$ is the zero vector, then $R(\mathbf{0}) = 0$. As a consequence, the zero sequence is an output sequence of any primitive FSR. The attribute *primitive* for the FSR's under discussion is justified for the following reason: A linear feedback shift register is primitive (in the sense of Definition 1) if and only if its characteristic polynomial is a primitive polynomial over $\mathbb{F}_q$ (see Lidl and Niederreiter [12, Chap. 8] and Golomb [6]). In the literature, the nonzero output sequences of a primitive LFSR are also called *m-sequences*, *maximal period sequences*, or *pseudo-noise sequences*.

Any two nonzero output sequences σ and τ of some primitive n-stage FSR over $\mathbb{F}_q$ are shifted versions of each other. That is, $\tau = T^d\sigma$ for some $0 \leq d \leq q^n - 2$. There are exactly

$$q^{-n}(q!)^{q^{n-1}}$$

primitive n-stage FSR's over $\mathbb{F}_q$ (see Van Aardenne-Ehrenfest and De Bruijn [1]) of which only $\phi(q^n - 1)/n$ are linear as there are that many primitive polynomials over $\mathbb{F}_q$ of degree n. Here ϕ is Euler's function.

Primitive LFSRs over $\mathbb{F}_q$ of length n can readily be constructed as long as primitive polynomials over $\mathbb{F}_q$ of degree n are available. The design of sparse primitive NLFSRs is a challenging task. By employing a small amount of theory and a huge amount of computing power at the present time it is possible to construct binary primitive NLFSRs in the lower thirties. For instance, the feedback function

$$F(x_0, x_1, \ldots, x_{31}) = x_0 + x_2 + x_6 + x_7 + x_{12} + x_{17} + x_{20} + x_{27} + x_{30}$$
$$+ x_3 x_9 + x_{12} x_{15} + x_4 x_5 x_{16}$$

defines a binary primitive NLFSR of length 32. Recently, a synchronous stream cipher based solely on binary primitive NLFSRs was proposed by Gammel, Göttfert and Kniffler [5]. See also [4].

3   Periodicity and Linear Complexity

Let σ be a sequence of elements of $\mathbb{F}_q$, and let f be a nonzero polynomial over $\mathbb{F}_q$. We call the sequence $\tau = f(T)\sigma$ a *linearly filtered* sequence derived from σ. The polynomial f is referred to as the *filter polynomial*.

Lemma 3. *Let $\sigma = (s_i)_{i=0}^{\infty}$ be a periodic sequence in V with minimal polynomial $m_\sigma \in \mathbb{F}_q[x]$, and let f be a nonzero polynomial over $\mathbb{F}_q$. The sequence $\tau = f(T)\sigma$ is periodic and its minimal polynomial is given by $m_\tau = m_\sigma / \gcd(m_\sigma, f)$.*

Proof. The assertion holds if σ is the zero sequence. Otherwise, the minimal polynomial of σ has positive degree. Let $\rho \in V$ be the impulse response sequence for the polynomial m_σ and consider $h/m_\sigma \in \mathbb{F}_q(x)$, the reduced rational generating function of σ. By Lemma 2, we have $\sigma = h(T)\rho$. It follows that $\tau = f(T)[h(T)\rho] = (fh)(T)\rho$. Let u be the uniquely determined polynomial over $\mathbb{F}_q$ with $fh \equiv u \bmod m_\sigma$ and $\deg(u) < \deg(m_\sigma)$. Since $m_\sigma(T)\rho = \mathbf{0}$, it follows that $\tau = u(T)\rho$. Another application of Lemma 2 shows that u/m_σ represents the generating function of τ. By reducing u/m_σ to lowest terms, we obtain the minimal polynomial of τ as $m_\tau = m_\sigma / \gcd(m_\sigma, u)$. However, as $u \equiv fh \bmod m_\sigma$, and since the polynomials h and m_σ are relatively prime, it follows that $\gcd(m_\sigma, u) = \gcd(m_\sigma, f)$. $\qquad\square$

Lemma 4. *Let $\sigma = (s_i)_{i=0}^{\infty}$ be a nonzero output sequence of a primitive n-stage FSR over $\mathbb{F}_q$. Then the minimal polynomial $m_\sigma \in \mathbb{F}_q[x]$ of σ is the product of distinct monic irreducible polynomials over $\mathbb{F}_q$ whose degrees divide n. The polynomials x and $x - 1$ do not divide the minimal polynomial m_σ.*

Proof. Let $r = \text{per}(\sigma)$. The n-tuples $\mathbf{s}_i = (s_i, s_{i+1}, \ldots, s_{i+n-1})$, $0 \leq i \leq r-1$, run exactly through all nonzero vectors of $\mathbb{F}_q^n$. Therefore, each nonzero element of $\mathbb{F}_q$ occurs exactly q^{n-1} times among the first coordinates of those n-tuples. It follows that $s_0 + s_1 + \cdots + s_{r-1} = 0$. Since σ is periodic with period r, we get

$$s_{i+r-1} + \cdots + s_{i+1} + s_i = 0 \quad \text{for } i = 0, 1, \ldots,$$

which means that

$$c(x) = x^{r-1} + x^{r-2} \cdots + x + 1 = \frac{x^r - 1}{x - 1} \in \mathbb{F}_q[x] \tag{3}$$

is a characteristic polynomial of σ. We have $c(1) = r \cdot 1 = (q^n - 1) \cdot 1 = -1 \neq 0$. Thus the element 1 is not a root of $c(x)$, nor is 0. Consequently, the minimal polynomial $m_\sigma(x)$, which divides $c(x)$, is neither divisible by $x - 1$ nor by x. Equation (3) implies that $m_\sigma(x)$ divides $x^{r+1} - x$, which, since $r + 1 = q^n$, is the product of all monic irreducible polynomials over $\mathbb{F}_q$ whose degrees divide n. $\square$

Let g be a monic polynomial over $\mathbb{F}_q$ of positive degree n. Consider the set $S(g)$ of all periodic sequences in V that have g as a characteristic polynomial: $S(g) = \{\sigma \in V : g(T)\sigma = \mathbf{0}\}$. The set $S(g)$ is a T-invariant, n-dimensional subspace of the vector space V. It follows that for all $\sigma \in S(g)$ and for all $f \in \mathbb{F}_q[x]$, the linearly filtered sequence $\tau = f(T)\sigma$ is in $S(g)$. The vector space $S(g)$ has a particularly simple structure if the polynomial g is primitive. In this case, $S(g) = \{\mathbf{0}, \sigma, T\sigma, \ldots, T^{r-1}\sigma\}$, where σ is an arbitrary sequence of $S(g)$ with a nonzero initial state vector and $r = q^n - 1$. It follows that the linearly filtered output sequence of a primitive LFSR is always a shifted version of the original output sequence. As linearly filtering primitive LFSR-output sequences does not produce "new" sequences we exclude the LFSR case from the following investigations.

Theorem 1. *Let $\sigma = (s_i)_{i=0}^\infty$ be a nonzero output sequence of a primitive n-stage NLFSR over $\mathbb{F}_q$, and let f be a nonzero polynomial over $\mathbb{F}_q$. Write f in the form*

$$f(x) = c\,x^a (x-1)^b f_1(x)^{e_1} \cdots f_s(x)^{e_s},$$

where $c \in \mathbb{F}_q^* = \mathbb{F}_q \setminus \{0\}$, $f_1, \ldots, f_s$ are distinct monic irreducible polynomials in $\mathbb{F}_q[x]$ none of which is equal to x or $x-1$, $e_1, \ldots, e_s$ are positive integers, a and b are nonnegative integers. Let $f_{i_1}, \ldots, f_{i_k}$ be all polynomials of $\{f_1, \ldots, f_s\}$ whose degrees divide n. Then the linear complexity $L(\tau)$ of the sequence $\tau = f(T)\sigma$ satisfies

$$L(\sigma) - \sum_{j=1}^k \deg(f_{i_j}) \leq L(\tau) \leq L(\sigma), \tag{4}$$

where $L(\sigma)$ denotes the linear complexity of σ and an empty sum has the value 0. In particular we have

$$L(\sigma) - \deg(f) \leq L(\tau) \leq L(\sigma). \tag{5}$$

Proof. By Lemma 4, the canonical factorization of m_σ in $\mathbb{F}_q[x]$ has the form $m_\sigma = \prod_{l=1}^{t} h_l$, where for all $l = 1, \ldots, t$, we have $\deg(h_l)$ divides n and $h_l(x) \neq x, x - 1$. Hence $\gcd(m_\sigma, f)$ divides $g = \prod_{j=1}^{k} f_{i_j}$, where an empty product has the value 1. It follows that the degree of $m_\sigma / \gcd(m_\sigma, f)$ is lower bounded by $\deg(m_\sigma) - \deg(g)$ and upper bounded by $\deg(m_\sigma)$. An application of Lemma 3 completes the proof. $\qquad\square$

Corollary 1. *Let $\sigma \in V$ be as in Theorem 1. If the canonical factorization of the filter polynomial $f \in \mathbb{F}_q[x]$ over $\mathbb{F}_q$ contains only irreducible factors equal to x or $x - 1$, or whose degrees do not divide n, then, for $\tau = f(T)\sigma$, we have $m_\tau = m_\sigma$, $L(\tau) = L(\sigma)$, and $\mathrm{per}(\tau) = \mathrm{per}(\sigma) = q^n - 1$.*

Proof. By the provisions above and Lemma 4, we infer that $\gcd(m_\sigma, f) = 1$. Therefore, according to Lemma 3, $m_\tau = m_\sigma$. This implies $L(\tau) = L(\sigma)$ and $\mathrm{per}(\tau) = \mathrm{ord}(m_\tau) = \mathrm{ord}(m_\sigma) = \mathrm{per}(\sigma) = q^n - 1$. $\qquad\square$

We are in particular interested in applying filter functions f whose degrees are smaller than the lengths of the respective NLFSR's. From a practical point of view this is the most interesting case (think of a hardware implementation of the shift register).

Corollary 2. *Let n be a prime and let σ be a nonzero output sequence of a primitive n-stage NLFSR over $\mathbb{F}_q$. If f is a nonzero polynomial over $\mathbb{F}_q$ with $\deg(f) < n$ and $\tau = f(T)\sigma$, then*

$$L(\tau) \geq L(\sigma) - \min(q - 2, n - 1).$$

If additionally, $f(c) \neq 0$ for all $c \in \mathbb{F}_q \setminus \{0, 1\}$, then $m_\tau = m_\sigma$, so that $L(\tau) = L(\sigma)$ and $\mathrm{per}(\tau) = \mathrm{per}(\sigma) = q^n - 1$.

Proof. Since n is prime, the canonical factorization of m_σ in $\mathbb{F}_q[x]$ can contain only irreducible polynomials of degree 1 or n. Hence the assertion follows from Lemma 3 and Theorem 1. $\qquad\square$

Proposition 1. *Let $\sigma = (s_i)_{i=0}^{\infty}$ be a nonzero output sequence of a primitive n-stage NLFSR over $\mathbb{F}_q$. Let $f \in \mathbb{F}_q[x]$ be a nonzero polynomial with $\deg(f) < n$, and let $\tau = f(T)\sigma$. If the minimal polynomial m_σ is divisible by at least one primitive polynomial $h \in \mathbb{F}_q[x]$ of degree n, then $\mathrm{per}(\tau) = q^n - 1$.*

Proof. By Lemma 4, $m_\sigma = h_1 \cdots h_t$, where $h_1, \ldots, h_t \in \mathbb{F}_q[x]$ are distinct monic irreducible polynomials whose degrees divide n. It follows that $\mathrm{ord}(h_j)$ divides $q^n - 1$ for $1 \leq j \leq t$. Let us assume that h_1 is primitive and has degree n. Since $0 \leq \deg(f) < n$, $m_\tau = m_\sigma / \gcd(m_\sigma, f)$ still contains the primitive polynomial h_1. Let—after a possible rearrangement of factors—the canonical factorization of m_τ be given by $m_\tau = h_1 \cdots h_s$, where $s \leq t$. As $\mathrm{ord}(h_1) = q^n - 1$, we obtain $\mathrm{per}(\tau) = \mathrm{ord}(m_\tau) = \mathrm{lcm}(\mathrm{ord}(h_1), \ldots, \mathrm{ord}(h_s)) = q^n - 1$. $\qquad\square$

Corollary 3. *Let $\sigma, \tau \in V$ and $f \in \mathbb{F}_q[x]$ be as in Proposition 1. If the linear complexity of σ satisfies $L(\sigma) \geq q^n - 1 - \phi(q^n - 1)$, where ϕ is Euler's function, then $\mathrm{per}(\tau) = q^n - 1$.*

Proof. The rth cyclotomic polynomial over $\mathbb{F}_q$ has degree $\phi(r)$ and, for $r = q^n - 1$, is the product of all (monic) primitive polynomials in $\mathbb{F}_q[x]$ having degree n (see [12]). Thus, if $L(\sigma) = \deg(m_\sigma) \geq r - \phi(r)$, at least one primitive polynomial $h \in \mathbb{F}_q[x]$ of degree n must be present in the canonical factorization of m_σ over $\mathbb{F}_q$. $\qquad\square$

The concept of linearly filtering nonlinear shift register sequences is, figuratively speaking, the mirror image of the concept of nonlinearly filtering linear shift register sequences. Consider an n-stage primitive feedback shift register over $\mathbb{F}_q$. Let us assume first that the shift register is linear. Let $\tau = (t_i)_{i=0}^\infty$ be an arbitrary periodic sequence of elements of $\mathbb{F}_q$ whose period divides $q^n - 1$. (The period might be equal to $q^n - 1$.) Assume that the LFSR is loaded with a nonzero initial state and that $\sigma = (s_i)_{i=0}^\infty$ is the corresponding output sequence. The n-tuples $\mathbf{s}_i = (s_i, s_{i+1}, \dots, s_{i+n-1})$, $0 \leq i \leq q^n - 2$, encompass all nonzero vectors of $\mathbb{F}_q^n$. We define a function $F : \mathbb{F}_q^n \to \mathbb{F}_q$ by setting $F(\mathbf{s}_i) = t_i$ for $0 \leq i \leq q^n - 2$, and $F(\mathbf{0}) = 0$. Unless τ is a shifted version of σ, the function F will be nonlinear. From the definition of F, it is immediate that, if we apply the filter function F to the n stages of the LFSR, we will obtain the sequence τ. In other words,

$$\tau = F(\sigma, T\sigma, \dots, T^{n-1}\sigma).$$

Thus we have shown: All periodic sequences of elements of $\mathbb{F}_q$ whose periods divide $q^n - 1$, can be obtained out of a given primitive n-stage LFSR over $\mathbb{F}_q$ by applying suitable (in general nonlinear) filter functions to the stages of the shift register.

A primitive n-stage LFSR over $\mathbb{F}_q$ can be regarded as a primitive feedback shift register over $\mathbf{F}_q$ whose nonzero output sequence has *minimum* linear complexity n. The counterpart is a primitive n-stage feedback shift register over $\mathbb{F}_q$ whose nonzero output sequence has *maximum* linear complexity $q^n - 2$. Such a shift register must necessarily be nonlinear. Let us denote the nonzero output sequence of such an n-stage NLFSR by $\sigma = (s_i)_{i=0}^\infty$. Since the linear complexity of σ is $q^n - 2$, the sequences $\sigma, T\sigma, \dots, T^{q^n-3}\sigma$ are linearly independent over $\mathbb{F}_q$. Together with the constant sequence $(1, 1, \dots)$, they form a basis of the $\mathbb{F}_q$-vector space consisting of all periodic sequences of elements of $\mathbb{F}_q$ whose periods divide $q^n - 1$. Let τ be any such sequence with the restriction that the minimal polynomial of τ is not divisible by $x - 1$. Then there exist uniquely determined scalars $a_i \in \mathbb{F}_q$, $0 \leq i \leq q^n - 3$, such that

$$\tau = a_0\sigma + a_1T\sigma + \cdots + a_{q^n-3}T^{q^n-3}\sigma.$$

In other words, $\tau = f(T)\sigma$ with a uniquely determined polynomial $f \in \mathbb{F}_q[x]$ of degree $< q^n - 2$.

Thus we have shown: All periodic sequences of elements of $\mathbb{F}_q$ whose periods divide $q^n - 1$ and whose minimal polynomials are not divisible by $x - 1$, can be obtained out of a given primitive maximum-linear-complexity n-stage NLFSR over $\mathbb{F}_q$ by linearly filtering the output sequence of the shift register.

4   Distribution Properties

Let σ be a nonzero output sequence of a primitive NLFSR over $\mathbb{F}_q$, and let f be a nonzero polynomial over $\mathbb{F}_q$. We will show that up to a slight aberration for the zero element the elements of $\mathbb{F}_q$ are equidistributed in the sequence $\tau = f(T)\sigma$, provided that the degree of the filter polynomial is not too large. Moreover, all possible strings of elements of $\mathbb{F}_q$ up to a certain length (which depends on the degree of the applied filter polynomial f) appear almost equally often within a full portion of the period of τ. If $f(x) = x^e g(x)$ with $e \geq 0$ and $g \in \mathbb{F}_q[x]$ with $g(0) \neq 0$, then the sequence $f(T)\sigma$ is a shifted version of the sequence $g(T)\sigma$. Therefore, w.l.o.g. we can restrict our attention to filter polynomials f which are not divisible by x.

Theorem 2. Let $\sigma = (s_i)_{i=0}^{\infty}$ a nonzero output sequence of a primitive n-stage NLFSR over $\mathbb{F}_q$. Let $f \in \mathbb{F}_q[x]$ with $f(0) \neq 0$ and $0 \leq \deg(f) = k \leq n - 1$. Let $\tau = (t_i)_{i=0}^{\infty} = f(T)\sigma$. For $1 \leq m \leq n - k$ and $\mathbf{b} = (b_1, \ldots, b_m) \in \mathbb{F}_q^m$, let $N(\mathbf{b})$ be the number of $i \in \{0, 1, \ldots, r - 1\}$ for which $(t_i, t_{i+1}, \ldots, t_{i+m-1}) = \mathbf{b}$. Then

$$N(\mathbf{b}) = \begin{cases} q^{n-m} & \text{for } \mathbf{b} \neq \mathbf{0}, \\ q^{n-m} - 1 & \text{for } \mathbf{b} = \mathbf{0}. \end{cases}$$

Proof. By assumption, $f(x) = a_0 + a_1 x + \cdots + a_k x^k$ with $a_0 a_k \neq 0$. Thus

$$t_i = a_0 s_i + a_1 s_{i+1} + \cdots + a_k s_{i+k} \quad \text{for } i = 0, 1, \ldots.$$

Let $\mathbf{b} = (b_1, \ldots, b_m) \in \mathbb{F}_q^m$ be fix. Consider the system of m linear equations in n unknowns $x_0, x_1, \ldots, x_{n-1}$, given by

$$\sum_{j=0}^{k} a_j x_{j+h} = b_{h+1}, \quad h = 0, 1, \ldots, m - 1. \tag{6}$$

Let A be the matrix of coefficients of the corresponding homogeneous system of linear equations, so that

$$A = \begin{pmatrix} a_0 & a_1 & \ldots & a_k & 0 & 0 & 0 & \ldots & 0 \\ 0 & a_0 & a_1 & \ldots & a_k & 0 & 0 & \ldots & 0 \\ \vdots & \vdots & \ddots & \ddots & & \ddots & & & \vdots \\ 0 & 0 & \ldots & a_0 & a_1 & \ldots & a_k & \ldots & 0 \end{pmatrix}.$$

Then A is an $m \times n$ matrix over $\mathbb{F}_q$ of rank m, since $a_0 \neq 0$. If $\mathbf{b} \neq \mathbf{0}$ then the augmented matrix $A' = (A, \mathbf{b}^t)$, which is the $m \times (n + 1)$ matrix whose first n columns are the columns of the matrix A and whose last column is the transpose of $\mathbf{b}$, has also rank m. Hence the system of linear equations in (6) has q^{n-m} distinct solution vectors $(x_0, \ldots, x_{n-1}) \in \mathbb{F}_q^n$.

If $\mathbf{b} = \mathbf{0}$, then the zero vector of $\mathbb{F}_q^n$ is one of the q^{n-m} solution vectors of the system (6). As i runs through $0, 1, \ldots, r - 1$, all nonzero n-tuples occur among $\mathbf{s}_0, \mathbf{s}_1, \ldots, \mathbf{s}_{r-1} \in \mathbb{F}_q^n$, so that $N(\mathbf{0}) = q^{n-m} - 1$. If $\mathbf{b} \neq \mathbf{0}$, then all q^{n-m} solution vectors of (6) are nonzero and thus occur among $\mathbf{s}_0, \mathbf{s}_1, \ldots, \mathbf{s}_{r-1}$, so that $N(\mathbf{b}) = q^{n-m}$. $\qquad\square$

5   A Running Key Generator

Consider k primitive binary NLFSR's of pairwise relatively prime lengths $n_1, \ldots, n_k$. For each $j = 1, \ldots, k$, the jth NLFSR is endowed with a modifiable linear feedforward logic described by a certain collection C_j of binary polynomials of degrees less than n_j. The key loading algorithm has to perform two tasks:

1. It loads each of the k NLFSR's with a nonzero initial state vector;
2. For each $j = 1, \ldots, k$, it selects a filter polynomial f_j from the set C_j.

Let σ_j be the binary sequence the jth NLFSR would produce, due to the chosen initial state vector, without filtering. Then $\tau_j = f_j(T)\sigma_j$ is the sequence after applying the chosen filter polynomial f_j. The linearly filtered sequences $\tau_1, \ldots, \tau_k$ are combined termwise by a Boolean combining function $F : \mathbb{F}_2^k \to \mathbb{F}_2$ to produce the keystream $\omega = F(\tau_1, \ldots, \tau_k)$.

The internal state of the running key generator at time t consists of the binary contents of the $n_1 + \cdots + n_k$ memory cells of the whole shift register bundle. The output function of the running key generator depends on the k-tuple $(f_1, \ldots, f_k)$ of filter polynomials that were chosen during key loading. A well designed key loading algorithm guarantees that each possible combination $(f_1, \ldots, f_k)$ of the filter polynomials will occur with the same probability if the secret key is chosen at random from the uniform distribution. The number N of different combinations of the filter polynomials is given by

$$N = \prod_{j=1}^{k} |C_j|,$$

where $|C_j|$ denotes the cardinality of the set C_j. Thus the running key generator has N different output functions and, therefore, can generate N translation distinct keystream sequences. The latter is a desired property by an information theoretical analysis of keystream generators carried out by Jansen [8, Chap. 7]. The rest of this section is devoted to the derivation of the minimal polynomial of the keystream sequence ω.

First we recall some results of Selmer [22, Chap. 4]. Let $f, g, \ldots, h \in \mathbb{F}_q[x]$ be nonconstant polynomials without multiple roots in their respective splitting fields over $\mathbb{F}_q$ and with nonzero constant terms. Then $f \vee g \vee \cdots \vee h$ is defined to be the monic polynomial whose roots are the distinct products $\alpha\beta \cdots \gamma$, where α is a root of f, β a root of g, and γ a root of h. The polynomial $f \vee g \vee \cdots \vee h$ is again a polynomial over the ground field $\mathbb{F}_q$. This follows from the fact that all conjugates (over $\mathbb{F}_q$) of a root of $f \vee g \vee \cdots \vee h$ are roots of $f \vee g \vee \cdots \vee h$.

Lemma 5. *Let $f, g, \ldots, h \in \mathbb{F}_q[x]$ be polynomials over $\mathbb{F}_q$ without multiple roots and with nonzero constant terms. The polynomial $f \vee g \vee \cdots \vee h \in \mathbb{F}_q[x]$ is irreducible if and only if the polynomials $f, g, \ldots, h$ are all irreducible and of pairwise relatively prime degrees. In this case,*

$$\deg(f \vee g \vee \cdots \vee h) = \deg(f) \deg(g) \cdots \deg(h).$$

If $\sigma, \tau, \ldots, \upsilon$ are periodic sequences in V with irreducible minimal polynomials $f, g, \ldots, h \in \mathbb{F}_q[x]$ of pairwise relatively prime degrees and with $f(0)g(0) \cdots h(0) \neq 0$, then $f \vee g \vee \cdots \vee h$ is the minimal polynomial of the product sequence $\sigma \tau \cdots \upsilon$.

Proof. See Selmer [22, Chap. 4].

Lemma 6. *For each $j = 1, \ldots, k$, let σ_j be a periodic sequence in V with minimal polynomial $m_j \in \mathbb{F}_q[x]$. If the polynomials $m_1, \ldots, m_k$ are pairwise relatively prime, then the minimal polynomial of the sum $\sigma = \sigma_1 + \cdots + \sigma_k$ is equal to the product $m_1 \cdots m_k$. Conversely, let σ be a periodic sequence in V whose minimal polynomial $m \in \mathbb{F}_q[x]$ is the product of pairwise relatively prime monic polynomials $m_1, \ldots, m_k \in \mathbb{F}_q[x]$. Then, for each $j = 1, \ldots, k$, there exists a uniquely determined periodic sequence σ_j with minimal polynomial $m_j \in \mathbb{F}_q[x]$ such that $\sigma = \sigma_1 + \cdots + \sigma_k$.*

Proof. A proof of the first part of the lemma can be found on page 426 in Lidl and Niederreiter [12]. To prove the second part, let $h/m \in \mathbb{F}_q(x)$ be the generating function of σ in the sense of Lemma 1. Let

$$\frac{h}{m} = \frac{h_1}{m_1} + \cdots + \frac{h_k}{m_k} \tag{7}$$

be the partial fraction decomposition of h/m. By the comments following Lemma 1, $\deg(h) < \deg(m)$ and $\gcd(h, m) = 1$. This implies $\deg(h_j) < \deg(m_j)$ and $\gcd(h_j, m_j) = 1$ for $1 \leq j \leq k$. The rational functions h_j/m_j correspond to uniquely determined periodic sequences $\sigma_j \in V$ with minimal polynomials m_j. Equation (7) implies that $\sigma = \sigma_1 + \cdots + \sigma_k$. $\qquad\square$

Lemma 7. *Let $S, T, \ldots, U$ be pairwise relatively prime integers greater than 1. Let $\sigma = (s_n)_{n=0}^{\infty}$, $\tau = (t_n)_{n=0}^{\infty}$, $\ldots$, $\upsilon = (u_n)_{n=0}^{\infty}$ be periodic binary sequences of periods $\mathrm{per}(\sigma) = 2^S - 1$, $\mathrm{per}(\tau) = 2^T - 1$, $\ldots$, $\mathrm{per}(\upsilon) = 2^U - 1$, respectively. Assume that the canonical factorizations over $\mathbb{F}_2$ of the minimal polynomials of $\sigma, \tau, \ldots, \upsilon$ are*

$$m_\sigma = \prod_{i=1}^{s} f_i, \quad m_\tau = \prod_{j=1}^{t} g_j, \quad \ldots, \quad m_\upsilon = \prod_{k=1}^{u} h_k. \tag{8}$$

Then the minimal polynomial of the product sequence $\sigma \tau \cdots \upsilon = (s_n t_n \cdots u_n)_{n=0}^{\infty}$ is given by

$$m_{\sigma\tau\cdots\upsilon} = \prod_{i=1}^{s} \prod_{j=1}^{t} \cdots \prod_{k=1}^{u} (f_i \vee g_j \vee \cdots \vee h_k). \tag{9}$$

In fact, (9) represents the canonical factorization of the minimal polynomial of $\sigma \tau \cdots \upsilon$ over $\mathbb{F}_2$.

Proof. It suffices to carry out the details of the proof for the product of two sequences σ and τ. The general statement then follows by induction. Consider the canonical factorization of the minimal polynomials m_σ and m_τ in (8). By hypothesis, $r = \text{per}(\sigma) = 2^S - 1$ which implies that m_σ divides $x^r - 1$. Recall that $x(x^r - 1) = x^{2^S} - x \in \mathbb{F}_2[x]$ is the product of all irreducible binary polynomials whose degrees divide S. It follows that the irreducible factors $f_1, \ldots, f_s$ are distinct and that $\deg(f_i)$ divides S for $1 \le i \le s$. Similarly, the irreducible polynomials $g_1, \ldots, g_t$ are distinct and $\deg(g_j)$ divides T for $1 \le j \le t$. Furthermore, the first-degree irreducible polynomial $p(x) = x$ does not occur among the polynomials $f_1, \ldots, f_s$ and $g_1, \ldots, g_t$. By Lemma 6, the sequences σ and τ possess unique representations

$$\sigma = \sum_{i=1}^{s} \sigma_i \quad \text{and} \quad \tau = \sum_{j=1}^{t} \tau_j,$$

where σ_i is a binary periodic sequence with minimal polynomial f_i for $1 \le i \le s$, and τ_j is a binary periodic sequence with minimal polynomial g_j for $1 \le j \le t$. It follows that

$$\sigma\tau = \sum_{i=1}^{s} \sum_{j=1}^{t} \sigma_i \tau_j.$$

By hypothesis, $\gcd(S, T) = 1$. It follows that for each $i \in \{1, \ldots, s\}$ and $j \in \{1, \ldots, t\}$, the corresponding irreducible polynomials f_i and g_j have relatively prime degrees. Invoking Lemma 5, we conclude that for each $i \in \{1, \ldots, s\}$ and $j \in \{1, \ldots, t\}$, the sequence $\sigma_i \tau_j$ has the irreducible minimal polynomial $f_i \vee g_j \in \mathbb{F}_2[x]$.

As will be shown below, the irreducible polynomials $f_i \vee g_j$, $1 \le i \le s$, $1 \le j \le t$, are distinct. Another application of Lemma 6 shows that the minimal polynomial of $\sigma\tau$ has the form

$$m_{\sigma\tau} = \prod_{i=1}^{s} \prod_{j=1}^{t} (f_i \vee g_j). \tag{10}$$

It remains to show that the polynomials $f_i \vee g_j$, $1 \le i \le s$, $1 \le j \le t$, are distinct. To see this, let f_i and f_i' be any two factors from the canonical factorization of m_σ, and let g_j and g_j' be any two factors from the canonical factorization of m_τ. Assume to the contrary that the two irreducible polynomials $f_i \vee g_j$ and $f_i' \vee g_j'$ are equal. Note that two irreducible polynomials over the finite field $\mathbb{F}_q$ are equal if and only if they have a common root (in some extension field of $\mathbb{F}_q$). Let γ be a common root of the polynomials $f_i \vee g_j$ and $f_i' \vee g_j'$. Then we can write γ in the form

$$\gamma = \alpha\beta = \alpha'\beta', \tag{11}$$

where α, β, α', and β' are roots of the polynomials f_i, g_j, f_i', and g_j', respectively. Since α is a root of the irreducible polynomial f_i, we have $\alpha \in \mathbb{F}_{2^{\deg(f_i)}}$, which is a subfield of $\mathbb{F}_{2^S}$, as $\deg(f_i)$ divides S. Similarly, we conclude that $\alpha' \in \mathbb{F}_{2^S}$ and $\beta, \beta' \in \mathbb{F}_{2^T}$. From (11) we obtain $\alpha/\alpha' = \beta'/\beta$. Clearly, $\alpha/\alpha' \in \mathbb{F}_{2^S}$ and $\beta'/\beta \in \mathbb{F}_{2^T}$. Since S and T are relatively prime we have $\mathbb{F}_{2^S} \cap \mathbb{F}_{2^T} = \mathbb{F}_2$, so that $\alpha/\alpha' = \beta'/\beta = 1$. Hence $\alpha = \alpha'$ and $\beta = \beta'$. This implies $f_i = f_i'$ and $g_j = g_j'$. $\qquad\square$

Theorem 3. *Let* $\sigma_1, \ldots, \sigma_k$ *be nonzero output sequences of* $k \geq 1$ *primitive binary NLFSR's of pairwise relatively prime lengths* $n_1, \ldots, n_k$. *Let the canonical factorization of the minimal polynomial of* σ_j *over* $\mathbb{F}_2$ *be given by*

$$m_{\sigma_j} = \prod_{i_j=1}^{d_j} h_{i_j} \quad \text{for } 1 \leq j \leq k.$$

Let $F : \mathbb{F}_2^k \to \mathbb{F}_2$ *be an arbitrary Boolean combining function with algebraic normal form*

$$F(x_1, \ldots, x_k) = a_0 + \sum_{1 \leq i \leq k} a_i x_i + \sum_{1 \leq i < j \leq k} a_{ij} x_i x_j + \cdots + a_{12 \ldots k} x_1 x_2 \cdots x_k.$$

Consider the linearly filtered sequences $\tau_j = f_j(T)\sigma_j$ *for* $1 \leq j \leq k$ *and the combined sequence* $\omega = F(\tau_1, \ldots, \tau_k)$. *If for* $j = 1, \ldots, k$, *the applied filter polynomial* $f_j \in \mathbb{F}_2[x]$ *does not contain any irreducible factors* $\neq x, x - 1$ *whose degrees divide* n_j, *then the minimal polynomial of* ω *is given by*

$$
\begin{aligned}
m_\omega = (x-1)^{a_0} &\left(\prod_{i_1=1}^{d_1} h_{i_1} \right)^{a_1} \left(\prod_{i_2=1}^{d_2} h_{i_2} \right)^{a_2} \cdots \left(\prod_{i_k=1}^{d_k} h_{i_k} \right)^{a_k} \\
\cdot &\left(\prod_{i_1=1}^{d_1} \prod_{i_2=1}^{d_2} (h_{i_1} \vee h_{i_2}) \right)^{a_{12}} \cdots \left(\prod_{i_{k-1}=1}^{d_{k-1}} \prod_{i_k=1}^{d_k} (h_{i_{k-1}} \vee h_{i_k}) \right)^{a_{k-1,k}} \\
\cdots &\left(\prod_{i_1=1}^{d_1} \prod_{i_2=1}^{d_2} \cdots \prod_{i_k=1}^{d_k} (h_{i_1} \vee h_{i_2} \vee \cdots \vee h_{i_k}) \right)^{a_{12 \ldots k}}.
\end{aligned}
\tag{12}
$$

Proof. By Corollary 1, $m_{\tau_j} = m_{\sigma_j}$ for $1 \leq j \leq k$. We have

$$\omega = a_0 + \sum_{1 \leq i \leq k} a_i \tau_i + \sum_{1 \leq i < j \leq k} a_{ij} \tau_i \tau_j + \cdots + a_{12 \ldots k} \tau_1 \tau_2 \ldots \tau_k. \tag{13}$$

For each summand we know the corresponding minimal polynomial from Lemma 7. It remains to show that the minimal polynomials of the individual summands are pairwise relatively prime. Lemma 6 then yields the presented formula for the minimal polynomial m_ω.

Consider any two different summands in the sum in (13). There exists at least one τ_l that appears in one of the two summands but not in the other. Consider the minimal polynomials of the two summands. By Corollary 1, we have $m_{\tau_l} = m_{\sigma_l}$, and therefore

$$m_{\tau_l} = \prod_{i_l=1}^{d_l} h_{i_l}.$$

According to Lemma 4, the irreducible factors h_{i_l} satisfy: (i) $\deg(h_{i_l})$ divides n_l; (ii) $\deg(h_{i_l}) \geq 2$. Consider the minimal polynomials of the two summands. By

Lemma 5, and since $\gcd(n_i, n_j) = 1$ for $i \neq j$, it follows that all irreducible factors in the canonical factorization of the minimal polynomial of the summand that does not contain τ_l have degrees relatively prime to n_l. On the other hand, each irreducible polynomial in the canonical factorization of the minimal polynomial of the summand that contains τ_l is of the form $(h_{(.)} \vee \cdots \vee h_{i_l})$, and its degree is a multiple of $\deg(h_{i_l})$, by Lemma 5. Therefore, the degree of the polynomial cannot be relatively prime to n_l. It follows that the minimal polynomials of any two summands in (13) are relatively prime. □

By taking the degrees of both sides in the formula (12), we can express the linear complexity of ω in terms of the linear complexities of the sequences $\sigma_1, \ldots, \sigma_k$.

A typical value for the linear complexity of a nonzero output sequence of a primitive binary n-stage NLFSR seems to be the maximum possible value $2^n - 2$. This is supported by extensive computer investigations of ours.

Corollary 4. *Assume that the underlying primitive binary NLFSR's are such that the linear complexities of the nonzero output sequences σ_j attain the maximum possible values $L(\sigma_j) = 2^{n_j} - 2$ for $1 \leq j \leq k$. Assume that the jth NLFSR is initialized with any nonzero vector of $\mathbb{F}_2^{n_j}$. Let for each j, the applied filter polynomial run through all $2^{n_j} - 1$ nonzero polynomials of $\mathbb{F}_2[x]$ with $0 \leq \deg(f_j) < n_j$. Then the linear complexities of the corresponding possible output sequences ω of the running key generator all satisfy*

$$F(2^{n_1} - n_1 - 1, \ldots, 2^{n_k} - n_k - 1) \leq L(\omega) \leq F(2^{n_1} - 2, \ldots, 2^{n_k} - 2).$$

Proof. By Corollary 3, we have $\mathrm{per}(\tau_j) = 2^{n_j} - 1$ for all nonzero $f_j \in \mathbb{F}_2[x]$ with $0 \leq \deg(f_j) < n_j$, $1 \leq j \leq k$. By Theorem 1, we have $L(\tau_j) \geq L(\sigma_j) - \deg(f_j) \geq 2^{n_j} - n_j - 1$. The assertion now follows from Theorem 3. □

6   Maximum Order Complexity

A sequence that is obtained by randomly choosing a string of r elements of $\mathbb{F}_q$ which is then repeated ad infinitum to produce a periodic sequence of $\mathbb{F}_q^\infty$ is expected to have maximum order complexity $2\lceil \log_q(r) \rceil$ (see Jansen [8]). By computer calculations we found that the mean value of the maximum order complexity of linearly filtered nonzero output sequences of primitive binary n-stage NLFSR's is close to the ideal value $2n$, provided that the applied filter polynomial f satisfies $2 \leq \deg(f) < n$.

Table 1 displays for a primitive binary NLFSR of length 12 the maximum order complexity values of its linearly filtered sequences $\tau = f(T)\sigma$. The applied filter polynomial $f \in \mathbb{F}_2[x]$ ranges over all binary polynomials with $f(0) = 1$ and $\deg(f) \leq 11$. Table 2 gives the mean values and standard deviations of the maximum order complexities of linearly filtered nonzero output sequences of primitive binary NLFSR's of different lengths. The lengths n of the NLFSR's vary in the range $4 \leq n \leq 23$. For $n = 4, 5, 6$ all binary primitive NLFSR's were taken into account. For each larger value of n, at least 300 randomly selected primitive binary NLFSR's were considered. In each considered n-stage NLFSR the applied filter polynomial f runs through all binary nonzero polynomials with $\deg(f) \leq n - 1$.

Table 1. Minimum, maximum, and average values of the maximum order complexity for linearly filtered nonzero output sequences of a primitive binary 12-stage NLFSR

$\deg(f)$	Min.	Max.	Average	$\deg(f)$	Min.	Max.	Average
0	12	12	12.000	6	20	30	23.375
1	20	20	20.000	7	19	31	23.578
2	19	27	23.000	8	20	30	23.352
3	21	30	24.750	9	19	34	23.203
4	19	29	23.750	10	19	35	23.322
5	19	28	23.375	11	19	34	23.332

Table 2. Mean value and standard deviation of the maximum order complexity for linearly filtered nonzero output sequences of primitive binary n-stage NLFSR's

n	Mean value	Std. Dev.	n	Mean value	Std. Dev.
4	5.99	1.64	14	27.64	2.91
5	8.16	2.28	15	29.65	2.63
6	10.55	2.36	16	31.83	3.06
7	12.77	2.39	17	33.58	2.40
8	15.03	2.95	18	35.57	2.30
9	17.37	3.82	19	37.80	2.72
10	19.35	3.08	20	40.18	2.80
11	21.42	2.75	21	42.42	3.39
12	23.62	3.10	22	43.42	2.01
13	25.74	3.40	23	45.49	2.26

Acknowledgement

The authors thank the anonymous referees for their useful comments and suggestions. Special thanks go to Lothrop Mittenthal and Johannes Mykkeltveit for insightful discussions at WCC 2005 in Bergen, and to Øyvind Ytrehus and Tor Helleseth for making the workshop a very special event.

References

1. T. van Aardenne-Ehrenfest and N. G. de Bruijn: Circuits and trees in oriented linear graphs, *Simon Steven* **28**, 203–217 (1951).
2. Z.-D. Dai and J.-H. Yang: Linear complexity of periodically repeated random sequences, *Advances in Cryptology — EUROCRYPT '91* (D. W. Davies, ed.), Lecture Notes in Computer Science, vol. 547, pp. 168–175, Springer-Verlag, 1991.

3. A. Fúster-Sabater and P. Caballero-Gil: On the linear complexity on nonlinearly filtered PN-sequences, *Advances in Cryptology — ASIACRYPT '94* (J. Pieprzyk and R. Safavi-Naini, eds.), Lecture Notes in Computer Science, vol. 917, pp. 80–90, Springer-Verlag, 1995.

4. B. M. Gammel, R. Göttfert, and O. Kniffler: Status of Achterbahn and tweaks, *SASC 2006—Stream Ciphers Revisited* (Leuven, Belgium, February 2-3, 2006), Workshop Record, pp. 302–315.

5. B. M. Gammel, R. Göttfert, and O. Kniffler: An NLFSR-based stream cipher, *IEEE International Symposium on Circuits and Systems — ISCAS 2006* (Island of Kos, Greece, May 21-24, 2006).

6. S. W. Golomb: *Shift Register Sequences*, Aegean Park Press, Laguna Hills, CA, 1982.

7. E. J. Groth: Generation of binary sequences with controllable complexity, *IEEE Trans. Inform. Theory* **IT-17**, 288–296 (1971).

8. C. J. A. Jansen: *Investigations On Nonlinear Streamcipher Systems: Construction and Evaluation Methods*, Ph.D. Thesis, Technical University of Delft, Delft, 1989.

9. E. Key: An analysis of the structure and complexity of nonlinear binary sequence generators, *IEEE Trans. Inform. Theory* **IT-22**, 732–736 (1976).

10. D. Laksov: Linear recurring sequences over finite fields, *Math. Scand.* **16**, 181–196 (1965).

11. C. C. Y. Lam and G. Gong: A lower bound for the linear span of filtering sequences, Workshop Record of *The State of the Art of Stream Ciphers* (Brugge, Oct. 2004), pp. 220–233.

12. R. Lidl and H. Niederreiter: *Finite Fields*, Encyclopedia of Mathematics and Its Applications, vol. 20, Addison-Wesley, Reading, Mass., 1983. (Now Cambridge Univ. Press.)

13. J. L. Massey and S. Serconek: A Fourier transform approach to the linear complexity of nonlinearly filtered sequences, *Advances in Cryptology — CRYPTO '94* (Y. G. Desmedt, ed.), Lecture Notes in Computer Science, vol. 839, pp. 332–340, Springer-Verlag, 1994.

14. W. Meidl and H. Niederreiter: On the expected value of the linear complexity and the k-error linear complexity of periodic sequences, *IEEE Trans. Inform. Theory* **48**, 2817–2825 (2002).

15. J. Mykkeltveit: Nonlinear recurrences and arithmetic codes, *Information and Control* **33**, 193-209 (1977).

16. J. Mykkeltveit, M.-K. Siu, and P. Tong: On the cycle structure of some nonlinear shift register sequences, *Information and Control* **43**, 202–215 (1979).

17. H. Niederreiter: Cryptology—The mathematical theory of data security, *Prospects of Mathematical Science* (T. Mitsui, K. Nagasaka, and T. Kano, eds.), pp. 189–209, World Sci. Pub., Singapore, 1988.

18. H. Niederreiter: *Random Number Generation and Quasi-Monte Carlo Methods*, CBMS-NFS Regional Conference Series in Applied Mathematics, vol. 63, SIAM, Philadelphia, PA, 1992.

19. H. Niederreiter: Sequences with almost perfect linear complexity profile, *Advances in Cryptology — EUROCRYPT '87* (D. Chaum and W. L. Price, eds.), Lecture Notes in Computer Science, vol. 304, pp. 37–51, Springer-Verlag, Berlin, 1985.

20. K. G. Paterson: Root counting, the DFT and the linear complexity of nonlinear filtering, *Designs, Codes and Cryptography* **14**, 247–259 (1998).

21. R. A. Rueppel: *Analysis and Design of Stream Ciphers*, Springer-Verlag, 1986.
22. E. S. Selmer: *Linear Recurrence Relations over Finite Fields*, Department of Mathematics, Univ. of Bergen, 1966.
23. T. Siegenthaler, R. Forré, and A. W. Kleiner: Generation of binary sequences with controllable complexity and ideal r-tupel distribution, *Advances in Cryptology — EUROCRYPT '87* (D. Chaum and W.L. Price, eds.), Lecture Notes in Computer Science, vol. 304, pp. 15–23, Springer-Verlag, 1988.
24. M. Willett: The minimum polynomial for a given solution of a linear recursion, *Duke Math. J.* **39**, 101–104 (1972).

Realizations from Decimation Hadamard Transform for Special Classes of Binary Sequences with Two-Level Autocorrelation

Nam Yul Yu and Guang Gong

Department of Electrical and Computer Engineering
University of Waterloo, Waterloo, Ontario, Canada
nyyu@engmail.uwaterloo.ca, ggong@calliope.uwaterloo.ca

Abstract. In an effort to search for a new binary two-level autocorrelation sequence, the decimation-Hadamard transform (DHT) based on special classes of known binary sequences with two-level autocorrelation is investigated. In the second order DHT of a binary generalized Gordon-Mills-Welch (GMW) sequence, we show that there exist realizations which can be theoretically determined by the second order DHT in its subfield. Furthermore, we show that complete realizations of any binary two-level autocorrelation sequence with respect to a quadratic residue (QR) sequence by the second order DHT are theoretically determined.

1 Introduction

Recently, Gong and Golomb developed a new method to study and search for two-level autocorrelation sequences for both binary and non-binary cases [6]. This method is iteratively to apply two operations: decimation and the Hadamard transform based on general orthogonal functions, referred to as the *decimation-Hadamard transform (DHT)*. Basically, it was inspired from Dillon and Dobbertin's work [2] where the Hadamard transform was used for the analysis of a new two-level autocorrelation sequence. The r-th order iterative DHT can transform one class of two-level autocorrelation sequences into another inequivalent class of such sequences, a process called *realization* [6]. Using the second order iterative DHT and starting with a single binary m-sequence, Gong and Golomb verified that one can obtain all the known two-level autocorrelation sequences of period $2^n - 1$ which have no subfield factorization for odd $n \leq 17$ [6].

In this paper, the DHT based on binary generalized GMW sequences and quadratic residue sequences is investigated. The binary generalized GMW sequence has the trace representation of an orthogonal function from $\mathbb{F}_{2^n}$ to $\mathbb{F}_2$ which is a composition of a component orthogonal function from $\mathbb{F}_{2^m}$ to $\mathbb{F}_2$ and a trace function, where $\mathbb{F}_{2^m}$ is a subfield of $\mathbb{F}_{2^n}$ [7] [12]. In the DHT of the sequence, we show that there exist the realizations which can be theoretically determined by the realizations of a sequence corresponding to the component orthogonal function in the subfield. In the realizations, we note that the DHT

Ø. Ytrehus (Ed.): WCC 2005, LNCS 3969, pp. 371–385, 2006.

of a binary generalized GMW sequence in the finite field is inherited from the DHT of its binary component sequence in the subfield.

In addition, using special properties of QR sequences, the realizations of any binary two-level autocorrelation sequence with respect to a QR sequence by the second order DHT are discussed. We show that the complete realizations can be theoretically determined and a valid realization of any binary two-level autocorrelation sequence with respect to a QR sequence is either a self-realization or a QR sequence.

This paper is organized as follows. In Section 2, we give some preliminary reviews of concepts and notations on sequences that we will use in this paper. In Section 3, the realizations of the binary generalized GMW sequences by the second order DHT are investigated. Mathematical proofs and experimental results are provided. In Section 4, the realizations of any binary two-level autocorrelation sequence based on a QR sequence are investigated. In Section 5, concluding remarks are given.

2   Preliminaries

In this section, we present some preliminary reviews on concepts and notations about sequences that we will frequently use in this paper. The following notation will be used throughout this paper.

- $\mathbb{Z}$ is the integer ring, $\mathbb{Z}_m$ the ring of integers modulo m, and $\mathbb{Z}_m^* = \{r \in \mathbb{Z}_m | r \neq 0\}$.
- $\mathbb{F}_Q = GF(Q)$ is the finite field with Q elements and $\mathbb{F}_Q^*$ the multiplicative group of $\mathbb{F}_Q$.
- For positive integers n and m, let $m|n$. The trace function from $\mathbb{F}_{2^n}$ to $\mathbb{F}_{2^m}$ is denoted by $Tr_m^n(x)$, i.e.,

$$Tr_m^n(x) = x + x^{2^m} + \cdots + x^{2^{m(\frac{n}{m}-1)}}, \quad x \in \mathbb{F}_{2^n},$$

or simply as $Tr(x)$ if $m = 1$ and the context is clear.

2.1   Correspondence Between Periodic Sequences and Functions from $\mathbb{F}_{2^n}$ to $\mathbb{F}_2$

Let $\mathcal{S}$ be the set of all binary sequences with period $t|(2^n - 1)$ and $\mathcal{F}$ be the set of all functions from $\mathbb{F}_{2^n}$ to $\mathbb{F}_2$. For any function $f(x) \in \mathcal{F}$, $f(x)$ can be represented as

$$f(x) = \sum_{i=1}^{r} Tr_1^{n_i}(A_i x^{t_i}), \quad A_i \in \mathbb{F}(2^{n_i})$$

where t_i is a coset leader of a cyclotomic coset modulo $2^{n_i} - 1$, and $n_i|n$ is the size of the cyclotomic coset containing t_i. For any sequence $\underline{a} = \{a_i\} \in \mathcal{S}$, there exists $f(x) \in \mathcal{F}$ such that $a_i = f(\alpha^i)$, $i = 0, 1, \cdots$, where α is a primitive element of $\mathbb{F}_{2^n}$. Then, $f(x)$ is called a *trace representation* of $\underline{a}$. ($\underline{a}$ is also referred to as an r-term sequence.)

2.2 Autocorrelation

The autocorrelation of $\underline{a}$ is defined by

$$C_{\underline{a}}(\tau) = \sum_{i=0}^{t-1}(-1)^{a_{i+\tau}+a_i}, \quad 0 \le \tau \le t-1 \tag{1}$$

where τ is a phase shift of the sequence $\underline{a}$ and the indices are computed modulo t, the period of $\underline{a}$. If $\underline{a}$ has period $2^n - 1$ and

$$C_{\underline{a}}(\tau) = \begin{cases} -1, & \text{if } \tau \not\equiv 0 \bmod 2^n - 1 \\ 2^n - 1, & \text{if } \tau \equiv 0 \bmod 2^n - 1, \end{cases}$$

then we say that the sequence $\underline{a}$ has an *(ideal) 2-level autocorrelation function.*

2.3 Hadamard Transform and the Inverse Transform

Let $f(x)$ be a polynomial function from $\mathbb{F}_{2^n}$ to $\mathbb{F}_2$. With a trace function $Tr(x)$ from $\mathbb{F}_{2^n}$ to $\mathbb{F}_2$, the Hadamard transform of $f(x)$ is defined by

$$\widehat{f}(\lambda) = \sum_{x \in \mathbb{F}_{2^n}} (-1)^{Tr(\lambda x)+f(x)}, \quad \lambda \in \mathbb{F}_{2^n}.$$

The inverse formula is given by

$$\chi(f(\lambda)) = \frac{1}{2^n} \sum_{x \in \mathbb{F}_{2^n}} (-1)^{Tr(\lambda x)} \widehat{f}(x), \quad \lambda \in \mathbb{F}_{2^n}.$$

2.4 Orthogonal Function

Let $f(x)$ be a function from $\mathbb{F}_{2^n}$ to $\mathbb{F}_2$ with $f(0) = 0$. If

$$C_f(\lambda) = \sum_{x \in \mathbb{F}_{2^n}} (-1)^{f(\lambda x)+f(x)} = \begin{cases} 0, & \text{if } \lambda \ne 1 \\ 2^n, & \text{if } \lambda = 1 \end{cases}$$

for $\lambda \in \mathbb{F}_{2^n}$, then we say that $f(x)$ is *orthogonal over* $\mathbb{F}$. Orthogonal function is a trace representation of a two-level autocorrelation sequence [6]. If $f(x)$ is a trace representation of $\underline{a}$ and autocorrelation function of $\underline{a}$ defined in (1) is $C_{\underline{a}}$, then

$$C_{\underline{a}}(\tau) = -1 + C_f(\lambda)$$

where $\lambda = \alpha^\tau \in \mathbb{F}_{2^n}^*$.

2.5 Decimation-Hadamard Transform (DHT)

Let $u(x)$ be orthogonal over $\mathbb{F}_2$ and $f(x)$ be a function from $\mathbb{F}_{2^n}$ to $\mathbb{F}_2$. For an integer $v \in \mathbb{Z}_{2^n-1}^*$, we define

$$\widehat{f}_u(v)(\lambda) = \sum_{x \in \mathbb{F}_{2^n}} (-1)^{u(\lambda x)+f(x^v)}, \quad \lambda \in \mathbb{F}_{2^n}.$$

Then, $\widehat{f}_u(v)(\lambda)$ is called *the first-order decimation-Hadamard transform (DHT)* of $f(x)$ *with respect to* $u(x)$, the first order DHT for short. With this notation, let $t \in \mathbb{Z}_{2^n-1}^*$. Then,

$$\widehat{f}_u(v,t)(\lambda) = \sum_{y \in \mathbb{F}_{2^n}} (-1)^{u(\lambda y)} \widehat{f}_u(v)(y^t) = \sum_{x,y \in \mathbb{F}_{2^n}} (-1)^{u(\lambda y)+u(y^t x)+f(x^v)} \quad (2)$$

is called the *second order decimation-Hadamard transform of* $f(x)$ *(with respect to* $u(x)$*), the second order DHT for short*. In DHT, the Hadamard transform is generalized by the use of the orthogonal function $u(x)$ instead of $Tr(x)$.

If $\widehat{f}_u(v,t)(\lambda) \in \{\pm 2^n\}$ for all λ in $\mathbb{F}_{2^n}$, the function $c(x)$ from $\mathbb{F}_{2^n}$ to $\mathbb{F}_2$ determined by

$$(-1)^{c(\lambda)} = \frac{1}{2^n} \widehat{f}_u(v,t)(\lambda),$$

is called a *realization* of $f(x)$ with respect to $u(x)$, and (v,t) is called a *realizable pair* [6].

3   Realizations on Binary Generalized GMW Sequences

In this section, the decimation-Hadamard transform based on the binary generalized Gordon-Mills-Welch (GMW) sequences is investigated.

Let n be a composite integer, m a proper factor of n, and $h(x)$ an orthogonal function from $\mathbb{F}_{2^m}$ to $\mathbb{F}_2$. For k with $\gcd(k, 2^n - 1) = 1$, a binary generalized GMW sequence $\underline{a} = \{a_i\}$ is defined by an evaluation of $f(x)$ at α^i [4], where α is a primitive element in $\mathbb{F}_{2^n}$ and $f(x)$ is given by

$$f(x) = h(x) \circ Tr_m^n(x^k) = h\left(Tr_m^n(x^k)\right).$$

Here, $f(x)$ is an orthogonal function from $\mathbb{F}_{2^n}$ to $\mathbb{F}_2$. In particular, if $h(x) = Tr_1^m(x^v)$ for v with $\gcd(v, 2^m - 1) = 1$ and $v \neq 1$, then the evaluation of $f(x)$ is a GMW sequence [7] [12]. For more details of GMW sequences, see [10] and [5].

For orthogonal functions $h(x), e(x)$ and $g(x)$ from $\mathbb{F}_{2^m}$ to $\mathbb{F}_2$, let $g(x)$ be a realization of $h(x)$ with respect to $e(x)$ by the second order DHT in $\mathbb{F}_{2^m}$, i.e.,

$$(-1)^{g(\mu^c)} = \frac{1}{2^m} \cdot \widehat{h}_e(a,b)(\mu) = \frac{1}{2^m} \sum_{x,y \in \mathbb{F}_{2^m}} (-1)^{e(\mu y)+e(y^b x)+h(x^a)}$$

or equivalently,

$$\sum_{x \in \mathbb{F}_{2^m}} (-1)^{e(\mu^b x)+h(x^a)} = \sum_{x \in \mathbb{F}_{2^m}} (-1)^{e(\mu x)+g(x^c)} \quad (3)$$

for $\mu \in \mathbb{F}_{2^m}$. In this realization, (a,b) is called a *realizable pair* of $h(x)$ with respect to $e(x)$ [6]. In this paper, we also use a triple (a,b,c) to indicate the realization including the decimation value of $g(x)$. From now on, the triple is called a *realizable triple*.

In Gong and Golomb's work [6], it is determined that if (v, t) is a realizable pair of $h(x)$ with respect to $e(x)$, there are at most six realizable pairs related to this pair for the case of $e(x) = h(x)$. In the following, we consider the result in case of $e(x) \neq h(x)$, i.e., asymmetric case.

Lemma 1. *Let $(v, t, 1)$ be a realizable triple of $h(x)$ with respect to $e(x)$ which realizes $g(x)$ by the second order DHT in $\mathbb{F}_{2^m}$, where $e(x) \neq h(x)$. Then, there exists another realizable triple $(-vt, t^{-1}, -t^{-1})$ of $h(x)$ with respect to $e(x)$ which realizes $g(x)$.*

Proof. If $(v, t, 1)$ and (a, b, c) are realizable triples of $h(x)$, then

$$2^m \cdot (-1)^{g(\mu)} = \sum_{x, y \in \mathbb{F}_{2^m}} (-1)^{e(\mu y) + e(y^t x) + h(x^v)}$$

$$= \sum_{z, w \in \mathbb{F}_{2^m}} (-1)^{e(\mu^{c^{-1}} z) + e(z^b w) + h(w^a)}.$$

Here, (a, b, c) can be a realizable triple if and only if there exists a variable change from (x, y) to (w, z) in the function $e(x)$ such that the above equality is true. In this case, only two kinds of variable changes are possible for $e(x) \neq h(x)$, i.e.,

i) $x^v = w^a$, $y^t x = \mu^{c^{-1}} z$, $\mu y = z^b w$ and ii) $x^v = w^a$, $y^t x = z^b w$, $\mu y = \mu^{c^{-1}} z$.

A nontrivial realizable triple (a, b, c) can be obtained only from i), and we can easily check $(a, b, c) = (-vt, t^{-1}, -t^{-1})$. Thus, $(-vt, t^{-1}, -t^{-1})$ is a realizable triple related to $(v, t, 1)$ realizable triple.   □

In the following, we show the main theorem on the second order DHT of the binary generalized GMW sequences.

Theorem 1. *Let n be a composite integer and m a proper factor of n. Let $(v, t, 1)$ be a realizable triple of $h(x)$ with respect to $e(x)$ which realizes $g(x)$ in $\mathbb{F}_{2^m}$. In other words,*

$$\frac{1}{2^m} \widehat{h}_e(v, t)(\mu) = (-1)^{g(\mu)}, \quad \mu \in \mathbb{F}_{2^m}$$

where $h(x)$, $g(x)$ and $e(x)$ are orthogonal functions from $\mathbb{F}_{2^m}$ to $\mathbb{F}_2$, respectively. Let $f(x), u(x)$ and $c(x)$ be orthogonal functions from $\mathbb{F}_{2^n}$ to $\mathbb{F}_2$ defined by

$$f(x) = h(x^v) \circ Tr_m^n(x), \quad u(x) = e(x) \circ Tr_m^n(x), \quad c(x) = g(x) \circ Tr_m^n(x)$$

where v is a decimation factor in $\mathbb{Z}_{2^m-1}^*$ with $\gcd(v, 2^m - 1) = 1$. Then, there exists a realizable triple $(s^{-1}, -s, s)$ of $f(x)$ with respect to $u(x)$ which realizes $c(x)$ by the second order DHT in $\mathbb{F}_{2^n}$, where $s \equiv -t^{-1} \pmod{2^m - 1}$. Precisely,

$$\widehat{f}_u(s^{-1})(\lambda^{-s}) = \widehat{c}_u(s)(\lambda), \quad \lambda \in \mathbb{F}_{2^n}$$

or equivalently,

$$\frac{1}{2^n} \widehat{f}_u(s^{-1}, -s)(\lambda) = (-1)^{c(\lambda^s)}, \quad \lambda \in \mathbb{F}_{2^n}.$$

Proof. Let's consider a decimation of the first order DHT of $f(x)$ with respect to $u(x)$ by the decimation pair $(s^{-1}, -s)$. Then,

$$
\begin{aligned}
\widehat{f_u}(s^{-1})(\lambda^{-s}) &= \sum_{x \in \mathbb{F}_{2^n}} (-1)^{u(\lambda^{-s}x)+f(x^{s^{-1}})} \\
&= \sum_{x \in \mathbb{F}_{2^n}} (-1)^{e(Tr_m^n(\lambda^{-s}x))+h((Tr_m^n(x^{s^{-1}}))^v)} \\
&= \sum_{\theta \in \mathbb{F}_{2^n}} (-1)^{e(Tr_m^n(\theta^s))+h((Tr_m^n(\lambda\theta))^v)}
\end{aligned}
$$

where $\lambda^{-s}x = \theta^s$. By decomposition of $\theta = \sigma\epsilon$ with $\sigma \in \mathbb{F}_{2^m}$, we have

$$
\begin{aligned}
\widehat{f_u}(s^{-1})(\lambda^{-s}) &= \sum_{\epsilon \in \Psi} \sum_{\sigma \in \mathbb{F}_{2^m}^*} (-1)^{e(\sigma^a Tr_m^n(\epsilon^s))+h(\sigma^v(Tr_m^n(\lambda\epsilon))^v)} + 1 \\
&= \sum_{\epsilon \in \Psi} \sum_{\sigma \in \mathbb{F}_{2^m}} (-1)^{e(\sigma^a Tr_m^n(\epsilon^s))+h(\sigma^v(Tr_m^n(\lambda\epsilon))^v)} - d + 1
\end{aligned}
$$

where $d = (2^n - 1)/(2^m - 1)$, $s \equiv a \pmod{2^m - 1}$, and $\Psi = \{1, \alpha, \alpha^2, \cdots, \alpha^{d-1}\}$ where α is a primitive element in $\mathbb{F}_{2^n}$. Let

$$
\delta_\epsilon = \sum_{\sigma \in \mathbb{F}_{2^m}} (-1)^{e(\sigma^a Tr_m^n(\epsilon^s))+h(\sigma^v(Tr_m^n(\lambda\epsilon))^v)}.
$$

With $(\zeta, \mu) = (Tr_m^n(\epsilon^s), Tr_m^n(\lambda\epsilon))$ and the orthogonality of $h(x)$ and $e(x)$, we obtain

$$
\delta_\epsilon = \begin{cases} 0, & \text{if } (\zeta, \mu) = (0, *) \text{ or } (*', 0) \\ 2^m, & \text{if } (\zeta, \mu) = (0, 0) \\ \delta'_\epsilon, & \text{otherwise} \end{cases}
$$

where both $*$ and $*'$ are nonzero elements in $\mathbb{F}_{2^m}$ and δ'_ϵ is defined for ϵ in $\Gamma = \{\epsilon \in \Psi | \zeta \neq 0 \text{ and } \mu \neq 0)\}$. Furthermore, we can express δ'_ϵ as follows.

$$
\delta'_\epsilon = \sum_{\rho \in \mathbb{F}_{2^m}} (-1)^{e\left(\rho^a \frac{Tr_m^n(\epsilon^s)}{(Tr_m^n(\lambda\epsilon))^a}\right)+h(\rho^v)} = \sum_{w \in \mathbb{F}_{2^m}} (-1)^{e\left(w \frac{Tr_m^n(\epsilon^s)}{(Tr_m^n(\lambda\epsilon))^a}\right)+h(w^{va^{-1}})}
$$

where $\rho = \sigma Tr_m^n(\lambda\epsilon)$ and $w = \rho^a$. With $\eta = \frac{Tr_m^n(\lambda\epsilon)}{(Tr_m^n(\epsilon^s))^{a-1}}$, we get

$$
\delta'_\epsilon = \sum_{w \in \mathbb{F}_{2^m}} (-1)^{e(\eta^{-a}w)+h(w^{va^{-1}})}.
$$

If $(v, t, 1)$ is a realizable triple of $h(x)$ with respect to $e(x)$ which realizes $g(x)$, then $(-vt, t^{-1}, -t^{-1})$ is also a realizable triple from Lemma 1. Thus, $(va^{-1}, -a, a)$

is a realizable triple for $a \equiv -t^{-1} \pmod{2^m - 1}$ if $(v, t, 1)$ is a realizable triple. From (3),

$$\delta'_\epsilon = \sum_{w \in \mathbb{F}_{2^m}} (-1)^{e(\eta^{-a}w) + h(w^{va^{-1}})} = \sum_{w \in \mathbb{F}_{2^m}} (-1)^{e(\eta w) + g(w^a)}$$

$$= \sum_{w \in \mathbb{F}_{2^m}} (-1)^{e\left(w \frac{Tr_m^n(\lambda\epsilon)}{(Tr_m^n(\epsilon^s))^{a-1}}\right) + g(w^a)} = \sum_{y \in \mathbb{F}_{2^m}} (-1)^{e(yTr_m^n(\lambda\epsilon)) + g(y^a Tr_m^n(\epsilon^s))}$$

where $y = \dfrac{w}{(Tr_m^n(\epsilon^s))^{a-1}}$. Finally,

$$\hat{f}_u(s^{-1})(\lambda^{-s}) = \sum_{\epsilon \in \Psi} \delta_\epsilon - d + 1 = \sum_{\epsilon \in \Gamma} \delta'_\epsilon + N \cdot 2^m - d + 1$$

$$= \sum_{\epsilon \in \Gamma} \sum_{y \in \mathbb{F}_{2^m}} (-1)^{e(yTr_m^n(\lambda\epsilon)) + g(y^a Tr_m^n(\epsilon^s))} + N \cdot 2^m - d + 1$$

$$= \sum_{\epsilon \in \Psi} \sum_{y \in \mathbb{F}_{2^m}} (-1)^{e(yTr_m^n(\lambda\epsilon)) + g(y^a Tr_m^n(\epsilon^s))} - d + 1$$

$$= \sum_{\epsilon \in \Psi} \sum_{y \in \mathbb{F}_{2^m}} (-1)^{e(Tr_m^n(\lambda y\epsilon)) + g(Tr_m^n((y\epsilon)^s))} - d + 1$$

$$= \sum_{z \in \mathbb{F}_{2^n}} (-1)^{e(Tr_m^n(\lambda z)) + g(Tr_m^n(z^s))}, \quad (z = y\epsilon)$$

$$= \sum_{z \in \mathbb{F}_{2^n}} (-1)^{u(\lambda z) + c(z^s)} = \hat{c}_u(s)(\lambda)$$

where N is the number of elements for $(\zeta, \mu) = (0,0)$ in Ψ, and $c(x) = g(x) \circ Tr_m^n(x)$. $\qquad\square$

Fig. 1 describes the relation of the orthogonal functions in Theorem 1 by the second order DHT. In Fig. 1, $c(x)$, a realization of $f(x)$ in $\mathbb{F}_{2^n}$ is determined by the extension of $g(x)$, a realization of $h(x)$ in $\mathbb{F}_{2^m}$, where $f(x) = h(x^v) \circ Tr_m^n(x)$

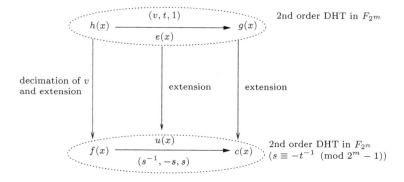

Fig. 1. Relation of orthogonal functions in Theorem 1

Table 1. Complete theoretical determination of realizations of GMW sequences for $n = 10$ ($h(x) = Tr_1^5(x)$)

v	$f(x)$	t	$g(x)$	$(s^{-1}, -s, s)$	$c(x)$
3	3, 17	1	3	(23,221,89), (29,35,247), (61,109,151), (85,343,85), (89,125,23), (91,101,215), (151,47,61), (215,157,91), (247,95,29), (511,1,511)	3, 17
		3	1, 5, 7	(7,73,439), (19,53,175), (25,367,41), (59,13,191), (107,239,49), (149,115,103), (205,383,5), (245,119,71), (379,83,235), (479,181,173)	1, 5, 7, 9, 19, 25, 69
		5	11	(13,59,79), (53,19,251), (73,7,127), (83,379,37), (115,149,347), (119,245,43), (181,479,17), (239,107,167), (367,25,223), (383,205,179)	11, 13, 21, 73
5	5, 9	1	5	(23,221,89), (29,35,247), (61,109,151), (85,343,85), (89,125,23), (91,101,215), (151,47,61), (215,157,91), (247,95,29), (511,1,511)	5, 9
		3	7	(7,73,439), (19,53,175), (25,367,41), (59,13,191), (107,239,49), (149,115,103), (205,383,5), (245,119,71), (379,83,235), (479,181,173)	7, 19, 25, 69
7	7, 19, 25, 69	1	7	(23,221,89), (29,35,247), (61,109,151), (85,343,85), (89,125,23), (91,101,215), (151,47,61), (215,157,91), (247,95,29), (511,1,511)	7, 19, 25, 69
		11	5	(5,179,205), (41,223,25), (49,167,107), (71,43,245), (103,347,149), (173,17,479), (175,251,19), (191,79,59), (235,37,379), (439,127,7)	5, 9
11	11, 13, 21, 73	1	11	(23,221,89), (29,35,247), (61,109,151), (85,343,85), (89,125,23), (91,101,215), (151,47,61), (215,157,91), (247,95,29), (511,1,511)	11, 13, 21, 73
		7	3	(17,173,181), (37,235,83), (43,71,119), (79,191,13), (127,439,73), (167,49,239), (179,5,383), (223,41,367), (251,175,53), (347,103,115)	3, 17
		11	1, 5, 7	(5,179,205), (41,223,25), (49,167,107), (71,43,245), (103,347,149), (173,17,479), (175,251,19), (191,79,59), (235,37,379), (439,127,7)	1, 5, 7, 9, 19, 25, 69
15	15, 23, 27, 29, 77, 85, 89, 147	1	15	(23,221,89), (29,35,247), (61,109,151), (85,343,85), (89,125,23), (91,101,215), (151,47,61), (215,157,91), (247,95,29), (511,1,511)	15, 23, 27, 29, 77, 85, 89, 147

represents a binary generalized GMW sequence with period $2^n - 1$. Furthermore, the corresponding realizable triples $(s^{-1}, -s, s)$ are determined by the realizable triple $(v, t, 1)$ in the subfield. In the DHT of a binary generalized GMW sequence in a finite field, we see that there exist the realizations and realizable triples which are theoretically determined by the realizations and realizable triples of a binary component sequence in its subfield. In the realizations, therefore, the DHT in a finite field is inherited from the DHT in its subfield in terms of a binary generalized GMW sequence.

Tables 1 and 2 show complete lists of realizations and realizable triples determined from Theorem 1 for the binary GMW and generalized GMW sequences for $n = 10$, respectively. In each case, both $e(x)$ and $u(x)$ represent m-sequences with period 31. Note that $u(x)$ can be any function whose subfield factorization is possible. The value of s in each realizable triple is a coset leader satisfying $s \equiv -t^{-1} \pmod{2^m - 1}$ and $\gcd(s, 2^n - 1) = 1$. The numbers in each function column under the label of a function represent trace exponents of the function.

Table 2. Complete theoretical determination of realizations of generalized GMW sequences for $n = 10$ $(h(x) = Tr_1^5(x + x^5 + x^7))$

v	$f(x)$	t	$g(x)$	$(s^{-1}, -s, s)$	$c(x)$
1, 5, 7	1, 5, 7, 9, 19, 25, 69	1	1, 5, 7	(23,221,89), (29,35,247), (61,109,151), (85,343,85), (89,125,23), (91,101,215), (151,47,61), (215,157,91), (247,95,29), (511,1,511)	1, 5, 7, 9, 19, 25 69
		3	11	(7,73,439), (19,53,175), (25,367,41), (59,13,191), (107,239,49), (149,115,103), (205,383,5), (245,119,71), (379,83,235), (479,181,173)	11, 13, 21, 73
		11	3	(5,179,205), (41,223,25), (49,167,107), (71,43,245), (103,347,149), (173,17,479), (175,251,19), (191,79,59), (235,37,379), (439,127,7)	3, 17
3, 11, 15	3, 17, 11, 13, 21, 73, 15, 23, 27, 29, 77, 85, 89, 47	1	3, 11, 15	(23,221,89), (29,35,247), (61,109,151), (85,343,85), (89,125,23), (91,101,215), (151,47,61), (215,157,91), (247,95,29), (511,1,511)	3, 17, 11, 13, 21, 73, 15, 23, 27, 29, 77, 85, 89, 47

In Table 1, $h(x) = Tr_1^5(x)$. Thus, $f(x) = Tr_1^5(x^v) \circ Tr_5^{10}(x)$ represents a binary GMW sequence with period 1023 for each v with $\gcd(v, 31) = 1$ and $v \neq 1$. Table 1 shows that 10 realizable triples in each realization are determined in $\mathbb{F}_{2^{10}}$. Those exactly match the experimental results of the second order DHT of $f(x)$ with respect to $u(x)$ in $\mathbb{F}_{2^{10}}$. From Table 1, we note that all binary GMW sequences and one binary generalized GMW sequence can be realized by the second order DHT of the binary GMW sequences and those realizations are theoretically determined by the realizations in the subfield $\mathbb{F}_{2^5}$. In Table 2, $h(x) = Tr_1^5(x + x^5 + x^7)$. Thus, $f(x) = h(x^v) \circ Tr_5^{10}(x)$ represents a binary generalized GMW sequence with period 1023 for each v with $\gcd(v, 31) = 1$. It is shown that 10 realizable triples in each realization are determined and all binary generalized GMW sequences can be realized by the second order DHT of the binary generalized GMW sequences, which matches the experimental results.

In the experiments of the second order DHT of the binary GMW and generalized GMW sequences for $n = 10$, we interestingly observed that there are no other realizations than the ones from Tables 1 and 2.

4   Realizations on Quadratic Residue (QR) Sequences

In this section, we study the realization of binary two-level autocorrelation sequences with respect to QR sequences by the second order DHT.

4.1   Basic Properties of QR Sequences

A QR sequence $\underline{q} = \{q_i\}$ with period $p \equiv 3 \pmod 4$ is defined by

$$q_i = \begin{cases} 1, & \text{if } i = 0 \pmod p \\ 0, & \text{if } i = \text{QR} \pmod p \\ 1, & \text{if } i = \text{QNR} \pmod p. \end{cases} \tag{4}$$

where 'QR' and 'QNR' represent quadratic residue and non-residue, respectively. For more details of quadratic residues, see [9]. Similarly, we can consider another distinct class of a QR sequence $\mathbf{q}' = \{q_i'\}$ with the same period.

$$q_i' = \begin{cases} 1, & \text{if } i = 0 \pmod{p} \\ 1, & \text{if } i = QR \pmod{p} \\ 0, & \text{if } i = QNR \pmod{p}. \end{cases} \tag{5}$$

The QR sequences with period p have two-level autocorrelation if and only if $p \equiv 3 \pmod 4$ [3]. Also, it has been known that there are only two cyclically distinct QR sequences with the same period, i.e., one is $\mathbf{q} = \{q_i\}$ in (4) and the other is $\mathbf{q}^{(d)} = \{q_{di}\}$ where d is QNR and $\mathbf{q}^{(d)} = \mathbf{q}'$ in (5).

Any QR sequence has its own trace representation [11] [1]. Let $p = 2^n - 1$. If the trace representation of the QR sequence $\mathbf{q}$ is $u(x)$, then the trace representation of $\mathbf{q}'$ is $u'(x) = u(x^d)$ for any QNR d in $\mathbb{Z}_p^*$. As the QR sequence is a two-level autocorrelation sequence for $p \equiv 3 \pmod 4$, both trace representations $u(x)$ and $u'(x)$ are orthogonal functions, respectively. In this paper, the trace representation of a QR sequence is called a *quadratic residue (QR) function*.

The cross-correlation of two distinct QR sequences with period $2^n - 1$ can be derived by using a similar way in [8]. This is stated as follows.

Proposition 1. *Let $\underline{\mathbf{a}} = \{a_i\}$ and $\underline{\mathbf{b}} = \{b_i\}$ be two shift distinct QR sequences with period $2^n - 1$ and their trace representations $u(x)$ and $u'(x)$ (or $u'(x)$ and $u(x)$), respectively. The cross-correlation of these two QR sequences has three values as shown below,*

$$C_{\underline{\mathbf{a}},\underline{\mathbf{b}}}(\tau) = \sum_{i=0}^{2^n-2} (-1)^{a_i + b_{i+\tau}} = \begin{cases} -2^n + 3, & \text{if } \tau = 0 \\ 3, & \text{if } \tau = QR \text{ (or } QNR) \\ -1, & \text{if } \tau = QNR \text{ (or } QR). \end{cases}$$

From the auto- and cross-correlation property of QR sequences, the Hadamard transform of a QR function with respect to itself or its distinct QR function is easily derived.

Lemma 2. *The Hadamard transform of $u(x)$ with respect to $g(x) = u(x^d)$ is defined by*

$$\widehat{u}_g(y) = \sum_{x \in \mathbb{F}_{2^n}} (-1)^{u(x) + g(yx)} = \sum_{x \in \mathbb{F}_{2^n}} (-1)^{u(x) + u(y^d x^d)}.$$

If d is QR, then

$$\widehat{u}_g(y) = \widehat{u}_u(y) = \begin{cases} 2^n, & \text{if } y = 1 \\ 0, & \text{otherwise.} \end{cases}$$

Otherwise,

$$\widehat{u}_g(y) = \widehat{u}_{u'}(y) = \begin{cases} -2^n + 4, & \text{if } y = 1 \\ 4, & \text{if } y = \alpha^i \text{ for } QR \text{ (or } QNR) \text{ } i \\ 0, & \text{if } y = 0 \text{ or } \alpha^i \text{ for } QNR \text{ (or } QR) \text{ } i \end{cases}$$

where α is a primitive element in $\mathbb{F}_{2^n}$.

Proof. If d is QR, then $g(x) = u(x^d) = u(x)$. Thus, the result follows from the fact that $u(x)$ is orthogonal. If d is QNR, on the other hand, then $g(x) = u(x^d) = u'(x)$. Since $\widehat{u}_{u'}(y) = C_{\underline{a}, \underline{b}}(\tau) + 1$, where $y = \alpha^\tau$ for $y \neq 0$, the result follows from Proposition 1.   □

4.2   Realizations of Binary Two-Level Autocorrelation Sequences with Respect to QR Sequences

Let $f(x)$ be an orthogonal function and $u(x)$ a QR function from $\mathbb{F}_{2^n}$ to $\mathbb{F}_2$. In the second order DHT of $f(x)$ with respect to $u(x)$ defined by (2), if $\lambda = 0$,

$$\widehat{f_u}(v, t)(0) = 2^n \tag{6}$$

from [6]. For λ in $\mathbb{F}_{2^n}^*$, we have

$$
\begin{aligned}
\widehat{f_u}(v, t)(\lambda) &= \sum_{x, z \in \mathbb{F}_{2^n}} (-1)^{u(z) + u(\lambda^{-t} z^t x) + f(x^v)} \quad (\lambda y = z) \\
&= \sum_{y, z \in \mathbb{F}_{2^n}} (-1)^{u(z) + u(y^t z^t) + f((\lambda y)^{vt})} \quad (\lambda^{-t} x = y^t) \\
&= \sum_{y \in \mathbb{F}_{2^n}} (-1)^{f((\lambda y)^{vt})} \sum_{z \in \mathbb{F}_{2^n}} (-1)^{u(z) + u(y^t z^t)} \\
&= \sum_{y \in \mathbb{F}_{2^n}} (-1)^{f((\lambda y)^{vt})} \widehat{u}_g(y)
\end{aligned}
\tag{7}
$$

where (v, t) is a decimation pair and $g(x) = u(x^t)$. First of all, we consider the second order DHT in (7) when t is QR.

Lemma 3. *Let $f(x)$ be an orthogonal function and $u(x)$ a QR function from $\mathbb{F}_{2^n}$ to $\mathbb{F}_2$, respectively. With a decimation pair of (v, t), if t is QR, the realization of $f(x)$ with respect to $u(x)$ by the second order DHT is a self-realization. Precisely,*

$$\widehat{f_u}(v, t)(\lambda) = 2^n \cdot (-1)^{f(\lambda^{vt})}$$

for λ in $\mathbb{F}_{2^n}$.

Proof. In (7), $g(x) = u(x^t) = u(x)$ if t is QR. From Lemma 2, $\widehat{u}_g(y)$ has nonzero value 2^n only at $y = 1$, and zero at all other y's in $\mathbb{F}_{2^n}$. Therefore, the result follows from (6) and (7).   □

When t is QNR, on the other hand, $\widehat{u}_g(y)$ becomes the Hadamard transform of $u(x)$ with respect to $u'(x)$. In this case, we firstly consider the case where $f(x)$ is not a QR function.

Lemma 4. *Let $f(x)$ be an orthogonal function which is not a QR function and $u(x)$ a QR function from $\mathbb{F}_{2^n}$ to $\mathbb{F}_2$, respectively. If t is QNR, then the second order DHT of $f(x)$ with respect to $u(x)$ with a decimation pair (v, t) does not produce any realization for any v.*

In order to prove Lemma 4, we need the following property of the orthogonal function $f(x)$.

Lemma 5. *If $f(x)$ is an orthogonal function from $\mathbb{F}_{2^n}$ to $\mathbb{F}_2$ where $2^n - 1$ is prime, then $f(1) = 1$.*

Proof. Let $\{a_i\}$ be a sequence represented by $f(x)$. Since $f(x)$ is orthogonal, $\{a_i\}$ is balanced with 2^{n-1} 1's and $2^{n-1} - 1$ 0's in one period. Furthermore, $\{a_i\}$ satisfies the coset-constant property [4], i.e., $a_{2i} = a_i$. For a prime $p = 2^n - 1$, all nonzero cosets modulo p have the same size n, and $\{a_i\}$ is constant with 0 or 1 on a coset. This gives $\frac{p-1}{2} = 2^{n-1} - 1$ 1's and $\frac{p-1}{2}$ 0's. Thus, $a_0 = f(1) = 1$ in order to obtain 2^{n-1} 1's.   □

Proof (Proof of Lemma 4). In the second order DHT given in (7), $\widehat{f_u}(v,t)(\lambda)$ should be $\pm 2^n$ for any λ in $\mathbb{F}_{2^n}^*$ if it is a valid realization [6]. To prove Lemma 4, therefore, it is sufficient to show that $\widehat{f_u}(v,t)(1)$ can be neither 2^n nor -2^n when t is QNR.

On the contrary, assume $\widehat{f_u}(v,t)(1) = \pm 2^n$ when $f(x)$ is not a QR function and t is QNR. Let δ and ρ be the numbers of QR and QNR indices satisfying $f(\alpha^{ivt}) = 0$ in a period of the sequence corresponding to $f(x)$, i.e.,

$$\delta = |\{i|\ f(\alpha^{ivt}) = 0 \text{ and } i \text{ is QR in } \mathbb{Z}_{2^n-1}^*\}|,$$
$$\rho = |\{i|\ f(\alpha^{ivt}) = 0 \text{ and } i \text{ is QNR in } \mathbb{Z}_{2^n-1}^*\}|.$$

From the balance property of $f(x)$,

$$\delta + \rho = 2^{n-1} - 1. \tag{8}$$

From Lemma 2 and Lemma 5,

$$\widehat{f_u}(v,t)(1) = \sum_{y \in \mathbb{F}_{2^n}} (-1)^{f(y^{vt})}\, \widehat{u}_{u'}(y)$$
$$= (-1)^{f(1)}(-2^n + 4) + 4\delta - 4(2^{n-1} - 1 - \delta) \tag{9}$$

where we assume $\widehat{u}_{u'}(y) = 4$ at $y = \alpha^i$ for QR i. If we assume that $\widehat{u}_{u'}(y) = 4$ at $y = \alpha^i$ for QNR i, then we have ρ instead of δ in the above, which does not change the final result.

Meanwhile, $f(\alpha^{ivt})$ should be constant on each coset from the coset-constant property of its corresponding sequence. Since each coset has the same size n and corresponds to either QR or QNR, the difference between numbers of QR and QNR indices of i satisfying $f(\alpha^{ivt}) = 0$ should be divisible by n, i.e., $|\delta - \rho| = kn$ for some integer k. From (9), $\delta = 2^{n-2}$ or 0 if $\widehat{f_u}(v,t)(1) = \pm 2^n$. In case of $\delta = 0$, $\rho = 2^{n-1} - 1$ from (8). Then, $|\delta - \rho| = 2^{n-1} - 1$ is divisible by n if n is odd prime. It means that $f(\alpha^{ivt})$ is just a QR sequence and $f(x)$ is a QR function. In case of $\delta = 2^{n-2}$ and $\rho = 2^{n-2} - 1$, on the other hand, $|\delta - \rho| = 1$ cannot be divided by n. With such values of δ and ρ, $f(\alpha^{ivt})$ might have different values on the same coset, which violates the coset-constant property. Thus, the case of $\delta = 2^{n-2}$ and $\rho = 2^{n-2} - 1$ is impossible.

For a QNR t, therefore, $\widehat{f_u}(v, t)(1)$ can be $\pm 2^n$ only if $f(x)$ is a QR function, which contradicts our assumption. Hence, if $f(x)$ is not a QR function, $\widehat{f_u}(v, t)(\lambda)$ cannot have a valid realization when t is QNR. □

From Lemma 4, there exist no realizations of a non-QR function $f(x)$ with respect to a QR function $u(x)$ when a decimation factor t is QNR. In the proof of Lemma 4, however, the realization of a QR function $f(x)$ may exist even though t is QNR. In this case, the realization depends on another decimation factor v.

Lemma 6. *Let $f(x)$ and $u(x)$ be the same QR functions from $\mathbb{F}_{2^n}$ to $\mathbb{F}_2$, i.e., $f(x) = u(x)$. If t is QNR, then the second order DHT of $f(x)$ with respect to $u(x)$ with a decimation pair (v, t) produces $u(x)$ or no realization depending on whether v is QR or QNR. In other words,*

$$\widehat{f_u}(v, t)(\lambda) = \begin{cases} 2^n \cdot (-1)^{u(\lambda)}, & \text{if } v \text{ is QR} \\ \text{no realization}, & \text{if } v \text{ is QNR} \end{cases}$$

for λ in $\mathbb{F}_{2^n}$.

Proof. If $f(x) = u(x)$, then (7) becomes

$$\widehat{f_u}(v, t)(\lambda) = \sum_{z \in \mathbb{F}_{2^n}} (-1)^{u(z)} \sum_{y \in \mathbb{F}_{2^n}} (-1)^{u((\lambda y)^{vt}) + u(z^t y^t)}.$$

If v is QR, then $u((\lambda y)^{vt}) = u((\lambda^t y^t)^v) = u(\lambda^t y^t)$. Thus,

$$\widehat{f_u}(v, t)(\lambda) = \sum_{z \in \mathbb{F}_{2^n}^*} (-1)^{u(z)} \sum_{x \in \mathbb{F}_{2^n}} (-1)^{u(x) + u(\lambda^t z^{-t} x)} = \sum_{z \in \mathbb{F}_{2^n}^*} (-1)^{u(z)} \, \widehat{u}_u(\lambda^t z^{-t})$$

where $x = z^t y^t$. Since $u(x)$ is orthogonal, $\widehat{u}_u(\lambda^t z^{-t})$ has nonzero value 2^n only at $\lambda z^{-1} = 1$. Combined with (6), therefore,

$$\widehat{f_u}(v, t)(\lambda) = 2^n \cdot (-1)^{u(\lambda)}.$$

If v is QNR, on the other hand, then $u(x^v)$ and $u(x)$ correspond to two distinct QR sequences. Thus,

$$\widehat{f_u}(v, t)(\lambda) = \sum_{z \in \mathbb{F}_{2^n}^*} (-1)^{u(z)} \sum_{x \in \mathbb{F}_{2^n}} (-1)^{u(x) + u(\lambda^{vt} z^{-vt} x^v)}$$

$$= \sum_{z \in \mathbb{F}_{2^n}^*} (-1)^{u(z)} \, \widehat{u}_{u'}((\lambda z^{-1})^t) \tag{10}$$

where $x = z^t y^t$. If $\widehat{f_u}(v, t)(\lambda)$ in (10) is evaluated at $\lambda = 1$, then

$$\widehat{f_u}(v, t)(1) = (-1)^{u(1)} \widehat{u}_{u'}(1) + \sum_{j \in \Theta} (-1)^{u(\alpha^j)} \cdot \widehat{u}_{u'}(\alpha^{-jt})$$

$$+ \sum_{j \in \Theta^c} (-1)^{u(\alpha^j)} \cdot \widehat{u}_{u'}(\alpha^{-jt}) = 3 \cdot 2^n - 8$$

Table 3. Realizations of $f(x)$ with respect to a QR function $u(x)$

(v, t)	(QR, QR)	(QR, QNR)	(QNR, QR)	(QNR, QNR)
$f(x) = u(x)$	$u(x)$	$u(x)$	$u'(x)$	None
$f(x) = u'(x)$	$u'(x)$	None	$u(x)$	$u(x)$
Other $f(x)$	$f(x^{vt})$	None	$f(x^{vt})$	None

where $\Theta = \{j \in \mathbb{Z}_{2^n-1}^* | j \text{ is QR}\}$ and $\Theta^c = \{j \in \mathbb{Z}_{2^n-1}^* | j \text{ is QNR}\}$. Since $\widehat{f_u}(v,t)(1) \neq \pm 2^n$, it is enough to show that there exists no realization of $f(x) = u(x)$ when both v and t are QNR.   □

Lemma 7. *Let $f(x)$ and $u(x)$ be distinct QR functions from $\mathbb{F}_{2^n}$ to $\mathbb{F}_2$, i.e., $f(x) = u'(x)$. If t is QNR, then the second order DHT of $f(x)$ with respect to $u(x)$ with a decimation pair (v,t) is given by*

$$\widehat{f_u}(v,t)(\lambda) = \begin{cases} 2^n \cdot (-1)^{u(\lambda)}, & \text{if } v \text{ is QNR} \\ no\ realization, & \text{if } v \text{ is QR} \end{cases}$$

for λ in $\mathbb{F}_{2^n}$.

Proof. This result follows by applying the similar procedure of the proof of Lemma 6.   □

From Lemma 3, 4, 6, and 7, we have the main theorem on the realizations of any binary two-level autocorrelation sequence with respect to a QR sequence.

Theorem 2. *Let $u(x)$ and $u'(x)$ be QR functions representing distinct QR sequences with period $2^n - 1$ and $f(x)$ be an orthogonal function from $\mathbb{F}_{2^n}$ to $\mathbb{F}_2$. In the second order DHT of $f(x)$ with respect to $u(x)$, the realizations of $f(x)$ are completely determined by $f(x)$ and its decimation pair (v,t) as listed in Table 3.*

In Table 3, each entry under (QR, QR) or the other three columns is the realization of $f(x)$ by the corresponding pair. For example, if $(v,t) = $ (QR, QR) and $f(x)$ is not a QR function, then (v,t) realizes $f(x^{vt})$, a self-realization. If $(v,t) = $ (QR, QNR) and $f(x) = u'(x)$, then the entry 'None' represents that (v,t) does not produce any realization.

From Theorem 2 and Table 3, we note that the complete realizations of any binary two-level autocorrelation sequence with respect to a QR sequence are theoretically determined, and a valid realization is either a self-realization or a QR sequence.

5   Conclusion

The second order DHT of special classes of binary two-level autocorrelation sequences has been investigated. Firstly, we showed that in the second order DHT of a binary generalized GMW sequence in a finite field, there exist realizations and corresponding realizable triples which can be theoretically determined by the realizations and realizable triples of its component sequence in the subfield. In the realizations, the DHT in a finite field is inherited from the DHT in its subfield in terms of a binary generalized GMW sequence. Secondly, we showed that the complete realizations of any binary two-level autocorrelation sequence with respect to a QR sequence can be theoretically determined, and a valid realization is either a self-realization or a QR sequence.

Acknowledgment

The authors' research is supported by NSERC (Natural Sciences and Engineering Research Council of Canada) Grant RGPIN 227700-00.

References

1. Dai, Z. D., Gong, G., Song, H. Y.: Trace representation and linear complexity of binary e-th residue sequences. Proceedings of International Workshop on Coding and Cryptography (WCC2003), Versailles, France. (2003) 121-133
2. Dillon, J. F., Dobbertin, H.: New cyclic difference sets with Singer parameters. Finite Fields and Their Applications 10. (2004) 342-389
3. Golomb, S. W.: Shift Register Sequences. Holden-day, Oakland, CA (1967). Revised edition: Aegean Park Press, Laguna Hills, CA (1982)
4. Golomb, S. W., Gong, G.: Signal Design for Good Correlation - for Wireless Communication, Cryptography and Radar. Cambridge University Press (2005)
5. Gong, G.: q-ary cascaded GMW sequences. IEEE Transactions on Information Theory. 42(1). (1996) 263-267
6. Gong, G., Golomb, S. W.: The decimation-Hadamard transform of two-level autocorrelation sequences. IEEE Transactions on Information Theory, 48(4). (2002) 853-865
7. Gordon, B., Mills, W. H., Welch, L. R.: Some new difference sets. Canadian J. Math. 14(4). (1962) 614-625
8. Gottesman, S. R., Grieve, P. G., Golomb, S. W.: A class of pseudonoise-like pulse compression codes. IEEE Transactions on Aerospace and Electronic Systems. 28(2). (1992) 355-362
9. Ireland, K., Rosen, M.: A Classical Introduction to Modern Number theory (2nd ed.). Springer-Verlag, New York (1991)
10. Klapper, A., Chan, A. H., Goresky, M.: Cascaded GMW sequences. IEEE Transactions on Information Theory. 39(1). (1993) 177-183
11. No, J. S., Lee, H. K., Chung, H., Song, H. Y., Yang, K.: Trace representation of Legendre sequences of mersenne prime period. IEEE Transactions on Information Theory. 42(6). (1996) 2254-2255
12. Scholtz, R. A., Welch, L. R.: GMW sequences. IEEE Transactions on Information Theory. 30(3). (1984) 548-553

Frequency/Time Hopping Sequences with Large Linear Complexities*

Yun-Pyo Hong and Hong-Yeop Song

CITY - Center for Information Technology of Yonsei University
Coding and Information Theory Lab.
Department of Electrical and Electronic Engineering, Yonsei University
134 Shinchon-dong Seodaemun-gu, Seoul, Korea, 120-749
{yp.hong, hy.song}@coding.yonsei.ac.kr

Abstract. In this paper, we discuss some methods of constructing frequency/time hopping (FH/TH) sequences over $GF(p^k)$ by taking successive k-tuples of given sequences over $GF(p)$. We are able to characterize those p-ary sequences whose k-tuple versions now over $GF(p^k)$ have the maximum possible linear complexities (LCs). Next, we consider the FH/TH sequence generators composed of a combinatorial function generator and some buffers. We are able to characterize the generators whose output FH/TH sequences over $GF(p^k)$ have the maximum possible LC for the given algebraic normal form.

1 Introduction

In a peer-to-peer frequency/time hopping (FH/TH) spread spectrum communication system, an interceptor may try to synthesize the entire FH/TH pattern from some frequency/time slots successively observed. That is, the interceptor may try to synthesize the linear feedback shift register (LFSR) [1][2] that can generate the next slots of the FH/TH pattern using, say, Berlekamp-Massey (BM) algorithm [3] over a finite field.

Let L be the linear complexity (LC) [4][5] of an FH/TH sequence. When the interceptor observes successive $2L$ frequency/time slots, he can successfully synthesize the next frequency/time slots as long as the same FH/TH sequence is used. Therefore, from the view point of the system designers, the system should change from one FH/TH sequence to another before $2L$ slots of the sequence are used, and the LC of the FH/TH sequences in use should be as large as possible.

Note that any FH/TH sequences are non-binary in general since there are usually more than 2 frequency/time slots available. In fact, an FH/TH communication systems using a few hundreds, or even a few thousands frequency/time slots are common in practice. It is well-known that the number of frequency/time slots affects directly the processing gain [2] of the FH/TH spread spectrum communication systems, at the price of the hardware complexity. Therefore, it is

---

* This work was supported by grant No.(R01-2003-000-10330-0) from the Basic Research Program of the Korea Science and Engineering Foundation.

Ø. Ytrehus (Ed.): WCC 2005, LNCS 3969, pp. 386–396, 2006.
© Springer-Verlag Berlin Heidelberg 2006

necessary to design non-binary sequences (i) with "large" LC, and (ii) over "large" alphabet, but (iii) with "little" increase in the hardware complexity.

In this paper, we consider the simple way of constructing a non-binary (p^k-ary) sequence T over a large alphabet from a given (p-ary) sequence S over a small alphabet, simply reading its successive k-tuples. By increasing the parameter k, one may obtain a sequence over as large alphabet as one wishes. We believe that this method is so simple to construct a p^k-ary sequence compared with a construction over $GF(p^k)$ because the multiplications over $GF(p^k)$ is much more complex than those over $GF(p)$ in the LFSR constructions which is general methods in the hardware systems. In this view point, there will be no significant increase in the complexity in actual hardware design. Therefore, this method satisfies the last two conditions listed in the previous paragraph.

On the other hand, we have to be very careful in analyzing the LC of the new sequences, including the definition of the LC of T over k-tuples over $GF(p)$ which is not a field any more. One way to solve this problem is to interpret the k-tuples over $GF(p)$ as elements of $GF(p^k)$. In this case, it is not much surprising to observe that two different basis may result in two different LC of T (now over $GF(p^k)$), and hence, the LC of T depends on the choice of basis (of $GF(p^k)$ over $GF(p)$).

We are here trying to rule out any possibility that the decrease in its LC using some other basis than that used in the design might help the intercepter to track the FH/TH sequence, assuming that the FH/TH sequence T with its LC equal to L (using the basis used in the design process) is used for the duration of $2L-1$ slots.

Given any one basis, it is clear that the LC of T is at most that of S. We are able to characterize those p-ary sequences S whose k-tuple versions T now over $GF(p^k)$ have the same minimal polynomials [4][5] as S, and therefore, the same LC as S (that is the maximum possible), for any choice of basis of $GF(p^k)$ over $GF(p)$. This leads to the construction of p^k-ary sequences with minimal polynomials essentially over $GF(p)$.

We apply the above characterization into two sequences with as large as possible period when the number of registers, r, is given: binary de Bruijn sequences of period 2^r [6] and p-ary m-sequences of period $p^r - 1$.

We consider the FH/TH sequence generators composed of a combinatorial function generator [7] and some buffers. We are able to characterize the FH/TH sequence generators which guarantee that a combinatorial function sequences, S, over $GF(p)$ have the maximum possible LC for the given algebraic normal form and that k-tuple versions T of S now over $GF(p^k)$ have the same minimal polynomials as S, and therefore, the same LC as S (that is the maximum possible) for any choice of basis of $GF(p^k)$ over $GF(p)$.

2 Constructions of Sequences over $GF(p^k)$ with Minimal Polynomials over $GF(p)$

Let $GF(q)$ be the finite field with q elements, and let p be a prime. Consider a given sequence $S = \{s_n | n = 0, 1, 2, ...\}$ over $GF(p)$. Let k be a positive integer, and define a new sequence (an FH/TH sequence) $T(k, S) = \{t_n | n = 0, 1, 2, ...\}$ based on S by the following:

$$t_n = (s_n, \ s_{n-1}, \ \ldots, \ s_{n-k+1}) \ . \tag{1}$$

Then, it is clear that the sequence $T(k, S)$ is over $GF(p)^k$, the k-tuple vector space over $GF(p)$. By using some but fixed basis such as a simple polynomial basis given by

$$\{\alpha^{k-1}, \ \alpha^{k-2}, \ \ldots, \ \alpha, \ 1\}, \tag{2}$$

where α is a primitive element of $GF(p^k)$, one can regard the sequence $T(k, S)$ being over a field $GF(p^k)$. This is a straightforward and simple way of enlarging the size of alphabet over which a sequence is.

Proposition 1. *The LFSR that generates a sequence $S = \{s_n\}$ over $GF(p)$ also generates $T(k, S)$ over $GF(p^k)$ as defined in (1) regardless of the choice of basis. The converse holds provided that the characteristic polynomial [4][5] that generates T over $GF(p^k)$ is essentially over $GF(p)$.*

Proof. Obvious.   □

Example 1. A ternary sequence S with period 26 is given by

$$0\ 0\ 1\ 1\ 1\ 0\ 2\ 1\ 1\ 2\ 1\ 0\ 1\ 0\ 0\ 2\ 2\ 2\ 0\ 1\ 2\ 2\ 1\ 2\ 0\ 2\ 0\ 0 \ \ldots .$$

Then the sequences $T(3, S)$ and $T(4, S)$ according to (1) are given by the following:

$T(3, S) = 000 \ 000 \ 100 \ 110 \ 111 \ 011 \ 201 \ 120 \ 112 \ 211 \ \cdots,$

$T(4, S) = 0002 \ 0000 \ 1000 \ 1100 \ 1110 \ 0111 \ 2011 \ 1201 \ 1120 \ 2112 \ \cdots .$

Note that both T's as well as S are generated by the LFSR shown in Fig. 1 with connection coefficients over $GF(3)$.

Proposition 1 does not guarantee that the LFSR for $T(k, S)$ over $GF(p^k)$, $k \geq 2$, is necessarily the shortest possible even if it is the shortest for S over $GF(p)$, but that the LC of $T(k, S)$ is at most that of S. In fact, the shortest LFSR for $T(k, S)$ over $GF(p^k)$, $k \geq 2$, (and hence the LC of T) cannot be uniquely determined unless a basis of $GF(p^k)$ is fixed. Following example shows this.

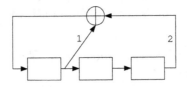

Fig. 1. The LFSR generating S and T's of Example 1

Example 2. (a) A binary sequence S_1 with period 63 is given by

110010000011111110101001001001101010111011011011101001111110010... .

The LC of S_1 over $GF(2)$ is 62, but that of $T(3, S_1)$ over $GF(2^3)$ is 60 with respect to any polynomial basis as in (2). (b) A binary sequence S_2 with period 63 is given by

010111111100110000011011111101010011111100011001110100101001011... .

The LC of $T(3, S_2)$ over $GF(2^3)$ is 55 or 53 with respect to the polynomial basis as in (2) using $x^3 + x + 1$ or $x^3 + x^2 + 1$, respectively.

A question at this point is the following: is it possible that the shortest LFSR that generates S over $GF(p)$ is indeed the shortest LFSR that generates $T(k, S)$ over $GF(p^k)$ with respect to some basis of $GF(p^k)$ over $GF(p)$ for $k \geq 2$? If it is possible to characterize such p-ary sequences S, then $T(k, S)$ over $GF(p^k)$ has the same minimal polynomial as S and hence it is over $GF(p)$.

Lemma 1. [4] (i) *The minimal polynomial of a sequence over $GF(q)$ divides any characteristic polynomial of the LFSR that generates the sequence over $GF(q)$. Therefore, it is uniquely determined up to the multiplication by a constant.* (ii) *An irreducible polynomial over $GF(q)$ of degree d remains irreducible over $GF(q^k)$ if and only if k and d are relatively prime.*

Theorem 1. *Let the minimal polynomial $C(x)$ of $S = \{s_n\}$ over $GF(p)$ be given by $C(x) = \prod_{i \in I}(f_i(x))^{m_i}$ for some irreducible polynomials $f_i(x)$ of degree d_i over $GF(p)$, some positive integers m_i, and some index set I. Let $T(k, S)$ over $GF(p^k)$ be defined as in (1) with respect to some but fixed basis for $k \geq 1$. Then,* (i) *the shortest LFSR that generates S is also the shortest LFSR that generates $T(k, S)$ over $GF(p^k)$, and therefore, their LCs are same, if k and d_i are relatively prime for all $i \in I$. Furthermore,* (ii) *it is also the shortest LFSR of $T(k, S)$ over $GF(p^m)$, and therefore, their LCs are same, for any $m \geq k$ such that m and d_i are relatively prime for all $i \in I$.*

Proof. (i) The LFSR with $C(x)$ also generate $T(k, S)$ over $GF(p^k)$ by Proposition 1. Suppose that the degree of $C(x)$ is not the least for $T(k, S)$. Then the shortest LFSR with characteristic polynomial $C'(x)$ exists and $C'(x)$ divides $C(x)$ by Lemma 1(i). $C'(x) = \prod_{i \in I}(f_i(x))^{s_i}$, where s_i is a non-negative integer, $0 \leq s_i \leq m_i$ for all $i \in I$, and $\sum_{i \in I} s_i < \sum_{i \in I} m_i$ by Lemma 1(ii). On the other hand, the polynomial $C'(x) = \prod_{i \in I}(f_i(x))^{s_i}$ is over $GF(p)$, and Proposition 1 (the converse part) implies that $C'(x)$ is also a characteristic polynomial for S over $GF(p)$ which is a desired contradiction. (ii) Furthermore, if we regard each term of $T(k, S)$ over $GF(p^m)$ for any $m \geq k$ such that m and d_i are relatively prime by inserting so many 0's at some fixed positions, all the previous arguments will be similarly applied.   □

The converse of Theorem 1 is not generally true. We are able to construct p^k-ary FH/TH sequences as in Theorem 1 whose LC are the same as the original (that is the maximum possible) with respect to any basis from p-ary sequences. Thus, if the p-ary sequences have large LC, the resulting FH/TH sequences have the same large LC as the original with respect to any basis. We would like to emphasize the following two cases to which Theorem 1 applies.

Corollary 1. (i) *For a p-ary m-sequence S of period $p^r - 1$ with p a prime, the shortest LFSR that generates S is also the shortest LFSR that generates $T(k, S)$ over $GF(p^k)$ as defined in (1) with respect to any basis if k is relatively prime to r. Furthermore, it is also the shortest LFSR of $T(k, S)$ over $GF(p^m)$ for any $m \geq k$ which is relatively prime to r.* (ii) *If a binary sequence S has a period 2^r (for example, binary de Bruijn sequences), then the shortest LFSR that generates S is also the shortest LFSR that generates $T(k, S)$ over $GF(2^k)$ as defined in (1) for any positive integer k. Furthermore, it is also the shortest LFSR of $T(k, S)$ over $GF(2^m)$ for any $m \geq k$.*

Proof. (i) Obvious. (ii) We note that the minimal polynomial $C(x)$ of a binary sequence S with period 2^r is of the form $(1 + x)^\tau$ for some positive integer τ [6]. □

For a binary de Bruijn sequence, S, with period 2^r and large LC which is at least $2^{r-1} + r$ [6], $T(k, S)$ over $GF(2^k)$ as defined in (1) has the same large LC as S by Corollary 1(ii). In addition, the symbol distribution of the $T(k, S)$ in one period is uniform, that is any symbol of the $T(k, S)$ appears exactly 2^{r-k} times, $r \geq k$, in one period. In reality, the finite field of characteristic 2 would be a good choice for the algebraic structure of FH/TH sequences because the computations over characteristic 2 are most efficiently implemented as hardware systems and the usual practice follows this idea. In above three points, $T(k, S)$ from binary de Bruijn sequences would be good candidates for FH/TH sequences in a peer-to-peer FH/TH spread spectrum communication system.

Example 3. A binary sequence S with period 16 is given by

$$0\ 0\ 0\ 0\ 1\ 0\ 1\ 1\ 1\ 1\ 1\ 1\ 0\ 1\ 0\ 0\ \dots\ .$$

An 8-ary sequence $T(3, S)$ with $k = 3$ over $GF(8)$ becomes

$$000\ 000\ 000\ 000\ 100\ 010\ 101\ 110\ 111\ 111\ 111\ 111\ \dots\ .$$

An 8-ary sequence $T'(3, S)$ over $GF(16)$ becomes

$$0000\ 0000\ 0000\ 0000\ 0100\ 0010\ 0101\ 0110\ 0111\ 0111\ 0111\ 0111\ \dots\ .$$

Here, the symbol 0 is padded at the leftmost position of the every term of $T(3, S)$, and the resulting 4-tuples are regarded as the elements of $GF(16)$. A 16-ary sequence $T(4, S)$ becomes

$$0001\ 0000\ 0000\ 0000\ 1000\ 0100\ 1010\ 1101\ 1110\ 1111\ 1111\ 1111\ \dots\ .$$

All these sequences have the same minimal polynomial and the corresponding LFSR is shown in Fig. 2.

Remark 1. Some interesting discussions are given in [8] and [9] which are methods of constructing p^k-ary m-sequences using several p-ary m-sequences of the same period. We note that the resulting m-sequences over $GF(p^k)$ do not have the same minimal polynomial as the component p-ary m-sequences. In [9], for example, if the minimal polynomial $C(x)$ of the component p-ary m-sequence over $GF(p)$ has degree kn, then the minimal polynomial of resulting p^k-ary m-sequence over $GF(p^k)$ has degree n, and in fact, it is a factor of $C(x)$ over $GF(p^k)$.

Remark 2. Some interesting discussions are given in [10] which establish a lower bound on the LC of a multisequence over $GF(q^k)$ in terms of the joint LC of its k element sequences of period N over $GF(q)$. We note that he characterize the period, N, of which the LC of a multisequence is the same as the joint LC of element sequences.

Now, let $U = \{u_n | n = 0, 1, 2, ...\}$ be a p-ary k-tuple FH/TH sequence in general. In order to determine its minimal polynomial and therefore, LC of U over $GF(p^k)$, we need to fix one basis for BM algorithm. Following theorem characterizes those U which do not need this.

Theorem 2. *Let $U = \{u_n | n = 0, 1, 2, ...\}$ be a p-ary k-tuple sequence in general, where $u_n = (u_n^{(1)}, u_n^{(2)}, \ldots, u_n^{(k)})$. Let a basis of $GF(p^k)$ over $GF(p)$ be fixed, and the minimal polynomial $C(x)$ of U over $GF(p^k)$ using BM algorithm be determined to be of the form $\prod_{i \in I}(f_i(x))^{m_i}$, where $f_i(x)$ are irreducible polynomials of degree d_i over $GF(p)$, m_i are positive integers, and I is some index set. Then, $C(x)$ is a uniquely determined minimal polynomial of U over $GF(p^k)$ regardless of the choice of basis, if k and d_i are relatively prime for all $i \in I$. Furthermore, $C(x)$ is the unique minimal polynomial of U over $GF(p^m)$ for any $m \geq k$ using any basis such that m and d_i are relatively prime for all $i \in I$.*

Proof. Suppose $C'(x)$ is the corresponding minimal polynomial of U now over $GF(p^k)$ with respect to another basis. Then, $C'(x)$ must divide $C(x)$ over $GF(p^k)$ by Lemma 1(i), since $C(x)$ also generates U over $GF(p^k)$ with respect to another basis. Using the same arguments as in the proof of Theorem 1, we have a contradiction unless $C'(x) = C(x)$.   □

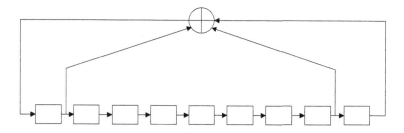

Fig. 2. The shortest LFSR generating S and three T's of Example 3

3   Frequency/Time Hopping Sequence Generators for Large Linear Complexities

We pay attention to the construction of S over $GF(p)$ with large LC. When $S^{(i)} = \{s_n^{(i)} | n = 0, 1, 2, ...\}$, $i = 1, 2, \ldots, N$, are sequences over $GF(p)$, a termwise product sequence $S = \prod_{i=1}^{N} S^{(i)} = \{s_n | n = 0, 1, 2, ...\}$ over $GF(p)$ based on $S^{(i)}$, $i = 1, 2, \ldots, N$, is defined as

$$s_n = \prod_{i=1}^{N} s_n^{(i)} \quad (multiplication \ in \ GF(p)) . \tag{3}$$

It is well-known that the LC of a termwise product sequence defined above is at most the product of the LCs of multiplied sequences.

Lemma 2. [5] *Let $Y = \{y_n\}$ and $Z = \{z_n\}$ be sequences over $GF(p)$ with some irreducible minimal polynomials $C_Y(x)$ and $C_Z(x)$ of degree l and m, respectively. If l and m are relatively prime, then $S = YZ$ over $GF(p)$ as defined in (3) has the irreducible minimal polynomial of degree $l \times m$.*

Corollary 2. *Let $S = YZ$ be a sequence over $GF(p)$ as constructed in Lemma 2. If $l \times m$ and k are relatively prime, then $T(k, S)$ over $GF(p^k)$ as defined in (1) has the same minimal polynomial as S.*

Proof. It is obvious by Lemma 2 and Theorem 1.   □

Example 4. The irreducible minimal polynomial of Y and Z over $GF(2)$ is $C_Y(x) = x^4 + x + 1$ and $C_Z(x) = x^3 + x + 1$, respectively. The irreducible minimal polynomial of $S = YZ$ over $GF(2)$ as defined in (3) is $x^{12} + x^9 + x^5 + x^4 + x^3 + x + 1$ whose degree is $12 = 3 \times 4$ because $gcd(3, 4) = 1$. $T(k, S)$ over $GF(2^k)$ as defined in (1) has the same minimal polynomial as S for k relatively prime to 12.

We consider the general case of Lemma 2, that is the case of termwise product sequences based on arbitrary number of sequences with general minimal polynomials composed of irreducible factors.

Lemma 3. [5] *Let $S^{(i)}$, $i = 1, 2, \ldots, N$, be sequence over $GF(p)$ with a minimal polynomial $C_{S^{(i)}}(x)$ of degree $M^{(i)}$, that divides $x^{p^{m^{(i)}} - 1} - 1$ for some $m^{(i)}$ and contains no linear factor. For any pair of distinct roots, α and β, of $C_{S^{(i)}}(x)$, $i = 1, 2, \ldots, N$, $\alpha\beta^{-1} \notin GF(p)$. If $m^{(i)}$, $i = 1, 2, \ldots, N$, are pairwise relatively prime, then $S = \prod_{i=1}^{N} S^{(i)}$ over $GF(p)$ as defined in (3) has the minimal polynomial of degree $M = \prod_{i=1}^{N} M^{(i)}$.*

The above lemma characterizes those LFSRs whose termwise product sequence has the maximum possible LC, that is the product of the LCs of multiplied sequences. We note that $\alpha\beta^{-1}$ never be in $GF(p)$ for any pair of distinct roots, α and β, of a minimal polynomial $C_{S^{(i)}}(x)$, $i = 1, 2, \ldots, N$, for the case of $p = 2$.

Corollary 3. *Let $S = \prod_{i=1}^{N} S^{(i)}$ be a sequence over $GF(p)$ as constructed in Lemma 3. If $\prod_{i=1}^{N} m^{(i)}$ and k are relatively prime, then $T(k, S)$ over $GF(p^k)$ as defined in (1) has the same minimal polynomial as S.*

Proof. Let $C_S(x)$ be the minimal polynomial of S, then the degree of any irreducible factor of $C_S(x)$ is of the form $\prod_{i=1}^{N} r^{(i)}$, where $r^{(i)} | m^{(i)}$, by Lemma 3 and Theorem 1 completes the proof. ☐

Example 5. The minimal polynomial of Y over $GF(2)$ is $C_Y(x) = x^3 + x^2 + 1$ that divides $x^{2^3-1} - 1$ and $C_Y(1) = 1$. The minimal polynomial of Z over $GF(2)$ is $C_Z(x) = x^6 + x^3 + x^2 + x + 1$ that divides $x^{2^4-1} - 1$ and $C_Z(1) = 1$. The minimal polynomial of $S = YZ$ over $GF(2)$ as defined in (3) is $x^{18} + x^{14} + x^{12} + x^{11} + x^{10} + x^9 + x^6 + x^4 + x^3 + x^2 + 1$ whose degree is $18 = 3 \times 6$ because $\gcd(3, 4) = 1$. $T(k, S)$ over $GF(2^k)$ as defined in (1) have the same minimal polynomial as S for k relatively prime to 3×4.

Now, we consider the FH/TH sequence generator composed of a combinatorial function generator [7] and k buffers shown in Fig. 3. Let a combinatorial function sequence, S, over $GF(p)$ by a combinatorial function, f, (that would make S have large LC) be represented in the algebraic normal form given by

$$
\begin{aligned}
s_n &= f(s_n^{(1)}, s_n^{(2)}, \ldots, s_n^{(N)}) \\
&= a_0 + \sum_{i=1}^{N} a_i s_n^{(i)} + \sum_{i=1}^{N} \sum_{j=i+1}^{N} a_{ij} s_n^{(i)} s_n^{(j)} + \ldots + a_{12\ldots N} s_n^{(1)} s_n^{(2)} \ldots s_n^{(N)},
\end{aligned}
\tag{4}
$$

where $S^{(i)}$, $i = 1, 2, \ldots, N$, are sequences over $GF(p)$ and the coefficients of f are elements of $GF(p)$. We note that the algebraic normal form as defined in (4) cannot represent all combinatorial functions. The maximum possible LC of a combinatorial function sequence, S, for the given algebraic normal form is given by

$$
M = F(M^{(1)}, M^{(2)}, \ldots, M^{(N)}),
\tag{5}
$$

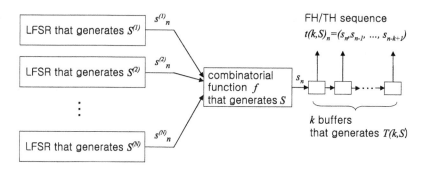

Fig. 3. Frequency/Time hopping sequence generators for large linear complexities

where $F(M^{(1)}, M^{(2)}, \ldots, M^{(N)})$ is defined as (4) with a coefficient being 0 if it is 0 or 1 otherwise and $M^{(i)}$ is the LC of $S^{(i)}$, $i = 1, 2, \ldots, N$, and operations of F are over the integers.

R. A. Rueppel characterize those LFSRs such that a combinatorial function sequence, S, has the maximum possible LC for the given algebraic normal form [5]. In the previous section, we characterize those p-ary sequences, S, whose k-tuple versions, $T(k, S)$, now over $GF(p^k)$ have the maximum possible LCs. In this view point, we focus on the relations between the above two characterizations. We are able to characterize those LFSRs such that a resulting k-tuple sequence (an FH/TH sequence), $T(k, S)$, has the maximum possible LC, M as defined in (5). That is, we are able to construct FH/TH sequences with large LCs by the generators shown in Fig. 3.

Lemma 4. [5] *Let $S^{(i)}$, $i = 1, 2, \ldots, N$, be sequences over $GF(p)$ with minimal polynomials $C_{S^{(i)}}(x)$ of degree $M^{(i)}$, that divide $x^{p^{m^{(i)}} - 1} - 1$ for some $m^{(i)}$ and contain no linear factor. For any pair of distinct roots, α and β, of $C_{S^{(i)}}(x)$, $i = 1, 2, \ldots, N$, $\alpha\beta^{-1} \notin GF(p)$. If $m^{(i)}$, $i = 1, 2, \ldots, N$ are pairwise relatively prime, then S over $GF(p)$ as defined in (4) has the minimal polynomial of degree M as defined in (5) for the given algebraic normal form, f.*

Corollary 4. *Let S be a sequence over $GF(p)$ as constructed in Lemma 4. If $\prod_{i=1}^{N} m^{(i)}$ and k are relatively prime, then $T(k, S)$ over $GF(p^k)$ as defined in (1) has the same minimal polynomial as S.*

Proof. Let $C_S(x)$ be the minimal polynomial of S, then the degree of any irreducible factor of $C_S(x)$ is of the form $\prod_{i=1}^{N} r^{(i)}$, where $r^{(i)} | m^{(i)}$, by Lemma 4 and Theorem 1 completes the proof. □

Example 6. The minimal polynomial of X, Y, Z over $GF(2)$ is $C_X(x) = x^6 + x^5 + x^4 + x^3 + x^2 + x + 1$, $C_Y(x) = x^6 + x^3 + x^2 + x + 1$, $C_Z(x) = x^{10} + x^8 + x^7 + x^5 + x^3 + x^2 + 1$ that divides $x^{2^3 - 1} - 1$, $x^{2^4 - 1} - 1$, $x^{2^5 - 1} - 1$ respectively and contains no linear factor. The minimal polynomial of S over $GF(2)$ defined by $s_n = f(x_n, y_n, z_n) = 1 + x_n + y_n + z_n + x_n y_n + y_n z_n + z_n x_n + x_n y_n z_n$ as (4) is of degree $539 = M(6, 6, 10) = 1 + 6 + 6 + 10 + 6 \cdot 6 + 6 \cdot 10 + 10 \cdot 6 + 6 \cdot 6 \cdot 10$ as defined in (5) because 3, 4, and 5 are pairwise relatively prime. $T(k, S)$ over $GF(2^k)$ as defined in (1) have the same minimal polynomial as S for k relatively prime to $3 \cdot 4 \cdot 5$. For example, $T(7, S)$ is a 128-ary FH/TH sequence whose LC is 539.

We believe that FH/TH sequences as constructed in Corollary 4 must be a good candidates of FH/TH patterns in a peer-to-peer FH/TH spread spectrum communication system for the following good reasons: (i) with "large" LC, and (ii) over "large" alphabet, but (iii) with "little" increase in the hardware complexity.

4 Concluding Remarks

We believe that the finite field of characteristic 2 would be a good choice for the algebraic structure of FH/TH sequences because the computations over characteristic 2 are most efficiently implemented as hardware systems and the usual practice follows this idea.

We have tried several other options but failed to extract any further reasonable behavior of non-binary FH/TH sequences over $GF(p^k)$ whose minimal polynomial and therefore, LC are uniquely determined regardless of the choice of basis other than those given in Theorem 1. Theorem 2 is slightly more general in that the p-ary k-tuple FH/TH sequences are not necessarily constructed as a k-tuple version of a p-ary sequence.

We note that Corollary 4 characterize those FH/TH sequence generators such that a combinatorial function sequence, S, and a resulting k-tuple sequence (an FH/TH sequence), $T(k, S)$, has the maximum possible LC for any given algebraic normal form, f, to resist the only BM attack. So, it is proper that we use the algebraic normal form, f, that has desired cryptographic properties such as correlation immunity, resiliency, nonlinearity, and propagation [7][11][12][13] to resist other attacks than the BM attack.

We note that the sequence terms of $T(k, S)$ are highly correlated with each other because t_n is the right shifted version of t_{n-1} with the only new leftmost component. This correlation between consecutive terms must be a vulnerable point to some other attacks. But, Theorem 1 and all corollaries in this paper also apply equally well to $T(k, S)$ defined by

$$t_n = (s_{n-\sigma(0)}, s_{n-\sigma(1)}, \ldots, s_{n-\sigma(k-1)}), \tag{6}$$

where σ is any permutation on $\{0, 1, \ldots, k-1\}$. A further generalization is also possible by using any integers instead of $\sigma(i)$ for each i. Therefore, we are able to solve the correlation problem between consecutive terms by the above method.

References

1. S. W. Golomb, *Shift Register Sequences*, Revised Edition, Aegean Park Press, Laguna Hills, CA 92654, 1982.
2. M. K. Simon, J. K. Omura, R. A. Scholtz, and B. K. Levitt, *Spread Spectrum Communications Handbook*, Revised Edition, McGraw-Hill, Inc., 1994.
3. J. L. Massey, "Shift-Register Synthesis and BCH decoding," *IEEE Transactions on Information Theory*, vol. IT-15, no. 1, pp. 122-127, Jan. 1969.
4. R. Lidl and H. Niederreiter, *Finite Fields*, Second Edition, Encyclopedia of Mathematics and Its Applications, vol. 20, Cambridge University Press, 1997.
5. R. A. Rueppel, *Analysis and Design of Stream Ciphers*, Springer-Verlag, 1986.
6. A. H. Chan, R. A. Games, and E. L. Key, "On the Complexities of de Bruijn Sequences," *Journal of Combinatorial Theory*, Series A 33, pp. 233-246, 1982.
7. S. W. Golomb and G. Gong, *Signal Design for Good Correlation for Wireless Communication, Cryptography, and Radar*, Cambridge University Press, 2005.
8. W. J. Park and J. J. Komo, "Relationships Between m-Sequences over $GF(q)$ and $GF(q^m)$," *IEEE Transactions on Information Theory*, vol. 35, no. 1, pp. 183-186, Jan. 1989.
9. G. Gong and G. Z. Xiao, "Synthesis and Uniqueness of m-Sequences over $GF(q^n)$ as n-Phase Sequences over $GF(q)$," *IEEE Transactions on Communications*, vol. 42, no. 8, pp. 2501-2505, Aug. 1994.

10. W. Meidl, "Discrete Fourier Transform, Joint Linear Complexity and Generalized Joint Linear Complexity of Multisequences," *Lecture Notes in Computer Science*, vol. 3486, pp. 101-112, Mar. 2005.
11. T. Siegenthaler, "Correlation-Immunity of Nonlinear Combining Functions for Cryptographic Applications," *IEEE Transactions on Information Theory*, vol. IT-30, no. 5, pp. 776-780, Sep. 1984.
12. W. Meier and O. Staffelbach, "Nonlinearity Criteria for Cryptographic Functions," *Lecture Notes in Computer Science*, vol. 434, pp. 549-562, 1990.
13. B. Preneel, W. V. Leekwijck, and L. V. Linden "Propagation Characteristics of Boolean Functions," *Lecture Notes in Computer Science*, vol. 473, pp. 161-173, 1990.

One and Two-Variable Interlace Polynomials: A Spectral Interpretation

Constanza Riera and Matthew G. Parker

Depto. de Álgebra, Facultad de Matemáticas, Universidad Complutense de Madrid,
Avda. Complutense s/n, 28040 Madrid, Spain
criera@mat.ucm.es
Selmer Centre, Inst. for Informatikk, Høyteknologisenteret i Bergen,
University of Bergen, Bergen 5020, Norway
matthew@ii.uib.no
http://www.ii.uib.no/~{}matthew

Abstract. We relate the one- and two-variable *interlace polynomials* of a graph to the spectra of a quadratic boolean function with respect to a strategic subset of local unitary transforms. By so doing we establish links between graph theory, cryptography, coding theory, and quantum entanglement. We establish the form of the interlace polynomial for certain functions, provide new one and two-variable interlace polynomials, and propose a generalisation of the interlace polynomial to hypergraphs. We also prove conjectures from [15] and equate certain spectral metrics with various evaluations of the interlace polynomial.

1 Introduction

The *interlace polynomial* was introduced by Arratia, Bollobás and Sorkin [2,3], as a variant of Tutte and Tutte-Martin polynomials [6]. They defined the interlace polynomial of a graph G, $q(G)$, by means of a recurrence formula, involving *local complementation* (LC) of the graph. Aigner and van der Holst, in [1], generalised the concept by means of a related interlace polynomial, $Q(G)$, and showed a new and easier way of constructing both polynomials $q(G)$ and $Q(G)$ using a matrix approach. They conclude that the polynomial $q(z)$, when evaluated at $z = 1$, gives the number of induced subgraphs of G with an odd number of perfect matchings (including the empty set), and that $Q(z)$, when evaluated at $z = 2$, gives the number of (general) induced subgraphs with an odd number of (general) perfect matchings, "general" meaning here that loops are allowed to be part of the matching.

In [4], Arratia, Bollobas and Sorkin defined an extension of the interlace polynomial q, defined by themselves in [3], to a new polynomial $q(x, y)$. Here we propose a similar extension of Q as defined by Aigner and Van der Holst in [1] to a new polynomial $Q(x, y)$. We also propose the HN-interlace polynomial Q^{HN} and its corresponding two-variable extension, $Q^{HN}(x, y)$. Also, we define the IN-interlace polynomial $Q^{IN}(x, y)$.

A main goal of this paper is to re-state the problem of constructing an interlace polynomial for a graph as a problem in transform theory. To be precise, the

Ø. Ytrehus (Ed.): WCC 2005, LNCS 3969, pp. 397–411, 2006.
© Springer-Verlag Berlin Heidelberg 2006

interlace polynomial of a graph summarises the spectra of the *Boolean function* associated with that graph, where the spectra are computed w.r.t. (with respect to) a certain well-chosen set of *Local Unitary* (LU) transforms. This re-statement allows us to propose new interlace polynomials, as mentioned above, to suggest new applications for these polynomials, and even to extend the problem in a natural way to hypergraphs. We focus on LU transforms which are formed from tensor products of the matrices I, H, and N, where,

$$H = \frac{1}{\sqrt{2}} \begin{pmatrix} 1 & 1 \\ 1 & -1 \end{pmatrix}, \qquad \text{is the } \textit{Walsh-Hadamard} \text{ kernel,}$$

$$N = \frac{1}{\sqrt{2}} \begin{pmatrix} 1 & i \\ 1 & -i \end{pmatrix}, i^2 = -1, \qquad \text{is the } \textit{Negahadamard} \text{ kernel,}$$

and I is the 2×2 identity matrix.

Definition 1. *The set of 3^n LU transforms, $\{I, H, N\}^n$, is the set comprising all transforms, U, of the form $U = \prod_{j \in \mathbf{R_I}} I_j \prod_{j \in \mathbf{R_H}} H_j \prod_{j \in \mathbf{R_N}} N_j$, each of size $2^n \times 2^n$, where, say, $V_j = I \otimes \ldots \otimes I \otimes V \otimes I \otimes \ldots \otimes I$, with V in the jth position, '$\otimes$' means the tensor product of matrices, and where the sets $\mathbf{R_I}, \mathbf{R_H}$, and $\mathbf{R_N}$, partition the set of vertices $\{0, \ldots, n-1\}$ [1].*

Transform subsets such as $\{I, H\}^n$ ...etc are then defined in the obvious way.

Define the n vertex graph, G, by its $n \times n$ adjacency matrix, Γ. We identify G with a quadratic Boolean function $p(x_0, x_1, \ldots, x_{n-1})$, where $p(\mathbf{x}) = \sum_{i<j} \Gamma_{ij} x_i x_j$ [20]. This identification allows us to interpret $q(G, 1)$ as the number of *flat spectra* of $p(\mathbf{x})$ w.r.t. the transform set, $\{I, H\}^n$, and $Q(G, 2)$ as the number of flat spectra of $p(\mathbf{x})$ w.r.t. the transform set $\{I, H, N\}^n$.

In section 3 we re-define the interlace polynomials q and Q using the modified adjacency matrix of the graph w.r.t. $\{I, H\}^n$ and $\{I, H, N\}^n$, respectively, as defined in [20, 21], and use them to compute the interlace polynomial of the clique (complete graph), and clique-line-clique.

In section 4 we define a new interlace polynomial, Q^{HN}, that summarises spectra w.r.t. $\{H, N\}^n$ in the same way that the interlace polynomials q and Q do with their respective sets. Our motivation for relating the concept of interlace polynomial to $\{H, N\}^n$ is that this set is related to the *Peak-to-Average Power Ratio* (PAR) w.r.t. both one and multi-dimensional continuous *Discrete Fourier Transforms*, and hence to problems in telecommunications and physics for tasks such as channel-sounding, spread-spectrum, and synchronization [19]. We compute Q^{HN} for the clique, line, and clique-line-clique functions. The polynomial Q^{HN} is also the basis for constructing Q for recursive structures.

By Glynn [11], a self-dual *quantum error correcting code (QECC)* $[[n, 0, d]]$ corresponds to a graph on n vertices, this being a so-called *graph state* [13] which may be assumed to be connected if the code is indecomposable. It is shown there that two graphs, G and G', give equivalent self-dual quantum codes if and only if they are LC-equivalent [2] (see definition 9). In this case, G and G'

---

[1] For instance, if $n = 4$, $\mathbf{R_I} = \{1\}$, $\mathbf{R_H} = \{0, 3\}$, and $\mathbf{R_N} = \{2\}$, then $U = H \otimes I \otimes N \otimes H$, where U is a 16×16 unitary matrix.

[2] Referred to as "Vertex-Neighbour-Complementation" (VNC)-equivalent in [11].

also map to GF(4) additive codes with identical weight distributions [7]. As the interlace polynomial, Q, is LC-invariant [1], it is also an invariant of the corresponding QECC. This result implies that Q is invariant under the application of certain LU *transforms* to the multipartite quantum state associated with the QECC [18], for it turns out that LC-equivalence for graph states can be characterised by LU-transformation via the set of transforms $\{I, H, N\}^n$ [20]. Therefore Q can be used to summarise some important properties of an associated quantum graph state. More specifically, an analysis of the spectra of a Boolean function provides measures of *entanglement* of the associated quantum multipartite state, as defined by the QECC and/or its associated quadratic Boolean function [13, 18, 20].

In section 5 we provide spectral interpretations of interlace polynomials, and generalise to hypergraphs, i.e. to Boolean functions of algebraic degree greater than 2. We prove conjectures proposed by Parker in [15] related to the line function (path graph) and its affine offsets. In [12, 17], the *Multivariate Merit Factor (MMF)* and *Clifford Merit Factor (CMF)* are defined, these being inverse measures of the energy of the Boolean function w.r.t. $\{H, N\}^n$ and $\{I, H, N\}^n$ respectively. By proving that the *power spectrum* of a quadratic Boolean function w.r.t. $\{I, H, N\}^n$ is always one or two-valued, we show that MMF and CMF can be derived from $Q^{HN}(4)$ and $Q(4)$, respectively.

In section 6 we propose the new two-variable interlace polynomial $Q(x, y)$, and derive some lemmas.

Our spectral approach allows us to interpret the interlace polynomial as a descriptor for some of the spectral characteristics of a Boolean function, with application to classical cryptography. For instance, one often wants to approximate a Boolean function, p, of n variables, by a simpler function. One way to do this is to use an *annihilator* Boolean function, g, such that $gp \approx 0$ or $g(p+1) \approx 0$. A generalisation of this is to look for a function g such that $g(p + a) \approx 0$, where a is a low degree Boolean function. These approximations provide a generalised measure of *probabilistic algebraic immunity* for a function, p. In particular, in the context of transforms w.r.t. $\{I, H\}^n$, with $\mathbf{R_I}$ and $\mathbf{R_H}$ integer sets that partition $\{0, 1, \ldots, n - 1\}$, then $g = \prod_{j \in \mathbf{R_I}} (x_j + c_j)$ is a degree $|\mathbf{R_I}|$ Boolean function of $|\mathbf{R_I}|$ variables, $c_j \in \text{GF}(2)$, and $a = d + \sum_{j \in \mathbf{R_H}} x_j$ is a degree-one Boolean function of $|\mathbf{R_H}|$ variables, with $d \in \text{GF}(2)$. In this case the transform spectra w.r.t. $\{I, H\}^n$ quantify the accuracy of all possible (g, a) pairs of the above form w.r.t. the approximation $g(p+a) \approx 0$. The spectra, in turn, are summarised by the interlace polynomial, q. Similarly, Q can be used to summarise the accuracy of approximations w.r.t. $\{I, H, N\}^n$, where a is now an affine function from $\text{GF}(2)^{|\mathbf{R_H}|+|\mathbf{R_N}|} \rightarrow \mathcal{Z}_4$. Q can also be used to assess the block cipher attack scenario where one has full read/write access to a subset of plaintext bits and access to all ciphertext bits [9]. Using similar arguments, Q^{HN} summarises all possible $\mathcal{Z}_4$-linear approximations to a Boolean function, i.e. w.r.t. $\{H, N\}^n$ [16]. Thus, the spectra w.r.t. $\{I, H, N\}^n$ or its subsets tell us more about the Boolean function, p, than is provided by just the spectrum w.r.t. the *Walsh-Hadamard transform* (WHT), and such spectra are conveniently summarised by their respective interlace polynomials. As seen in [20], just an enumeration of the flat spectra of a function w.r.t. $\{I, H, N\}^n$ or its subsets provides a good measure of the 'strength' of the function in various contexts.

2   Definitions and Notation

We recapitulate here some definition and results of [20, 21]:

Definition 2. *[23] A Boolean function* $p(\mathbf{x}) : GF(2)^n \to GF(2)$ *is bent iff* $P = 2^{-n/2}(\bigotimes_{i=0}^{n-1} H)(-1)^{p(\mathbf{x})}$ *has a flat spectrum, or, in other words, if* $P = (P_\mathbf{k}) \in \mathbb{C}^{2^n}$ *is such that* $|P_\mathbf{k}| = 1 \ \forall \ \mathbf{k} \in GF(2)^n$.

If the function is quadratic, we associate to it a simple non-directed n-vertex graph, and in this case a flat spectrum is obtained iff Γ, the $n \times n$ adjacency matrix of the graph, has maximum rank mod 2 [14]. In [20], we generalised this concept, considering not only the Walsh-Hadamard transform $\bigotimes_{i=0}^{n-1} H$, but the complete set of unitary transforms w.r.t. $\{I, H, N\}^n$. We studied there the number of flat spectra of a function w.r.t. $\{I, H, N\}^n$, or in other words the number of unitary transforms $U \in \{I, H, N\}^n$ such that $P_U = (P_{U,\mathbf{k}}) \in \mathbb{C}^{2^n}$ has $|P_{U,\mathbf{k}}| = 1 \ \forall \ \mathbf{k} \in GF(2)^n$, where

$$(P_{U,\mathbf{k}}) = U(-1)^{p(\mathbf{x})} = (\prod_{j \in \mathbf{R_I}} I_j \prod_{j \in \mathbf{R_H}} H_j \prod_{j \in \mathbf{R_N}} N_j)(-1)^{p(\mathbf{x})} . \qquad (1)$$

We also considered the number of flat spectra w.r.t. some subsets of $\{I, H, N\}^n$, namely $\{H, N\}^n$ (when $\mathbf{R_I} = \emptyset$) and $\{I, H\}^n$ (when $\mathbf{R_N} = \emptyset$). We proved there that a quadratic Boolean function will have a flat spectrum w.r.t. a transform in $\{I, H, N\}^n$ iff a certain modification of its adjacency matrix, Γ, concretely the matrix resultant of the following actions, has maximal rank mod 2:

- for $i \in \mathbf{R_I}$, we erase the ith row and column of Γ.
- for $i \in \mathbf{R_N}$, we substitute 0 for 1 in position $[i, i]$, i.e. we assign $\Gamma_{ii} = 1$.
- for $i \in \mathbf{R_H}$, we leave the ith row and column of Γ unchanged.

This modified adjacency matrix is also helpful to compute the interlace polynomial of a graph.

3   The Interlace Polynomial

We define polynomials q and Q, equivalently to definitions offered in [1], but relate the interlace polynomial with the spectra of a graph w.r.t. $\{I, H\}^n$ and $\{I, H, N\}^n$.

Definition 3. *The interlace polynomial* q *of a graph* G *in* n *variables is*

$$q_n(G; z) = \sum_{U \in \{I, H\}^n} (z - 1)^{co(\Gamma_U)} , \qquad (2)$$

where co(Γ_U) stands for the corank of the modified adjacency matrix of the graph w.r.t. the transform $U \in \{I, H\}^n$, Γ_U, obtained by erasing from the adjacency matrix of the graph the rows and columns whose indices are in $\mathbf{R_I}$ (see [20]).

Remark: $q(G; 1)$ is the number of flat spectra of the function w.r.t. $\{I, H\}^n$.

Definition 4. *The line function (or path graph), $p_l(\mathbf{x})$ is*

$$p_l(\mathbf{x}) = \sum_{j=0}^{n-2} x_j x_{j+1} + \mathbf{c} \cdot \mathbf{x} + d , \tag{3}$$

where $\mathbf{x}, \mathbf{c} \in GF(2)^n$, $\mathbf{x} = (x_0, \ldots, x_{n-1})$, and $d \in GF(2)$.

Definition 5. *The clique function (complete graph) is*

$$p_c(\mathbf{x}) = \sum_{0 \leq i < j \leq n-1} x_i x_j , \tag{4}$$

where $\mathbf{x} = (x_0, \ldots, x_{n-1}) \in GF(2)^n$.

Remark: [3] The interlace polynomial q for the path graph satisfies, for $n \geq 2$, $q_n(z) = q_{n-1}(z) + z q_{n-2}(z)$, with $q_1(z) = 1, q_2(z) = 2z$; for the complete graph, $q_n(z) = 2^{n-1} z$.

Definition 6. *The n-clique-line-m-clique is*

$$p_{n,m}(\mathbf{x}) = \sum_{0 \leq i < j \leq n-1} x_i x_j + x_{n-1} x_n + \sum_{n \leq i < j \leq n+m-1} x_i x_j , \tag{5}$$

where $\mathbf{x} = (x_0, \ldots, x_{n+m-1}) \in GF(2)^{n+m}$.

We consider the clique-line-clique function to be structurally interesting due to the results on page 31 of [8] which tend to suggest that the best QECCs (and most entangled graph states) have a graph in their LC-orbit which can be described as a *nested-clique graph* - more generally, *nested-regular graph*. So an investigation of the clique-line-clique structure is an attempt to understand these graphical structures more.

Lemma 1. *For the n-clique-line-m-clique (5), $q_n(z) = 3 \cdot 2^{n+m-4} z^2 + 2^{n+m-3} z$.*

Definition 7. *The interlace polynomial Q of a graph G in n variables is*

$$Q_n(G; z) = \sum_{V \in \{I, H, N\}^n} (z - 2)^{co(\Gamma_V)} , \tag{6}$$

where co(Γ_V) means the corank of the modified adjacency matrix of the graph w.r.t. $V \in \{I, H, N\}^n$, Γ_V, obtained by erasing the rows and columns whose indices are in $\mathbf{R_I}$, as before, and then substituting 0 by $v_i \in GF(2)$ in the diagonal, in those indices $i \in \mathbf{R_H} \cup \mathbf{R_N}$, where $v_i = 1$ iff $i \in \mathbf{R_N}$ (see [20]).

Remark: $Q(G; 2)$ is the number of flat spectra of the function w.r.t. $\{I, H, N\}^n$.

Remark: The formula of Q for the path graph is found in [1].

Lemma 2. *For the complete graph (4),*

$$Q_{n+1}(z) = 2Q_n(z) + z^n, \ n \geq 2, \ with \ Q_1(z) = z \ .$$

The closed form is $Q_n = 2^{n-1}(z-1) + (z-2)^{-1}(z^n - 2^n)$.

Remark: When $z = 2$, we get $(n+1)2^{n-1}$, the number of flat spectra for the complete graph w.r.t. $\{I, H, N\}^n$ [21].

Lemma 3. *For the n-clique-line-m-clique (5), when n, $m \geq 3$, the interlace polynomial Q is:*

$$Q_{n,m}(z) = 2^{n+m-2} - 2^{n+m-4}z + 3 \cdot 2^{n+m-4}z^2 + z^{n-1}2^{m-2}(z-1)$$

$$+ z^{m-1}2^{n-2}(z-1) + \frac{3 \cdot 2^{m-1}z + z^{m-1} - 2^m}{z-2}(z^{n-1} - 2^{n-1})$$

$$+ \frac{3 \cdot 2^{n-1}z + z^{n-1} - 2^n}{z-2}(z^{m-1} - 2^{m-1})$$

$$+ \frac{z+4}{(z-2)^2}(z^{n-1} - 2^{n-1})(z^{m-1} - 2^{m-1}) \ .$$

4   The HN-Interlace Polynomial

We now define an interlace polynomial related to the set $\{H, N\}^n$ as q and Q were related to the sets $\{I, H\}^n$ and $\{I, H, N\}^n$ respectively.

Definition 8. *The HN-interlace polynomial for a graph G in n variables is*

$$Q_n^{HN}(G; z) = \sum_{W \in \{H, N\}^n} (z - 2)^{co(\Gamma_W)} \ , \tag{7}$$

where $co(\Gamma_W)$ means the corank of the modified adjacency matrix of the graph w.r.t. $W \in \{H, N\}^n$, Γ_W, obtained by substituting 0 by $v_i \in GF(2)$ in the diagonal, in those indices $i \in \mathbf{R_H} \cup \mathbf{R_N}$, where $v_i = 1$ iff $i \in \mathbf{R_N}$ (see [20]).

Remark: $Q^{HN}(G; 2)$ is the number of flat spectra of the function w.r.t. $\{H, N\}^n$.

Lemma 4. *The HN-interlace polynomial for the path graph (3) is*

$$Q_{n+1}^{HN}(z) = 2^n - Q_n^{HN}(p_l; z), \ with \ Q_1^{HN}(p_l; z) = z - 1 \ .$$

In closed form,

$$Q_n^{HN}(z) = \frac{1}{3} \left(2^n + (-1)^{n-1}\right) z + (-1)^n \ .$$

Lemma 5. *For the complete graph (4),*

$$Q_{n+1}^{HN}(z) = Q_n^{HN}(z) + (z-1)^n + (-1)^n(z-3), \quad \text{with } Q_1^{HN}(z) = z-1 \ .$$

In closed form,

$$Q_n^{HN}(z) = \begin{cases} 1 + (z-2)^{-1}((z-1)^n - 1), & \text{for } n \text{ even} \\ z-2 + (z-2)^{-1}((z-1)^n - 1), & \text{for } n \text{ odd} \end{cases}$$

Remark: When $z = 2$, we get $n + \frac{1+(-1)^n}{2}$, the number of flat spectra for the complete graph w.r.t. $\{H, N\}^n$, as seen in [21].

Lemma 6. *For the n-clique-line-m-clique (5), the HN-interlace polynomial is*

$$Q_{n,m}^{HN}(z) = -2 + 6\chi_n\chi_m + 3\chi_{n+1}\chi_{m+1} + (2 - 2\chi_n\chi_m - \chi_{n+1}\chi_{m+1})z$$
$$+ \frac{z+1}{(z-2)^2}\left((z-1)^{n-1} - 1\right)\left((z-1)^{m-1} - 1\right)$$
$$+ \frac{z+1+z\chi_m - 3\chi_m}{z-2}\left((z-1)^{n-1} - 1\right)$$
$$+ \frac{z+1+z\chi_n - 3\chi_n}{z-2}\left((z-1)^{m-1} - 1\right) \ ,$$

where $\chi_k = \frac{1 + (-1)^k}{2}$.

5 Spectral Interpretations of the One-Variable Interlace Polynomial

In definition 7 in section 3, the interlace polynomial Q was related to the set of transforms $\{I, H, N\}^n$. We now give further spectral interpretations of q, Q, and Q^{HN}. This allows us to extend the interlace concept to hypergraphs (or Boolean functions of higher degree than two). Given a graph G with adjacency matrix Γ, its *complement* is defined to be the graph with adjacency matrix $\Gamma + I + \mathbf{1}$ (mod 2), where I is the identity matrix and $\mathbf{1}$ is the all-ones matrix.

Definition 9. *[6,11,13] The action of Local Complementation (LC) on a graph G at vertex v is defined as the graph transformation obtained by replacing the subgraph $G[\mathcal{N}(v)]$ (i.e., the induced subgraph of the neighbourhood of the vth vertex of G) by its complement.*

Theorem 1. *[1] The interlace polynomial Q is invariant under LC.*

Proof. From definition 7 and [20], one can show that Q is invariant w.r.t. $\{I, H, N\}^n$. But, as seen in [20], this set defines the LC operation.

Definition 10. *[2,4] The action of pivot on a graph, G, at two connected vertices, u and v, (i.e. where G contains the edge uv), is given by $LC(v)LC(u)LC(v)$ - that is the action of LC at vertex v, then vertex u, then vertex v again.*

Theorem 2. *[2] The interlace polynomial q is invariant under pivot.*

Proof. By considering definition 3 it is possible to show that q is invariant w.r.t. $\{I, H\}^n$. One can then show that pivot can be defined by $\{I, H\}^n$ [22].

Theorem 3. *The corank of the modified adjacency matrix is*

$$co(\Gamma_U) = \log_2(\max_{\mathbf{k}} |P_{U,\mathbf{k}}|^2) \ ,$$

where $P_{U,\mathbf{k}}$ are the entries of P_U as defined in (1).

Proof. We prove the theorem for $U \in \{H, N\}^n$, as the case for $U \in \{I, H, N\}^n$ then follows trivially. First, we must recall the autocorrelation of a boolean function $p(\mathbf{x})$ w.r.t $\{H, N\}^n$:

$$A_{\mathbf{k}} = \sum_{\mathbf{x} \in GF(2)^n} (-1)^{p(\mathbf{x}) + p(\mathbf{x}+\mathbf{k}) + \sum_{i=0}^{n-1} \chi_{\mathbf{R_N}}(i) k_i (x_i + 1)} \ ,$$

where $\mathbf{k} = (k_0, k_1, \ldots, k_{n-1}) \in GF(2)^n$, and $\chi_{\mathbf{R_N}}(i)$ is the characteristic function of $\mathbf{R_N}$, i.e,

$$\chi_{\mathbf{R_N}}(i) = \begin{cases} 1, & i \in \mathbf{R_N} \\ 0, & i \notin \mathbf{R_N} \end{cases}$$

We use extensively the *Wiener-Kinchine* property

$$\begin{pmatrix} A_{0\ldots0} \\ A_{0\ldots1} \\ \vdots \\ A_{1\ldots1} \end{pmatrix} \begin{array}{c} U \\ \longrightarrow \\ \longleftarrow \\ U^{-1} \end{array} \begin{pmatrix} |P_{0\ldots0}|^2 \\ |P_{0\ldots1}|^2 \\ \vdots \\ |P_{1\ldots1}|^2 \end{pmatrix} \tag{8}$$

Let $co(\Gamma_U) = c$. Then, from [20], we can deduce that exactly 2^c of the autocorrelation values $A_{\mathbf{k}}$ are different from zero, and furthermore that, for those $\mathbf{k}$'s, $A_{\mathbf{k}} = \pm 2^n$. Clearly, $A_{0\ldots0} = 2^n$.

We differentiate two cases. First, let $U = H \otimes \cdots \otimes H$. Then, in U there always exists a row i with entries in ± 1 ordered in such a way that, when multiplying by $(A_{0\ldots0}, A_{0\ldots1}, \ldots, A_{1\ldots1})^T$, we get $2^{n/2} 2^c$. By (8), this is $|P_{\mathbf{k}}|^2$, for some $\mathbf{k}$. Then, after normalization, we get 2^c. Clearly, this value is the maximum value that we can obtain, so the theorem is true for $U = H \otimes \cdots \otimes H$.

Now, let any $U \in \{H, N\}^n$ except $H \otimes \cdots \otimes H$. By (8), we can obtain the autocorrelation vector from the power spectrum as $(A_{0\ldots0}, A_{0\ldots1}, \ldots, A_{1\ldots1})^T = U^{-1}(|P_{0\ldots0}|^2, |P_{0\ldots1}|^2, \ldots, |P_{0\ldots0}|^2)^T$. Because of the shape of N, in U^{-1} half of the rows have purely imaginary entries. Since both the $|P_{\mathbf{k}}|^2$'s and the $A_{\mathbf{k}}$'s are real, we have that the corresponding $A_{\mathbf{k}}$'s must be equal to zero. Trivially, the rows in U^{-1} that have purely imaginary entries correspond to the columns in U with purely imaginary entries, the rest being real. Now, as in the previous case, we can always find a row i such that, when multiplying by $(A_{0...0}, A_{0...1}, \ldots, A_{1...1})^T$, we get $2^{n/2}2^c$, and this is the maximum value we can get.

Definition 11. *[18, 16, 10, 8] The* Peak-to-Average Power Ratio [3] *of a vector $s \in \mathbb{C}^{2^n}$, with respect to a set of $2^n \times 2^n$ unitary transforms* **T**, *is*

$$PAR_{\mathbf{T}}(s) = 2^n \max_{\substack{U \in \mathbf{T} \\ \mathbf{k} \in \mathbb{Z}_2^n}} (|P_{U,\mathbf{k}}|^2), \quad \text{where } P_U = (P_{U,\mathbf{k}}) = Us \in \mathbb{C}^{2^n} . \quad (9)$$

Corollary 1. *Let $p(\mathbf{x})$ be a quadratic Boolean function, and let $s = (-1)^{p(\mathbf{x})}$. Then, by theorem 3, the logarithm (base 2) of the Peak-to-Average Power Ratio of s, $\log_2(PAR_{\mathbf{T}}(s))$, is equal to the degree of the interlace polynomial q, Q^{HN}, or Q, for $\mathbf{T} = \{I, H\}^n$, $\{H, N\}^n$ or $\{I, H, N\}^n$, respectively.*

Remark: It follows, trivially, that $\deg(q) \leq \deg(Q)$ and $\deg(Q^{HN}) \leq \deg(Q)$. [4]

Lemma 7. *Let G be a graph which is the union of two disjoint graphs, G_1 and G_2, in n and m variables respectively. Then, $q^{n+m}(G; z) = q^n(G_1; z)q^m(G_2; z)$, $Q^{n+m}(G; z) = Q^n(G_1; z)Q^m(G_2; z)$ and $Q_{n+m}^{HN}(G; z) = Q_n^{HN}(G_1; z)Q_m^{HN}(G_2; z)$.*

It follows from corollary 1 and lemma 7 that PAR is *multiplicative* on the union of disjoint graphs.

The "GDJ sequences", defined in [15], can be identified, without loss of generality, with the path graph. Using a result of [21], we prove Conjectures 1 – 3 of [15].

Lemma 8. *(Conjecture 1 of [15]) PAR_H of the path graph is 1.0 for even n and 2.0 for odd.*

Proof. The proof for the number of flat spectra w.r.t. the path graph [21] tells us that

$$D_n = v_0 D_{n-1} + D_{n-2} \bmod 2 , \quad (10)$$

where D is the determinant of the generic modified adjacency matrix of the line on n variables w.r.t. $\{H, N\}^n$. As in this case $v_i = 0$ for all i, we get that $D_n = D_{n-2}$, mod 2. Expanding for n even, $D_n = D_2 = 1$; for n odd, $D_n = D_1 = 0$. From the proof of Q^{HN} for the path graph (lemma 4), we know that the rank of the matrix cannot be lower than $n - 1$.

---

[3] PAR_{IHN} can be used as a lower bound on PAR_l (where PAR_l is the PAR w.r.t. the infinite set of local unitary transforms - see [18, 8]) and therefore as an upper bound on the *geometric measure* $1 - \Gamma_{\max}^2$ (after normalisation), because $PAR_l = 2^n(1 - \Gamma_{\max}^2)$. The geometric measure is an *entanglement monotone* for quantum states (see [25, 24]).

[4] It also follows from previous comments that $n - \deg(Q)$ can be used as an upper bound on the log form of the geometric measure of quantum entanglement, $E_{\log_2}$, as defined in [24], where $E_{\log_2} \leq n - \deg(Q)$.

Lemma 9. *(Conjecture 2 of [15])* PAR_N *of the path graph is 1.0 for* $n \neq 2$ *mod 3 and 2.0 for* $n = 2$ *mod 3.*

Proof. From (10), and as in this case $v_i = 1$ for all i, we get that $D_n = D_{n-1} + D_{n-2}$ mod 2. It is clear that $D_1 = 1$, $D_2 = 0$ and $D_3 = D_2 + D_1 = 1$. For $n > 3$, $D_n = D_{n-1} + D_{n-2} = D_{n-2} + D_{n-3} + D_{n-2} = D_{n-3}$. Expanding the argument, when $n = 0$ mod 3, $D_n = D_3 = 1$; when $n = 1$ mod 3, $D_n = D_1 = 1$; when $n = 2$ mod 3, $D_n = D_2 = 0$.

Corollary 1. *(Conjecture 3 of [15]) ¿From lemmas 8 and 9 it follows that* PAR_H *and* PAR_N *of the path graph are both 1.0 for* n *even,* $n \neq 2$ *mod 3.*

Lemma 10. $PAR_N(s) = |q(-1)|$. *Furthermore, for quadratics,* PAR_N *is pivot-invariant.*

Proof. In [5] and [1] it is shown that $q(-1) = (-1)^r 2^{n-r}$, where r is the rank of $\Gamma + I$. From [20] and the results above it follows that $PAR_N(s) = |q(-1)|$. The last part follows from [22], as we prove there that the pivot orbit lies within $\{I, H\}^n$, and q is invariant w.r.t. this set.

Theorem 4. *Let* $p(\mathbf{x})$ *be a quadratic Boolean function. Let* $s = (-1)^{p(\mathbf{x})}$, *and let* $U \in \mathbf{T}$, *where* $\mathbf{T} = \{I, H, N\}^n$ *or one of its subsets. Then, the* power spectrum $|P_U|^2 = (|P_{U,\mathbf{k}}|^2)$, *where* $P_U = (P_{U,\mathbf{k}}) = Us \in \mathbb{C}^{2^n}$ *is the spectrum of* p *under* U, *is either flat (one-valued) or two-valued. Furthermore, if it is two-valued, one of the values is 0 and the other value is equal to* $2^{co(\Gamma_U)}$.

Proof. We prove that the power-spectrum is one or two-valued w.r.t. $\{H, N\}^n$ as the case for $\{I, H, N\}^n$ then follows trivially. Firstly, we characterise the possible sets of spectral values produced via the action of the transforms H_0 and N_0 on any Boolean function. Then we show that, for a quadratic, the subsequent actions of H_1 or N_1 on these partial spectra produce identically-structured sets of values for the power spectra which can be one or two-valued with one value equal to zero. Further action by H or N on the remaining tensor positions leaves the structure of these sets invariant. The evaluation to the corank follows from theorem 3.

Definition 12. *(see [8]) An* independent set (IS) *of a graph* G *is a subset of the set of vertices* V *such that no two vertices in the subset are adjoint.*

Lemma 11. $PAR_{IH} = 2^{\max|IS|}$.

Proof. $log_2(PAR_{IH})$ is, as we saw in theorem 3, the maximal value of the corank of the modified adjacency matrix over all transforms in $\{I, H\}^n$. But the corank is maximal when the graph has been completely separated, and its value will tell us the least possible number of fixings we have to do to get a completely disjoint graph. But this is exactly the maximal size independent set, $\max|IS|$, that is, the maximal number of variables that such a graph can have.

Corollary 2. $deg(q) = \max|IS|$.

Proof. By corollary 1 and lemma 11.

Furthermore:

Theorem 5. *[8] If the maximum independent set over all graphs in the LC orbit of the graph G has size $\lambda(G)$, then all functions corresponding to graphs in the orbit will have $PAR_{IHN} = 2^{\lambda(G)}$.*

Corollary 3. $deg(Q) = \lambda(G)$.

Proof. By corollary 1 and theorem 5.

Definition 13. *[12,17] The* Multivariate Merit Factor (MMF) *and the* Clifford Merit Factor (CMF) *are* $MMF = \dfrac{4^n}{2\sigma}$, *and* $CMF = \dfrac{6^n}{2E}$, *where*

$$2\sigma = \sum_{\substack{U \in \{H,N\}^n \\ \mathbf{k} \in \mathbb{Z}_2^n}} |P_{U,\mathbf{k}}|^4 - 4^n, \qquad 2E = \sum_{\substack{U \in \{I,H,N\}^n \\ \mathbf{k} \in \mathbb{Z}_2^n}} |P_{U,\mathbf{k}}|^4 - 6^n .$$

Corollary 4. $MMF = \dfrac{4^n}{2^n Q^{HN}(4) - 4^n}$, *and* $CMF = \dfrac{6^n}{2^n Q(4) - 6^n}$.

Proof. By theorems 3 and 4, and the fact that $\sum_{\mathbf{k}} |P_{U,\mathbf{k}}|^2 = 2^n$.

σ and E are derived from their respective L_4-norms, (e.g. L_4-norm$_{IHN} = (2^n Q(4))^{\frac{1}{4}}$). We can generalise the result to express the L_p norms in terms of the interlace polynomials.

Lemma 12. *The L_p-norms w.r.t. $\{I,H,N\}^n$, $\{H,N\}^n$, and $\{I,H\}^n$ for all $1 \leq p < \infty$, are,*

$$L_p\text{-}norm_{IHN} = (2^n Q(2^{\frac{p-2}{2}} + 2))^{\frac{1}{p}},$$
$$L_p\text{-}norm_{HN} = (2^n Q^{HN}(2^{\frac{p-2}{2}} + 2))^{\frac{1}{p}}$$
$$L_p\text{-}norm_{IH} = (2^n q(2^{\frac{p-2}{2}} + 1))^{\frac{1}{p}},$$

respectively.

Theorem 4, together with theorem 3, tell us that, for quadratics, the interlace polynomial encapsulates much of the information about the spectrum. But for higher degree Boolean functions (i.e. hypergraphs), the number of values of the spectrum grows with the number of variables, and concretely, for each function, with the number of variables we have to fix to get a quadratic function. So, for higher-degree functions, we lose information by just considering the maximum of the spectrum - we require a more detailed generalisation of the interlace polynomial. We defer the complete solution of this problem to future work but offer an initial generalisation to hypergraphs below from which, by theorem 3, we can still compute the number of flat spectra and the PAR, and that preserves the property $Q(G) = Q(G_1)Q(G_2)$, if G_1 and G_2 are disjoint hypergraphs:

Definition 14. *The interlace polynomial[5] of a hypergraph is*

$$Q = \sum_{U \in \{I,H,N\}^n} (z-2)^{\log_2(\max_{\mathbf{k}} |P_{U,\mathbf{k}}|^2)}$$

---

[5] Note that, in general, it will not be really a polynomial, because some of the exponents might be non-integer, and even irrational. In some cases, though, they are rational, so we can, by multiplying by a certain $(z-2)^l$, get a polynomial.

6   Two-Variable Interlace Polynomials

6.1   Interlace Polynomial $q(x, y)$

We offer a definition of $q(x, y)$ equivalent to the one proposed in [4]:

Definition 15. *The 2-variable interlace polynomial $q(G; x, y)$ of a graph G in n variables is defined as*

$$q(G; x, y) = \sum_{U \in \{I, H\}^n} (x - 1)^{rk(\Gamma_U)} (y - 1)^{co(\Gamma_U)} , \qquad (11)$$

where $co(\Gamma_V)$ and $rk(\Gamma_V)$ stand respectively for corank and rank of the modified adjacency matrix of the graph w.r.t. $V \in \{I, H, N\}^n$, Γ_V, obtained by erasing the rows and columns whose indices are in $\mathbf{R_I}$ (see definition 3).

Remark: $q(2, y) = q(y)$. Therefore, $deg(q(2, y)) = log_2(\text{PAR}_{IH})$.

Lemma 13. *$q(x, 1)$ gives the number of flat spectra of the function w.r.t. $\{I, H\}^n$ partitioned according to their weight in I's. Furthermore, $n - deg(q(x, 1))$ is the least number of fixings that we have to do to get a flat spectrum.*

Proof. From the definition

$$q(x, 1) = \sum_{U, co(\Gamma_U) = 0} (x - 1)^{rk(\Gamma_U)} .$$

Now, when $co(\Gamma_U) = 0$, it is clear that the matrix has full rank, and therefore that $rk(\Gamma_U) = n - |R_I|$. Thus, $q(x, 1)$ tells you where to locate the flat spectra, and the degree is maximal when the number of fixings is minimal.

Lemma 14. *$q(1, y)$ gives the number of independent sets partitioned according to their size (i.e. according to their weight in I's). Furthermore, $deg(q(1, y))$ gives the maximal size of an independent set.*

Proof.

$$q(1, y) = \sum_{U, rk(\Gamma_U) = 0} (y - 1)^{co(\Gamma_U)} .$$

But the only subgraph such that $rk(\Gamma_U) = 0$ is the empty graph. Thus, $q(1, y)$ tells us how to separate totally the graph and how many fixings we have to make to do so, and as $co(\Gamma_U) = n - |R_I|$, the degree of $q(1, y)$ tells us the least number of fixings we have to do to get a completely disjoint graph; i.e., how many variables can have such a graph.

Lemma 15. *$deg(q(2, y)) = deg(q(1, y))$.*

Proof. As can be deduced from [4], the degree of $q(1, y)$ is equal to $\max|IS|$. The lemma follows by taking into account lemma 11.

Remark: $q(x, y)$ gives us the bentness of the function. That is, if x^n appears in $q(x, y)$, then the function is bent. Otherwise, it is not bent.

Lemma 16. *The following equality holds:*

$$q_x(G; 1, 1) = \#\{U : rk(\Gamma_U) = 1, co(\Gamma_U) = 0\} = 0 \ ,$$

where the subindex means derivative w.r.t. x.

Proof. The first equality follows trivially. The second equality follows from the fact that $rk(U) + co(U) = dim(U)$, and that any modified adjacency matrix (w.r.t. $\{I, H\}^n$) of dimension 1 is the 1×1 matrix (0), which has rank 0 and corank 1.

6.2   Interlace Polynomial $Q(x, y)$

Definition 16. *The 2-variable interlace polynomial $Q(G; x, y)$ of a graph G in n variables is defined as*

$$Q(G; x, y) = \sum_{V \in \{I, H, N\}^n} (x - 2)^{rk(\Gamma_V)} (y - 2)^{co(\Gamma_V)} \ , \qquad (12)$$

where $co(\Gamma_V)$ and $rk(\Gamma_V)$ stand respectively for the corank and rank of the modified adjacency matrix of the graph w.r.t. $V \in \{I, H, N\}^n$, Γ_V, obtained by erasing the rows and columns whose indices are in $\mathbf{R_I}$, as before, and then substituting 0 by $v_i \in GF(2)$ in those indices $i \in \mathbf{R_H} \cup \mathbf{R_N}$, where $v_i = 1$ iff $i \in \mathbf{R_N}$ (see definition 7).

Remark: $Q(3, y) = Q(y)$ as defined in (6).

Lemma 17. $Q(2, y) = q(1, y - 1)$.

Proof. Clearly,

$$Q(2, y) = \sum_{V, rk(\Gamma_V) = 0} (y - 2)^{co(\Gamma_V)} \ .$$

The only subgraph such that $rk(\Gamma_V) = 0$ is the empty graph. Moreover, $rk(\Gamma_V) = 0$ iff $\mathbf{R_N} = \emptyset$. Thus, $Q(2, y) = \sum_{V \in \{I, H\}^n, rk(\Gamma_V) = 0} (y - 2)^{co(\Gamma[S])} = q(1, y - 1)$.

Lemma 18. *The following equalities hold:*

$$Q_x(G; 2, 2) = \#\{V : rk(\Gamma_V) = 1, co(\Gamma_V) = 0\} = n \ ,$$
$$Q_{x,y}(G; 2, 2) = \#\{V : rk(\Gamma_V) = 1, co(\Gamma_V) = 1\} = \#edges = \# \ terms \ p(\mathbf{x}) \ ,$$

where the subindex means derivative w.r.t. the corresponding variable.

Proof. For both equations, the first equality follows trivially. For the first equation, the second equality comes from the fact that the modified adjacency matrix cannot have rank 1 and corank 0 unless in the case $V = N_j$ for some j, so there are n possibilities. For the second equation, as all the modified adjacency matrices are symmetric, then both the rank and the corank being 1 can only happen in the case $\Gamma_V = \begin{pmatrix} 1 & 1 \\ 1 & 1 \end{pmatrix}$; that is, when $V = N_i N_j$ for each edge ij. Now, the number of edges is precisely the number of terms of the quadratic boolean function $p(\mathbf{x})$.

7 Conclusions

We have shown that one and two-variable interlace polynomials can be used to summarise many of the spectral properties of quadratic boolean functions with respect to a special subset of tensor transforms. We also derived interlace polynomials for the clique and clique-line-clique functions. We then defined the HN-interlace polynomial, and derived its form for the clique, the line, and the clique-line-clique functions. We proved some conjectures of [15], and presented other spectral interpretations of the interlace polynomial. We also generalised the interlace polynomial to hypergraphs.

References

1. M. Aigner and H. van der Holst, "Interlace Polynomials", *Linear Algebra and its Applications*, **377**, pp. 11–30, 2004.
2. R. Arratia, B. Bollobas, and G. B. Sorkin, "The Interlace Polynomial: a new graph polynomial", *Proc. 11th Annual ACM-SIAM Symp. on Discrete Math.*, pp. 237–245, 2000.
3. R. Arratia, B. Bollobas, and G. B. Sorkin, "The Interlace Polynomial of a Graph", *J. Combin. Theory Ser. B*, **92**, 2, pp. 199–233, 2004. http://arxiv.org/pdf/math/0209045 , v2, 13 Aug. 2004.
4. R. Arratia, B. Bollobas and G. B. Sorkin, "Two-Variable Interlace Polynomial", *Combinatorica*, **24**, 4, pp. 567–584, 2004. http://arxiv.org/pdf/math/0209054, v3, 13 Aug. 2004.
5. P. N. Balister, B. Bollobas, J. Cutler and L. Pebody, "The Interlace Polynomial of Graphs at -1", *Europ. J. Combinatorics*, **23**, pp. 761–767, 2002.
6. A. Bouchet, "Tutte-Martin Polynomials and Orienting Vectors of Isotropic Systems", *Graphs Combin.*, **7**, pp. 235–252, 1991.
7. A. R. Calderbank, E. M. Rains, P. W. Shor and N. J. A. Sloane, "Quantum Error Correction Via Codes Over GF(4)," *IEEE Trans. on Inform. Theory*, **44**, pp. 1369–1387, 1998, http://xxx.soton.ac.uk/pdf/quant-ph/?9608006.
8. L. E. Danielsen, "On Self-Dual Quantum Codes, Graphs, and Boolean Functions," *Master's Thesis*, Selmer Centre, Inst. for Informatics, University of Bergen, Bergen, Norway, March 2005. http://arxiv.org/pdf/quant-ph/0503236.
9. L. E. Danielsen, T. A. Gulliver and M. G. Parker, "Aperiodic Propagation Criteria for Boolean Functions," *Accepted for Inform. Comput.*, Sept. 2005. http://www.ii.uib.no/ matthew/apcpaper.pdf.

10. L. E. Danielsen and M. G. Parker, "Spectral Orbits and Peak-to-Average Power Ratio of Boolean Functions with respect to the $\{I, H, N\}^n$ Transform", *SETA'04, Sequences and their Applications, Seoul*, Proceedings of SETA04, Lecture Notes in Computer Science, LNCS 3486, 2005. http://xxx.soton.ac.uk/ps/cs.IT/0504102.
11. D. G. Glynn, "On Self-Dual Quantum Codes and Graphs", *Submitted to the Electronic Journal of Combinatorics*, http://homepage.mac.com/dglynn/quantum_files/Personal3.html, April 2002.
12. T. A. Gulliver and M. G. Parker, "The Multivariate Merit Factor of a Boolean Function", IEEE ITSOC Information Theory Workshop 2005 on Coding and Complexity, Rotorua, New Zealand, 29th Aug. - 1st Sept., 2005. http://www.ii.uib.no/~matthew/NZRecursionsCamera1.pdf.
13. M. Hein, J. Eisert and H. J. Briegel, "Multi-Party Entanglement in Graph States", *Phys. Rev. A*, **69**, 6, 2004. http://xxx.soton.ac.uk/pdf/quant-ph/0307130.
14. F. J. MacWilliams and N. J. A. Sloane, **The Theory of Error-Correcting Codes**, Amsterdam: North-Holland, 1977.
15. M. G. Parker, "Constabent Properties of Golay-Davis-Jedwab Sequences", *ISIT2000, Sorrento, Italy*, June, 2000. http://www.ii.uib.no/~matthew/BentGolayISIT.ps.
16. M. G. Parker, "Generalised S-Box Nonlinearity", *NESSIE Public Document - NES/DOC/UIB/WP5/020/A*, https://www.cosic.esat.kuleuven.ac.be/nessie/reports/phase2/SBoxLin.pdf, 11 Feb., 2003.
17. M. G. Parker, "Univariate and Multivariate Merit Factors", *SETA'04, Sequences and their Applications, Seoul, Proceedings of SETA04, Lecture Notes in Computer Science, LNCS 3486, Springer-Verlag, 2005*, http://www.ii.uib.no/~matthew/seta04-mf.pdf.
18. M.G. Parker and V. Rijmen, "The Quantum Entanglement of Binary and Bipolar Sequences", short version in *Sequences and Their Applications*, Discrete Mathematics and Theoretical Computer Science Series, Springer-Verlag, 2001, long version at http://xxx.soton.ac.uk/ps/quant-ph/?0107106 or http://www.ii.uib.no/~matthew/BergDM3.ps, June 2001.
19. M. G. Parker and C. Tellambura, "A Construction for Binary Sequence Sets with Low Peak-to-Average Power Ratio", *Technical Report No 242, Dept. of Informatics, University of Bergen, Norway*, http://www.ii.uib.no/publikasjoner/texrap/ps/2003-242.ps, Feb. 2003.
20. C. Riera and M. G. Parker, "Generalised Bent Criteria for Boolean Functions", accepted for IEEE Trans Inform. Theory, July, 2005. http://xxx.soton.ac.uk/pdf/cs.IT/0502049.
21. C. Riera, G. Petrides and M. G. Parker, "Generalised Bent Criteria for Boolean Functions (II)". http://xxx.soton.ac.uk/pdf/cs.IT/0502050.
22. C. Riera and M. G. Parker, "On Pivot Orbits of Boolean Functions", *Proceedings of the Fourth International Workshop on Optimal Codes and Related Topics (OC 2005), Pamporovo, Bulgaria*, June 2005. http://www.ii.uib.no/~matthew/octalk4.ps.
23. O. S. Rothaus, "On Bent Functions", *J. Comb. Theory*, **20A**, pp. 300–305, 1976.
24. T. Wei, M. Ericsson, P. M. Goldbart, and W. J. Munro, "Connections between relative entropy of entanglement and geometric measure of entanglement", *Quantum Information and Computation*, **4**, pp. 252–272, 2004. http://xxx.soton.ac.uk/pdf/quant-ph/0405002.
25. T. Wei, and P. M. Goldbart, "Geometric measure of entanglement and applications to bipartite and multipartite quantum states", *Physical Review*, **A 68**, 2003. http://xxx.soton.ac.uk/pdf/quant-ph/0307219.

Improved Bounds on Weil Sums over Galois Rings and Homogeneous Weights

San Ling[1] and Ferruh Özbudak[2]

[1] Division of Mathematical Sciences
School of Physical and Mathematical Sciences
Nanyang Technological University
Block 5 Level 3, 1 Nanyang Walk, Singapore 637616, Republic of Singapore
lingsan@ntu.edu.sg
[2] Department of Mathematics, Middle East Technical University
İnönü Bulvarı, 06531, Ankara, Turkey
ozbudak@metu.edu.tr

Abstract. We generalize a recent improvement for the bounds of Weil sums over Galois rings of characteristic p^2 to Galois rings of any characteristic p^l. Our generalization is not as strong as for the case p^2 and we indicate the reason. We give a class of homogeneous weights, including the homogeneous weight defined by Constantinescu and Heise, and we show their relations. We also give an application of our improvements on the homogeneous weights of some codewords.

1 Introduction

The Weil-Carlitz-Uchiyama bound on exponential sums over finite fields is a well-known result that has found many useful applications in coding theory and sequence design [An, B-M, He, Si, T]. An analog of this bound in the case of Galois rings was obtained by Kumar *et al.* in [K-H-C]. An improved bound for a related Weil-type exponential sum over Galois rings of characteristic 4 was later obtained by Helleseth *et al.* [H-K-M-S], leading to sharp lower bounds on the minimum distance of the binary Kerdock and Delsarte-Goethals codes. Recently, an analog of this improved upper bound was obtained for Galois rings of characteristic p^2 (cf. [L-O]), for any prime p, along with some applications [L-O2].

In this paper, we generalize the results of [L-O] to Galois rings of any characteristic p^l and explore their applications. While the general result is not as strong as for the case of characteristic p^2, the reason for this weakness is identified and, when $l \geq 3$, improvements upon the general result may be obtained in some special cases. A new class of homogeneous weights, which includes the homogeneous weight introduced by Constantinescu and Heise [C-H], is given and their relations studied. Just as the homogeneous weight of Constantinescu and Heise is naturally connected to exponential sums, so are the new homogeneous weights introduced in this paper. We also give an application of our improvements on the homogeneous weights of some codewords. The improvement obtained in this

Ø. Ytrehus (Ed.): WCC 2005, LNCS 3969, pp. 412–426, 2006.
© Springer-Verlag Berlin Heidelberg 2006

paper permits to establish new bounds on homogeneous weights of certain code-words.

The paper is organized as follows. In Section 2 we introduce the new class of homogeneous weights and study their relations. We consider the generalization of the results of [L-O] and the applications in Section 3. We fix the following conventions throughout the paper:

- p: a prime number
- $m \geq 2$, $l \geq 2$: positive integers
- $n = p^m - 1$
- $\mathbb{N}$: the set of natural numbers
- $\mathbb{Z}$: the ring of integers
- $\mathbb{F}_p$, $\mathbb{F}_{p^m}$: finite fields of cardinality p and p^m
- $\mathrm{tr}_m : \mathbb{F}_{p^m} \to \mathbb{F}_p$: trace map from $\mathbb{F}_{p^m}$ onto $\mathbb{F}_p$
- $\mathbb{Z}_p$: the ring of p-adic integers
- $\mathbb{Q}_p$: the field of p-adic numbers
- K_m: the unramified extension of $\mathbb{Q}_p$ having all n-th roots of unity with $[K_m : \mathbb{Q}_p] = m$
- $\mathrm{Tr}_{K_m/\mathbb{Q}_p} : K_m \to \mathbb{Q}_p$: trace map from K_m onto $\mathbb{Q}_p$
- $\Gamma = \{0\} \cup \{$all n-th roots of unity in $K_m\}$
- $\mathfrak{O}_{K_m}$: the ring of integers of K_m
- ξ: a primitive n-th root of unity in K_m
- $\mathrm{GR}(p^l, m)$: Galois ring of characteristic p^l with cardinality p^{lm}
- $\mathbb{Z}_{p^l}$: the ring of integers modulo p^l
- $\mathrm{Tr}_m : \mathrm{GR}(p^l, m) \to \mathbb{Z}_{p^l}$: trace map from $\mathrm{GR}(p^l, m)$ onto $\mathbb{Z}_{p^l}$
- $\eta : \mathfrak{O}_{K_m} \to \mathfrak{O}_{K_m}/p\mathfrak{O}_{K_m} \cong \mathbb{F}_{p^m}$: reduction modulo p map in $\mathfrak{O}_{K_m}$
- $\eta_l : \mathfrak{O}_{K_m} \to \mathfrak{O}_{K_m}/p^l\mathfrak{O}_{K_m} \cong \mathrm{GR}(p^l, m)$: reduction modulo p^l map in $\mathfrak{O}_{K_m}$
- $\Gamma_m = \eta_l(\Gamma)$: Teichmüller set in $\mathrm{GR}(p^l, m)$
- $\beta = \eta_l(\xi)$: primitive n-th root of unity in $\mathrm{GR}(p^l, m)$
- $\omega = \eta(\xi)$: primitive n-th root of unity in $\mathbb{F}_{p^m}$
- $\mu : \mathrm{GR}(p^l, m) \to \mathrm{GR}(p^l, m)/p\mathrm{GR}(p^l, m) \cong \mathbb{F}_{p^m}$: reduction modulo p map in $\mathrm{GR}(p^l, m)$

Note that $\eta = \mu \circ \eta_l$.

2 Homogeneous Weights

In this section we introduce a class of homogeneous weights and we give a relation among them.

For each $u \in \mathbb{Z}_{p^l}$ and $0 \leq t \leq l - 1$, we have

$$\sum_{x \in p^t \mathbb{Z}_{p^l}} e^{\frac{2\pi i u x}{p^l}} = \begin{cases} p^{l-t} & \text{if } u \in p^{l-t}\mathbb{Z}_{p^l} \\ 0 & \text{if } u \notin p^{l-t}\mathbb{Z}_{p^l}. \end{cases}$$

Therefore for $1 \leq t \leq l - 1$

$$\sum_{x \in \mathbb{Z}_{p^l} \setminus p^t \mathbb{Z}_{p^l}} e^{\frac{2\pi i u x}{p^l}} = \begin{cases} p^l - p^{l-t} & \text{if } u = 0 \\ -p^{l-t} & \text{if } u \in p^{l-t}\mathbb{Z}_{p^l} \setminus \{0\} \\ 0 & \text{if } u \notin p^{l-t}\mathbb{Z}_{p^l} \end{cases}$$

and

$$p^{l-1} - p^{l-t-1} - \frac{1}{p} \sum_{x \in \mathbb{Z}_{p^l} \setminus p^t \mathbb{Z}_{p^l}} e^{\frac{2\pi i u x}{p^l}} = \begin{cases} 0 & \text{if } u = 0 \\ p^{l-1} & \text{if } u \in p^{l-t}\mathbb{Z}_{p^l} \setminus \{0\} \\ p^{l-1} - p^{l-t-1} & \text{if } u \notin p^{l-t}\mathbb{Z}_{p^l}. \end{cases}$$

Definition 1. *For $1 \leq t \leq l - 1$, the weight $w_{(l,t)}$ on $\mathbb{Z}_{p^l}$ is defined as*

$$w_{(l,t)}(u) = \begin{cases} 0 & \text{if } u = 0 \\ p^{l-1} & \text{if } u \in p^{l-t}\mathbb{Z}_{p^l} \setminus \{0\} \\ p^{l-1} - p^{l-t-1} & \text{if } u \notin p^{l-t}\mathbb{Z}_{p^l}. \end{cases}$$

Note that

$$w_{(l,1)}(u) = \begin{cases} 0 & \text{if } u = 0 \\ p^{l-1} & \text{if } u \in p^{l-1}\mathbb{Z}_{p^l} \setminus \{0\} \\ p^{l-1} - p^{l-2} & \text{if } u \notin p^{l-1}\mathbb{Z}_{p^l} \end{cases}$$

and

$$w_{(l,l-1)}(u) = \begin{cases} 0 & \text{if } u = 0 \\ p^{l-1} & \text{if } u \in p\mathbb{Z}_{p^l} \setminus \{0\} \\ p^{l-1} - 1 & \text{if } u \notin p\mathbb{Z}_{p^l}. \end{cases}$$

The weight $w_{(l,1)}$ is the homogeneous weight defined by Constantinescu and Heise [C-H].

For $l \geq 3$ and $1 \leq i \leq l - 2$, since $p^{l-i}\mathbb{Z}_{p^l} \subseteq p^{l-(i+1)}\mathbb{Z}_{p^l}$ we have

$$w_{(l,i+1)}(u) - w_{(l,i)}(u) = \begin{cases} 0 & \text{if } u \in p^{l-i}\mathbb{Z}_{p^l} \\ p^{l-i-1} & \text{if } u \in p^{l-(i+1)}\mathbb{Z}_{p^l} \setminus p^{l-i}\mathbb{Z}_{p^l} \\ p^{l-i-1} - p^{l-i-2} & \text{if } u \notin p^{l-(i+1)}\mathbb{Z}_{p^l}. \end{cases} \qquad (1)$$

Definition 2. *For $1 \leq t \leq l - 1$ and $u \in \mathbb{Z}_{p^l}$, let $u^{(t)} \in \{0, 1 \ldots, p^t - 1\}$ be the integer such that $u^{(t)} \equiv u \mod p^t$. By convention we define $u^{(l)} = u$. Note that $u^{(t)} \in \mathbb{Z}_{p^t}$.*

We observe that for $1 \leq j \leq l - 1$ we have $u \in p^j\mathbb{Z}_{p^l} \iff u^{(j)} = 0$; and for $2 \leq j \leq l$ we have $u \in p^{j-1}\mathbb{Z}_{p^l} \iff u^{(j)} \in p^{j-1}\mathbb{Z}_{p^j}$.

Definition 3. *For $l \geq 3$, $2 \leq t \leq l - 1$ and $1 \leq j \leq t - 1$, the weight $w_{(t,j)}$ on $\mathbb{Z}_{p^l}$ is defined as*

$$w_{(t,j)}(u) = w_{(t,j)}(u^{(t)}),$$

where $w_{(t,j)}(u^{(t)})$ is determined using Definition 1 applied for $\mathbb{Z}_{p^t}$.

For $l \geq 3$, $1 \leq i \leq l - 2$ and $u \in \mathbb{Z}_{p^l}$, using (1) and Definition 3 we obtain

$$w_{(l,i+1)}(u) = w_{(l,i)}(u) + w_{(l-i,1)}(u). \qquad (2)$$

Theorem 1. *Let $u \in \mathbb{Z}_{p^l}$. For $2 \le t \le l$ we have*

$$w_{(t,t-1)}(u) = w_{(t,1)}(u) + w_{(t-1,1)}(u) + \cdots + w_{(2,1)}(u). \tag{3}$$

Moreover for $l \ge 3$ and $1 \le t \le l - 2$ we have

$$w_{(l,t)}(u) = w_{(l,l-1)}(u) - w_{(l-t,l-t-1)}(u). \tag{4}$$

Proof. First we prove (3). The case $t = 2$ holds trivially. For $t = 3$, by (2) we have $w_{(3,2)}(u) = w_{(3,1)}(u) + w_{(2,1)}(u)$. For $t \ge u$, by a repeated application of (2) $t - 2$ times, we obtain

$$
\begin{aligned}
w_{(t,t-1)}(u) &= w_{(t,t-2)}(u) + w_{(2,1)}(u) \\
&= w_{(t,t-3)}(u) + w_{(3,1)}(u) + w_{(2,1)}(u) \\
&\vdots \\
&= w_{(t,1)}(u) + w_{(t-1,1)}(u) + \cdots + w_{(2,1)}(u).
\end{aligned}
$$

It remains to consider (4). For $l \ge 3$ and $1 \le t \le l - 2$ let $i = l - 2 - t$. We prove (4) by induction on i for $0 \le i \le l - 3$. For $i = 0$, we have $t = l - 2$ and by (2)

$$
\begin{aligned}
w_{(l,l-1)}(u) &= w_{(l,l-2)}(u) + w_{(2,1)}(u), \\
&= w_{(l,t)}(u) + w_{(l-t,l-t-1)}(u),
\end{aligned}
$$

and hence (4) holds. For $l \ge 4$ and $0 \le i \le l - 4$, assume that (4) holds for i, or equivalently $t = l - 2 - i$. Now we complete the proof by showing that (4) holds for $i + 1$, or equivalently for $t - 1$. By (3) we have

$$
\begin{aligned}
w_{(l-t+1,l-t)}(u) &= w_{(l-t+1,1)}(u) + w_{(l-t,1)}(u) + \cdots + w_{(2,1)}(u), \\
w_{(l-t,l-t-1)}(u) &= w_{(l-t,1)}(u) + \cdots + w_{(2,1)}(u),
\end{aligned}
$$

and hence

$$w_{(l-t+1,l-t)}(u) = w_{(l-t+1,1)}(u) + w_{(l-t,l-t-1)}(u). \tag{5}$$

Adding $w_{(l,t-1)}(u)$ to both sides of (5) we get

$$
\begin{aligned}
w_{(l-t+1,l-t)}(u) + w_{(l,t-1)}(u) &= w_{(l-t+1,1)}(u) + w_{(l,t-1)}(u) \\
&\quad + w_{(l-t,l-t-1)}(u).
\end{aligned} \tag{6}
$$

Note that by (2)

$$w_{(l-t+1,1)}(u) + w_{(l,t-1)}(u) = w_{(l,t)}(u). \tag{7}$$

Therefore by (6) and (7) we have

$$w_{(l-t+1,l-t)}(u) + w_{(l,t-1)}(u) = w_{(l,t)}(u) + w_{(l-t,l-t-1)}(u). \tag{8}$$

By the induction hypothesis at i we have

$$w_{(l,t)}(u) + w_{(l-t,l-t-1)}(u) = w_{(l,l-1)}(u). \tag{9}$$

Using (8) and (9) we complete the proof. $\qquad\square$

Remark 1. Using Theorem 1, for $l \ge 2$ and $1 \le t \le l - 1$, each $w_{(l,t)}(u)$ can be written as a combination of $w_{(2,1)}(u), w_{(3,1)}(u), \ldots, w_{(l,1)}(u)$ using only integer coefficients.

3   An Improvement of Bounds for Weil Sums

In this section we generalize the improvement of [L-O] to any characteristic p^l and we indicate the reason why our generalization is not as strong as the case of characteristic p^2.

Throughout this section the homogeneous weight $w_{(l,1)}$ of Section 2 will also be denoted by w_{hom}. The Frobenius map Frob on $\text{GR}(p^l, m)$ sends $u_0 + pu_1 + \cdots + p^{l-1}u_{l-1} \in \text{GR}(p^l, m)$ with $u_0, \ldots, u_{l-1} \in \Gamma_m$ to $u^p + pu_1^p + \cdots + p^{l-1}u_{l-1}^p \in \text{GR}(p^l, m)$. It is extended to the map on the polynomial ring $\text{GR}(p^l, m)[x]$ as $\text{Frob}\left(\sum_{i=0}^t g_i x^i\right) = \sum_{i=0}^t \text{Frob}(g_i) x^{ip}$. We recall that a polynomial $g(x) \in \text{GR}(p^l, m)[x]$ is called *non-degenerate* if it cannot be written in the form $g(x) = \text{Frob}(h(x)) - h(x) + u$, where $h(x) \in \text{GR}(p^l, m)[x]$ and $u \in \text{GR}(p^l, m)$.

Let $f(x) \in \text{GR}(p^l, m)[x]$ be a non-degenerate polynomial and let $\hat{f}(x) = f(x) - f(0) \in \text{GR}(p^l, m)[x]$. Let

$$c = \left(0, \text{Tr}_m(\hat{f}(\beta)), \text{Tr}_m(\hat{f}(\beta^2)), \ldots, \text{Tr}_m(\hat{f}(\beta^n))\right) \in \mathbb{Z}_{p^l}^{p^m},$$

and for $\lambda \in \mathbb{Z}_{p^l}$ let $c(\lambda) = (\lambda, \ldots, \lambda) + c \in \mathbb{Z}_{p^l}^{p^m}$. Note that if $\text{Tr}_m(f(0)) = \alpha$, then

$$c(\alpha) = \left(\text{Tr}_m(f(0)), \text{Tr}_m(f(\beta)), \text{Tr}_m(f(\beta^2)), \ldots, \text{Tr}_m(f(\beta^n))\right).$$

For $i = 0, \ldots, p^l - 1$ and $\lambda \in \mathbb{Z}_{p^l}$, let N_i denote the number of coordinates of c equal to i and let $N_i(\lambda)$ denote the number of coordinates of $c(\lambda)$ equal to i. By convention we set $N_{p^l} = N_0$. We first prove some technical lemmas.

Lemma 1. *For $0 \leq \lambda \leq p^l - 1$, let $\bar{\lambda} = \lambda^{(l-1)}$, which is defined in Definition 2. We have*

$$\sum_{i=1}^{p^l-1} N_i(\lambda) = p^m - N_{p^l - \lambda}, \tag{10}$$

$$\sum_{i=1}^{p-1} N_{ip^{l-1}}(\lambda) = \left(\sum_{i=1}^{p-1} N_{ip^{l-1} - \bar{\lambda}}\right) + N_{p^l - \bar{\lambda}} - N_{p^l - \lambda}, \tag{11}$$

$$w_{\text{hom}}(c(\lambda)) \equiv p^{l-2}\left(\sum_{i=1}^{p} N_{ip^{l-1} - \bar{\lambda}}\right) - p^{l-1}N_{p^l - \lambda} \mod p^m \tag{12}$$

and if $\bar{\lambda} = 0$,

$$w_{\text{hom}}(c(\lambda)) \equiv -p^{l-1}N_{p^l - \lambda} - p^{l-2}\sum_{\substack{i=1 \\ p^{l-1} \nmid i}}^{p^l-1} N_i \mod p^m. \tag{13}$$

Proof. Note that (10) follows from the observations $\sum_{i=0}^{p^l-1} N_i(\lambda) = p^m$ and $N_0(\lambda) = N_{p^l-\lambda}$. As $\lambda \equiv \bar{\lambda} \mod p^{l-1}$ we have

$$\sum_{i=0}^{p-1} N_{ip^{l-1}}(\lambda) = \sum_{i=0}^{p-1} N_{ip^{l-1}}(\bar{\lambda}).\tag{14}$$

Using (14), $N_0(\lambda) = N_{p^l-\lambda}$ and $N_0(\bar{\lambda}) = N_{p^l-\bar{\lambda}}$ we prove (11). We have

$$w_{\text{hom}}(c(\boldsymbol{\lambda})) = (p^{l-1} - p^{l-2}) \sum_{i=1}^{p^l-1} N_i(\lambda) + p^{l-2} \sum_{\substack{i=1 \\ p^{l-1}|i}}^{p^l-1} N_i(\lambda).\tag{15}$$

Using (10) and (11) we get

$$w_{\text{hom}}(c(\boldsymbol{\lambda})) \equiv -(p^{l-1} - p^{l-2})N_{p^l-\lambda} + p^{l-2}\left(\sum_{i=1}^{p-1} N_{ip^{l-1}-\bar{\lambda}}\right)$$

$$+p^{l-2}N_{p^l-\bar{\lambda}} - p^{l-2}N_{p^l-\lambda} \mod p^m$$

$$\equiv p^{l-2}\left(\sum_{i=1}^{p} N_{ip^{l-1}-\bar{\lambda}}\right) - p^{l-1}N_{p^l-\lambda} \mod p^m.$$

In order to prove the remaining item, we assume that $\bar{\lambda} = 0$. From (12) we have

$$w_{\text{hom}}(c(\boldsymbol{\lambda})) \equiv p^{l-2}\left(\sum_{i=1}^{p} N_{ip^{l-1}}\right) - p^{l-1}N_{p^l-\lambda} \mod p^m.\tag{16}$$

We prove (13) using (16) and $\sum_{i=1}^{p} N_{ip^{l-1}} + \sum_{\substack{i=1 \\ p^{l-1}\nmid i}}^{p^{l-1}} N_i = \sum_{i=1}^{p^l} N_i = p^m$. □

Lemma 2. *We have* $\sum_{\lambda \in \mathbb{Z}_{p^l}} w_{\text{hom}}(c(\boldsymbol{\lambda})) = p^l(p^{l-1} - p^{l-2})p^m$ *and hence*

$$w_{\text{hom}}(c) \equiv -\sum_{\lambda=1}^{p^l-1} w_{\text{hom}}(c(\boldsymbol{\lambda})) \mod p^m.$$

Proof. We have

$$w_{\text{hom}}(c(\boldsymbol{\lambda})) = (p^{l-1} - p^{l-2})(p^m - N_0(\lambda)) + p^{l-2}\sum_{i=0}^{p-1} N_{ip^{l-1}}(\lambda)\tag{17}$$
$$-p^{l-2}N_0(\lambda).$$

Moreover the following identities also hold:

$$\sum_{\lambda \in \mathbb{Z}_{p^l}} (p^m - N_0(\lambda)) = p^m p^l - \sum_{\lambda \in \mathbb{Z}_{p^l}} N_0(\lambda) = p^m(p^l - 1),\tag{18}$$

$$\sum_{\lambda \in \mathbb{Z}_{p^l}} \sum_{i=0}^{p-1} N_{ip^{l-1}}(\lambda) = \sum_{i=0}^{p-1} \sum_{\lambda \in \mathbb{Z}_{p^l}} N_{ip^{l-1}}(\lambda) = pp^m, \tag{19}$$

and

$$\sum_{\lambda \in \mathbb{Z}_{p^l}} N_0(\lambda) = p^m. \tag{20}$$

Using (17), (18), (19) and (20) we get

$$\sum_{\lambda \in \mathbb{Z}_{p^l}} w_{\mathrm{hom}}(\mathbf{c}(\lambda)) = (p^{l-1} - p^{l-2})p^m(p^l - 1) + p^{l-2}pp^m - p^{l-2}p^m$$

$$= p^l(p^{l-1} - p^{l-2})p^m. \qquad \square$$

Lemma 3. *Let* $F(x) \in \mathbb{Q}_p[x]$ *be an integer-valued polynomial of degree at most* $p^l - 1$. *The corresponding map*

$$\widetilde{F} : \mathbb{Z}_p/p^l\mathbb{Z}_p \to \mathbb{Z}_p/p\mathbb{Z}_p \cong \mathbb{F}_p$$
$$\alpha + p^l\mathbb{Z}_p \mapsto F(\alpha) + p\mathbb{Z}_p$$

is well defined. Moreover for any $c \in \mathbb{Z}_p$, *we have*

$$\eta(F(c)) = \widetilde{F}(\eta_l(c)).$$

Proof. As in the proof of [L-O, Lemma 2.9], we may assume $F(x) = \binom{x}{r}$, where $1 \leq r \leq p^l - 1$, and in this case for $\gamma \in \mathbb{Z}_p$ we have

$$F(x + p^l\gamma) - F(x) = \binom{p^l\gamma}{1}\binom{x}{r-1} + \cdots + \binom{p^l\gamma}{r-1}\binom{x}{1} + \binom{p^l\gamma}{r}.$$

For $1 \leq j \leq p^l - 1$, writing

$$\binom{p^l\gamma}{j} = \frac{p^l\gamma(p^l\gamma - 1)\cdots(p^l\gamma - j + 1)}{j!} = p^{\beta(j,\gamma)}u,$$

where $u \in \mathbb{Z}_p \setminus p\mathbb{Z}_p$ is a unit, it suffices to prove that $\beta(j,\gamma) \geq 1$ in order to prove that $\widetilde{F}$ is well-defined.

Note that $j! = p^{\epsilon(j)}u_1$, where $u_1 \in \mathbb{Z}_p \setminus p\mathbb{Z}_p$ is a unit and $\epsilon(j) = \lfloor j/p \rfloor + \lfloor j/p^2 \rfloor + \cdots + \lfloor j/p^{l-1} \rfloor$. Furthermore, $p^l\gamma(p^l\gamma - 1)\cdots(p^l\gamma - j + 1) = p^{\delta(j,\gamma)}u_2$, where $u_2 \in \mathbb{Z}_p \setminus p\mathbb{Z}_p$ is a unit and

$$\delta(j,\gamma) \geq l + \left\lfloor \frac{j-1}{p} \right\rfloor + \left\lfloor \frac{j-1}{p^2} \right\rfloor + \cdots + \left\lfloor \frac{j-1}{p^{l-1}} \right\rfloor.$$

Since $\left\lfloor \frac{j-1}{p^i} \right\rfloor - \left\lfloor \frac{j}{p^i} \right\rfloor \geq -1$, it follows that

$$\beta(j,\gamma) = \delta(j,\gamma) - \epsilon(j) \geq 1.$$

For the final statement, for any $c \in \mathbb{Z}_p$, note that

$$\widetilde{F}(\eta_l(c)) = \widetilde{F}(c + p^l\mathbb{Z}_p) = F(c) + p\mathbb{Z}_p = \eta(F(c)). \qquad \square$$

For $0 \le j \le p^{l-1} - 1$, let

$$F_j(x) = \binom{x + j}{p^{l-1}}.$$

Note that F_j is an integer-valued polynomial of degree p^{l-1}. For $0 \le j \le p^{l-1}-1$, using Lemma 3 we define the $\mathbb{F}_p$-vector $\widetilde{F}_j(c)$ of length p^m obtained by applying $\widetilde{F}_j$ to c componentwise. For $i = 0, \ldots, p-1$ and $0 \le j \le p^{l-1} - 1$, let $N_i(\widetilde{F}_j(c))$ denote the number of coordinates of $\widetilde{F}_j(c)$ equal to i.

Lemma 4. *For integers $1 \le i \le p - 1$ and $0 \le j \le p^{l-1} - 1$, we have*

$$N_i(\widetilde{F}_j(c)) = N_{ip^{l-1}-j} + N_{ip^{l-1}-j+1} + \cdots + N_{ip^{l-1}-j+(p^{l-1}-1)}.$$

Proof. Using Lucas's theorem on binomial coefficients (cf. [Lu]) we obtain for $0 \le x_0, \ldots, x_{l-1}, x_l \le p - 1$ that

$$\binom{x_0 + x_1 p + \cdots + x_{l-1}p^{l-1} + x_l p^l}{p^{l-1}} \equiv x_{l-1} \mod p.$$

Hence for $0 \le x_0, \ldots, x_{l-1} \le p - 1$ and $0 \le j \le p^{l-1} - 1$ we have in modulo p

$$F_j(x_0 + x_1 p + \cdots + x_{l-1}p^{l-1}) \equiv \binom{x_0 + x_1 p + \cdots + x_{l-1}p^{l-1} + j}{p^{l-1}} \tag{21}$$

$$\equiv \begin{cases} x_{l-1} & \text{if } x_0 + \ldots x_{l-2}p^{l-2} + j \le p^{l-1} - 1, \\ x_{l-1} + 1 & \text{if } x_0 + \ldots x_{l-2}p^{l-2} + j \ge p^{l-1}. \end{cases}$$

The proof follows from (21). $\qquad\qquad\square$

Using Lemma 4, let M be the $((p-1)p^{l-1}) \times (p^l - 1)$ matrix with coefficients from $\{0, 1\} \subseteq \mathbb{Z}$ such that

$$\begin{bmatrix} N_1(\widetilde{F}_0(c)) \\ \vdots \\ N_{p-1}(\widetilde{F}_0(c)) \\ \vdots \\ N_1(\widetilde{F}_{p^{l-1}-1}(c)) \\ \vdots \\ N_{p-1}(\widetilde{F}_{p^{l-1}-1}(c)) \end{bmatrix} = M \cdot \begin{bmatrix} N_1 \\ N_2 \\ \vdots \\ N_{p^l-1} \end{bmatrix}. \tag{22}$$

By reordering the rows of M we obtain the $(p-1)p^{l-1} \times (p^l - 1)$ matrix A

$$A = \begin{bmatrix} 1\,1 \ldots 1 & & & \\ & 1 \ldots 1\,1 & & \\ & & \ddots & \\ & & & 1 \ldots 1 \end{bmatrix},$$

where the nonzero entries in each row consist of p^{l-1} consecutive 1's and they are shifted by one position to the right in the next row.

Let C and B be the $p^{l-1} \times (p^{l-1} - 1)$ and $(p-1)p^{l-1} \times (p^{l-1} - 1)$ matrices

$$C = \begin{bmatrix} -I_{p^{l-1}-1} \\ 1 \cdots 1 \end{bmatrix}, \quad B = \begin{bmatrix} C \\ C \\ \vdots \\ C \end{bmatrix},$$

where $I_{p^{l-1}-1}$ is the identity matrix of size $p^{l-1} - 1$. Using similar operations as in the proof of [L-O, Lemma A.1], we obtain that the row space of A over $\mathbb{Q}$ is equal to the row space of $\begin{bmatrix} I_{(p-1)p^{l-1}} & B \end{bmatrix}$ over $\mathbb{Q}$ and any linear combination of the rows of $\begin{bmatrix} I_{(p-1)p^{l-1}} & B \end{bmatrix}$ with coefficients from $\mathbb{Z}$ is a linear combination of the rows of A with coefficients from $\mathbb{Z}$.

In the next proposition, for $\lambda \in \mathbb{Z}_{p^l}$ we relate the homogeneous weight $w_{\text{hom}}(c(\lambda))$ of $c(\lambda)$ to some corresponding p-ary codewords.

Proposition 1. *For $\lambda \in \mathbb{Z}_{p^l}$, there exists $a_{(j,i)}(\lambda) \in \mathbb{Z}$ with $0 \leq i \leq p-1$ and $0 \leq j \leq p^{l-1} - 1$ such that*

$$w_{\text{hom}}(c(\lambda)) \equiv p^{l-2} \sum_{j=0}^{p^{l-1}-1} \sum_{i=0}^{p-1} a_{(j,i)}(\lambda) N_i(\widetilde{F}_j(c)) \mod p^m.$$

Proof. Using Lemma 2, it is enough to prove it for $1 \leq \lambda \leq p^l - 1$. Therefore we assume that $\lambda \neq 0$. Let $0 \leq \lambda_0, \ldots, \lambda_{l-1} \leq p-1$ and $0 \leq \bar{\lambda} \leq p^{l-1} - 1$ be the integers such that

$$\lambda = \lambda_0 + \lambda_1 p + \cdots + \lambda_{l-2} p^{l-2} + \lambda_{l-1} p^{l-1}, \text{ and}$$
$$\bar{\lambda} = \lambda_0 + \lambda_1 p + \cdots + \lambda_{l-2} p^{l-2}.$$

First we consider the case $\bar{\lambda} \neq 0$. Let $1 \leq \gamma \leq p^{l-1} - 1$ and $0 \leq j \leq p-1$ be the integers given by $\gamma = p^{l-1} - \bar{\lambda}$ and $j = p - 1 - \lambda_{l-1}$. We have

$$\sum_{i=1}^{p} N_{ip^{l-1}-\bar{\lambda}} = N_\gamma + N_{\gamma+p^{l-1}} + \cdots + N_{\gamma+(p-1)p^{l-1}} \text{ and} \tag{23}$$

$$N_{p^l - \lambda} = N_{\gamma + jp^{l-1}}. \tag{24}$$

Let $\{f_1, \ldots, f_{(p-1)p^{l-1}}\}$ and $\{e_1, \ldots, e_{p^l-1}\}$ be the standard bases of the vector spaces $\mathbb{Q}^{(p-1)p^{l-1}}$ and $\mathbb{Q}^{p^l-1}$. Using (12), (23) and (24) we obtain that

$$w_{\text{hom}}(c(\lambda)) \equiv p^{l-2} \left(e_\gamma + e_{\gamma+p^{l-1}} + \cdots + e_{\gamma+(p-1)p^{l-1}} - p e_{\gamma+jp^{l-1}} \right)$$
$$\cdot \begin{bmatrix} N_1 \\ \vdots \\ N_{p^l-1} \end{bmatrix} \mod p^m. \tag{25}$$

It follows from the definition of the matrix B that we have

$$\boldsymbol{f}_{\gamma+jp^{l-1}} \cdot \left[I_{(p-1)p^{l-1}} \; B\right] = \boldsymbol{e}_{\gamma+jp^{l-1}} - \boldsymbol{e}_{(p-1)p^{l-1}+\gamma}. \tag{26}$$

Then we have

$$\begin{aligned}
&\boldsymbol{e}_\gamma + \boldsymbol{e}_{\gamma+p^{l-1}} + \cdots + \boldsymbol{e}_{\gamma+(p-1)p^{l-1}} - p\boldsymbol{e}_{\gamma+jp^{l-1}} \\
&\equiv \left(\boldsymbol{f}_\gamma + \boldsymbol{f}_{\gamma+p^{l-1}} + \cdots + \boldsymbol{f}_{\gamma+(p-2)p^{l-1}} - p\boldsymbol{f}_{\gamma+jp^{l-1}}\right) \\
&\quad \cdot \left[I_{(p-1)p^{l-1}} \; B\right].
\end{aligned} \tag{27}$$

For $\bar{\lambda} \neq 0$, the proof follows from (22), (25) and (27).

Next we consider the case $\bar{\lambda} = 0$. Let $1 \leq j \leq p-1$ be the integer given by $p^l - \lambda = jp^{l-1}$. Using (13) we obtain that

$$\begin{aligned}
w_{\mathrm{hom}}(\boldsymbol{c}(\boldsymbol{\lambda})) &\equiv -p^{l-2}\Big((\boldsymbol{e}_1 + \cdots + \boldsymbol{e}_{p^{l-1}-1}) + (\boldsymbol{e}_{p^{l-1}+1} + \cdots + \boldsymbol{e}_{2p^{l-1}-1}) \\
&\quad + \cdots + (\boldsymbol{e}_{(p-1)p^{l-1}+1} + \cdots + \boldsymbol{e}_{(p-1)p^{l-1}+p^{l-1}-1}) \\
&\quad + p\boldsymbol{e}_{jp^{l-1}}\Big) \\
&\quad \cdot \begin{bmatrix} N_1 \\ \vdots \\ N_{p^l-1} \end{bmatrix} \quad \bmod p^m.
\end{aligned} \tag{28}$$

From the definition of the matrix B we have

$$\begin{aligned}
&\boldsymbol{f}_{jp^{l-1}} \cdot \left[I_{(p-1)p^{l-1}} \; B\right] \\
&= \boldsymbol{e}_{jp^{l-1}} + \left(\boldsymbol{e}_{(p-1)p^{l-1}+1} + \cdots + \boldsymbol{e}_{(p-1)p^{l-1}+p^{l-1}-1}\right).
\end{aligned} \tag{29}$$

Using (26) and (29) we get

$$\begin{aligned}
&\Big((\boldsymbol{e}_1 + \cdots + \boldsymbol{e}_{p^{l-1}-1}) + (\boldsymbol{e}_{p^{l-1}+1} + \cdots + \boldsymbol{e}_{2p^{l-1}-1}) \\
&\quad + \cdots + (\boldsymbol{e}_{(p-1)p^{l-1}+1} + \cdots + \boldsymbol{e}_{(p-1)p^{l-1}+p^{l-1}-1}) \\
&\quad + p\boldsymbol{e}_{jp^{l-1}}\Big) \\
&\equiv \Big((\boldsymbol{f}_1 + \cdots + \boldsymbol{f}_{p^{l-1}+1}) + (\boldsymbol{f}_{p^{l-1}+1} + \cdots + \boldsymbol{f}_{2p^{l-1}-1}) \\
&\quad + \cdots + (\boldsymbol{f}_{(p-2)p^{l-1}+1} + \cdots + \boldsymbol{f}_{(p-2)p^{l-1}+p^{l-1}-1}) \\
&\quad + p\boldsymbol{f}_{jp^{l-1}}\Big) \\
&\quad \cdot \left[I_{(p-1)p^{l-1}} \; B\right] \quad \bmod p^m.
\end{aligned} \tag{30}$$

We complete the proof using (22), (28) and (30). $\qquad \square$

Let $\hat{a}_0(x), \hat{a}_1(x), \ldots, \hat{a}_{l-1}(x) \in \Gamma[x]$ be the polynomials defined by

$$\hat{f}(x) = \hat{a}_0(x) + p\hat{a}_1(x) + \cdots + p^{l-1}\hat{a}_{l-1}(x).$$

Let $I_0, \ldots, I_{l-1} \subseteq \mathbb{N} \setminus \{0\}$ be the supports of the polynomials $\hat{a}_0(x), \ldots, \hat{a}_{l-1}(x)$ respectively. For $0 \leq t \leq l-1$, let $I_{t,m} \subseteq \mathbb{N} \setminus \{0\}$ be the set

$$I_{t,m} = \{p^u i \mid i \in I_t,\ 0 \leq u \leq m-1\}. \tag{31}$$

Let A_m be the subset of $\mathbb{N} \setminus \{0\}$ defined as

$$A_m = \bigcup_{k=1}^{p^{l-1}} \{\alpha_1 + \cdots + \alpha_k \mid \alpha_1, \ldots, \alpha_k \in I_{0,m} \cup \cdots \cup I_{l-1,m}\}. \tag{32}$$

Let $A_{m,0}$ be a complete set of representatives of the distinct p-cyclotomic cosets of A_m modulo n. For $0 \le i \le l-1$, let $\bar{a}_i \in \mathfrak{O}_{K_m}[x]$ such that $\eta_l(\bar{a}_i(x)) = \hat{a}_i(x)$ and the support of $\bar{a}_i(x)$ is the same as the support of $\hat{a}_i(x)$. Let $\bar{f}(x) = \bar{a}_0(x) + p\bar{a}_1(x) + \cdots + p^{l-1}\bar{a}_{l-1}(x)$.

Proposition 2. *For $0 \le j \le p^{l-1} - 1$ and $i \in A_{m,0}$, there exists $\gamma_{(j,i)} \in \mathfrak{O}_{K_m}$ such that*

$$\eta\left(F_j\left(\mathrm{Tr}_{K_m/\mathbb{Q}_p}(\bar{f}(\xi^u))\right)\right) = \mathrm{tr}_m\left(\sum_{i \in A_{m,0}} \eta\left(\gamma_{(j,i)}\right)\omega^{iu}\right)$$

for each $u = 1, \ldots, n$.

Proof. For $0 \le j \le p^{l-1} - 1$, let

$$H_j(x) = F_j(\mathrm{Tr}_{K_m/\mathbb{Q}_p}(\bar{f}(x))).$$

Since $H_j(0) = 0$ and $H_j(\alpha) \in \mathbb{Z}_p$ for each $\alpha \in \Gamma$, using [L-O, Proposition 2.10] we obtain $\gamma_{(j,i)} \in \mathfrak{O}_{K_m}$ for $i \in A_{m,0}$ such that

$$F_j\left(\mathrm{Tr}_{K_m/\mathbb{Q}_p}(\bar{f}(\xi^u))\right) = \mathrm{Tr}_{K_m/\mathbb{Q}_p}\left(\sum_{i \in A_{m,0}} \gamma_{(j,i)}\xi^{iu}\right).$$

The proof follows by applying η to the both sides. $\square$

Let the weight function $w_p : \mathbb{N} \to \mathbb{N}$ be defined as the function sending $a \in \mathbb{N}$ to the sum of its digits of the representation of a in base p. Let

$$W = \max\{ w_p(\alpha_1) + \cdots + w_p(\alpha_{p^{l-1}}) \\ \mid \alpha_1, \ldots, \alpha_{p^{l-1}} \in I_{0,m} \cup \cdots \cup I_{l-1,m}\} \text{ and} \tag{33}$$

$$l_m = \left\lceil \frac{m}{W} \right\rceil - 1, \quad h_m = \left\lfloor \frac{m}{W} \right\rfloor.$$

Now we state some known results for which we refer the reader to [L-O] and the references therein. Let $a_0(x), \ldots, a_{l-1}(x) \in \Gamma[x]$ be the polynomials defined by

$$f(x) = a_0(x) + pa_1(x) + \cdots + p^{l-1}a_{l-1}(x).$$

Recall that the weight function D_f of the polynomial $f(x)$ is

$$D_f = \max\{p^{l-1} \deg a_0(x), p^{l-2} \deg a_1(x), \ldots, \deg a_{l-1}(x)\}.$$

There exists a multiset $\mathcal{W}$ of complex numbers such that $|\mathcal{W}| \leq D_f - 1$ and for any $\gamma \in \mathcal{W}$ we have $|\gamma| = p^{m/2}$ and for any $s \geq 1$

$$\sum_{a \in \mathbb{Z}_{p^l} \backslash \{0\}} \sum_{x \in \Gamma_{ms}} e^{2\pi i \frac{\mathrm{Tr}_{ms}(f(x))}{p^l}} = -\sum_{\gamma \in \mathcal{W}} \gamma^s. \tag{34}$$

There exists an algebraic function field $E/\mathbb{F}_{p^m}$ of genus g with L-polynomial $L_E(x) = \prod_{i=1}^{2g}(1 - \theta_i x)$ such that for any $s \geq 1$

$$\sum_{a \in \mathbb{Z}_{p^l} \backslash \{0\}} \sum_{x \in \Gamma_{ms}} e^{2\pi i \frac{\mathrm{Tr}_{ms}(af(x))}{p^l}} = -\sum_{i=1}^{2g} \theta_i^s. \tag{35}$$

There also exists an algebraic function field $E_{l-1}/\mathbb{F}_{p^m}$ of genus g_{l-1} such that E_{l-1} is a subfield of E and the roots of its L-polynomial $L_{E_{l-1}}(x) = \prod_{i=1}^{2g_{l-1}}(1 - \psi x)$ satisfy

$$\sum_{a \in p^{l-1}\mathbb{Z}_{p^l} \backslash \{0\}} \sum_{x \in \Gamma_{ms}} e^{2\pi i \frac{\mathrm{Tr}_{ms}(f(x))}{p^l}} = -\sum_{i=1}^{2g_{l-1}} \psi_i^s \tag{36}$$

for any $s \geq 1$.

Now we are ready to give our main result.

Theorem 2. *We have*

$$\left| \sum_{a \in \mathbb{Z}_{p^l} \backslash p^{l-1}\mathbb{Z}_{p^l}} \sum_{x \in \Gamma_m} e^{2\pi i \frac{\mathrm{Tr}_m(af(x))}{p^l}} \right| \leq p^{lm+l-1} \left\lfloor \frac{p^{hm}(g - g_{l-1}) \left\lfloor 2p^{\frac{m}{2}-hm} \right\rfloor}{p^{lm+l-1}} \right\rfloor,$$

and

$$\left| \sum_{a \in \mathbb{Z}_{p^l} \backslash p^{l-1}\mathbb{Z}_{p^l}} \sum_{x \in \Gamma_m} e^{2\pi i \frac{\mathrm{Tr}_m(af(x))}{p^l}} \right| \leq p^{lm+l-1} \left\lfloor \frac{p^{hm}\frac{p^l-p^{l-1}}{2}(D_f - 1)\left\lfloor 2p^{\frac{m}{2}-hm} \right\rfloor}{p^{lm+l-1}} \right\rfloor.$$

Proof. By definition of w_{hom}, for $\alpha \in \mathbb{Z}_{p^l}$ we have

$$w_{\mathrm{hom}}(\alpha) = p^{l-2}(p - 1) - \frac{1}{p} \sum_{a \in \mathbb{Z}_{p^l} \backslash p^{l-1}\mathbb{Z}_{p^l}} e^{2\pi i \frac{a\alpha}{p^l}}. \tag{37}$$

Let $\lambda = \mathrm{Tr}_m(f(0))$. From (37) we have

$$w_{\mathrm{hom}}(c(\lambda)) = p^{l-2}(p - 1)p^m - \frac{1}{p} \sum_{a \in \mathbb{Z}_{p^l} \backslash p^{l-1}\mathbb{Z}_{p^l}} \sum_{x \in \Gamma_m} e^{2\pi i \frac{\mathrm{Tr}_m(af(x))}{p^l}}. \tag{38}$$

For $0 \leq j \leq p^{l-1}-1$ and $i \in A_{m,0}$, let $\gamma_{(i,j)} \in \mathfrak{O}_{K_m}$ be obtained using Proposition 2 such that

$$\eta\left(F_j\left(\mathrm{Tr}_{K_m/\mathbb{Q}_p}(\bar{f}(\xi^u))\right)\right) = \mathrm{tr}_m\left(\sum_{i \in A_{m,0}} \eta\left(\gamma_{(j,i)}\right)\omega^{iu}\right)$$

for each $u = 1, \ldots, n$. For $0 \leq j \leq p^{l-1} - 1$, let $h_j(x) = \sum_{i \in A_{m,0}} \eta \left(\gamma_{(j,i)} \right) x^i \in$ $\mathbb{F}_{p^m}[x]$. It follows from Lemma 3 that $\widetilde{F}_j(c) = (0, \mathrm{tr}_m(h_j(\omega)), \ldots, \mathrm{tr}_m(h_j(\omega^n)))$ for $0 \leq j \leq p^{l-1} - 1$. Using [L-O, Proposition 2.4] we obtain that $N_i(\widetilde{F}_j(c))$ is divisible by p^{lm} for $1 \leq i \leq p - 1$ and $0 \leq j \leq p^{l-1} - 1$. As $m \geq l_m$, by Proposition 1 and (38) we get

$$p^{l_m+l-1} \left| \sum_{a \in \mathbb{Z}_{p^l} \backslash p^{l-1} \mathbb{Z}_{p^l}} \sum_{x \in \Gamma_m} e^{2\pi i \frac{\mathrm{Tr}_m(af(x))}{p^l}} \right. \tag{39}$$

and hence for any positive integer s, the sum in (39) is divisible by $p^{h_m s}$. Using (34), (35), (36) and similar arguments as in the proof of [L-O, Corollary 3.6], we obtain

$$\left| \sum_{a \in \mathbb{Z}_{p^l} \backslash p^{l-1} \mathbb{Z}_{p^l}} \sum_{x \in \Gamma_m} e^{2\pi i \frac{\mathrm{Tr}_m(af(x))}{p^l}} \right| \leq p^{h_m} (g - g_{l-1}) \left\lfloor 2p^{\frac{m}{2} - h_m} \right\rfloor, \tag{40}$$

and

$$\left| \sum_{a \in \mathbb{Z}_{p^l} \backslash p^{l-1} \mathbb{Z}_{p^l}} \sum_{x \in \Gamma_m} e^{2\pi i \frac{\mathrm{Tr}_m(af(x))}{p^l}} \right| \leq p^{h_m} \frac{p^l - p^{l-1}}{2} (D_f - 1) \left\lfloor 2p^{\frac{m}{2} - h_m} \right\rfloor. \tag{41}$$

We complete the proof using (39), (40) and (41). $\qquad \square$

Remark 2. The definition of W in (33) and the divisibility parameters l_m and h_m depend directly on Proposition 2 and the "representing" set A_m. It is desirable to have a small representing set in the proposition such that we can reduce W (and hence increase l_m and h_m). This is possible in a canonical way for all polynomials in the case $l = 2$ due to an interesting reduction (see the proof of [L-O, Proposition 2.11]). Therefore the set A_m of [L-O, Definition 2.1] is much "smaller" than the set A_m of (32). It seems difficult to find such a canonical reduction for the general case $l \geq 3$.

For $l \geq 3$, in some special cases, we can improve Theorem 2.

Example 3. For $p = 2$ and $l = 3$, assume that

$$\hat{f}(x) = \hat{a}_0(x) + 4\hat{a}_2(x) \in \mathrm{GR}(8, m)[x],$$

that is $\hat{a}_1(x) = 0$. Let I_0 and I_2 be the supports of $\hat{a}_0(x)$ and $\hat{a}_2(x)$ respectively. Let $I_{0,m}$, $I_{2,m}$ be the sets defined in (31) and B_m be the subset of $\mathbb{N} \setminus \{0\}$ given by

$$B_m = (4I_{0,m} \cup 3I_{0,m} \cup 2I_{0,m} \cup I_{0,m}) \cup I_{2,m},$$

where $tI_{0,m} = \{\alpha_1 + \cdots + \alpha_t | \alpha_1, \ldots, \alpha_t \in I_{0,m}\}$ for $1 \leq t \leq 4$. Using a similar reduction as in the proof of [L-O, Proposition 2.11], we observe that a version of Proposition 2 obtained by changing A_m to B_m holds as well. Let

$$\widetilde{W} = \max \left\{ 4 \max\{w_p(i) | i \in I_0\}, \ \max\{w_p(i) | i \in I_2\} \right\},$$

$\tilde{l}_m = \lceil \frac{m}{\widetilde{W}} \rceil - 1$ and $\tilde{h}_m = \lfloor \frac{m}{\widetilde{W}} \rfloor$. Therefore we get an improved version of Theorem 2 by replacing l_m and h_m by $\tilde{l}_m$ and $\tilde{h}_m$ respectively.

For $\boldsymbol{u} = (u_1, \ldots, u_n) \in \mathbb{Z}_{p^l}^n$, let the weight $w_{(l,1)}(\boldsymbol{u})$ of $\boldsymbol{u}$ be defined as

$$w_{(l,1)}(\boldsymbol{u}) = \sum_{i=1}^{n} w_{(l,1)}(u_i).$$

Corollary 1. *Let* $f(x) \in \mathrm{GR}(p^l, m)[x]$ *be a non-degenerate polynomial and*

$$\boldsymbol{u} = \left(\mathrm{Tr}_m(f(w)), \mathrm{Tr}_m(f(w^2)), \ldots, \mathrm{Tr}_m(f(w^n))\right) \in \mathbb{Z}_{p^l}^n.$$

For the weight $w_{(l,1)}(\boldsymbol{u})$ *we have*

$$\left| w_{(l,1)}(\boldsymbol{u}) - n \left(p^{l-1} - p^{l-2}\right) - \theta \right| \leq p^{l_m + l - 2} \left\lfloor \frac{p^{h_m}(g - g_{l-1}) \left\lfloor 2p^{\frac{m}{2} - h_m} \right\rfloor}{p^{l_m + l - 1}} \right\rfloor,$$

and

$$\left| w_{(l,1)}(\boldsymbol{u}) - n \left(p^{l-1} - p^{l-2}\right) - \theta \right| \leq p^{l_m + l - 2} \left\lfloor \frac{p^{h_m} \frac{p^l - p^{l-1}}{2}(D_f - 1) \left\lfloor 2p^{\frac{m}{2} - h_m} \right\rfloor}{p^{l_m + l - 1}} \right\rfloor,$$

where

$$\theta = \begin{cases} p^{l-1} - p^{l-2} & \text{if } \mathrm{Tr}_m(f(0)) = 0, \\ -p^{l-2} & \text{if } \mathrm{Tr}_m(f(0)) \in p^{l-1}\mathbb{Z}_{p^l} \setminus \{0\}, \\ 0 & \text{if } \mathrm{Tr}_m(f(0)) \notin p^{l-1}\mathbb{Z}_{p^l}. \end{cases}$$

Remark 3. Using the methods of this section and Theorem 1, we can also obtain analogous bounds of Corollary 1 for the homogeneous weights $w_{(l,t)}$ with $2 \leq t \leq l - 1$.

Acknowledgements

The research of the first author is partially supported by NTU Research Grant No. M48110000.

The research of the second author is partially supported by the Turkish Academy of Sciences in the framework of Young Scientists Award Programme (F.Ö./TÜBA-GEBIP/2003-13).

References

[An]   D.R. Anderson, "A new class of cyclic codes", *SIAM J. Appl. Math.*, vol. 16, pp. 181-197, 1968.

[B-M]   I.F. Blake and J.W. Mark, "A note on complex sequences with low correlations", *IEEE Trans. Inform. Theory*, vol. 28, no. 5, pp. 814-816, September 1982.

[C-H]   I. Constantinescu and T. Heise, "A metric for codes over residue class rings of integers", *Problemy Peredachi Informatsii*, vol. 33, no. 3, pp. 22-28, 1997.

[He]   T. Helleseth, "On the covering radius of cyclic linear codes and arithmetic codes", *Disc. Appl. Math.*, vol. 11, pp. 157-173, 1985.

[H-K-M-S]   T. Helleseth, P.V. Kumar, O. Moreno and A.G. Shanbhag, "Improved estimates via exponential sums for the minimum distance of $\mathbb{Z}_4$-linear trace codes", *IEEE Trans. Inform. Theory*, vol. 42, no. 4, pp. 1212-1216, July 1996.

[K-H-C]   P.V. Kumar, T. Helleseth and A.R. Calderbank, "An upper bound for Weil exponential sums over Galois rings with applications", *IEEE Trans. Inform. Theory*, vol. 41, no. 2, pp. 456-468. March 1995.

[L-O]   S. Ling and F. Özbudak, "An improvement on the bounds of Weil exponential sums over Galois rings with some applications", *IEEE Trans. Inform. Theory*, vol. 50, no. 10, pp. 2529-2539, 2004.

[L-O2]   S. Ling and F. Özbudak, "Improved p-ary codes and sequence families from Galois rings of characteristic p^2", *SIAM J. Discrete Math.*, vol. 19, no. 4, pp. 1011-1028, 2006.

[Lu]   E. Lucas, "Sur les congruences des nombres eulériens et des coefficients différentiels des fonctions trigonométriques suivant un module premier", *Bull. Soc. Math. France*, vol. 6, pp. 49-54, 1878.

[Si]   V.M. Sidelnikov, "On mutual correlation of sequences", *Soviet Math. Dokl.*, vol. 12, no. 1, pp. 197-201, 1971.

[T]   A. Tietäväinen, "On the covering radius of long BCH codes", *Disc. Appl. Math.*, vol. 16, pp. 75-77, 1987.

Locally Invertible Multivariate Polynomial Matrices

Ruben G. Lobo, Donald L. Bitzer, and Mladen A. Vouk

North Carolina State University, Raleigh NC 27695, USA
rglobo@ncsu.edu, bitzer@ncsu.edu, vouk@ncsu.edu

Abstract. A new class of rectangular zero prime multivariate polynomial matrices are introduced and their inverses are computed. These matrices are ideal for use in multidimensional systems involving input-output transformations. We show that certain multivariate polynomial matrices, when transformed to the sequence space domain, have an invertible subsequence map between their input and output sequences. This invertible subsequence map can be used to derive the polynomial inverse matrix together with a set of pseudo-inverses. All computations are performed using elementary operations on the ground field without using any polynomial operations.

1 Introduction

Multivariate polynomial matrices play a fundamental role in the theory of linear multidimensional systems. They have a wide range of applications in circuits, systems, controls, signal processing and recently in the theory of multidimensional convolutional codes [1,2,3,4,5,6,7,8]. Matrices that have inverses, are of particular interest because many problems in multidimensional systems can be formulated as finitely generated modules over a multivariate polynomial ring. Three notions of primeness, namely, zero prime, minor prime and left factor prime arise when describing the structure of polynomial matrices. Among these, it is well known [2] that only zero prime matrices have inverses.

In this paper, we introduce a new class of rectangular zero prime multivariate polynomial matrices and derive their inverses. Algorithms for the inversion of zero prime matrices using Gröbner basis techniques, interpolation and Fourier transforms exist [4,9]. However, we are motivated by the fact that in the design of many applications involving multidimensional input-output transformations, one is free to choose the transformation matrix [4,6,7]. In such situations, it is highly desirable to be able to select matrices with solutions (inverses) such that they satisfy the desired primeness properties. Our method allows for explicit construction of invertible multidimensional transformations.

We show that certain multivariate polynomial matrices introduced in [10], when transformed to the sequence space domain, have an invertible subsequence map between their input and output sequences. We define these as "locally

Ø. Ytrehus (Ed.): WCC 2005, LNCS 3969, pp. 427–441, 2006.
© Springer-Verlag Berlin Heidelberg 2006

invertible" and prove that they are zero prime. We use the invertible subsequence map to derive the polynomial inverse matrix together with a set of pseudo-inverses (inverses with delay) and show that these computations can be performed using elementary operations on the ground field without using any polynomial operations.

2 Preliminaries

Let $R = \mathbb{F}_q[z_1, \ldots, z_m]$ be a polynomial ring in m variables over a finite field $\mathbb{F}$ with q elements. Let $G(z) \in R^{k \times n}$ be a rectangular polynomial matrix with k rows and n columns, $(k < n)$, having elements in R. We use the short form notation z to represent the m variables $z_1, \ldots, z_m$.

Definition 1. *Let $G(z) \in R^{k \times n}$ be of rank k. Let $\mathcal{M}$ be the set of full-size minors of $G(z)$ and let $\mathcal{I}$ denote the ideal generated by $\mathcal{M}$. The matrix $G(z)$ is said to be (a) left factor prime LFP if whenever $G(z)$ is factored as a product $G(z) = T(z)G_1(z)$ with $T(z) \in R^{k \times k}$ and $G_1(z) \in R^{k \times n}$, then determinant of $T(z)$ is a unit in R, (b) minor prime MP if the elements of $\mathcal{M}$ have no common divisors in R except for units, and (c) zero prime ZP if $\mathcal{I} = R$.*

These primeness notions apply as follows [2].

Theorem 1. *For $m = 1$, $ZP \equiv MP \equiv LFP$. For $m = 2$, $ZP \not\equiv MP \equiv LFP$. For $m \geq 3$, $ZP \not\equiv MP \not\equiv LFP$. Always, $ZP \Rightarrow MP \Rightarrow LFP$.*

Definition 2. *A delay-free polynomial inverse of $G(z)$ is a matrix $G(z)^{-1} \in R^{n \times k}$ such that $G(z)G(z)^{-1} = I$, where I is the identity matrix. A matrix $G(z)^{-1}_{(d_1, \ldots, d_m)} \in R^{n \times k}$ is a polynomial inverse with delay or pseudo-inverse for $G(z)$ if $G(z)G(z)^{-1}_{(d_1, \ldots, d_m)} = z_1^{d_1} \cdots z_m^{d_m} I$. The exponent $d_i > 0$ is considered to be the delay of the z_ith variable.*

The proof of the following theorem using the Quillen-Suslin theorem and the Cauchy-Binet formula can be found in [3].

Theorem 2. *A matrix $G(z) \in R^{k \times n}$ has a delay-free polynomial inverse if and only if it is ZP.*

It is well known, see for example [7,8], that there is an $\mathbb{F}$-isomorphism between the multivariate polynomial ring R and the multidimensional (m-D) finite sequence space

$$S = \{\omega : \mathbb{N}^m \to \mathbb{F}_q\}, \tag{1}$$

where elements of $\mathbb{F}$ are attached to the coordinates $(i_1, \ldots i_m)$ of the m-D positive integer lattice $\mathbb{N}^m$. Here ω has finite support, that is $\omega(i_1, \ldots i_m) = 0$ for all but finitely many points $(i_1, \ldots i_m) \in \mathbb{N}^m$. The isomorphism is defined as

$$\psi : S \longrightarrow R \\ \omega \mapsto \sum_{(i_1, \ldots, i_m)} \omega(i_1, \ldots, i_m) z_1^{i_1} \cdots z_m^{i_m}. \tag{2}$$

The coordinates of $\mathbb{N}^m$ are associated with monomials of R via the correspondence $(i_1, \ldots, i_m) \leftrightarrow z_1^{i_1} \cdots z_m^{i_m}$ and an element of $\mathbb{F}$ at the coordinate $(i_1, \ldots, i_m)$ becomes the coefficient of the monomial $z_1^{i_1} \cdots z_m^{i_m}$.

When there is more than one element of $\mathbb{F}$ attached to each coordinate of $\mathbb{N}^m$, we have the $\mathbb{F}$-isomorphism

$$\psi^n : \quad \begin{array}{c} S^n \\ (\omega_1, \ldots, \omega_n) \end{array} \begin{array}{c} \longrightarrow \\ \mapsto \end{array} \begin{array}{c} R^n \\ (\psi(\omega_1), \ldots, \psi(\omega_n)) \end{array}. \tag{3}$$

Let $G(z) \in R^{k \times n}$ be a polynomial matrix having elements $g(z)_x^{(y)} \in R$. The m-D sequence space representation of $G(z)$ using the transformation

$$\psi^{-1}\left(g(z)_x^{(y)}\right) = g_x^{(y)}, \tag{4}$$

gives us the sequence generator $G \in S^{k \times n}$ with elements $g_x^{(y)} \in S$.

$$G(z) = \begin{bmatrix} g(z)_1^{(1)}, & \ldots, & g(z)_1^{(n)} \\ & \vdots & \\ g(z)_k^{(1)}, & \ldots, & g(z)_k^{(n)} \end{bmatrix} \quad \xrightarrow{\psi^{-1}} \quad G = \begin{pmatrix} g_1^{(1)}, & \ldots, & g_1^{(n)} \\ & \vdots & \\ g_k^{(1)}, & \ldots, & g_k^{(n)} \end{pmatrix} \tag{5}$$

Operations in R involving polynomial multiplication can be replaced by discrete convolution in S. Let M_i be the largest exponent of the z_ith variable among the monomials of the polynomial entries $g(z)_x^{(y)}$ of $G(z)$. The sequence $g_x^{(y)}$ of G will be of length $M_i + 1$ along the ith dimension of the sequence space. The output sequence $v \in S^n$ corresponding to an input sequence $u \in S^k$ is given by

$$v = u * G \tag{6}$$

$$v^{(1)} = u^{(1)} * g_1^{(1)} + \cdots + u^{(k)} * g_k^{(1)}$$

$$\vdots$$

$$v^{(n)} = u^{(1)} * g_1^{(n)} + \cdots + u^{(k)} * g_k^{(n)}.$$

The sequences $v^{(1)}, \ldots, v^{(n)}$ are multiplexed to form the output sequence $v \in S^n$. The convolution operation $u^{(x)} * g_x^{(y)}$ implies that for all $(i_1, \ldots, i_m) \geq 0$,

$$v_{(i_1, \ldots, i_m)}^{(y)} = \sum_{l_m=0}^{M_m} \cdots \sum_{l_1=0}^{M_1} u_{((i_1-l_1), \ldots, (i_m-l_m))}^{(x)} g_{x(l_1, \ldots, l_m)}^{(y)}, \tag{7}$$

where addition and multiplication is performed in $\mathbb{F}_q$ and

$$u_{((i_1-l_1), \ldots, (i_m-l_m))}^{(x)} \triangleq 0 \quad \forall \, i_r < l_r.$$

For the isomorphism $\psi : S \to R$, since the transformation is discrete convolution in the domain S and polynomial multiplication in the range R, the law of composition is $\psi(\omega_1 * \omega_2) = \psi(\omega_1)\psi(\omega_2)$.

3 Local Invertibility

In this section we show that certain polynomial matrices of the form $G(z) \in R^{k \times n}$, when transformed to their sequence space representation $G \in S^{k \times n}$, have a subsequence map between their input (S^k) and output (S^n) sequence spaces. We prove that the sequence generator G has an inverse if the subsequence map is invertible. We use the inverse subsequence map to obtain the inverse sequence generator $G^{-1} \in S^{n \times k}$, which can then be transformed using the map ψ into the inverse polynomial matrix $G(z)^{-1} \in R^{n \times k}$.

3.1 Sequence Ordering

A m-D sequence space S is defined (1) as having finite support with elements of $\mathbb{F}$ attached to the coordinates $(i_1, \ldots, i_m)$ of the m-D positive integer lattice $\mathbb{N}^m$. There is no restriction on the way in which elements of $\mathbb{F}$ are *ordered* at each coordinate of the lattice. For values of $m > 1$, one has the option of ordering the input symbols in S^k and the output symbols in S^n along the m axes of their respective sequence spaces. This gives rise to the possibility of having different choices of sequence space ordering. For a given polynomial matrix $G(z) \in R^{k \times n}$, since input sequences in S^k have k symbols and output sequences in S^n have n symbols of $\mathbb{F}$ attached to each coordinate of their respective lattices, the rate $\frac{k}{n}$ is a natural candidate to predefine sequence space ordering. We formalize this notion of sequence space ordering based on the rate $\frac{k}{n}$ as follows.

Definition 3. *For a given polynomial matrix $G(z) \in R^{k \times n}$, if the rate $\frac{k}{n}$ is specified using the notation*

$$\frac{k}{n} = \prod_{i=1}^{m} \frac{k_i}{n_i}, \tag{8}$$

then, $\prod_{i=1}^{m} k_i$ defines the ordering of the input sequence and $\prod_{i=1}^{m} n_i$ defines the ordering of the output sequence. The input sequence is ordered by attaching k_i symbols, and the output sequence is ordered by attaching n_i symbols, to each coordinate of the lattice $\mathbb{N}^m$ along the ith dimension of their respective sequence spaces.

3.2 Subsequence Mapping

Since we are dealing with rectangular matrices ($n > k$), the output sequence $v \in S^n$ obtained from $v = u * G$, will be larger (have more symbols) than the input sequence $u \in S^k$. Local invertibility finds an *invertible map*, if one exists, between *subsequences* of the input and output sequence.

We begin this construction by letting $G(z) \in R^{k \times n}$ be any polynomial matrix, not necessarily ZP, not even of full row rank. Let the rate $\frac{k}{n}$ be factored as $\prod_{i=1}^{m} \frac{k_i}{n_i}$ to predefine a sequence ordering of $\prod_{i=1}^{m} k_i$ in S^k and $\prod_{i=1}^{m} n_i$ in S^n as described in Definition 3. Let $G \in S^{k \times n}$ be the equivalent sequence generator of $G(z)$. The condition $u_{((i_1 - l_1), \ldots, (i_m - l_m))}^{(x)} \triangleq 0 \ \forall \ i_r < l_r$ in (7) implies that the boundaries of the input sequence $u \in S^k$ have to be padded with $k_i M_i$

zeros in the ith dimension. Now consider the production of output symbols in each dimension during the convolution operation as shown in Table 1. The first $k_i(M_i + 1)$ input symbols (which include $k_i M_i$ padded zeros and k_i valid input symbols) produce n_i output symbols in the ith dimension. In the next iteration

Table 1. Subsequence Mapping

Iteration	Input	Output
1	$k_i(M_i + 1)$	n_i
2	$k_i(M_i + 2)$	$2n_i$
3	$k_i(M_i + 3)$	$3n_i$
	$\cdots$	
j	$k_i(M_i + j)$	jn_i

k_i additional input symbols produce n_i additional output symbols and so on. If we require the number of input and output symbols to be equal, we set

$$k_i(M_i + j) = jn_i, \qquad (9)$$

for the jth iteration. Solving for j and substituting back in $k_i(M_i + j)$ or jn_i gives us

$$w_i = \frac{k_i n_i M_i}{n_i - k_i}. \qquad (10)$$

The quantity w_i is called the *subsequence mapping length* and is interpreted as follows: Even though output sequences are always larger than their corresponding input sequences, given a reference coordinate of $(l_1, \ldots, l_m)$, a subsequence of w_i symbols starting at l_i of an input sequence in S^k will map to a subsequence of w_i symbols starting at l_i of its corresponding output sequence in S^n along the ith dimension. The total number of symbols in this equal sized subsequence map between the input and output sequence is

$$w = \prod_{i=1}^{m} w_i. \qquad (11)$$

For example we have row vector subsequences of size w_1 in 1-D, rectangular subsequences of size $w_1 \times w_2$ in 2-D, cuboid subsequences of size $w_1 \times w_2 \times w_3$ in 3-D, orthotope (hypercube) subsequences of size $\prod_{i=1}^{m} w_i$ in m-D and so on.

Remark 1. If the parameters n_i, k_i and M_i are such that w_i in (10) is not a positive integer, then a subsequence map of equal length will not exist. This requirement restricts the use of this technique to matrices whose rate $\frac{k}{n}$ can be factored such that $n_i > k_i$ in each dimension. Furthermore, even if $n_i > k_i$, the factorization should be such $(n_i - k_i) \mid k_i$, or $(n_i - k_i) \mid n_i$, or $(n_i - k_i) \mid M_i$.

We will now proceed to find out if this subsequence map is invertible. Consider a set of w, m-D standard basis input subsequences $b_r \in S^k$, each of size $\prod_{i=1}^{m} w_i$

with order $\prod_{i=1}^{m} k_i$. Let $\hat{g}_r \in S^n$ be the corresponding m-D output subsequences of equal size $\prod_{i=1}^{m} w_i$ with order $\prod_{i=1}^{m} n_i$ obtained using $b_r * G$.

$$S^k \longrightarrow S^n$$
$$b_r \longmapsto \hat{g}_r \tag{12}$$

An inverse subsequence map will exist only if (12) is injective. Since S is isomorphic to R, linear independence of sequences in S implies that they are linearly independent both $\mathbb{F}$-linearly and with respect to shifts. Here, we are only looking for symbol-wise linear independence "within" the subsequence map. This can easily be done by serializing the m-D output subsequence map into 1-D sequences and performing elementary row or column operations on them in $\mathbb{F}$. A m-D sequence can be serialized (or unfolded) into a 1-D sequence in many ways. For example it can be unfolded in row-major form, column-major form or by following the path traced by a space-filling curve as it passes through the points of the m-D lattice and so on. The choice of the unfolding technique is not important as long as it is bijective and one is consistent when folding (17) the 1-D sequence back into a m-D sequence. We denote unfolding with the following bijective operator.

$$\mathcal{U}_f : \mathbb{N}^m \longrightarrow \mathbb{N}^1 \tag{13}$$

The inverse subsequence map is found as follows: Each $\hat{g}_r$ (12) of size $\prod_{i=1}^{m} w_i$ and order $\prod_{i=1}^{m} n_i$ is unfolded into a 1-D sequence $\mathbf{g}_r$ of length w and order n.

$$\mathcal{U}_f(\hat{g}_r) = \mathbf{g}_r \tag{14}$$

A $w \times w$ reduced encoding matrix $\hat{G}$ is constructed by using each $\mathbf{g}_r$ as a row.

$$\hat{G} = [\mathbf{g}_1, \ldots, \mathbf{g}_w]^T \tag{15}$$

The inverse of $\hat{G}$ (if one exists) can be found using elementary operations in $\mathbb{F}$.

$$\hat{G}^{-1} = [\mathbf{p}_1, \ldots, \mathbf{p}_w]^T \tag{16}$$

The rows of the *inverse reduced encoding matrix* $\hat{G}^{-1}$ represent the inverse subsequence map. Each 1-D sequence $\mathbf{p}_r$ of length w and order k is folded back into a m-D subsequence $\hat{p}_r$ of size $\prod_{i=1}^{m} w_i$ and order $\prod_{i=1}^{m} n_i$ using

$$\mathcal{U}_f^{-1}(\mathbf{p}_r) = \hat{p}_r , \tag{17}$$

to obtain the inverse subsequence map

$$S^n \longrightarrow S^k$$
$$c_r \longmapsto \hat{p}_r \tag{18}$$

The $c_r \in S^n$ (18) are m-D standard basis output subsequences, each of size $\prod_{i=1}^{m} w_i$ with order $\prod_{i=1}^{m} n_i$.

Definition 4. *A polynomial matrix $G(z) \in R^{k \times n}$ is said to be locally invertible if it has a nonsingular reduced encoding matrix.*

Since the subsequence map (12) is constructed using a standard basis as input, it can be used to obtain an output subsequence corresponding to any input subsequence. Let $u_i \in \mathbb{F}$ be the w symbols that make up an input subsequence $\hat{u} \in S^k$ of size $\prod_{i=1}^{m} w_i$. The standard basis input subsequences in (12) can be used to represent $\hat{u}$ as

$$\hat{u} = u_1 b_1 + u_2 b_2 + \cdots + u_w b_w. \tag{19}$$

The corresponding output subsequence $\hat{v} \in S^n$ of size $\prod_{i=1}^{m} w_i$ is given by

$$\hat{v} = u_1 \hat{g}_1 + u_2 \hat{g}_2 + \cdots + u_w \hat{g}_w. \tag{20}$$

Let $v_i \in \mathbb{F}$ be the w symbols that make up $\hat{v}$. The standard basis output subsequences in (18) can be used to represent $\hat{v}$ as

$$\hat{v} = v_1 c_1 + v_2 c_2 + \cdots + v_w c_w. \tag{21}$$

Now, the input subsequence $\hat{u}$ can be recovered from the output subsequence $\hat{v}$ using the inverse subsequence map. That is,

$$\hat{u} = v_1 \hat{p}_1 + v_2 \hat{p}_2 + \cdots + v_w \hat{p}_w. \tag{22}$$

We started this construction by letting $G(z)$ be any polynomial matrix. Next we show that if the reduced encoding matrix is nonsingular then $G(z)$ is not only of full row rank but also ZP.

Theorem 3. *Let $G \in S^{k \times n}$ be a sequence generator with an invertible subsequence map of size $\prod_{i=1}^{m} w_i$ between its input and output sequences. Then, G has a delay-free sequence inverse $G^{-1} \in S^{n \times k}$, such that $G * G^{-1} = I \in S^{k \times k}$.*

Proof. It is sufficient to show that every input sequence $u \in S^k$ can be recovered free of delay from its corresponding unique output sequence $v = u * G \in \mathcal{C} \subset S^n$.

Since G has an subsequence map of size $\prod_{i=1}^{m} w_i$, any input sequence $u \in S^k$ can be constructed from overlapping subsequences $\hat{u} \in S^k$ of size $\prod_{i=1}^{m} w_i$ that are shifted by k_i symbols in the ith dimension. The corresponding output subsequences $\hat{v} \in S^n$ of size $\prod_{i=1}^{m} w_i$ (obtained from (20)) can be overlapped by shifting by n_i symbols in the same dimension to construct the output sequence $v \in \mathcal{C}$.

If the sequence generator map is not injective, then there exist sequences $u_1 \neq u_2 \in S^k$ such that $u_1 * G = u_2 * G = v \in \mathcal{C}$. If this is true, then there exist input subsequences $\hat{u}_1 \neq \hat{u}_2$ of size $\prod_{i=1}^{m} w_i$ that map to the same subsequence $\hat{v}$ of size $\prod_{i=1}^{m} w_i$ of the output sequence v. This is a contradiction to the statement that the subsequence map is invertible, and we conclude that the sequence generator map is injective.

Any output sequence v can be constructed from overlapping subsequences $\hat{v} \in S^n$ of size $\prod_{i=1}^{m} w_i$ that are shifted by n_i symbols in the ith dimension. The corresponding input subsequences $\hat{u} \in S^k$ of size $\prod_{i=1}^{m} w_i$ (obtained from (22)) can be overlapped by shifting by k_i symbols in the same dimension to reconstruct the input sequence $u \in S^k$. The injective mapping and delay-free inversion of output sequences using the invertible subsequence map of the sequence generator G completes the proof. $\square$

Theorem 3 shows that subsequence (local) invertibility implies global invertibility. From the $\mathbb{F}$-isomorphism defined in (2) follows immediately

Corollary 1. *A locally invertible polynomial matrix is ZP.*

Proof. If $G(z)$ is a locally invertible polynomial matrix, then it has an invertible subsequence map defined by its nonsingular reduced encoding matrix. Then from Theorem 3, equation (2), and Theorem 2 it follows that $G(z)$ is ZP.   □

3.3   The Reduced Encoding Matrix

The reduced encoding matrix defined in (15) can be formed by inspection. Since a standard basis subsequence is used as an input in the subsequence map (12), the output subsequences will just contain elements of the sequence generator. The structure of the reduced encoding matrix depends on the ordering of the sequence space (8) and the unfolding technique (13) used to form its rows. To see this we first need to establish the following representation.

When $k > 1$, the sequences $g_x^{(y)} \in S$ of G from (5) can be represented in a composite form, where, for a fixed y each $g_x^{(y)}$; $x = 1$ to k is multiplexed into a single sequence, k_i symbols at a time along the ith dimension.

$$G_c = \left\{ g^{(1)}, \ldots, g^{(n)} \right\} \tag{23}$$

The ordering of the composite sequences $g^{(y)} \in S^k$ is now consistent with the ordering $\prod_{i=1}^m k_i$ of the input sequence space S^k. Since the sequences $g_x^{(y)}$ are of length of $M_i + 1$ symbols in the ith dimension, the length of the composite sequence $g^{(y)}$ along that dimension is $L_i = k_i(M_i + 1)$.

Definition 5. *A sequence $s \in S^l$ with order $l = \prod_{i=1}^m l_i$ is said to be in* reverse composite form $\mathcal{R}(s)$ *when it is reversed l_i symbols at a time along each of its dimensions.*

Let $\zeta + \mathcal{R}(g^{(y)})$ be the symbol wise addition of an all-zero sequence $\zeta \in S^k$ of size $\prod_{i=1}^m w_i$ and the composite generator sequence $g^{(y)} \in S^k$ of size $\prod_{i=1}^m L_i$ in reverse composite form. Row-major unfolding of a sequence in S^k of finite size $\prod_{i=1}^m t_i$ with order $\prod_{i=1}^m k_i$ is defined as

$$\begin{aligned} \mathcal{U}_{fr} : \quad \mathbb{N}^m &\longrightarrow \mathbb{N}^1 \\ (i_1, \ldots, i_m) &\mapsto i_1 + \sum_{j=1}^{m-1} i_{j+1} \prod_{l=1}^{j} \frac{t_l}{k_l} \,. \end{aligned} \tag{24}$$

In equation (12) since a standard basis is used as input, the output subsequences $\hat{g}_r$ will just contain symbols of the sequences $g_x^{(y)}$. If the standard basis sequences $b_r \in S^k$ in (12) are labeled in row-major form and if the corresponding $\hat{g}_r$ in (14) are unfolded into 1-D sequences using row-major unfolding $\mathcal{U}_{fr}(\hat{g}_r) = \mathbf{g}_r$ to form "rows" of the matrix $\hat{G}$ as described in (15), then the resulting "columns" of $\hat{G}$ will be shifts of the 1-D sequences

$$\mathcal{U}_{fr}(\zeta + \mathcal{R}(g^{(y)})). \tag{25}$$

The values m, k_i and w_i determine the amount by which the columns are shifted and the reduced encoding matrix $\hat{G}$ will have an elegant nested structure as described below.

The $w \times w$ reduced encoding matrix $\hat{G}$ has $1 + \frac{w_m - L_m}{k_m}$ columns made up of the matrix $\hat{G}_{m-1}$, with the jth column shifted down by $(j-1)k_m \prod_{i=1}^{m-1} w_i$ rows.

$$
\hat{G} = \begin{bmatrix} \hat{G}_{m-1} & & & \\ \vdots & \hat{G}_{m-1} & & \\ & \vdots & \ddots & \\ & & \hat{G}_{m-1} & \\ & & & \vdots \end{bmatrix} \tag{26}
$$

Similarly, for values of x ranging from $(m-1)$ to 2, the matrix $\hat{G}_x$ consists of $1 + \frac{w_x - L_x}{k_x}$ columns made up of the matrix $\hat{G}_{x-1}$, with the jth column shifted down by $(j-1)k_x \prod_{i=1}^{x-1} w_i$ rows.

$$
\hat{G}_x = \begin{bmatrix} \hat{G}_{x-1} & & & \\ \vdots & \hat{G}_{x-1} & & \\ & \vdots & \ddots & \\ & & \hat{G}_{x-1} & \\ & & & \vdots \end{bmatrix} \tag{27}
$$

The matrix $\hat{G}_1$ has $1 + \frac{w_1 - L_1}{k_1}$ columns made up of the matrix $\hat{G}_0$, with the jth column shifted down by $(j-1)k$ rows.

$$
\hat{G}_1 = \begin{bmatrix} \hat{G}_0 & & & \\ \vdots & \hat{G}_0 & & \\ & \vdots & \ddots & \\ & & \hat{G}_0 & \\ & & & \vdots \end{bmatrix} \tag{28}
$$

The fundamental matrix $\hat{G}_0$ has n columns consisting of the 1-D sequences defined in (25). That is,

$$
\hat{G}_0 = \left[\mathcal{U}_{fr}\left(\zeta + \mathcal{R}(g^{(1)})\right), \ \ldots, \ \mathcal{U}_{fr}\left(\zeta + \mathcal{R}(g^{(n)})\right) \right]. \tag{29}
$$

To summarize, the $w \times w$ reduced encoding matrix $\hat{G}$ is made up of m nested column-matrices. The columns of these nested matrices are in turn just shifts of the 1-D sequences $\mathcal{U}_{fr}\left(\zeta + \mathcal{R}(g^{(y)})\right)$.

Remark 2. The property of local invertibility can be used to construct ZP polynomial matrices. For given values of n, k, m and M_i, if there exists a sequence space ordering of $\prod_{i=1}^{m} k_i$ and $\prod_{i=1}^{m} n_i$ that yields a positive integer valued subsequence mapping length of w_i along each dimension, then one can construct the fundamental matrix $\hat{G}_0$ shown in (29) using randomly generated 1-D vectors of the form $a^{(y)} = \mathcal{U}_{fr}(\zeta + \mathcal{R}(g^{(y)}))$. That is,

$$\hat{G}_0 = [a^{(1)}, \dots, a^{(n)}].$$

If the resulting reduced encoding matrix (26) inverts, then by Corollary 1 the polynomial matrix $G(z)$ obtained using

$$\mathcal{U}_{fr}^{-1}(a^{(y)}) \to \zeta + \mathcal{R}(g^{(y)}) \to \psi^k(g^{(y)}) \to g(z)_x^y,$$

will be ZP.

3.4   Polynomial Inverses

In Section 3.2 we saw that if the polynomial matrix $G(z) \in R^{k \times n}$ is locally invertible, then, the transformation from S^n to S^k can be performed using the inverse subsequence map. In this section we derive the inverse sequence generator $G^{-1} \in S^{n \times k}$ from the inverse reduced encoding matrix and transform it using the map ψ to obtain the inverse polynomial matrix $G(z)^{-1} \in R^{n \times k}$.

The subsequence inversion in (22) can be performed using matrix multiplication with the inverse reduced encoding matrix (16) as follows

$$\hat{u} = \mathcal{U}_f^{-1}(\mathcal{U}_f(\hat{v})\hat{G}^{-1}). \tag{30}$$

When the subsequence inversion is viewed as shown above, each symbol attached to a coordinate of the m-D input subsequence $\hat{u}$ is obtained by multiplying the 1-D output subsequence $\mathcal{U}_f(\hat{v})$ with a column $\mathbf{g}_r^{-1}$ of the inverse reduced encoding matrix.

$$\hat{G}^{-1} = [\mathbf{p}_1, \dots, \mathbf{p}_w]^T = [\mathbf{g}_1^{-1}, \dots, \mathbf{g}_w^{-1}] \tag{31}$$

The transformation using the inverse subsequence map produces a $(w_i - k_i)$ symbol overlap (see proof of Theorem 3) in each dimension between consecutive overlapping input m-D subsequences in the input sequence space S^k. The $w - k$ columns of $\hat{G}^{-1}$ that are responsible for this overlap right-shift the symbols of the input subsequence. The remaining k columns produce non-overlapping symbols and correspond to the inverse composite sequence generator

$$G_c^{-1} = \{g^{(1)^{-1}}, \dots, g^{(k)^{-1}}\}. \tag{32}$$

The columns of $\hat{G}^{-1}$ that represent G_c^{-1} depend on the choice of sequence space ordering and unfolding technique used in the construction of the reduced encoding matrix. For a given sequence space ordering, if the reduced encoding matrix $\hat{G}$ in (15) is constructed using row-major unfolding as described in Section 3.3, then, the last k columns of $\hat{G}^{-1}$ (31) will correspond to the k non-overlapping symbols of the m-D input subsequence generated during each step of the subsequence inversion. Each of these columns have to be folded back into m-D subsequences and then reversed to obtain the inverse composite generator sequences $g^{(y)^{-1}}$.

$$\mathcal{U}_f^{-1}(\mathbf{g}_r^{-1}) \to \mathcal{R}(g^{(y)^{-1}}) \to \psi^n(g^{(y)^{-1}}) \to g^{-1}(z)_y^{(x)}$$

The polynomial representation of G_c^{-1} using the isomorphism ψ^n gives us the *delay-free polynomial inverse* $G(z)^{-1} \in R^{n \times k}$ with elements $g^{-1}(z)_y^{(x)} \in R$.

$$G(z)G(z)^{-1} = I_{k \times k} \tag{33}$$

The remaining $w - k$ columns of $\hat{G}$ that are responsible for the overlap, produce right-shifts of k_i along each axis of the m-D input sequence space S^k and correspond to $\frac{w-k}{k}$ unique *pseudo-inverses* $G(z)_{(d_1,\ldots,d_m)}^{-1} \in R^{n \times k}$, such that

$$G(z)G(z)_{(d_1,\ldots,d_m)}^{-1} = z_1^{d_1} \cdots z_m^{d_m} I_{k \times k} , \tag{34}$$

with $0 \le d_i \le \frac{w_i - k_i}{k_i}$ and $\sum_{i=1}^{m} d_i \neq 0$. The exponent d_i can be considered as the *delay* in the ith dimension.

Example 1. Let $R = \mathbb{F}_2[z_1, z_2]$ and $G(z) \in R^{2 \times 6}$ with $M_1 = 2$ and $M_2 = 1$.

$$G(z) = \begin{bmatrix} z_1^2 z_2 & 0 & z_2 & z_1^2 & 1 & 0 \\ 0 & 1+z_1^2 z_2 & 0 & z_1 z_2 & z_1 & 1+z_1^2+z_2 \end{bmatrix}$$

The 2-D sequence generator $G \in S^{2 \times 6}$ transformed using ψ^{-1} is

$$G = \begin{pmatrix} 0\,0\,0 & 0\,0\,0 & 0\,0\,0 & 0\,0\,1 & 1\,0\,0 & 0\,0\,0 \\ 0\,0\,1, & 0\,0\,0, & 1\,0\,0, & 0\,0\,0, & 0\,0\,0, & 0\,0\,0 \\[4pt] 0\,0\,0 & 1\,0\,0 & 0\,0\,0 & 0\,0\,0 & 0\,1\,0 & 1\,0\,1 \\ 0\,0\,0, & 0\,0\,1, & 0\,0\,0, & 0\,1\,0, & 0\,0\,0, & 1\,0\,0 \end{pmatrix}$$

If we factor the rate $\frac{2}{6}$ as $\frac{1}{2} \times \frac{2}{3}$ to predefine an ordering of $\prod_{i=1}^{2} k_i = 1 \times 2$ in S^2 and $\prod_{i=1}^{2} n_i = 2 \times 3$ in S^3, then, the subsequence mapping lengths along i_1 and i_2 are

$$w_1 = \frac{1 \times 2 \times 2}{2-1} = 4, \qquad w_2 = \frac{2 \times 3 \times 1}{3-2} = 6.$$

The set of $w = 24$ standard basis input subsequences $b_r \in S^2$, each of size 4×6 with order 1×2 map to output subsequences $\hat{g}_r \in S^6$ of size 4×6 with order 2×3 as shown below.

b_1	$\hat{g}_1$	b_2	$\hat{g}_2$	b_3	$\hat{g}_3$	b_{24}	$\hat{g}_{24}$
1 0 0 0	10 00	0 0 0 0	01 00	0 1 0 0	00 10	0 0 0 0	00 00
0 0 0 0	00 00	1 0 0 0	00 00	0 0 0 0	00 00	0 0 0 0	00 00
0 0 0 0	00 00	0 0 0 0	00 00	0 0 0 0	00 00	0 0 0 0	00 00
0 0 0 0 $\mapsto$	00 00,	0 0 0 0 $\mapsto$	00 00,	0 0 0 0 $\mapsto$	00 00,	$\cdots$, 0 0 0 0 $\mapsto$	00 01
0 0 0 0	00 00	0 0 0 0	00 00	0 0 0 0	00 00	0 0 0 0	00 00
0 0 0 0	00 00	0 0 0 0	00 00	0 0 0 0	00 00	0 0 0 1	00 01

Each 2-D output subsequence is unfolded $\mathcal{U}_{fr}(\hat{g}_r) = \mathbf{g}_r$ to form a row of $\hat{G} = [\mathbf{g}_1, \ldots, \mathbf{g}_{24}]^T$. Because row-major unfolding is used, the resulting $\hat{G}$ shown in

```
⎡1 0 0 0 0 0                          ⎤    ⎡1 0 0 0 0 0 0 0 0 0 0 0 0 0 0 0 0 0 0 0 0 0 0 0⎤
⎢0 1 0 0 0 0                          ⎥    ⎢0 1 0 0 0 0 0 0 0 0 0 0 0 0 0 0 0 0 0 0 0 0 0 0⎥
⎢0 0 0 0 0 0 1 0 0 0 0 0              ⎥    ⎢0 0 0 1 0 0 0 0 0 0 0 0 0 0 0 0 0 0 0 0 0 0 0 0⎥
⎢0 0 0 1 0 0 0 1 0 0 0 0              ⎥    ⎢0 0 0 1 0 0 0 1 0 0 0 0 0 0 0 1 0 0 0 1 1 0 0 0⎥
⎢0 0 1 0 0 0 0 0 0 0 0 0              ⎥    ⎢0 0 0 0 0 0 1 0 0 1 0 0 0 1 0 0 0 1 0 0 1 1 0 0 1⎥
⎢0 0 0 0 0 1 0 0 0 1 0 0              ⎥    ⎢0 0 0 0 0 0 0 0 1 0 0 0 0 0 0 0 1 0 0 0 1 1 0⎥
⎢            0 0 1 0 0 0              ⎥    ⎢0 0 1 0 0 0 0 0 0 0 0 0 0 0 0 0 0 0 0 0 0 0 0 0⎥
⎢            0 0 0 0 0 1              ⎥    ⎢0 0 0 0 0 0 1 0 0 0 0 0 0 0 1 0 0 0 1 1 0 0 0⎥
⎢0 0 0 1 0 0          1 0 0 0 0 0     ⎥    ⎢0 0 0 0 0 1 0 0 0 0 0 0 0 0 0 0 0 0 0 0 0 0 0 0⎥
⎢0 0 0 0 0 1          0 1 0 0 0 0     ⎥    ⎢0 0 0 0 0 1 0 0 1 0 0 0 0 0 0 1 0 0 0 1 1 0⎥
⎢0 0 0 0 0 0 0 0 1 0 0 0 0 0 0 0 0 0 1 0 0 0 0 0⎥  ⎢0 1 0 0 0 0 0 0 1 0 0 0 1 0 0 0 0 1 0 0 1 1 0⎥
⎢0 0 0 0 1 0 0 0 0 0 0 1 0 0 0 1 0 0 0 1 0 0 0 0⎥  ⎢0 0 0 0 0 0 1 0 0 0 0 0 0 0 0 0 0 0 0 0 0 0 0⎥
⎢0 0 0 0 1 0 0 0 0 0 0 0 0 0 1 0 0 0 1 0 0 0 0 0⎥  ⎢0 0 0 1 0 0 1 1 0 0 0 0 0 0 1 0 0 0 1 1 0 0 0⎥
⎢0 1 0 0 0 1 0 0 0 0 1 0 0 0 0 0 0 1 0 0 0 1 0 0⎥  ⎢0 0 0 0 0 0 0 0 0 0 0 0 0 0 0 1 0 0 0 1 1 0⎥
⎢            0 0 0 0 1 0          0 0 1 0 0 0     ⎥  ⎢0 0 0 0 0 0 1 0 0 0 1 1 0 0 1 0 0 1 0 0 1 1 0 0 1⎥
⎢            0 1 0 0 0 1          0 0 0 0 0 1     ⎥  ⎢0 0 0 0 0 0 0 0 0 0 0 0 0 0 0 0 1 0 0 0 0 0 0⎥
⎢                     0 0 0 1 0 0                ⎥  ⎢0 0 0 0 0 0 0 0 0 0 0 0 0 0 0 0 0 0 0 1 0 0 0⎥
⎢                     0 0 0 0 0 1                ⎥  ⎢0 0 0 0 0 0 0 0 0 0 0 0 0 0 0 0 0 1 0 0 0 0 0⎥
⎢                     0 0 0 0 0 0 0 0 0 1 0 0    ⎥  ⎢0 0 0 0 0 1 0 0 1 1 0 0 0 0 0 1 0 0 0 1 1 0⎥
⎢                     0 0 0 0 1 0 0 0 0 0 0 1    ⎥  ⎢0 0 0 0 0 0 0 0 0 0 0 0 0 0 0 0 0 0 0 1 1 0 0 1⎥
⎢                     0 0 0 0 1 0 0 0 0 0 0 0    ⎥  ⎢0 1 0 0 0 0 0 0 1 0 0 0 1 1 0 0 0 1 0 0 1 1 0⎥
⎢                     0 1 0 0 0 1 0 0 0 0 1 0    ⎥  ⎢0 0 0 0 0 0 0 0 0 0 0 0 0 0 0 0 0 0 1 0 0 0 0 0⎥
⎢                              0 0 0 0 1 0       ⎥  ⎢0 0 0 0 0 0 0 0 0 0 0 0 0 0 0 0 0 0 0 0 1 0⎥
⎣                              0 1 0 0 0 1       ⎦  ⎣0 0 0 0 0 0 0 0 0 0 0 0 0 0 0 0 0 0 0 1 1 0 0 0⎦
```

(a) $\hat{G}_{24\times24}$: The rows are 1-D sequences in S^6 with order 2×6 and the columns are 1-D sequences in S^2 with order 1×2

(b) $\hat{G}_{24\times24}^{-1}$: The rows are 1-D sequences in S^2 with order 1×2 and the columns are 1-D sequences in S^6 with order 2×6

Fig. 1. The reduced encoding matrix and its inverse (*blanks denote zeros*)

Fig. 1(a) has the nested structure defined in Section 3.3. Since the reduced encoding matrix is nonsingular (the inverse is shown in Fig. 1(b)), the polynomial encoder $G(z)$ is locally invertible. The last $k = 2$ columns $\mathbf{g}_{23}^{-1}$ and $\mathbf{g}_{24}^{-1}$ of $\hat{G}^{-1} = [\mathbf{g}_1^{-1}, \ldots, \mathbf{g}_{24}^{-1}]$ correspond to the inverse composite sequence generator

$$G_c^{-1} = \{g^{(1)^{-1}}, \quad g^{(2)^{-1}}\} = \{\mathcal{R}(\mathcal{U}_{fr}^{-1}(\mathbf{g}_{23}^{-1})), \quad \mathcal{R}(\mathcal{U}_{fr}^{-1}(\mathbf{g}_{24}^{-1}))\}.$$

The delay-free inverse $G(z)^{-1}$ is obtained by folding these columns and reversing them $n_1 = 2$ and $n_2 = 3$ symbols at a time as shown in Fig. 2. The remaining $w - k = 22$ columns of $\hat{G}^{-1}$ produce forward shifts of $k_1 = 1$ and $k_2 = 2$ along the i_1 and i_2 dimensions of the input sequence space and correspond to $\frac{w-k}{k} = 11$ pseudo-inverses $G(z)_{(d_1,d_2)}^{-1}$ shown in Fig. 3.

$\mathcal{U}_{fr}^{-1}(\mathbf{g}_{23}^{-1})$ $\qquad$ $g^{(1)^{-1}}$ $\qquad\qquad$ $\mathcal{U}_{fr}^{-1}(\mathbf{g}_{24}^{-1})$ $\qquad$ $g^{(2)^{-1}}$

$$
\begin{array}{ccc}
00\ 00 & 00\ 00 & 10\ 01 \\
00\ 01 & 01\ 00 & 10\ 00 \\
01\ 10 & 10\ 01 & 10\ 00 \\
\end{array}
\;\xrightarrow{\psi^6}\;
\begin{bmatrix}
1 \\ z_1 \\ 1 \\ z_2 \\ 1+z_2 \\ z_1 z_2
\end{bmatrix}
,\qquad
\begin{array}{ccc}
00\ 00 & 00\ 00 & 01\ 00 \\
00\ 00 & 00\ 00 & 00\ 10 \\
10\ 00 & 00\ 10 & 00\ 00 \\
\end{array}
\;\xrightarrow{\psi^6}\;
\begin{bmatrix}
0 \\ 1 \\ z_1 \\ 0 \\ z_1 z_2 \\ 0
\end{bmatrix}
$$

$$
\begin{array}{ccc}
01\ 10 & 10\ 01 & 00\ 00 \\
00\ 10 & 10\ 00 & 01\ 00 \\
00\ 10 & 10\ 00 & 10\ 01 \\
\end{array}
\qquad\qquad
\begin{array}{ccc}
00\ 01 & 01\ 00 & 00\ 00 \\
10\ 00 & 00\ 10 & 00\ 00 \\
00\ 00 & 00\ 00 & 00\ 10 \\
\end{array}
$$

Fig. 2. Extracting the delay-free inverse from the inverse reduced encoding matrix

$G(z)^{-1}$ $G(z)^{-1}_{(1,0)}$ $G(z)^{-1}_{(2,0)}$ $G(z)^{-1}_{(3,0)}$ $G(z)^{-1}_{(0,1)}$ $G(z)^{-1}_{(1,1)}$

$$
\begin{bmatrix} 1 & 0 \\ z_1 & 1 \\ 1 & z_1 \\ z_2 & 0 \\ 1+z_2 & z_1z_2 \\ z_1z_2 & 0 \end{bmatrix}
\begin{bmatrix} z_1 & 1 \\ 1+z_2 & z_1 \\ z_1 & 1 \\ z_1z_2 & z_2 \\ z_1+z_1z_2 & z_2 \\ 1 & z_1z_2 \end{bmatrix}
\begin{bmatrix} 0 & z_1 \\ 0 & 1+z_2 \\ 1 & z_1 \\ 1 & z_1z_2 \\ z_2 & z_1z_2 \\ 0 & 1 \end{bmatrix}
\begin{bmatrix} 0 & 1 \\ 0 & z_1 \\ z_1 & 0 \\ z_1 & z_2 \\ z_1z_2 & 0 \\ 0 & z_1+z_1z_2 \end{bmatrix}
\begin{bmatrix} 0 & z_1 \\ 0 & z_2 \\ 1 & 0 \\ 0 & z_1z_2 \\ 0 & 0 \\ 0 & 0 \end{bmatrix}
\begin{bmatrix} 0 & 0 \\ 0 & 0 \\ z_1 & 1 \\ 0 & 0 \\ 0 & z_2 \\ 0 & 0 \end{bmatrix}
$$

$G(z)^{-1}_{(2,1)}$ $G(z)^{-1}_{(3,1)}$ $G(z)^{-1}_{(0,2)}$ $G(z)^{-1}_{(1,2)}$ $G(z)^{-1}_{(2,2)}$ $G(z)^{-1}_{(3,2)}$

$$
\begin{bmatrix} 1 & 0 \\ 0 & 0 \\ 0 & z_1 \\ 0 & 0 \\ 0 & z_1z_2 \\ 0 & 0 \end{bmatrix}
\begin{bmatrix} z_1 & 1 \\ 0 & 0 \\ 0 & 1 \\ 0 & z_2 \\ 0 & z_2 \\ 0 & z_1z_2 \end{bmatrix}
\begin{bmatrix} 0 & z_1 \\ 0 & z_2 \\ z_2 & z_1 \\ 0 & z_1z_2 \\ 0 & z_1z_2 \\ 0 & z_2 \end{bmatrix}
\begin{bmatrix} 0 & 1 \\ 0 & 0 \\ z_1z_2 & 0 \\ 0 & z_2 \\ 0 & 0 \\ 0 & 0 \end{bmatrix}
\begin{bmatrix} z_2 & z_1 \\ 0 & 0 \\ 0 & 0 \\ 0 & z_1z_2 \\ 0 & 0 \\ 0 & 0 \end{bmatrix}
\begin{bmatrix} z_1z_2 & 0 \\ 0 & z_1z_2 \\ 0 & 1 \\ 0 & 0 \\ 0 & 0 \\ 0 & z_2 \end{bmatrix}
$$

Fig. 3. The polynomial representation of the inverse reduced encoding matrix

Example 2. This example demonstrates the construction of a ZP polynomial matrix using local invertibility as suggested in Remark 2. Let $m = 3, k = 1, n = 8$ and $M_1 = M_2 = M_3 = 2$. In the interest of simplifying the illustration in Fig. 4 we assume $\mathbb{F} = \mathbb{F}_2$.

If we factor the rate $\frac{k}{n} = \frac{1}{8}$ as $\frac{1}{2} \times \frac{1}{2} \times \frac{1}{2}$ to predefine input sequence ordering as $\prod_{i=1}^{3} k_i = 1 \times 1 \times 1$ and output sequence ordering as $\prod_{i=1}^{3} n_i = 2 \times 2 \times 2$, then the subsequence mapping lengths along i_1, i_2 and i_3 are $w_1 = w_2 = w_3 = 4$.

Consider the reduced encoding matrix $\hat{G}_{64 \times 64}$ shown in Fig. 4 constructed using $n = 8$, randomly generated 1-D sequences of the form $a^{(y)} = \mathcal{U}_{fr}(\zeta + \mathcal{R}(g^{(y)}))$. Here ζ is a all-zero 3-D subsequence of size $\prod_{i=1}^{3} w_i = 4 \times 4 \times 4$ in S^1 and $g^{(y)}$ is a element of G_c of size $\prod_{i=1}^{3} L_i = 3 \times 3 \times 3$, where $L_i = k_i(M_i + 1) = 3$.

If $\hat{G}$ is nonsingular (it is in this case) the polynomial matrix $G(z) \in R^{1 \times 8}$ obtained from the 1-D sequences $a^{(y)}$ will be ZP.

$$a^{(1)} = [100(0)000(0)000(00000)110(0)000(0)000(00000)000(0)000(0)000]^T$$

The zeros in braces represent blanks in the reduced encoding matrix.

$\mathcal{U}_{fr}^{-1}(a^{(1)}) = \zeta + \mathcal{R}(g^{(1)})$ $\mathcal{R}(g^{(1)})$ $g^{(1)}$

$$
\begin{array}{ccc}
\begin{matrix} 1\,0\,0\,0 & 1\,1\,0\,0 & 0\,0\,0\,0 \\ 0\,0\,0\,0 & 0\,0\,0\,0 & 0\,0\,0\,0 \\ 0\,0\,0\,0, & 0\,0\,0\,0, & 0\,0\,0\,0 \\ 0\,0\,0\,0 & 0\,0\,0\,0 & 0\,0\,0\,0 \\ i_3 = 0 & i_3 = 1 & i_3 = 2 \end{matrix}
&
\begin{matrix} 1\,0\,0 & 1\,1\,0 & 0\,0\,0 \\ 0\,0\,0, & 0\,0\,0, & 0\,0\,0 \\ 0\,0\,0 & 0\,0\,0 & 0\,0\,0 \\ i_3 = 0 \; i_3 = 1 \; i_3 = 2 \end{matrix}
&
\begin{matrix} 0\,0\,0 & 0\,0\,0 & 0\,0\,0 \\ 0\,0\,0, & 0\,0\,0, & 0\,0\,0 \\ 0\,0\,0 & 0\,1\,1 & 0\,0\,1 \\ i_3 = 0 \; i_3 = 1 \; i_3 = 2 \end{matrix}
\end{array}
$$

$$\downarrow \psi^1$$

$$z_1 z_2^2 z_3 + z_1^2 z_2^2 z_3 + z_1^2 z_2^2 z_3^2$$

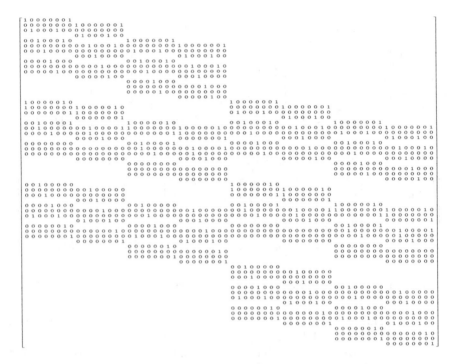

Fig. 4. 3-D reduced encoding Matrix $\hat{G}_{64 \times 64}$ (*blanks denote zeros*)

Since $k = 1$, there is no multiplexing involved and $g^{(y)} = g_x^{(y)}$.

$$G_c = \{g^{(1)}, \ldots, g^{(8)}\} = G = (g_1^{(1)}, \ldots, g_1^{(8)})$$

The sequence generator $G \in S^{1 \times 8}$ is transformed using the map ψ^1 to obtain the ZP polynomial matrix $G(z) \in R^{1 \times 8}$. Since $k = 1$, the inverse polynomial matrix $G(z)^{-1} \in R^{8 \times 1}$ is obtained from the last column of inverse reduced encoding matrix.

$$G(z) = \begin{bmatrix} z_1 z_2^2 z_3 + z_1^2 z_2^2 z_3 + z_1^2 z_2^2 z_3^2 \\ z_2 + z_2^2 z_3^2 \\ z_1^2 z_2^2 + z_1 z_2 z_3 + z_1^2 z_2 z_3 + z_1^2 z_2 z_3^2 \\ z_2^2 + z_2 z_3^2 \\ z_1^2 z_2 + z_2 z_3 + z_1^2 z_3^2 \\ z_2 + z_3^2 + z_2^2 z_3^2 \\ z_1^2 + z_1^2 z_2^2 z_3 + z_1^2 z_2 z_3^2 \\ 1 + z_1^2 z_2 z_3 + z_2^2 z_3 + z_1^2 z_2^2 z_3^2 \end{bmatrix}^T \qquad G(z)^{-1} = \begin{bmatrix} z_2 + z_3 + z_1 z_3 + z_1 z_2 z_3 \\ z_2 z_3 + z_1 z_3 + z_1 z_2 z_3 \\ z_1 + z_2 z_3 \\ z_1 + z_3 + z_1 z_3 + z_1 z_2 z_3 \\ z_1 z_2 \\ z_1 z_2 + z_2 z_3 + z_1 z_2 z_3 \\ z_1 z_2 z_3 \\ 1 + z_1 z_2 z_3 \end{bmatrix}$$

4   Conclusion

Locally invertible matrices introduced in this paper are a class of rectangular zero prime multivariate polynomial matrices whose polynomial inverse can be easily computed by performing elementary operations in the ground field. We have shown that for a given m-variate polynomial matrix with k rows and n columns, if the rate $\frac{k}{n}$ can be factored as $\prod_{i=1}^{m} \frac{k_i}{n_i}$ such that each w_i in (10) is a positive integer, then, there exists a subsequence map of size $\prod_{i=1}^{m} w_i$ between its input and output sequences. The bijectivity of this m-D subsequence map can be tested by unfolding the output subsequences to form a reduced encoding matrix. If the reduced encoding matrix is nonsingular then the polynomial matrix is shown to be zero prime. The polynomial inverse matrix and a set of pseudo-inverses are then derived from the inverse subsequence map using the inverse reduced encoding matrix. Since the reduced encoding matrix has a specific structure that depends on the parameters of the polynomial matrix, it can used as a powerful tool to construct invertible transformations of any dimension.

References

1. Bose, N.K., ed.: Multidimensional Systems: Progress, Directions and Open Problems. Reidel, Dordrecht, The Netherlands (1985)
2. Youla, D.C., Gnavi, G.: Notes on n-dimensional system theory. IEEE Trans. Circuits and Systems **26** (1979) 105–111
3. Youla, D., Pickel, P.: The Quillen-Suslin theorem and the structure of n-dimensional elementary polynomial matrices. IEEE Trans. Circuits and Systems **31**(6) (1984) 513–517
4. Park, H., Kalker, T., Vetterli, M.: Grobner bases and multidimensional FIR multirate systems. Journal of multidimensional systems and signal processing **8** (1997) 11–30
5. Fornasini, E., Valcher, M.E.: nD polynomial matrices with applications to multidimensional signal analysis. Multidimensional Systems and Signal Processing **8** (1997) 387–408
6. Fornasini, E., Valcher, M.E.: Algebraic aspects of 2D convolutional codes. IEEE Trans. Inform. Theory **IT-40**(4) (1994) 1068–1082
7. Weiner, P.A.: Basic properties of multidimensional convolutional codes. In: Codes, Systems and Graphical Models. IMA Volumes in Mathematics and Its Applications, 123. Springer-Verlag, New York (2001) 397–414
8. Luerssen, H.G., Rosenthal, J., Weiner, P.A.: Duality between multidimensional convolutional codes and systems. In Colonius, F., Helmke, U., Wirth, F., Prätzel-Wolters, D., eds.: Advances in Mathematical Systems Theory, A Volume in Honor of Diederich Hinrichsen. Birkhauser, Boston (2000) 135–150
9. Vologiannidis, S., Karampetakis, N.: Inverses of multivariable polynomial matrices by discrete fourier transforms. Multidimensional Systems and Signal Processing **15**(4) (2004) 341–361
10. Lobo, R., Bitzer, D., Vouk, M.: Inverses of multivariate polynomial matrices using discrete convolution. In: Proceedings of the 2005 International Workshop on Coding and Cryptography (WCC 2005), Bergen, Norway. (2005) 481–490

Author Index

Lecture Notes in Computer Science

For information about Vols. 1–3954

please contact your bookseller or Springer

Vol. 3995: G. Müller (Ed.), Emerging Trends in Information and Communication Security. XX, 524 pages. 2006.

Vol. 3994: V.N. Alexandrov, G.D. van Albada, P.M.A. Sloot, J. Dongarra (Eds.), Computational Science – ICCS 2006, Part IV. XXXV, 1096 pages. 2006.

Vol. 3993: V.N. Alexandrov, G.D. van Albada, P.M.A. Sloot, J. Dongarra (Eds.), Computational Science – ICCS 2006, Part III. XXXVI, 1136 pages. 2006.

Vol. 3992: V.N. Alexandrov, G.D. van Albada, P.M.A. Sloot, J. Dongarra (Eds.), Computational Science – ICCS 2006, Part II. XXXV, 1122 pages. 2006.

Vol. 3991: V.N. Alexandrov, G.D. van Albada, P.M.A. Sloot, J. Dongarra (Eds.), Computational Science – ICCS 2006, Part I. LXXXI, 1096 pages. 2006.

Vol. 3990: J. C. Beck, B.M. Smith (Eds.), Integration of AI and OR Techniques in Constraint Programming for Combinatorial Optimization Problems. X, 301 pages. 2006.

Vol. 3989: J. Zhou, M. Yung, F. Bao, Applied Cryptography and Network Security. XIV, 488 pages. 2006.

Vol. 3987: M. Hazas, J. Krumm, T. Strang (Eds.), Location- and Context-Awareness. X, 289 pages. 2006.

Vol. 3986: K. Stølen, W.H. Winsborough, F. Martinelli, F. Massacci (Eds.), Trust Management. XIV, 474 pages. 2006.

Vol. 3984: M. Gavrilova, O. Gervasi, V. Kumar, C.J. K. Tan, D. Taniar, A. Laganà, Y. Mun, H. Choo (Eds.), Computational Science and Its Applications - ICCSA 2006, Part V. XXV, 1045 pages. 2006.

Vol. 3983: M. Gavrilova, O. Gervasi, V. Kumar, C.J. K. Tan, D. Taniar, A. Laganà, Y. Mun, H. Choo (Eds.), Computational Science and Its Applications - ICCSA 2006, Part IV. XXVI, 1191 pages. 2006.

Vol. 3982: M. Gavrilova, O. Gervasi, V. Kumar, C.J. K. Tan, D. Taniar, A. Laganà, Y. Mun, H. Choo (Eds.), Computational Science and Its Applications - ICCSA 2006, Part III. XXV, 1243 pages. 2006.

Vol. 3981: M. Gavrilova, O. Gervasi, V. Kumar, C.J. K. Tan, D. Taniar, A. Laganà, Y. Mun, H. Choo (Eds.), Computational Science and Its Applications - ICCSA 2006, Part II. XXVI, 1255 pages. 2006.

Vol. 3980: M. Gavrilova, O. Gervasi, V. Kumar, C.J. K. Tan, D. Taniar, A. Laganà, Y. Mun, H. Choo (Eds.), Computational Science and Its Applications - ICCSA 2006, Part I. LXXV, 1199 pages. 2006.

Vol. 3979: T.S. Huang, N. Sebe, M.S. Lew, V. Pavlović, M. Kölsch, A. Galata, B. Kisačanin (Eds.), Computer Vision in Human-Computer Interaction. XII, 121 pages. 2006.

Vol. 3978: B. Hnich, M. Carlsson, F. Fages, F. Rossi (Eds.), Recent Advances in Constraints. VIII, 179 pages. 2006. (Sublibrary LNAI).

Vol. 3977: N. Fuhr, M. Lalmas, S. Malik, G. Kazai (Eds.), Advances in XML Information Retrieval and Evaluation. XII, 556 pages. 2006.

Vol. 3976: F. Boavida, T. Plagemann, B. Stiller, C. Westphal, E. Monteiro (Eds.), Networking 2006. Networking Technologies, Services, and Protocols; Performance of Computer and Communication Networks; Mobile and Wireless Communications Systems. XXVI, 1276 pages. 2006.

Vol. 3975: S. Mehrotra, D.D. Zeng, H. Chen, B.M. Thuraisingham, F.-Y. Wang (Eds.), Intelligence and Security Informatics. XXII, 772 pages. 2006.

Vol. 3973: J. Wang, Z. Yi, J.M. Zurada, B.-L. Lu, H. Yin (Eds.), Advances in Neural Networks - ISNN 2006, Part III. XXIX, 1402 pages. 2006.

Vol. 3972: J. Wang, Z. Yi, J.M. Zurada, B.-L. Lu, H. Yin (Eds.), Advances in Neural Networks - ISNN 2006, Part II. XXVII, 1444 pages. 2006.

Vol. 3971: J. Wang, Z. Yi, J.M. Zurada, B.-L. Lu, H. Yin (Eds.), Advances in Neural Networks - ISNN 2006, Part I. LXVII, 1442 pages. 2006.

Vol. 3970: T. Braun, G. Carle, S. Fahmy, Y. Koucheryavy (Eds.), Wired/Wireless Internet Communications. XIV, 350 pages. 2006.

Vol. 3969: Ø. Ytrehus (Ed.), Coding and Cryptography. XI, 443 pages. 2006.

Vol. 3968: K.P. Fishkin, B. Schiele, P. Nixon, A. Quigley (Eds.), Pervasive Computing. XV, 402 pages. 2006.

Vol. 3967: D. Grigoriev, J. Harrison, E.A. Hirsch (Eds.), Computer Science – Theory and Applications. XVI, 684 pages. 2006.

Vol. 3966: Q. Wang, D. Pfahl, D.M. Raffo, P. Wernick (Eds.), Software Process Change. XIV, 356 pages. 2006.

Vol. 3965: M. Bernardo, A. Cimatti (Eds.), Formal Methods for Hardware Verification. VII, 243 pages. 2006.

Vol. 3964: M. Ü. Uyar, A.Y. Duale, M.A. Fecko (Eds.), Testing of Communicating Systems. XI, 373 pages. 2006.

Vol. 3963: O. Dikenelli, M.-P. Gleizes, A. Ricci (Eds.), Engineering Societies in the Agents World VI. XII, 303 pages. 2006. (Sublibrary LNAI).

Vol. 3962: W. IJsselsteijn, Y. de Kort, C. Midden, B. Eggen, E. van den Hoven (Eds.), Persuasive Technology. XII, 216 pages. 2006.

Vol. 3960: R. Vieira, P. Quaresma, M.d.G.V. Nunes, N.J. Mamede, C. Oliveira, M.C. Dias (Eds.), Computational Processing of the Portuguese Language. XII, 274 pages. 2006. (Sublibrary LNAI).

Vol. 3959: J.-Y. Cai, S. B. Cooper, A. Li (Eds.), Theory and Applications of Models of Computation. XV, 794 pages. 2006.

Vol. 3958: M. Yung, Y. Dodis, A. Kiayias, T. Malkin (Eds.), Public Key Cryptography - PKC 2006. XIV, 543 pages. 2006.

Vol. 3956: G. Barthe, B. Grégoire, M. Huisman, J.-L. Lanet (Eds.), Construction and Analysis of Safe, Secure, and Interoperable Smart Devices. IX, 175 pages. 2006.

Vol. 3955: G. Antoniou, G. Potamias, C. Spyropoulos, D. Plexousakis (Eds.), Advances in Artificial Intelligence. XVII, 611 pages. 2006. (Sublibrary LNAI).